TEMPERATE FRUITS

Volume 1: Pome and Stone Fruits

The Editor

Professor (Dr.) Sisir Kumar Mitra worked (until 28.02.2014) at the Bidhan Chandra Krishi Viswavidyalaya, Mohanpur, West Bengal, India. He served as Head, Department of Fruits and Orchard Management (1996-2000), Dean, Faculty of Horticulture (2007-2009) and Dean, Post Graduate Studies (2009-2012) of the university. He did Post-doctoral research (1985-1986) at the Horticulture Research International, East Malling, UK as Commonwealth Fellow and worked as Visiting Professor at the Institute Coltivazoni Arboree, Italy, University of Lavras, Brazil, and Kyoto University, Japan, Adjunct Professor, UBKV, Cooch Behar, West Bengal. He has been conferred with Honorary Fellowship by the International Society for Horticultural Science (ISHS), Confederation of Horticulture Associations of India (CHAI) and Fellow Horticulture Society of India (HSI).

Professor Mitra guided 24 Ph.D. and 32 M.Sc. (Horticulture) students and has published 252 scientific papers, 74 popular articles, 11 books, 18 book chapters, 1 IBPGR Descriptors and 28 technical reports.

Served as Member of the International Scientific Committee, Chaired, Technical Sessions and delivered keynote/invited lecture in 43 International events. He was the Chairman, International Society for Horticultural Science (ISHS) Section, Tropical and Subtropical Fruits, (2010-2018). Presently he is serving as Board Member responsible for publication of ISHS, Science Editor, ISHS and Editor-in-Chief, Fruits: The International Journal of Tropical and Subtropical Horticulture. He is also the Chairman, Editorial Advisory Board of Acta Horticulturae and Chronica Horticulturae, Chairman, ISHS Working Group on Avocado, and Jackfruit and other Moraceae. He has been Convener of seven International Symposia and Editor of eight volumes of Acta Horticulturae.

He has served as Technical Advisor of the Board of Trustee of International Tropical Fruits Network of FAO, and as FAO Consultant. He served as Member, in many Research Advisory Committees (RAC) of ICAR Institutes and also a member of QRT. He is Advisor, CAB International, UK, and Member of Editorial Board of many professional journals. He is in the Board of Directors, of the West Bengal State Food Processing and Horticulture Development Corporation Limited, and Technical Advisor in many academic and

TEMPERATE FRUITS

Volume 1: Pome and Stone Fruits

S.K. Mitra, D.S. Rathore and T.K. Bose

Revised and Enlarged by

S.K. Mitra

2025

Daya Publishing House®

A Division of

Astral International Pvt. Ltd.

New Delhi – 110002

First Published 1990
Revised and enlarged Edition 2020
Reprinted, 2025

ISBN 978-93-89605-32-7 (HB)

Publisher's Note:

Every possible effort has been made to ensure that the information contained in this book is accurate at the time of going to press, and the publisher and author cannot accept responsibility for any errors or omissions, however caused. No responsibility for loss or damage occasioned to any person acting, or refraining from action, as a result of the material in this publication can be accepted by the editor, the publisher or the author. The Publisher is not associated with any product or vendor mentioned in the book. The contents of this work are intended to further general scientific research, understanding and discussion only. Readers should consult with a specialist where appropriate.

Every effort has been made to trace the owners of copyright material used in this book, if any. The author and the publisher will be grateful for any omission brought to their notice for acknowledgment in the future editions of the book.

Published by : **Daya Publishing House®**
A Division of
Astral International Pvt. Ltd.
– ISO 9001:2015 Certified Company –
4736/23, Ansari Road, Darya Ganj
New Delhi-110 002
Ph. 011-43549197, 23278134
E-mail: info@astralint.com
Website: www.astralint.com

Digitally Printed at : **Replika Press Pvt. Ltd.**

Acknowledgement

Many people have, in one way or another, helped in the production of this book; most were mentioned in the 'preface' in the first edition and I hope they will forgive me if I do not mention them all again. I express my gratitude to all the contributors of the first edition of book from where I worked to publish this revised and enlarged edition.

I like to thank Professor Eddo Rugini, University of Tuscia, Viterbo, Italy, Professor Sadiye Gozlekci, Akdeniz University, Turkey, Late Professor Norman Looney, Pacific Agri-Food Research Centre, Agriculture and Agri-Food Canada, Dr. Damiano Avanzato, International Horticulture Consultant, Rome, Italy, Dr. K.K. Pramanick, IARI Regional Station, Shimla, Himachal Pradesh, India, Professor B. Ghosh, Dr. S. Debnath, Dr. D. Majhi and Sri. S. Sau, Bidhan Chandra Krishi Viswavidyalaya, Mohanpur, West Bengal, India for support, information, photographs and ideas.

I express my sincere gratitude to my respected teacher Professor T.K. Bose who was also the editor of the first edition. He is the one behind this revised edition, felt the needs of students, teachers and researchers and encouraged me to work for compiling with updated information.

My special thanks to Dr. P.K. Pathak, KVK, W.B. University of Animal and Fishery Sciences, West Bengal, India for his editorial assistance without which possibly I could not have finished.

Professor S.K. Mitra

Preface to the Second Edition

This book was first published in 1991 and was well received by students, teachers and researchers in many universities and institutes in India and abroad. In the past 28 years (since its first publication), tremendous progress have been made in temperate fruit research, notably in developing varieties and rootstocks, canopy management, understanding the flowering physiology, storage technology, biological control of pests and diseases, resistance breeding, biotechnology, *etc.*, Revision of the book with updated information was considered necessary.

In the process of a comprehensive literature survey it was felt impossible to compile all the information in one volume. This enlarged edition on temperate fruits is, therefore, published in two volumes. The first volume consists of pome and stone fruits (apples, pears, peaches, plums, apricots and cherries) in more than 550 pages. The second volume deals with temperate nuts and berries (almonds, chestnuts, hazelnuts, pecans, pistachios, walnut, strawberries, currants, gooseberries, raspberries, blueberries and cranberries) in about 650 pages.

Apart from thorough revision with updated scientific information on various aspects of mentioned pome, stone, nuts and berries, substantial reorganization of the text has been made. Emphasis has been laid to include released cultivars and rootstocks, recent trends in propagation and orchard management, pruning and canopy management, nutrition and tissue analysis, water management, improved technology for harvesting, postharvest handling, storage and ripening. Special emphasis has been laid to include the development in breeding and improvement through biotechnological approaches like molecular markers, genetic transformation, genetic map and microbes for N and P input. Other information like organic production technology, protected cultivation and impact of climate change are included, where sufficient information is available.

Though the subject is vast, this two volumes book presents the concepts in a condensed, informative and lucid manner. It will be of immense help to the students, teachers and researchers of horticulture, plant science, plant physiology and plant protection.

Professor S.K. Mitra

Preface to the First Edition

The present title has been planned and designed to meet the long-felt need for a book covering major aspects of important temperate fruits. The considerable large volume of research done on these fruits and the wealth of information accumulated over the years made the publication of such a volume imperative. It thus becomes a useful textbook for the students of horticulture all over the world as well as important reference book for the researchers on this subject.

As a well-prepared and comprehensive textbook, the title incorporates recent literature, updated scientific and general information and important tables that in many cases, have not compiled before. As such, the book will make the readers conversant with the current researches and development in this field.

I am grateful to all the contributors for their efforts and care with which they prepared the different chapters of the book. I express my gratitude Dr. S.K. Mitra and Dr. D.S. Rathore, the co-editors who helped me immensely in editing as well as during printing process. I wish to record my appreciation to my colleagues-Prof. N. Mukherjee, Dr. B. Ghosh and also to Dr. B. Chowdhury of Horticulture and Allied Publishers, Mr. K. Dasgupta, Mrs. K. Neyogi and Mr. S. Bhowmick for their help.

This publication has been financed by United Bank of India, Old Court House Street Branch, Calcutta and I am thankful to Mr. J.V. Shetti, Chairman and Managing Director of the Bank for his interest and cooperation.

Calcutta *T.K.Bose*

August, 1991

Contents

Volume 1
Temperate Fruits: Pome and Stone Fruits

Contributors of Temperate Fruits: Pome and Stone Fruits (1990)

Kishore, D.K.

Indian Agricultural Research Institute, Regional Station, Amartara, Shimla 171004, Himachal Pradesh

Mitra, S.K.

Faculty of Agriculture, Bidhan Chandra Krishi Viswavidyalaya, Mohanpur, West Bengal 741 235,

Pandey, R.M.

Department of Agriculture and Coperation, Government of India, Krishi Bhavan, New Delhi 110 001.

Pathak, P.

Faculty of Agriculture, Bidhan Chandra Krishi Viswavidyalaya, Mohanpur, West Bengal 741 235,

Pathak, R.A.

Department of Horticulture, Narendra Dev University of Agriculture and Technology, Faizabad, Uttar Pradesh 224 229.

Pathak, R.K.

Department of Horticulture, Narendra Dev University of Agriculture and Technology, Faizabad, Uttar Pradesh 224 229.

Randhawa, S.S.

Indian Agricultural Research Institute, Regional Station, Amartara, Shimla 171004, Himachal Pradesh

Rathore, D.S.

National Bureau of Plant Genetic Resources, Regional Station, Phagli, Shimla, Himachal Pradesh 171 004

Sharma, Y.P.

Indian Agricultural Research Institute, Regional Station, Amartara, Shimla 171004, Himachal Pradesh

Contributors of Temperate Fruits: Nuts and Berries

Bist, H.S.

Department of Fruit Culture and Orchard Management, Dr. Y.S. Parmar University of Horticulture and Forestry, Solan 173230, Himachal Pradesh.

Chauhan, J.S.

Department of Fruit Culture and Orchard Management, Dr. Y.S. Parmar University of Horticulture and Forestry, Solan 173230, Himachal Pradesh.

Jindal, K.K.

Regional Horticulture Research Station, Dr. Y.S. Parmar University of Horticulture and Forestry, Mashobra, Shimla 171008, Himachal Pradesh.

Karkar, B.K.

Regional Horticulture Research Station, Dr. Y.S. Parmar University of Horticulture and Forestry, Mashobra, Shimla 171008, Himachal Pradesh.

Mehta, K.

Department of Fruit Culture and Orchard Management, Dr. Y.S. Parmar University of Horticulture and Forestry, Solan 173230, Himachal Pradesh.

Mitra, S.K.

Faculty of Agriculture, Bidhan Chandra Krishi Viswavidyalaya, Mohanpur, West Bengal 741 235,

Monastra, F.

Instituto Sperimentale per la Fruitticoltura, MAF, Rome, Italy

Pathak, P.

Faculty of Agriculture, Bidhan Chandra Krishi Viswavidyalaya, Mohanpur, West Bengal 741 235,

Rathore, D.S.

National Bureau of Plant Genetic Resources, Regional Station, Phagli, Shimla, Himachal Pradesh 171 004

Rugini, E.

Universita degli studi della Tuscia, Instituto di ortofloroarboricultura, 01100 Viterbo, Italy.

Sharma, S.D.

Regional Horticulture Research Station, Dr. Y.S. Parmar University of Horticulture and Forestry, Sharbo, Rekong Peo, Kinnaur, Himachal Pradesh.

Thakur, B.S.

Regional Horticulture Research Station, Dr. Y.S. Parmar University of Horticulture and Forestry, Kandaghat, Solan 172215, Himachal Pradesh.

Thakur, K.S.

Department of Postharvest Technology, Dr. Y.S. Parmar University of Horticulture and Forestry, Solan 173230, Himachal Pradesh.

Chapter 1

Apple

Apple, the premier table fruit of the world, has been under cultivation since time immemorial. It is a typical temperate tree fruit. Within the broad Temperate Zone, apple cultivation is concentrated in areas where the environment is particularly favourable. It gives the highest yield of good quality fruits in regions having long day hours with high light intensity and relatively warm days with cool nights and low relative humidity during the growing season. Today, only the best quality apples are profitable and it is essential to maximize the yield. High yields of good quality fruits can only be obtained when the growing inputs result in clean, well-structured orchards where fruiting and growth are correctly balanced (White, 1984).

Composition and Uses

An apple a day keeps the doctor away-this old adage focuses man's attention on the importance of apple in the daily diet. Apple has long been the staple fresh fruit in the temperate parts of the world. Eating apples is believed to reduce the incidence of dental caries, helps to control obesity and supply extra energy for heavy exercise. The average composition of apples (Paul and Southgate, 1978) is presented in Table 1.1.

Carbohydrates found in apple consist of sugars, dextrins, starch, hemicellulose, cellulose and pectic substances. The apple fruits have been found to contain appreciable amount of sorbitol (D-glucitol, the water-soluble alcohol of D-glucose) and various sugars (sucrose, glucose, fructose). At harvest, the sorbitol, sucrose, glucose and fructose contents in apple cv. Mclntosh were recorded as 8, 39, 6, 49,18, 37 and 48.90 mg/fresh weight, respectively (Taper and Liu, 1969). The acid content of fresh fruits helps to develop the natural flavour in apple juice. Apples have been found to have very strong antioxidant activity, inhibit cancer cell proliferation, decrease lipid oxidation, and lower cholesterol. Apples contain a variety of phytochemicals, including quercetin, catechin, phloridzin and chlorogenic acid, all of which are strong antioxidants. The phtochemical composition of apples however, varies greatly between different varieties (Boyer and Liu, 2004). Five quercetin and two phloretin glysoside were Isolated from apple skin. They were identified as quercetin, 3-*0*-galactoside (hyperin), 3-*0*-glucoside (isoquercitrin), 3-*0*-xyloside (reynoutrin), 3-*0*-arabinoside (avicularin) and 3-*0*-rhamnoside (quercitrin), phloretin glucoside (phlorizin) and phloretin xyloglucoside. The glycosides of phloretin were also found in apple flesh (Oleszek *et al.*, 1988). Although the major acids in apple are maleic and citric, Hulme (1958) stated that a variety of other acids (Table 1.2) are also present.

Table 1.1: Composition of Fresh Apple per 100 g Flesh without Core or Skin

Per 100g Flesh without core or skin			
Calories	37-46	Thiamine (mg)	0.04
Water (g)	84.3-85.6	Vitamin B6 (mg)	0.03
Sugars (g)	9.2-11.8	Folate (μg)	5
Fibre (g)	2.0-2.4	Pantothenate (mg)	0.1
Total N (g)	0.04-0.05	Biotin (μg)	0.3
Sodium (mg)	2	Vitamin C (mg)	3.15
Potassium (mg)	120		

Paul and Southgate (1978).

Table 1.2: Organic Acids in Apple Fruits*

Whole Fruit or juice of whole fruit	*Pulp*	*Peel*
Quinic	Maleic	Maleic
Glycolic	Citric	Citric
Succinic	Quinic	Quinic
Lactic	Shikimic	Citramalic
Galacturonic	Succinic	Shikimic
Citramalic	Glyceric	Glyceric
Mucic	α-Oxoglutaric	α-Oxoglutaric
	Pyruvic	Pyruvic
	Oxaloacetic	Oxaloacetic
	Glyoxlic	
	Isocitric	

* Hulme (1958).

The English cooking apple, Bramley's Seedling, has a white, firm-textured flesh, used in salads, sandwich filling, starters, snacks, main dishes, puddings, sweetmeats, jams, jellies, pickles and products include pie filling, slices, sauce and powder while the liquid products, juice and cider are based on fresh juice and stored concentrate. The utilization of apple as proposed by Beech (1983) is presented in Figure 1.1.

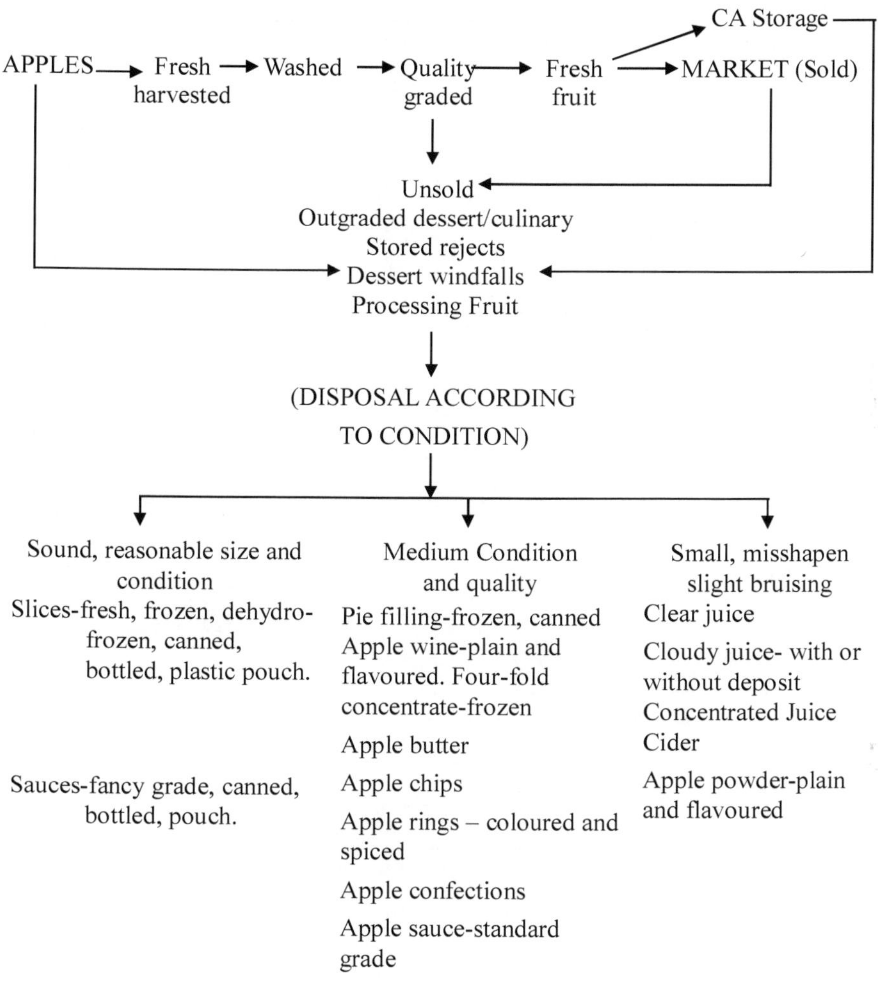

Figure 1.1: Utilization of the Apple.

Production

Apple is produced in all the continents of the world. The largest producer is China followed by United States and Turkey. The important apple producing countries in Europe are Poland, Italy, France, Russia, Ukraine, Germany. India is the second largest apple producer in Asia after China and ranks fifth in the world (Table 1.3)

Table 1.3: Apple Production in different Countries of the World (Production in 2012)

Rank	*Country*	*Production (MT)*	*Rank*	*Country*	*Production (MT)*
1.	China	37,001,601	29.	South Korea	3,94,596
2.	United States	4,110,050	30.	Mexico	3,75,045
3.	Turkey	2,889,000	31.	Syria	3,34,166
4.	Poland	2,877,340	32.	Australia	2,89,064
5.	India	2,203,400	33.	Moldova	2,81,849
6.	Italy	1,991,310	34.	Netherlands	281,200
7.	Iran	1,700,000	35.	Canada	269,837
8.	Chile	1,625,000	36.	Greece	251,000
9.	Russia	1,403,000	37.	Azerbaijan	234,771
10.	France	1,382,900	38.	Switzerland	227,106
11.	Brazil	1,403,000	39.	Belgium	220,400
12.	Argentina	1,250,000	40.	Tajikistan	218,300
13.	Ukraine	1,126,800	41.	United Kingdom	202,900
14.	Germany	972,405	42.	Portugal	186,000
15.	Uzbekistan	820,000	43.	Serbia	178,713
16.	South Africa	795,758	44.	Lebanon	155,000
17.	Japan	793,800	45.	Peru	147,114
18.	North Korea	785,000	46.	Kyrgyzstan	137,000
19.	Hungary	650,595	47.	Kazakistan	132,000
20.	Pakistan	620,000	48.	Israel	131,608
21.	Spain	558,900	49.	Macedonia	127,171
22.	Egypt	541,239	50.	Czeh Republic	118,709
23.	Belarus	511,479	51.	Tunisia	111,000
24.	Morocco	485,642	52.	Armenia	110,289
25.	Austria	471,420		**World (Total Production)**	**76,378,700**
26.	Romania	462,935			
27.	New Zealand	448,000			
28.	Algeria	397,529			

Note: Other apple producing countries produces less than 100,000 tonnes.

*FAOSTAT, FAO (December, 2014).

Table 1.4: Global Fresh Apple Imports (Top 10 importing countries in 2014)

County	*Volume (Million kg)*
Russia	1050
Germany	631
United Kingdom	437
Netherlands	269
Spain	241
Mexico	236
Canada	222
United States	208
India	205
Taiwan	158

Source: Global Trade Atlas (GTA Navigator).

Table 1.5: Fresh Apples per Capita Consumption in Selected Countries (2007)

Country	*Consumption (kg/year)*	*Country*	*Consumption (kg/year)*
Austria	25.23	China	5.48
Belgium	18.27	Japan	6.06
Bulgaria	4.31	Taiwan	31.68
Denmark	19.60	Turkey	11.69
France	16.12	Argentina	7.38
Germany	19.35	Australia	4.64
Hungary	16.68	Chile	9.58
Italy	20.38	New Zealand	14.57
United Kingdom	10.14	South Africa	4.11
Canada	12.56	Russia	8.11
Mexico	6.27		
United States	7.53		

Source: Belrose, Inc., World Apple Review, 2008.

Global apples exports by country totaled US$ 7,517,285,000 in 2014 which was 16.1 per cent more than in 2010, but decreased by 6.5 per cent compared with 2013.

The Top 15 Countries that Exported the Highest Dollar Value Worth of Apples (Exports by country during 2014)

Rank	*Country*	*Export Value US$*	*Percentage of Total Apple Exports*
1.	United States	1,088,369,000	14.5
2.	China	1,027,637,000	13.7
3.	Italy	975,630,000	13.0

Rank	Country	Export Value US$	Percentage of Total Apple Exports
4.	Chile	822,347,000	10.9
5.	France	719,954,000	9.6
6.	Poland	441,766,000	5.9
7.	New Zealand	433,160,000	5.8
8.	South Africa	353,652,000	4.7
9.	Netherlands	221,079,000	2.9
10.	Belgium	149,496,000	2.0
11.	Argentina	137,387,000	1.8
12.	Spain	107,562,000	1.4
13.	Belarus	86,537,000	1.2
14.	Japan	81,741,000	1.1
15.	Serbia	81,307,000	1.1

Source: Trade Map, International Trade Centre.

Origin and Distribution

The primary centre of origin of apple is thought to be the region which includes Asia Minor, the Caucasus, Soviet central Asia and Himalayan India and Pakistan (Wilcox, 1962). It has existed in Europe, both in wild and cultivated forms, since prehistoric times and have been well developed there at the beginning of Christian era (Shoemaker and Teskey, 1959). Records of apple cultivars are available as early as 100 BC (Magness, 1941). Cultivation of apple by Greeks and Romans seems to have been started from a few centuries BC, and as result of their travels and colonization, it was spread throughout Europe and Asia. Central Asia has been reported to contain the greatest diversity of *Malus*, and this area also appears to be the centre of origin of the domesticated apple (Janick and Moore, 1996). *M. siversii* which grows wild in Kazakhstan and Kyrgzstan, is thought to be the main progrenitor species (Pereira–Lorenzo *et al.*, 2009). *M. siversii* has very similarity with *M. domestica* in morphology and fruit flavour (Forsline, 1995). This species may have hybridized with *M. prunifolia, M. baccata,* and *M. sieboldi* to the East and with *M. turkmenorum* and *M. sylvestries* to the West. Subsequently, well-established apple cultivars were selected and introduced in to Europe and specially to the Mediterranean regions by the Romans (Juniper *et al.*, 1999). Analyses of nuclear rDNA and chloroplast DNA (cpDNA) sequences indicate that the domesticated apple is most closely related to series *Malus* species. Moreover, the occurrence of shared 18-bp duplication in the cpDNAs of wild and cultivated apple supports the close relationship between them (Harris *et al.*, 2002). It was introduced in America after it has been cultivated for more than 2000 years in Europe. It is not known when the apple was introduced in the cooler parts of India but evidence showed it as being scarce in Agra in 1632 (Temple, 1914).

Taxonomy

The genus *Malus* belongs to the subfamily Pomoideae and the family Rosaceae with a basic chromosome number of 17. The genus *Malus* has 25 species (Rehder, 1954) and several subspecies of so-called Crab apples, many of which are cultivated as ornamental trees for their attractive flower and fruits.

The cultivated apple has been earlier designated as *Malus pumila* (formerly *M. communis* or *Pyrus malus*). Some workers prefer to split this species in *M. domestica* (the cultivated apple), *M. pumila* (including var. *paradisiaca,* the paradise apple) and *M. sylvestris* (the wild Crab apple). However, *M. pumila* is considered to be the parent of most of our cultivated apples (Watkins, 1979). The cultivated apple is likely the result of interspecific hybridization and at present the binomal *Malus* x *domestica* has been generally accepted as the appropriate scientific name (Korban and Skirvin, 1984). The description of some principal *Malus* species are as follows:

M. floribunda **(Flowering Crab)**

It is a native of Japan. Trees are small and thorny. Flowers are highly ornamental, rose-red colour. Fruits are small in size and red in colour.

M. baccata **(Siberian Crab)**

The Siberian Crab grows wild from Siberia to Manchuria and the Himalayan region. Spreading trees, flowers white and showy. Fruits yellow or red firm and translucent.

M. angustifolia **(Narrow-leaved Crab-apple)**

It is distributed from Pennsylvania to Tennessee and Florida. Leaves are small, narrow and lanceolate. Flowers are pink in colour fruits small.

M. coronaria **(American Crab-apple)**

Trees are small, bushy and thorny. Flowers are large with persistent calyx. Fruits are greenish yellow, fragrant and waxy.

M. ioensis

Trees are small. Flowers large, fruits are green with light coloured spots.

M. soulardii

This is a natural hybrid of *M. sylvestris* and *M. ioensis.* Trees are small. Fruits are large with good aroma.

M. sylvestris **(Wild Crab-apple)**

Trees are large. There are several cultivars differing in form, size, colour and taste, It is considered to be native of western Asia and southeastern Europe.

M. sylvestris and *M. pumila* are considered as the major ancestral species of modern apples but *M. spectabilis, M. prattii, M. baccata, M. prunifolia* and *M. astracanica* act as significant additional contributors (Knight, 1963b).

Cytotaxonomic Background

The majority of apple cultivars are diploid with 2n = 34 and a genome of moderate size (1C = 2.25 pg which corresponds to approximately 1.5 x 10^9 bp) wheras some cultivars are triploid with 2n = 3x = 51 (Pereira-Loren *et al.*, 2009). The Maloideae subfamily probably has resulted from an ancient autopolyploidization of a 9 chromosome progenitor to 18 chromosomes. Then it was followed by a chromosome loss resulting in current 17-chromosome apple cultivars whereas other subfamilies are X = 7, 8 or 9 (Folta and Gardiner, 2009; Giovannoni, 2010). A list of diploid and triploid apple clones is presented below (Pratt *et al.*, 1978).

Diploids (34 Chromosomes)

Alamanka
Alexander
Allington Pippin
Almey
American Forestier
American Summer Pearrnain
Ananas Berzenicki
Anderson Jonathan
Anna
Arkad Zimnyi
Ashmead's Kernel
Autumn Arctic
Ballarat
Barbara Ann
Beauty
Bellefleur Rekord
Benoni
Black Crofton
Blanche Ames
Blaze
Bob White
Bottle Greening
Bramtot
Britemac
Brock
Burke Sweet
Cardinal
Carroll
Chehalis
Columnaris
Cowichan
Cranberry
Crimson Gold
Crittenden de Juane
Dochdianay
Dolgo
Duke of Clarence
Dunn's Seedling
Early Harvest
Eastman Sweet
Ebenezer Lambkin
Fall Russet
Frettingham Victoria
Freyberg
Fuji
Gala
Gano
Garland
Golden Hornet
Gooisbey
Hawaii
Henrietta Crosby

Henry F DuPont
Holiday
Honeygold
Hoover
Idared
Imperial McIntosh
Jay Darling
Jerseymac
Jonadel
Jonalicious
Julyred
Katherine
Kensib
King Cole
King Luscious
Lady Carrington
Large Yellow Siberian
Lyman Prolific
Magnolia Gold
Maiden Blush
Malus baccata
Malus halliana
Malus purpurea aldenhamensis
Marry Potter
Mingan
Mio
Mollie's Delicious
Monarch
Nipissing
Nova Easygro
Ortley
Ozrak Gold
Paducah
Parkman
Paulared
Pepin Shafrannyi
Pink Beauty
Pioneer Scarlet
PK-14
Poly Eades
Prairie Rose
Prima
Prime Gold
Prof. Sprenger
Quinte
Radiance
Raritan
Red Canada
Red Jacket
Red Thorle
Redflesh
Reineta Encarnada
Renet Bergamotyj
Roanoke
Rodney
Scotia
Shenandoah
Slavyanka
Snowdrift
Stark Blushing Golden
Stark Splendour
Stark Summer Delicious
Stirling
Sumatooka
Summer Scarlet
Summerred
Sungold
Tack Ben Davis

Thewgold	Toko
Twistbody Jersey	Veitch Scarlet
Virginiagold	Winter Gold

Triploips (51 chromosomes)

Bisbel Giant Winsap	Boro Belle de Boskoop
Boutielle de Liseaux	Charlottae
Colby Baldwin	Galbraith Baldwin
Knuthenborg Gravenstein	Rouge Belle de Boskoop

Apple Genome

A draft suquence of the apple genome has been released by the Italian-led international team of scientist in August 2010 (Velasco *et al.*, 2010). The genome of 'Golden Delicious' one of the most widely grown apple variety in the world, is the source of published sequence. The release marks a major break through in apple genetics and genomics. Several mysterious questions have been answered by the release, including the number of genes encoded by the apple genome, the origin of the 17 chromosomes of apple, and the origin of the domesticated apples (Xu, 2010). The apple genome sequencing project was initiated in 2007 involving 85 scientists from Italy, New Zealand, France and Belgium and was coordinated by Dr. Riccardo Velasco, a plant geneticist based at the E. Mach Foundation, Trento, Italy.

A strategy known as Whole-Genome Shotgum (WGS) was adopted in apple genome sequencing. Prior to sequencing, high molecular weight genomic (DNA cDNA molecule of several Mb) was isolated and fragmented into smaller pieces. The smaller pieces of DNA were used to construct 27 genomic libraries with size ranging from 500 bp to 135 kilo (1000) base pairs (kb) for sequencing. Both the Sanger Method based sequencers (GE Healthcare's Mega BACE 4500 and ABI 3730 XL DNA analyzer) and the G2 sequencers (Roche/454 Genome Sequencer) were used. Ultimately, a total of 39.2 million reads accounting for approximately 13 billion sequenced nucleotides were generated among which 26 per cent were by Sanger Method and 74 per cent by the Roche/454 Genome Sequencer (Velasco *et al.*, 2010). Assembling of the 39.2 million reads (based on identical sequences shared between individual reads) has led to 122, 146 contigs of an accumulative length of 604 Mb, covering 81.3 per cent of the genome with a new estimate of 742.3 Mb. Using 1,643 genetic markers, 528.3 Mb representing 103.076 contigs were anchored to 17 chromosomes. The anchored 528.3 Mb sequences are gene rich regions and contain more than 90 per cent of the apple gene assigned with chromosomes (Valasco *et al.*, 2010). However, 138.4 Mb (18.6 per cent of the genome) have not been determined with their corresponding chromosomal origins although most (98 per cent) are repetitive elements that make up nearly two thirds of apple genome. Analyses of the apple genome sequence find that the apple genome encoders 57,386 predicted genes. Importantly, nearly 20 per cent (11,444) of those predicted genes are apple specific as they do not show significant similarities with other plant genes (Velasco *et al.*, 2010). Discovery of the large number of apple specific genes strongly suggest that the apple genome is noticeably unique, supporting the distinctive patterns of apple as a tree fruit in growth, development and adaptation to the temperate environments. Several gene families are overpresented in apple genome, including those related to sorbitol (a type of sugar) metabolism, disease resistance, and biosynthesis of volatiles, aromatics, pigments and antioxidants. In addition, apple-specific groups of genes encoding MADS-box transcription factors which are important in regulating apple fruit growth and development were identified as well (Xu, 2010). To resolve the mysterious origin of the domesticated apple, 23 nucleic genes across 74 *Malus* accessions including *M. domestica* (12), *M. sieversii* (10), and *M. sylvestris* (21) were sequenced and compared. The comprehensive study supported the view that *M. domestica* was originated from *M. sieversii*. As the cultivated apple was domesticated from *M. sieversii* only around 4000 years ago, a relatively short time in plant evolution, the study proposes that two species merged into one (Velasco *et al.*, 2010).

Cultivars

All the new commercial cultivars of apple are derived from carefully planned crossing and selection programme with specific objectives. These new cultivars are generally more resistant to diseases and abiotic stress, more productive and have good storage ability. The different cultivars have been variously classified, as (i) early, (ii) mid-season and (iii) late; (i) dessert and (ii) culinary apples; and (i) green apples, (ii) red apples and (iii) apples of the Cox type. Listed below are a few dessert and culinary apples of UK.

Dessert Apples

Discovery
Merton Knave
Worcester
Fortune
St. Edmund's Pippin
Lord Lambourne
Ribston Pippin
Sunset
Malling Suntan (triploid)
Malling Kent
Orleans Reinette
Golden Delicious
Crispin (triploid)
George Cave
Miller's Seedling
James Grieve
Merton Worcester
Charles Ross
King of the Pippins
Egremont Russet
Cox's Orange Pippin
Kidd's Orange Red
Laxton's Superb
Spartan
Ashmead's Kernel
Late Orange

Culinary Apples

Early Victoria
Rev W Wilks

Grenadier	Golden Noble
Lane's Prince Albert	Newton Wonder
Bramley's (triploid)	Edward VII

In the USSR, the cultivars recommended for areas with high scab and powdery mildew pathogenicity are Antel, Azau, Koster, Khasan and Bagryanoe (Alekseeva, 1983). The cv. Renet Kichunova showed most resistance to scab in different ecological and geographical conditions in the USSR (Kochanov and Konovalov, 1986). The cvs. Orlik (early winter) Lobo (winter) and Spartan (late winter) were recommended for cultivation in Khar Kov province (Pavlenko, 1986). The cvs. Renet Burkhardta; Malinovka, Augustovskoe and Malinovoe Oberlyandskoe are regarded as high yielding (Nesterov, 1986), while Melba, Miltosh, Pepin Chernenko and Diana are winter-hardy cultivars of the USSR (Sidnon, 1986).

The Technical Commission for the evaluation of fruit cultivars has recommended 22 dessert apple cultivars for Switzerland (Anon., 1984-1985). They are classified into 4 groups: (i) main cultivars suitable for general large-scale commercial production (*e.g.*, Gravensteiner and Summerred), (ii) supplementary cultivars for local interest and for special purposes (*e.g.*, Julyred, Discovery, James Grieve and Primerouge); (iii) valuable new cultivars recommended for limited planting to obtain more information of their practical performance (*e.g.*, Vista Bella, Jerseymac), and (iv) older cultivars grown on a substantial scale but not recommended for new planting (*e.g.*, Klarapfel, Stark Earliest).

Yellow Transparent, Red Astrachan, Gravenstein and American Summer Pearmain are the old early cultivars of China (Zhaonian Guo, 1986). In Ching Mai Province of Thailand, the cv. Anna produced the largest fruit while the highest acid content was recorded in Einsherner, and harvest firmness was highest in Rome Beauty (Subhadrabandhu and Punsri, 1986) Blazek (2011) studied the climatic demands of apple cultivars in Czech Republic. He stated that 'Golden Delicious' is most suitable for growing in the warmest areas. 'Gala' could be grown in areas up to 400 masl. 'Gloster' is performing well in warm and medium altitudes. 'Angold' 'Sampion', 'Bohemia' and 'Rubin' an suitable for colder region of the country. Chadha (1978) recommended the following cultivars for different apple growing states of India.

Himachal Pradesh: King of Pippins, Red Junet, Tydeman's Worcester, Golden Delicious, McIntosh, Red Gold, Richared, Starking Delicious, Granny Smith, Yellow Newton.

Jammu and Kashmir: Benoni, Irish Peach, Cox's Orange Pippin, Kerry Pippin, Lal Cider, Razakwar, Ambri, Apirouge, Golden Delicious, Lal Ammbri, Maharajki, Red Delicious, Sunehari.

Uttar Pradesh: Early Shanburry, Fanny, Cortland, Golden Delicious, McInstosh, Red Delicious, Buckingham, Rymer.

According to fruiting time under the conditions prevailing in the apple growing areas of Kumaon and Garhwal hills of UP (India), Srivastava (1975) classified the cultivars as:

Apple Orchard

Bitter Pit

Columnar Apple Trees

Gala

Fire Blight

Discovery

Gloster 69

Greensleeves

Granny Smith

Golden Delicious

Jonagold

Kalei

Leaf Miner

Red Delicious

Red Fuji

Russeting in Apple

Watercore

Training System

Central Leader

Modified Cenreal Leader

Cordon

Espailer

Open Centre

Early: Beauty of Bath, Red Astrachan, Early Shanburry, Summer Golden Pippin, Fanny, Benoni, Devonshire Red Ouarrenden, Green Sweet.

Mid-season: King of Pippin, Red McIntosh, McIntosh, Jonathan, Spartan, Red Delicious, Royal Delicious, Cortland.

Late: Golden Delicious, Rymer, Esopus Spitzenberg, Peck's Pleasant, Buckingham.

The cultivars, Tropical Beauty and Parlin's Beauty, were judged best in respect of productivity and fruit quality and are recommended for cultivation under mild-hill conditions. (Gautam and Chauhan, 1986).

Brief descriptions of some commercially important apple cultivars of the world are presented below.

Delicious

The leading world cultivar 'Delicious' (also known as 'Red Delicious') was selected as a seedling in lowa, USA in 1980 and is characterized by a distinct oblong shape with lobes at the calyx end. It is dominated by a number of highly coloured strains, such as Starking Delicious, Red Chief, Ruby Red, Red Spur, Royal (Starking) Delicious, Richared Delicious *etc.* Popularity is declining in the United States and Europe, but it is still highly appreciated in South America and Asia. Consumers appear to prefer 'Fuji' to 'Delicious', but due to the extensive plantings, it will remain as a leading cultivar.

Golden Delicious

This yellow apple was selected as a seeding in West Virginia, USA in 1980. 'Golden Delicious' is the leading cultivar outside China due to enormous planting in Europe and its continued popularity in the Eastern United States. Clearly this apple has peaked, but will continue to dominate the apple picture because of its excellent quality for fresh and processed products and high productivity despite only moderate storage life. Flesh creamy-white, firm, crisp, sweet with blend of acidity.

Jonathan

Tree medium size, moderate vigour and a reliable cropper. Fruits are round to conical, medium to small and reddish in colour with good quality. Fruits are, however, susceptible to black Jonathan spots.

McIntosh

Tree vigorous, broadly diverging. Fruit medium, oblate round, skin smooth, shining carmine colour on a pale green background. Flesh white, develops red shade after storage, tender, crisp, sweet with a good acid blend, juicy and mild flavoured. The cultivar is, however, susceptible to scab.

Bramley's Seedling

Bramley's the best-known English cooking apple is ideal for stewing and especially for baking. Skin deep green, waxy, sometimes with a slight orange-red blush.

Fuji

This is the popular cultivar of China. It derives from a cross of two American apples 'Ralls Janet' and 'Delicious' and was introduced in 1962 by the Tohoku Horticulture Research Station in Japan. Its mild sweet flavour, fine texture and large fruit size has made 'Fuji' very popular especially for Asian and others preferring sweet apples.

Gala

This New Zealand Apple is the result of a cross between 'Kidd's Orange Red' and 'Golden Delicious'. This mid-summer apple, despite it's disease susceptibility, has rocketed into one of the leading apples as a result of its crisp flesh, sweet, aromatic, flavour and wide appeal. New Plantings are predominantly of the red sports.

Grany Smith

This cultivar was developed from a seedling fruiting in Australia in 1968. It has firm, tart fruit that broke the prejudice against green apples for fresh use. Late maturity has restricted production in Northern areas. It is widely appreciate for processing including cider, juice and cut apples. It performs very well in hot climates, and in areas with marginal winter chilling.

Jonagold

A triploid red apple with large size and good texture was released in Geneva, New York, USA from a cross between 'Golden Delicious' and 'Jonathan' and likely is based on an unreduced egg cell of 'Golden Delicious'. It rescued the European apple industry in 1980s. The red sport 'Jonagored' is widly planted.

Braeburn

A seedling discovered in New Zealand was introduced in 1952. It is very high quality apple but potential for expansion is limited because of its fruiting habit and it is subject to calcium-related storage problems.

Kent (Diploid) Cox X Jonathan

A high-yielding late maturing apple cultvar. The fruit is slightly conical with a dark red colour on a green background. Trees are medium in size and less susceptible to core flush and bitter pit than Cox. The cultivars Worcester, Golden Delicious, Greensleeves and Spartan are suitable pollinators of Kent.

Suntan (Triploid) Cox X Court Pendu Platt

It is very similar in appearance to Cox but with superior aromatic flavour to Cox and higher acidity. Fruits are orange-red flush and stripes on a greenish gold background. Trees are vigorous and spreading. Fruits are susceptible to bitter pit. It is pollinated by Kent.

Greensleeves (Diploid) James Grieve X Golden Delicious

The cultivar is a heavy cropper and gives good yields on spurs and one-year-old wood early in the life of the tree. Trees are semi-dwarf and flower profusely. Fruits are green-yellow, crisp, juicy, slightly acidic with attractive flavour. Greensleeves is a good pollenizer for Cox, Merton Knave, Worcester, and all these cultivars pollinate Greensleeves.

Jester (Diploid) Worcester X Starkspur Golden Delicious

Trees are moderately vigorous but compact with wide angled branches. It is very precocious and the fruits are bright red in colour. Fruits are crisp and juicy.

Jupiter (Triploid) Cox X Starking

Trees are more vigorous than Cox with wide branch angles, high yielding and with Cox flavour. Fruits are green-yellow with an orange-red flush and stripes. The cultivars, Discovery, Spartan, Goden Delicious, Gala, Greensleeves and Jester are suitable pollenizers of Jupiter.

Gloster 69

Cultivar was released from West Germany. Trees are upright and fairly vigorous. Fruits are oblong conical, ribbed and often flat sided. The flesh is creamy white, fine textured and juicy with sweet and pleasant flavour. It is compatible with Cox's Orange Pippin, Golden Delicious, Gala and Spartan, and itself a good pollenizer for Suntan. Gloster has a tendency to alternate bearing and occasionally to produce oversize fruits (leading to poor storage quality), and susceptibility to bruising and water core (Mantinger, 1987).

Cortland

Tree vigrous, diverging. Fruits are medium to small, round oblate, skin smooth, carmine streaks on a yellowish red background. Flesh is greenish white, firm, crisp, juicy with a mild flavour, Sweet in taste.

Spartan

A good eating cultivar. Fruit is firm, medium, conical, purplish red with white flesh, taste sweet with blend of acidity and good aroma, and stores extremely well.

Ambri

Ambri is the only indigenous variety grown in India. It is a late season variety (ripes in September-October), fruits has red streaks over a greenish yellow background. The fruit can be stored for 10 months in storage. Because of its excellent storage life, the variety has been largely used in India as one of the parent in breeding programme.

Prijedorska Zelenika

An old-established cultivar of Yugoslaiva and is thoroughly compatible with modern rootstocks (Marjanovic, 1983). The cultivar gives high yields on M.9, M. 26 or MM. 106 rootstocks than on wild Crab seedling stock. Fruits are roundish but somewhat elongated, weighing around 150 g, greenish yellow when ripe, with an attractive scattering of russet spots. The flesh is white, firm and juicy, acid-sweet in flavour, with a pleasant aroma. It is moderately resistant to drought and low temperature.

Stark Summer Treat

Developed from NJ109055 X Mollie's Delicious in the USA. This is a summer cultivar, ripening 3-4 weeks earlier than Red Delicious (Mehlenbacher *et al.*, 1985). The fruits are large and bright red, and should be picked daily as they ripen. The flesh is straw coloured and moderately firm. Dessert quality is good. McIntosh and Red Delicious are good pollenizers for Stark Summer Treat.

Himekami

This cultivar was bred at the Morioka Branch of the Fruit Research Station, Shimokuriyagawa, Japan and derived from pollination of Fuji by Jonathan (Horie, 1985). The red-skinned, crisp fruit wighs on an average 300g, matures in late September and keeps for 4-5 weeks. The trees are erect and vigorous, bearing regularly.

Shamrock

Shamrock originated from a cross between the irradiated McIntosh clone 10C8-43-1 and the Golden Delicious strain Starkspur. It is a green apple, nearly identical in appearance, texture and taste with Granny Smith, but matures 6 weeks earlier and has a more desirable growth havit. It was recommended for northern areas of Canada where Granny Smith does not mature before cold weather slows down growth (Lance and Mac-Donald, 1987)

Fiesta

This is a English apple cultivar (raised at East Malling Research Station), a cross between Cox and Idared. Fruits have yellow background, with a red lightly striped blush covering up to 90 per cent of the surface. They can be stored successfully in cold storage at 1.5-3°C until January and in CA storage until the end of May (Goddrie, 1987d). Fruits are larger and mature earlier than Cox (Lovelidge, 1987).

Red Elstar

This mutant apple cultivar was found in 1981 in a Dutch plantation of Elstar; the top portion of one tree bore fruits, which were earlier and more coloured than the normal fruits of mother plant (Visser and Verhaegh, 1987). Red Elstar was released for commercial cultivation in July, 1987. The fruit colour develops earlier and the intensity is more with a greater blush and fewer stripes, at the normal time of harvest.

Centenaria

This was selected from Golden Delicious x Valinhense (Rainha) at the Instituto Agronomico, Campinas (lAC) Brazil. It has a cold requirement of 150 h below 7.2°C. Fruits are large (220-250 g), yellow, and mature early (Anon., 1987a).

Marquesa

Marquesa, also from the same parents as Centenaria selected at lAC requires

120 h cold, The fruits weigh 150-220 g and resemble those of Red Delicious. Yield is good, when cross-pollination is adequate, although the cultivar is self-fertile, Maturity is earlier than Rainha (Anon., 1987a).

Meran

Meran is derived from a chance seedling of unknown parentage. It is more vigorous than Golden Delicious and because of its habit it is ideal for cultivation as spindle bushes (Mantinger *et al.*, 1986). Meran begins yielding from the second year. Fruit set tends to be heavy and fruit thinning is required to increase ovarall fruit size and to prevent alternate bearing. The fruits are firm and has higher acid content than Golden Delicious at harvest.

Fu Cui

Fu Cui was derived from the cross Chi Yang x Golden Delicious in 1962 in China (Zhaonian Guo, 1986) and was selected from a progeny of 73 seedlings. The flesh of Fu Cui is crisp, fine and tender with sweet and good flavour. It is highly productive and resistant to apple leaf spot and powdery mildew.

Romus-2

The cultivar Romus-2, released from Romania in 1984 has good resistance to scab, early maturity (mid-July onwards) high yield and good fruit quality. Fruits are medium sized (120 g), largely red and have a good flavour (Anon., 1987b).

Delbarestivale

This French dessert apple cultivar (alternative name Delcorf) is a cross between Jongrimes and Golden Delicious. It was raised by a nurseryman. M. Delbard at Malicorne, France. Suitable pollenizers include Akane. Red Delicious. Gala and Idared. The fruits are green-yellow in skin colour, with little blush, flesh creamy white and most are in the 70-80 mm size grade or larger (Goddrie. 1987a).

Rubinette

Rubinette, also known as Rafzubin, is a diploid chance seedling of Golden Delicious found in Switzerland; the presumed pollen parent was Cox's Orange (Kellerhals Hohn, 1987). The fruits are small to medium in size (60-70 mm diameter), weigh 120-150 g, orange-red, have a slight tendency to russeting. Fruits are tasty, juicy and firm, and have a high sugar and acid content. Maturity in Switzerland is in early October. Bearing is precocious and regular. However, because of its small fruit size it is not recommended for commercial planting in the Netherlands (Goddrie, 1987b).

Malychenkovskoe

This Soviet apple cultivar, bred at Volgograd from a cross between Boiken and Zhigulevskoe, has weak vigour and begins to fruit in the 3rd/4th year. The fruits are large (mean weight 160 g), uniform and yellowish-green with a red blush, juicy and have acid-sweet flesh. The mean yield is 17.18 t/ha and belonging to the early winter maturity group.

Table 1.7: Hybrids Developed in India

Hybrids	*Parentage*	*Important Characters*
Ambred	Red Delicious x Ambri	Good keeping quality; low incidence of powdery mildew, sooty blotch and scab
Ambstarking	Starking Delicious x Ambri	Tolerant to scab
Ambroyal	Royal Delicious x Ambri	Semi-dwarf tree, semi-spur type; good dessert quality
Ambrich	Richared X Ambri	Tree semi-dwarf, semi-spur spreading, Good desert quality, tolerant to apple scab.
Chaubattia Princess	Red Delicious x Early Shanbury	Early (matures in the last week of June), fruit deep red streaks on pale background, very sweet and good keeping quality
Chaubattia Anupam	Red Delicious x Early Shanbury	Tree vigorous and uprignt, ripens in the second week of July, flesh creamy-white, crispy
Lal Ambri	Red Delicious x Ambri	Red Coloured apple
Sunehari	Ambri x Golden Delicious	Good keeping quality, colour similar to Golden Delicious
Shalimar Apple 1	Sunhari x Prima	Mid-season variety, Sunhari in shape and Ambri in taste, resistant to apple scab
Shalimar Apple 2	Red Delicious x Ambri	Tree medium, wide spreading, mid-season variety, Attains maturity in 138 to 151 days after full bloom

Freedom

Originated as a cross made between NY 18492 and NY 49821-46. Freedom is highly resistant to *Venturia inaequalis* (Lamb *et al.*, 1985). The fruit is large in size and oblate in shape with firm, moderately coarse flesh. It is suitable as dessert and for processing as cider or juice.

Iwakami

Iwakami was selected from Fuji pollinated by Jonathan. The fruits are yellow-green with a bright red blush, medium size, weighing on an average 200-250 g, crisp and juicy with strong acidity. They keep well for 20 days in an ordinary store and 60 days in cold storage (Yoshida *et al.*, 1985).

Pinova

Pinova, a diploid, compatible with its parents (Clivia x Golden Delicious), was released from Pillnitz, GDR, in 1986. It is harvested in late October and stores well till June. Fruits with pleasant flavour are suitable for processing. It is fairly resistant to spring and winter frost, scales, mildew and fireblight. It comes early into bearing, but has a biennial tendency (Fischer and Patzold, 1986).

Majda

This is a Yugoslav apple cultivar, bred from a cross between Jonathan x Golden Noble. Growth is vigorous, high yielder and resistant to scab (Crnko and Marn,

1987a). Mean fruit weight is 175 g and the fruits are suitable for fresh consumption or processing. It ripens about 5 clays after Idared.

Taishan Gala

This is a large and red bud mutation of Royal Gala developed recently in China (Li-LinGuang *et al.*, 2012). The fruit is large, weighing 212.8 g on average. The quality is excellent, with 15.0 per cent soluble solids content. It is highly productive and an early ripening apple cultivar.

Qicefuhong

Qicefuhong is a spur-type bud mutant, selected from Fuji at China (Xu-GuiXuan *et al.*, 2012), It produces large and dark red fruit, high yielder, excellent fruit quality and a late-ripening cultivar

Hongan

Hongan was originated from an artificial cross between Fuji and Jonathan at Korea (Kwon-Sront, 2011). The fruit has light red skin on a greenish yellow ground and yellowish-white flesh. The fruit contains 13.7° Brix, resistant to bitter rot.

Kalei

Kalei is the result of a 20 years breeding program by the Queensland Government, Australia. This new apple is resistant to scab and tolerant to *Alternaria* disease and Western flower thrip. Kalei is well adapted to high density growing systems and can produce crops between 80 and 100 tonnes per hectar each year. The variety has been developed through controlled cross pollination.

In India, good numbers of hybrids have been developed in the last four decades. Some of them have excellent storage life, and some are scab resistant (Table 1.7).

Climate

Temperature

The apple is a typical temperate tree fruit and is generally grown in the temperate regions of the world. Within the broad temperate zone, production is concentrated in compact areas in which the environment is particularly favourable (Jackson, 1983). Apple succeeds best in regions where the trees experience uninterrupted winter rest, since the trees require certain minimum chilling period for growth and fruiting. Kronenberg (1979) suggested a chilling requirement of 1000 hours at below 7°C. However, some cultivars have low (250 hours) chilling requirement. Extreme cold temperature on the other hand caused freezing injury. The cvs. San Miguel and Cella, known for their very late flowering, have very high chilling requirements of 1400 and 1600 h, respectively, at below 7°C. In years of mild winter weather, when these requirements were not properly met, anomalies in flowering were observed (Tabuenca, 1983). However, there are low chilling apple varieties eg 'Anna' 'Princesa' which have low chilling requirement of 350 to 450 chill units for satisfactory flowering (Njuguna *et al.*, 2004, Nachtigall, 2004). Jackson and Hamer (1980) found a close correlation between high temperature in the pre-blossom period and low yield in Cox apples in the UK. In seasons of high temperature in February, March and April, Cox apples set poorly. Flowers developed under nigh temperature conditions are usually smaller and the flower buds are sensitive to frost at the green cluster stage. The spur leaves become small and thin (Miller *et al.*, 1986). However, higher temperature during fruit growth hastens fruit maturity. In the UK, Luton and Hamer (1983) reported that the harvest of cv. Cox was about 3.5 days earlier for each 1°C increase in mean temperature during June, July and August. In comparatively warmer climate in New Zealand, Cox matures within 3.5 months after blossoming compared with 5 months in UK (Denne, 1963). The climatic conditions have a marked effect on the rate of fruit development during the early part of the growing season resulting in generally softer textured apples at harvest. Work on Bramley indicated that high temperature in July may be associated with poor texture at harvest (Sharples, 1984b). Uota (1950) found a close association between mean night temperature and red colour development in McIntosh apples and low night temperatures during the last few weeks before harvest were particularly beneficial. High temperatures (around 76°F or 24.4°C) inhibit anthocyanin production. Anthocyanin synthesis was found most rapid in the cold, *i.e.*, at average temperatures of 46.5°F (8.1°C) and 60.5°F (15.8°C) during the night and day, respectively (Uota, 1952). Cool growing seasons, however, have been found to lead to a higher risk of core flush in Cox and McIntosh (Sharples, 1973) while fruits grown under high temperatures are more susceptible to superficial scald than those grown at lower temperatures (Sharples, 1975). Summer drought (complete lack of precipitation) delays senscence due to an increased demand of carbohydrate during autumn. The delayed enabled replenishment of root carbohydrate reserves, which is critical for next year's growth and fruiting (Naschitz *et al.*, 2014).

Death of trees following freezing injury is a major factor limiting apple cultivation (Jackson, 1983). Czynczyk and Holubowicz (1984) reported that with a soil temperature of -11°C at a depth of 5 cm, nearly 75 per cent of trees on M.9 rootstock were killed. The rootstocks, M.2, M.4 and M.9, proved frost susceptible. The cvs. Ontario, James Grieve and Berlepsch were reported to be more susceptible to frost than Stark Earliest, Klarapfel and Glockenapfel (Schumacher and Fankhauser, 1985). The winter of 1984-85 caused severe injuries to Finnish (Finland) apple orchards. In January, the temperature fell to –40°C and in February it stayed near 35° C for a long time. Young trees 1, 5 years or less, were found more susceptible to injuries than older trees. Trees exposed to hard north winds were most badly injured (Sako, 1986).

Sunshine

Abundant sunshine is important in growing apples since it is largely responsible for good colour development. Surveys of apples of several cultivars growing at elevations of 800 and 300 m in Palestine showed that the area of red colour on fruit surface was greater at higher altitude (Damast, 1949). The intensity of solar radiation determines the level of economic yield. Only the outermost zone of Cox trees was found to produce good coloured fruits compared with only a few smaller and green fruits on heavily shaded inner zones (Jackson, 1970). Solar radiation penetrating to all parts of the crown and providing 0.3 to 0.8 cal cm^2 $minute^1$ creates optimal conditions for productive leaf activity (Provorchenko, 1986).

Rainfall

Apple trees are particularly sensitive to low soil moisture supply (Jackson, 1983). Water stress during growing season reduces number and size of fruits, and increases June drop (Goode, 1975). High precipitation during the four weeks after bloom was found to be associated with a high intensity red colour development (Mowry, 1960-61). Application of water during the fortnight preceding harvest markedly improves fruit colour (Piolanti, 1964). However, rainfall during the later part of fruit growth evidently tends to promote russeting (Walter, 1967). Dalbro (1958) found that the incidence of russeting on Cox in different years was closely related to the number of rainy days; and in controlled experiments in which some trees were covered with plastic tents from two weeks after blossoming until harvest, the fruits remained quite smooth. Approximately 300 mm rainfall between April to September is considered sufficient for apple orchards in UK. Waterlogging for a considerable period of time was also found detrimental. Soil waterlogging imposed for a 6-week period in spring, summer or autumn reduced vegetative growth and fruit yield in Macspur trees (on M. 26). Shoot and trunk growth was reduced most by spring and summer waterlogging. Waterlogging during spring reduced the yield by 52 per cent (Olien, 1987).

Soil

The distribution of apple growing regions all over the world clearly indicates that apples can grow on a wide range of soils. However, a well-drained, slightly acidic (about pH 6.5-6.7) loam soil with good depth (45 cm or more) is considered ideal for apple cultivation. Planting in shallow soils is possible, but it will involve digging a pit not less than 45 cm deep and 75 cm square at each site to be filled with a good medium loam soil (Baker, 1983). In the UK, an ideal soil is one which is well-drained, of good structure and organic matter status and supplying about 150 mm of available water (White, 1984). Most of the apple soils are slightly to moderately acidic (pH 5.5 to 6.8), although apple is also grown on alkaline soils in the central Prairies of Canada, the plateau of central Mexico, and parts of the Galilee in Israel (Cummins and Aldwinckle, 1983).

Soils with heavy clay or compact subsoil should be avoided for planting apples as in clay soil, most roots slope downwards and grow in the subsoil. Growth of the trees, particularly those on M.26 rootstock, was adversely affected due to soil compaction (Slowik, 1970). A direct relationship was found between crop productivity and depth of friable soil layer. Increasing the friable soil depth from 50-60 to 120 cm was found to increase the yield of apple cv. Renet Simirenko on local Caucasian apple and Anis Kubanskii seedling of the USSR (Teren'ko and Kal'gina, 1988). In the valleys of NW Caucasus, USSR, the cv. Renet Simerenko give yields of 200-250 h kg/ha on alluvial meadow soils with a gravel layer at 60 cm, and 250-300 h kg/ha on a deeper soil (Kal'gina, 1988). Soil compaction effected growth and nutrient uptake by the plant. Ferree and Schmid (2001) reported that a bulk density of 1.5 g/cm^3 caused a decrease in dry weight of shoots, leaves and roots of apple cv. Melrose on different rootstocks tried. Trees on M-26 EMLA and MM. 106 EMLA were more affected by severe soil compaction. Compacted soil resulted in a decrease in leaf concentration of K and B and an increase in Mg and Mn.

The root growth of apple begins when the soil temperature reaches about 7°C while it shows a minimal growth at 1-4°C and limits growth above 28°C (Ferree and Carlson, 1987). The optimum temperature for root growth is around 18°C (Nightingale, 1935). Warm (5-19°C) and high (18-26°C) soil temperature could be achieved by using clear and black plastic covers over bare soil and with buried heat tapes under insulation and clear plastic (Hammond and Seeley, 1978). The best growth and productivity of apple are found on soils with good drainage and aeration. Apple trees can not tolerate wet feet, and waterlogging may cause root-rot and consequent death of trees. However, recent development in rootstock breeding showed the relative tolerance of M.13, M.16 (Cummins and Norton, 1974), M.7 and M.26 (Rom and Brown, 1979) rootstocks to excess moisture.

Rootstocks

All successful apple rootstocks, either seedling or clonal, are of the genus *Malus,* although the apple can grow and even come into bearing on pear (*Pyrus communis*) roots. In early days, French Crab seedlings (*M. sylvestris*) were widely used in the USA as rootstocks. Some of the Asiatic species of apples are apomictic and breed true from seeds even when pollinated by other species. These apomictic species have been used as rootstocks for commercial apple cultivars with the hope that they might be of value in controlling tree growth and age of fruiting (Sax, 1949). *Mallus hupehensis, M. toringoides. M. sargenti* and *M.sikkimensis* are apomictic (Sax, 1949, 1953). The seedlings are generally resistant to root-knot and root-lesion nematodes and moderately resistant to oak root fungus.

Since the clonal rootstocks are more uniform than sexually reproduced seedlings, one might expect greater uniformity of trees propagated on clonal stock. By the fifteenth century, budding and grafting techniques were developed and the use of rootstock became possible (Ferree and Carlson, 1987). The apple growers began to recognize that grafting also permitted the selection of specific rootstocks for their capacities to influence favourably the productivity of the scion cultivar and for their abilities to survive the edaphic, climatic and biotic components of the rootstock environment (Cummins and Aldwinckle, 1983). The selection of suitable rootstock of apples for a certain tract has been an important investigation and rootstock breeding programme was carried out simultaneously in different research institutes of the world. Numerous clonal, asexually propagated apple rootstocks have been developed for various purposes, An ideal rootstock should possess uniformity in vigour, resistance to diseases and pests without impairing the productivity and quality.

Characteristics of Rootstocks

Malling Rootstock

In 1912, the East Malling Research Station in England selected and developed a series of sixteen apple rootstocks which were named as Malling stock (M) and were numbered in Arabic numericals. Some of these rootstocks are very dwarfing while others are semi-dwarfing or vigorous. All these rootstocks are easily propagated by stooling and trench layering and some even by stem cuttings.

Dwarfing Rootstocks

M.9 (Jaune de Metz)

This is the most widely used dwarfing rootstock for apple cultivation, which orginated as a chance seedling. This rootstock can be identified in the nursery with large oblong leaves and pie-crust edges. Shoots are straight with reddish yellow wood. The major demerits of M.9 stock are its brittle roots and shallow root system. The roots are unable to perform the dual duty of holding the plant firmly to the ground as well as absorbing the nutrients from the soil and thus needs staking (Sinha *et al.*, 1982). This was, however, re-cloned as M.9 EMLA (East Malling - Long Ashton) which is free from all known viruses (Cutting and Montgomery, 1973). M.9 can be used as an interstock, but the dwarfing effect was found to be more when used as a rootstock than as an interstock.

M.27

Originally known as seedling 3431, it was raised at East Malling by Mark Tydeman in 1929 from a cross between M.9 and M.13. This is the most dwarfing rootstock now available in the market and a 15-year-old bush trees on M.27 is approximately half the size of those on M.9 (Preston, 1978). The leaves are large, leathery and puckered. The leaf margins are coarse but stems are smooth and feathery. Trees on M.27 are approximately 1.2 to 1.5 m in height and started cropping immediately in the second year of planting, depending on the cultivar. The triploid cultivars (Bramley's Seedling, Crispin), which are inherently vigorous, produce very dwarf trees on M.27 enabling them to grow very intensively in multi-row system. It is now possible to consider very high density planting with this rootstock as close as 0.5 m apart in rows 1.5 m apart planted in multi-row beds (Preston, 1978). M.27 can be propagated by stooling or by long hardwood leafless winter cuttings in January.

M.20

It originated as a rogue in a batch of M.9 at East Malling (Preston, 1978). The Cox tree on M.20 was found to produce similar tree size as on M.27. The disadvantage of M.20 rootstock is the emergence of large number of suckers from underground and thus unfit for high-density orcharding.

Semi-dwarfing Rootstocks

M.2

This is the traditional apple rootstock of moderate vigour used in UK and part of Europe. However, in view of the susceptibility of M.2 rootstock to woolly aphid infestation, it is unlikely that M.2 would ever replace Northern Spy as the recommended rootstock of moderate vigour for New Zealand (McKenzie, 1964). It is also susceptible to crown gall but resistant to crown rot. The leaves are evenly shaped and very close, with many lenticles on the stem. Trees on this rootstock which are usually propagated by stooling, comes into bearing early.

M.7

This rootstock produces trees larger than on M.9 and the semi-dwarfing rootstock; M.26. It has a stronger root system than M.9 and thus needs no staking. A tree on M.7 produces fruits early but the rootstock is susceptible to crown gall and crown rot. The leaf stalk is upright with lamina at right angles. The shape of the leaves is nearly circular and thin. This can easily be propagated by layering.

M.26

This is a cross between M.16 and M.9 and makes trees smaller than on M.7. It is suitable for high density planting. Sometimes it produces even smaller trees than on M.9. Carlson (1970) stated that Golden Delicious on M.26 is smaller than on M.9. It has good anchorage when planted 10-15 cm deeper in the orchard than on the nursery. The leaves are smooth with weavy margins, shoots are chocolate brown with a silvery sheen. It can easily be propagated by hardwood leafless winter cuttings (Howard *et al.,* 1984).

Vigorous Rootstocks

M.12

The trees on this rootstock are very vigorous, requiring wider spacing. They come into bearing late. The stocks are, although somewhat resistant to crown gall, not adapted to low soil moisture conditions. The root system is well anchored and scion forms a spreading crown. It is very difficult to propagate.

M.16

It is the most vigorous rootstock of Malling series. Trees on this root are well anchored and vigorous, produce fruits earlier than on seedling roots. It is susceptible to Collar-rot (Smith. 1955) and woolly aphid (Mckenzie, 1964). Multiplication of this rootstock is usually done by layering.

M.25

This rootstock is the result of a cross of Northern Spy and M.2 made at East Malling in the late 1920s. This is a very vigorous rootstock, makes trees similar in size to those on M.16. It is outstanding in its vigour and precocity. Where vigorous trees of Bramley's Seedlings are required this rootstock may be suitable on account of its precocity and vigour. In a nursery, it could be recognized by zigzag stems with small leaves and short petioles. The tips of the leaves are slivery and are closed together.

Malling-Merton Rootstocks

The rootstocks in the Malling series, although had a wide choice for tree vigour, have certain disadvantages, particularly their susceptibility to woolly aphids. To overcome this problem, the John Innes Horticultural Institute, Merton and East Malling Research Station jointly started breeding programme in 1928 with Malling series and Northern Spy, a cultivar known to be woolly aphid resistant. All these rootstocks which were bred in this series were designated as MM. (Malling-Merton) rootstocks. Most of the seedlings resulted from crossing the resistant Northern Spy with some of the Malling rootstocks, and those selected for trial were propagated as clones reported by Preston (1953) at the 13th International Horticulture Congress in London, 1952. The recommended MM stocks are listed below.

MM. 104 (Malling 2 X Northern Spy)

The trees on this rootstock are vigorous and have a well-developed root system. The rootstocks can easily be recognized by its zigzag stems and lenticles at the base of the shoots. Fully developed or mature leaves are upcupped.

MM. 106 (Malling I x Northern Spy)

This rootstock produces trees similar to those on M.7 in size and yield, and the yield of Cox on MM. 106 was found to be 40 per cent higher than on M.7, 15 years after planting. This stock produces a strong, well-anchored tree, can do without stakes, has a good number of roots, and is found to be free from woolly aphid and collar rot. The leaves of the rootstock are large, broad and leathery. The large stipules with one serration is found right to the shoot tip. The MM. 106 is an easy rooter in the stool beds and is a good rootstock for slight heavy to light soil.

MM. 109 (Malling 2 x Northern Spy)

Trees on MM. 109 are larger than those on M.2. The root system and yield are, however, similar to those of M.2. It can be propagated easily by stooling.

MM. 111 (Northern Spy x Merton 793)

This rootstock produces semi-dwarfing trees similar to those on M.2 but yield about 25 per cent more than on M.2. It is somewhat tolerant to drought. The leaves of the rootstock are normally upcupped and often misshapen. It is fairly easy to propagate by layering and hardwood cuttings.

Other Important Rootstocks

Northern Spy

This rootstock is resistant to woolly aphid (*Eriosoma lanigerum*) and is the traditional apple rootstock of New Zealand, but this clone has a major disadvantage. It fails to produce profitable trees on second-class soils (McKenzie, 1964). This rootstock has been used in several breeding programmes to meet the need of woolly aphid resistance on superior horticultural stocks.

Robusta No. 5

This was developed at Ottawa, Canada, as a selection of *M. robusta.* It is apparently resistant to fire blight and is very hardy. The trees on Robusta No.5 are similar to those on M.2. The propagation methods suitable for this rootstock are layering and cuttings.

P.12

This is a dwarfing rootstock developed from a cross between M.9. and common

Antonovka at The Research Institute of Pomology and Floriculture at Skierniewice, Poland. Trees on this rootstock are smaller than on M.9, and have coqsiderable resistance to crown rot. It can be propagated easily. The other important rootstocks in this P series are P.l, P.2, P.16 and P.18.

Bemali

A hybrid of Mank's Codlin x M.4 and the first selection from the apple rootstock breeding programme at Balsgard, Sweden (Cumins and Aldwinckle, 1982). Bemali is almost as dwarfing as M.26. It was selected for its ease of propagation, freedom from woolly aphid and excellent productivity (Olden, 1953).

Jork 9

The fruit research station at Jork, West Germany has released this selection from open pollinated M.9 seedlings (Tiemann and Dammann, 1981). J.9 is hardier than M.9 but less hardy than M.ll and is moderately susceptible to scab and powdery mildew. J.9 is considered suitable for the scion cultivars Gloster, Jamba and Boskoop Schmitzhubsh, but is less satisfactory than M.9 for Jonagold (Faby *et al.*, 1986).

Antonovkas

The Antonovka group has been used widely where winter hardiness is required (Hutchinson, 1974). Antonovka Kamenichka is resistant to fire blight; Antonovka Poltorofontanaja which has polygenic scab resistance but is very susceptible to fire blight (Cumins and Aldwinckle, 1982).

Ottawa 3 (0.3)

This was selected from Robin x M. 9 (Robin is a hardy Crab) at the Ottawa Research Station at Canada. Trees on 0.3 are highly productive (Spangelo *et al.*,1974) but are susceptible to fire blight and woolly apple aphids (Cummins and Aldwinckle, 1982).

MAC 9 (Mark)

It arose from open-pollinated seedlings of M.9 at Michigan State University. Trees on MAC 9 are similar in size to those on M.9 or M.26. Like M.9, MAC 9 can be propagated easily in the stoolbeds, and is better anchored. However it is very sensitive to drought (Fernandez *et al.*, 1997) and to crown gall.

Novole (PI 286613)

This is a Geneva selection from a set or open-pollinated seedlings of *M prunifolia* from Japan. Novole is a very vigorous stock and is easily propagated by cuttings (Cummins and Aldwinckle, 1982). It is resistant to crown rot, fire blight and tomato ringspot virus (Cummins and Aldwinckle, 1983).

62-396

Rootstock 62-396 (a cross between hybrid 13-14 and Budagovskii paradize) was developed in the USSR. It is readily propagated by vertical layers, starts shoot growth on stock-plants early in spring and is frost resistant (with roots undamaged by -16°C at 20 cm depth). Stockplant productivity is moderate, but considerably better than that from M.9 or Budagovskii paradize (Gul'Ko and Gul'Ko, 1985).

J-TE-G

A hybrid of M9 x Croncels, developed at Techbuzice, Czech Republic (Dvorak, 1988). Similar vigour to M.27, very high yield efficiency.

Voinesti 2

A hybrid of M.9 x Cretese, developed at Voinesti, Romania (Parnia *et al.*, 1997). Vigour similar to M.27 or P.22, good yield efficiency.

Supporter 1, Supporter 2 and Supporter 3

These three hybrids were developed at Dresden, Germany (Fischer, 1997, 1999). Supporter 1 is a hybrid of M.9 x *Malus baccata*; less vigorous than M.9, high yield efficiency. Supporter 2 developed from a cross of M.9 x *Malus micromalus*, is also less vigorous than M.9, high yielder. Supporter 3 (M.9 x *Malus micromalus*), vigour similar to M.9; high yield efficiency.

Interstock

In apple, dwarfing interstocks are used to avoid the undesirable root system of a dwarfing rootstock. The semi-dwarfing rootstock and interstock have shown some interesting relationship between stock and scion. Different *Malus* species such as M. *sikkimensis* (an apomictic species of India), hybrids of *M. sargenti* x *M. astracanica* are largely used as interstocks. *Malus sikkimensis* seedling rootstocks are compatible with McIntosh, Cortland, Baldwin, Stayman (Sax, 1953). The hybrid 33340 (*M. sargenti* x *M. astracanica*) seedling rootstocks are also compatible with McIntosh, Cortland, Baldwin, Stayman (Sax, 1953). The hybrid 33340 (*M.sargenti* x *M.astracanica*) seedling rootstocks are compatible with McIntosh and Cortland, although they are semi-dwarfing. The use of dwarfing interstock for apple is often accompanied by a swelling of the interstock and the diameter is larger than the scion or the trunk of the rootstock. This behaviour is particularly evident when Malling 7, 8 or 9 are used as dwarfing intersrock (Argles, 1937). Grubb (1939) stated that the long sections of dwarfing interstock have more dwarfing effect than short ones. Felius and Toorenaar (1959) and Carlson and Robitaile (1970) have reported that the effectiveness of dwarfing interstock is correlated with their length. Parry and Rogers (1972) stated that long interstocks resulted in 30-40 per cent less vigour than short interstocks, even with interstocks of clones more vigorous than the rootstock (Table 1.8). This is thought to be due mainly to the persistence of an initial check in the nursery caused by the method of working.

Lord *et al.* (1985) working with Empire apple recorded higher production on M.26, M.9/MM. 106 and M.27/MM. 106 than on M.9/MM.111, M.9 or M.27 and fruits from trees on M.27/MM. 111 entered into their climacteric later than those on M.26, M.27, M.9 or M.9/MM. 111.

Trees on M.9/MM.I06 (lighter soils) and M.9/MM.111 are the most common interstem combinations in Massachusetts. However, there are few plantations of

'C' series of interstem trees. These are four piece trees with a seedling root, K-14 winter- hardy trunk, a dwarfing interstem (C-6 or C-52), and the scion cultivar (Lord. *et al.*, 1983). The Polish series rootstocks, P.2 and P.22, show particular promise as dwarfing interstems (Cummins and Aldwinckle, 1983).

Table 1.8: Extension Growth of Cox on M.2 with Various Interstocks*

Interstock		*Extension Growth***	
Clone	*Length (in)*	*Nursery (1959) cm*	*Orchard (m) (1960/62)*
M.9	2	93	2.62
	14	44	1.91
M.26	2	146	2.53
	14	73	1.86
3428	2	89	2.61
	14	68	2.22
M.7	2	86	3 23
	14	44	2.31
M2	2	105	3 58
	14	49	2.21
Crab C	2	101	3.01
	14	49	2.14
3430	2	226	3.28
	14	122	2.70
Single worked	6	94	2.78
	8	93	2.66

** Nursery extension growth includes the length of feathers.

* Parry and Rogers (1972).

In Germany, Cox's Orange Pippin, Roter Boskoop Ley and Jonagold on MM. 106 and M.26 rootstocks and 15,20 or 25 cm long M.9 and M.27 interstocks were deep planted in 1982 with the interstocks projected 10 cm above ground level; control trees were grafted directly on M. 9 at 25 cm stem height. Trees with M. 9 interstocks were more vigorous and in the first year of harvesting produced higher yield than the control. Trees with M.27 interstocks were less vigorous and produced fewer shoots than the control (with the exception of Cox's Orange Pippin on MM. 106/M.27) and the yields in the 3rd to 5th year after planting were lower (Keipert, 1987a). The effect of M.9 and B.9 interstock on cropping of McIntosh, Starking and Bancroft grown on Antonovka and M.11 rootstocks were evaluated by Patzold (1986). In general, the longer the dwarfing interstock, the greater was its dwarfing effect and increase in yield of larger fruits. Very short interstocks (plates up to 10 cm) resulted in increased vigour and reduction in yield. All trees with interstocks grew more vigorously, were better anchored, gave higher yields per tree, but produced more suckers than trees grafted directly on M.9. Dwarfing interstocks (M.27 and virus-tested M.9) on more vigorous rootstocks (M.26 and MM. 106) reduced the growth of cvs. Gravensteiner, Boskoop, Jonagold and Glockenapfel than when grafted directly on virus-tested M. 9 (Riesen and Widmer, 1987).

Rootstock Adaptations

Soil

The rootstocks M.26, MM. 106 and M1.793 are very sensitive to waterlogging while M.13 and M.16 are relatively tolerant to this condition (Cummins and Norton, 1974). M.1 and M.13 are sensitive to drought (Tukey, 1965a) while M.7. M.9 and M.17 are relatively drought tolerant (Blasse, 1960). The rootstocks M.7, M.16, M.25. MM. 109 are relatively resistant to high soil temperature and M.1. M.2, M.9 and MM. 104 perform poorly at a temperature of 25°C or above (Gur *et al.*, 1976: Holubowicz *et al.*, 1982). The depth of rooting increases from sand to loam to clay soil. A shallow scaffold with vertically descending roots is seen in loam and sandy soil, while in clay, most roots slope downward and grow in the subsoil (Rogers and Vyvyan, 1934). In Japan, the apple cv. Fuji showed the best tree form and greatest yield on M.26 in deep Sekigami andosols with higher water retention. On shallow Shibani andosols, Fuji on MM. 106 and *M. prunifolia* showed the best growth, the later being early bearing and giving the best yields. The growth of trees on M.9 and M.26 was poor on showa soil (sand-dune regosols) but trees on MM. 106 and on *M. prunifolia* grew well and had higher yields (Matsui *et al.*, 1984a); on Masuda gray lowland soils and on Hiraka wet andosols, trees on M.26 showed the best growth (Matsui *et al.*, 1984b). In gravelly brown lowland soil condition, apple cv. Fuji grows well on M.9 Nagano (developed in Japan) rootstock (Tamai *et al.*, 2002).

Climate

The Malling Merton (MM) and Malling (M) rootstocks have proved winter hardy in North America. The rootstock M.26 was found more tolerant to low winter temperature than M.7. M.9. MM 104 or MM 106 (Wildung *et al.*, 1973). Ottawa 4, Ottawa 3, MM.106 and C-52 showed tolerance at -35°C (Holubowicz *et al.*, 1982) and Delicious on M.7 was found to survive at -44°C while on M.2 showed severe injury (Lapins, 1962). The rootstock M.7 proved frost resistance along with A.2 in Switzerland while M.2, M.4 and M.9 were frost susceptible (Schumacher and Fankhauser. 1985). The rootstock KSC.3 and KSC.6 (parents for both rootstocks Beautiful Arcade X Antonovka) developed in Canada are winter hardy (Prive and Embree, 1997).

Diseases and Pests

Most of the clonal rootstocks are highly susceptible to fire blight (*Erwinia amylovora*). A few, such as, P-18, Ottawa 11, are moderately resistant and Robusta 5 and Novole are claimed to be resistant (Cummins and Aldwinckle, 1982). Among the M and MM rootstocks M.26 and M.9 are very susceptible while MM. 106 and M.27 are moderately susceptible to fire blight.

The rootstocks, Novole, P.18, Bud.9, Ottawa 3, M.9, P.2. P.22 and M.27, are resistant to crown rot and M.l, M.16, and Northern Spy are resistant to crown gall. The trees on MM. 106 and MM.111 were found to be less damaged by black root rot

(Dozier *et al.*, 1974) than on MM. 104, Robusta 5, Novole, M.2, M.7, M.9, M.13, M.26, M.27, Ottawa 3 and rootstocks of P series (P.1, P.2, P.18 and P.22) were moderately resistant to powdery mildew (Cummins and Aldwinckle, 1982). Common latent virus tolerant rootstocks are A.2, MM.106, Bud.9, M.9 and Mark.9. The heat-treated and virus-tested rootstocks are provided by East Malling Research Station and Long Ashton Research Station through Nuclear Stock Association in England which were given number with EMLA suffix to M or MM. Those rootstocks appear to be a subclone of original ones and showed higher tree vigour due to removal of viruses. The M.9 EMLA showed about 50 per cent higher growth than the original M.9 (Ferree and Carlson, 1987).

The rootstocks Northern Spy, Robusta 5, Novole, MM.106, Bemali, *etc.*, are resistant to moderately resistant to woolly aphid, while A.2, P.18. Bud.9, M.26, Ottawa 3, M.9, M.27, Bud 54 -146 and Bud 57- 491 are susceptible to this pest (Table 1.9).

Table 1.9: Responses of Rootstock Candidates to Biotic Factors of Environment*

	Crown Rot	*Fire Blight*	*Powdery Mildew*	*White Root Rot*	*Woolly Aphids*	*Common Latent Viruses*	*TmRSV***
Vigorous							
Alnarp 2 (A2)	MS	VS	MS	NT	S	Tol.	NT
Robusta 5 (R5)	M	VR	MR	NT	VR	MS	G
Antonovka (KA)	NT	MS	NT	NT	MS	NT	NT
YP (Mb-4)	NT	VS	NT	NT	NT	NT	NT
Novole	R	R	MR	NT	MR	S	G
Semi-vigorous							
MM. 106	S	MS	S	MS	R	Tol.	VS
Bud. 54-118	NT	MS	NT	NT	S	NT	NT
Bud. 57-490	MR	M	S	NT	MS	NT	NT
1-48-46 (NCPI)	NT	NT	NT	NT	NT	NT	NT
P-18	VR	MR	M	NT	S	NT	NT
Ottawa 8	NT	M	M	NT	M	M	NT
Ottawa 11	NT	MR	M	MR	M	M	G
Semi-dwarf							
M.26	M	VS	M	NT	VS	MR	S
Bud. 9(RLP)	VR	S	MS	NT	S	To l.	NT
11-25-23 (NCPI)	NT	NT	NT	NT	NT	NT	
P-1	MR	MS	NT	NT	MS	NT	NT
Ottawa 3	R	S	M	NT	VS	M	NT
J9	NT	NT	MS	NT	NT	NT	NT
Bemali	NT	NT	NT	NT	MR	NT	NT

	Crown Rot	*Fire Blight*	*Powdery Mildew*	*White Root Rot*	*Woolly Aphids*	*Common Latent Viruses*	*TmRSV***
Dwarf							
M.9	R	VS	M	NT	S	Tal.	NT
1-47-1 (NCPI)	NT	NT	NT	NT	NT	NT	NT
M.20	NT	NT	NT	NT	NT	NT	NT
Mark (MAC-9)	NT	S	M	NT	MS	Tol.	MS
OAR-I	NT	MS	NT	NT	M	NT	NT
P-2	R	MS	NT	NT	MS	NT	S
P-22	R	MS	NT	NT	MS	NT	NT
PI-80 (Pillnitz)	NT	NT	NT	NT	NT	NT	NT
Super-dwarf							
M.27	R	S	M	NT	S	Tol.	NT
Bud. 54-146	NT	VS	MS	NT	S	NT	NT
Bud. 57-491	MS	S	MS	NT	S	NT	NT

**VS: Very Susceptible; S: Susceptible; MS: Moderately Susceptible; M: Intermediate; MR: Moderately Resistant; R: Resistant; VR: Very Resistant; NT: Not Tested.

* Complied from Cummins and Aldwinckle (1982).

Rootstock Relationship to Tree Size

The built-in size controlling mechanism of the rootstock affords a relatively cheap and permanent method of containing trees within a pre-determined size (Preston, 1978). Trees on M.9 and M.27 are small, those on M.25 are large and those on MM. 106 are intermediate in size. However, anomalies are also found with rootstocks within a similar vigour group with different scions. Lane's Prince Albert, for example, makes a larger tree at East Malling on M.1 than on M.2, the reverse being the case with Worcester Pearmain (Preston, 1958, 1959). Similarly, trees on MM.106 and M.7 are almost identical in size in trials at East Malling (Preston, 1966), whereas in the USA, MM. 106 makes a larger tree than does M.7 (Carlson, 1975). The height and spread of Empire apple on M.26, M.9/MM. 106, M. 9/MM.111, M.27/MM.106 and M 27/MM. 111 were similar and greater than those on M.9 or M.27 (Lord *et al.*, 1985). Vigorous cultivars such as Jonagold and Canada grow well on M.27 (Evequoz, 1985). In New Zealand, Golden Delicious trees on MM.101, MM. 104 and MM.105 were equivalent to those on M.16, while MM. 102 proved least vigorous. The trees on M.4, M.6 and MM.102 were poorly anchored but the root system of trees on MM. 104, MM.105 and MM. 106 were firm in the ground (McKenzie, 1985). The relative tree size of available and potential rootstocks are summarized in Table 1.10.

In India, M.9 proved to be the dwarfest of all stocks, whereas M.7 followed by MM. 106 was the most vigorous (Sinha *et al.*, 1982). McIntosh trees on Ottawa 8 were larger than on M.26. The trees on Bud 490 were larger than on Ottawa 8 and smaller than on Ottawa 11, but were less efficient than those on Ottawa 8 and

Table 1.10: Tree Size Controlling Apple Rootstocks

Very Vigorous	*Semi-vigorous*	*Semi-dwarf*	*Dwarf*	*Super-dwarf*
Alnarp 2 (A2)	M.2	M.7	M.B	M.27
Aruonovka (KA)	M.4	M.26	M.9	Bud. 54-146
YP (Mb-4)	MM. 106	Bud.9 (RLP)	M.20	Bud. 57-491
Novole	MM.111	H-25-23 (NCPI)	1-47-1 (NCPI)	MAC-39
M.I	Bud. 54-118	P. 1	MAC-9	
M.13	Bud. 57-490	P. 16	P 2	
M.25	1-48-46 (NCPI)	0.3	P. 22	
MAC-4	P. IR	Jork. 9 (J9)	KSC26	
MAC-24	0.1	Bemali	OAR I	
O.11	0.2	MAC-I		
KSC 3	KSC7	KSC28		
KSC6	KSC 11	Northern Spy		
Robusta 5	KSC25			
	Anis			
	Beautiful Arcade			

Malling Series: M; Malling Merton Series: MM; Polish Series: P; Russian Series: Bud; Michigan Apple Clone Series: MAC; Ottawa Series: O; Kentville Stock Clone Series: KSC; YP: Selected in Finland; Jork 9 (J 9): The Fruit Research Station at Jork, West Germany. OAR: Oregon Series; Novole: Geneva Selection.

on other dwarfing rootstocks like M.8, M.9, M.26, M.27 and Ottawa 3 (Ferree and Schmid, 1987).

Rootstock Relationship to Productivity

One of the most striking attributes of M.9 is its capacity to induce early fruiting (precocity) in scion cultivars. Even the very unprecocious Northern Spy begins bearing by the fourth year on M. 9, when properly trained (Cummins and Aldwinckle, 1983). Trees on M. 25 showed high level of fruit set (Preston, 1955). Preston (1966) compared fruiting response of Cox and Jonathan on 27 rootstocks for fifteen years. The accumulated crop yield in 15 years (Table 1.11) in two different sites, *i.e.*, Bearsted Soil Series (marginal soil without irrigation) and Malling Soil Series (sandy loam) showed that on both soils the yields from trees on MM. 106 and M. 7 were very similar although the trees on MM. 106 on the Bearsted Series were smaller than those on M. 7. The yields from Cox and Jonathan on MM. 103 and MM. 104 on both soils were as heavy as or heavier than those from trees on M. 4, which is well known for its capacity to induce heavy cropping in many scion cultivars. Both cultivars on MM. 116 yielded more fruits than on MM. 106 or M. 7, especially on the Bearsted Soil Series, but the trees on MM. 116, were slightly larger. Cox trees on MM. 111, which were similar in size as those on M.2, bore more heavily on both soils, but there was no appreciable difference between these two rootstocks with Jonathan. Cox on M. 25 out yielded those on M. 26 on the Malling Series while little difference was found with Jonathan. Bramley's Seedlings on M. 25 produced 64 per cent more fruit in 10 years than did on M. 16, although the trees

Table 1.11. Accumulated Crop Yield per Tree (lb) at 15 Years*

Malling Soil Series			*Bearsted Soil Series*		
Rootstock	*Cox*	*Jonathan*	*Rootstock*	*Cox*	*Jonathan*
Trial A			***Trial C***		
M.9	279	219	M.9	144	156
829	283	820	MM.102	302	129
Spy	568	566	Spy	323	337
MM.I06	621	632	829	324	752
M.7	654	588	M.7	354	348
MM. 105	733	895	MM.106	356	296
MM. 102	749	354	MM.101	447	586
MM.116	773	909	MM.116	477	604
MM.101	798	713	M.4	506	582
M.4	928	858	MM. 105	577	817
MM. 103	993	999	MM.104	594	797
MM.I04	1.165	988	MM. 103	721	749
Mean	712	712		427	513
Trial B			***Trial D***		
M.I	592	843	- MM. 108	123	545
MM.107	603	749	M.I	300	315
MM. 108	669	957	MM.112	383	593
MM.112	697	779	M.2	438	618
MM.26	779	1.026	M.4	453	683
MM.114	816	893	MM.115	547	792
M.2	868	772	I.D.V	564	865
MM.113	881	980	M.25	572	790
411	915	949	MM.114	581	779
MM.115	978	1.009	MM.107	602	1,002
MM.110	996	743	MM.109	614	943
I.D.V.	1.033	1.036	M.16	614	884
MM.109	1.049	1.078	MM.111	659	633
MM.111	1,060	796	MM.113	664	995
M.25	1,090	972	MM.110	769	833
M.4	1, 125	1.106	411	797	947
Mean	884	918		542	763

* Preston (1966).

were similar in size (Preston, 1978). On deep fertile soil in New Zealand, Golden Delicious on MM.106 produced the heaviest annual yield per tree, equivalent to 92 t/ha compared with trees on MM. 101, MM. 102, MM. 104, MM. 105, M. 4 and M. 7 (McKenzie, 1985) and MM. 106 was recommended for semi-intensive orchards on fertile soils. In USA, Redspur Delicious on MM. 106 gave higher yield than on

MM.111 and M.7 and exhibited greatest precocity while Golden Delicious on M. 7 generally performed best (Archbold *et al.*, 1987). In India, the cvs. Jonathan, Red Delicious and Rymer produced higher yield on Merton 7 and M. 13 than on M.2, Merton 793 and Crab C (Hari Om and Pathak, 1983). Cultivars 'Sinap Orlovskii' and 'Auksis' showed higher yield and bigger fruits on P 22 at Lativia (Rubauskis *et al.*, 2011). Skrzynski and PoniedZieck (2000) in their experiment conducted at Poland with apple cv. Jonagold on different rootstoks reported that trees on P.22 were 50 per cent of the size of those on M. 26 (after 7 growing seasons). Cumulative yield per tree after 6 harvests ranged from 54 to 66 Kg. The highest productivity index amounting 3.5 Kg/cm^2 was recorded on P.22 rootstock, whereas lowest on M.26 and P.14 (1.8 and 1.6 Kg/cm^2, respectively). The biggest fruits (200-210 g on average) were harvested from trees on P.14 and M.9 while the smallest fruits (190-200g) were recorded from trees on P.2 and M.26. In Belgium, apple cv. King Jonagold showed the highest yield on M.9 7337 (92.5 Kg/tree) followed by Bud 146 (84.9 Kg/tree) and J. TE. G (76.3 Kg/tree) (Vercammen, 2004).

Mineral Uptake

Trees of Cox's Orange Pippin on M. 9 showed higher concentration of Ca and Mn and low levels of Mg, B and Fe than on M. 5 or M. 12 (Roach, 1947). Awad and Kenworthy (1963) observed relatively high Ca in leaves of fruiting clones on M. 1 and M. 16, and lower in trees on M. 2, M. 5, M. 7 and M. 13, Mg was higher in M. 16 than on M. 1. Trees of Empire apple showed higher concentration of Mn in leaves on M. 27 than on M. 9, M. 26, and interstock/rootstock combinations of M.9/MM. 106, M.9/MM. 111, M. 27/MM 106 and M. 27/MM. 111 (Lord *et al.*, 1985). Trees on M. 2 and M. 5 are susceptible to K deficiency (Warne and Wallace, 1935). Scion cultivars may differ in their influence on levels of leaf N, K, Ca and Mg (Tukey *et al.*, 1962; Whitfield, 1964). Poling and Overly (1979) noted significant differences in the levels of Mg, Ca, Fe, Mn and Na in leaves of McIntosh apple scions grafted on 16 different rootstocks. Trees on MM. 106 rootstock were among the highest in leaf Mg and Ca, while those on M.4 were among the lowest in leaf Mg. Ca, Na and AI. (Table 1.12). Rostslocks showed significant effects on the mineral concentrations in the leaves of apple cv. Nagafu 6 Fuji. Leaves on P.22 had higher Ca concentration than those on M.7 EMLA for three years. Leaves on P.22 had the highest leaf Mg concentrations than on M.9, NAKBT 337 and M.7 EMLA (Ikjo and Fallahi, 2001)

Fruit Quality

Rootstocks influenced the fruit size and quality (Gernyseva, 1964; Walter. 1966). In Norway, apple cv. Summerred produced fruits of good size on Graham Jubilee and MM. 106 than on A2 (Redalen, 1987). The colouration of fruit, in general. is better on M.6 and M. 9 (Rogers, 1927; Hilkenbaumer, 1959). At any given harvesting date, Golden Delicious on M. 27 and M. 26 impart a less intense green colour to fruits whereas MM. 104, MM. l09a, M. 2a, M. 16 and M. 7a give rise to darker green fruits. Fruit colour is closely related to the degree of maturity, which also varies with tree vigour depending on the rootstock and soil and other environmental conditions (Villeneuve *et al.*, 1988). Teaotia *et al.* (1983) noted marked effect of different rootstocks on fruit quality of apple cvs. Jonathan, Red Delicious and

Table 1.12: Effects of 16 Apple Rootstocks on Mineral Element Composition of 'McIntosh' Scion Leaves. Means for 4 years. 1973-1976*

	Element Composition (dry wt)**									
Rootstock	N (per cent)	P (per cent)	K (per cent)	Ca (per cent)	Mg (per cent)	Fe (ppm)	Mn (ppm)	B (ppm)	Zn (pprn)	Cu (ppm)
M9	1.92	0.151	1.46	1.13abcd	0.178ABC	67B	74.OCD	31.7	22.9	11.6
M 26	1.93	0.141	1.45	1.01cd	0.191AB	76AB	I38.8A	25.0	20.9	8.4
M 9/MM 106	1.98	0.154	1.44	1.31a	0.168ABCD	69B	69.30	30.5	26.5	11.0
M 9/Ddlg	1.89	0.141	1.43	1.18 abc	0.186ABC	71B	84.6CD	31.7	29.3	11.4
M 25	2.00	0.192	1.63	1.03 cd	0.165ABCD	77AB	80.ICD	30.3	21.9	10.6
M7	1.98	0.153	1.51	1.17abcd	0.192AB	110A	72.2CD	28.8	25.7	10.8
MM 106	1.98	0.168	1.58	I.25ab	0.194A	80AB	76.6CD	32.9	25.6	12.1
M2	1.96	0.153	1.52	1.07 bcd	0.158ABCO	63B	68.00	34.1	30.4	9.1
M4	1.96	0.171	1.83	0.96d	0.148CD	82AB	88.0 CD	29.6	234	11.7
MM 102	1.93	0.172	1.59	1.08bcd	0.1430	76AB	125.0AB	30.5	25.6	10.2
MM 111	1.90	0.139	1.71	1.16abcd	0.187ABC	82AB	80.ICD	31.4	26.9	12.2
Alnarp 2	1.93	0.164	1.72	1.07bcd	0.149CD	I06A	84.8CD	29.2	24.0	10.3
MM 104	1.86	0.157	1.54	I.09 bcd	0.189AB	85AB	73.7CD	30.8	25.3	10.4
Robusta 5	1.99	0.168	I. 75	1.18 bcd	0.152BCD	93AB	101.4BC	33.2	24.3	10.5
Seedling	1.85	0.145	1.43	1.05 bcd	0.196A	85AB	85.3 CD	30.2	29.9	9.2
MM109	1.85	0.155	1.59	1. 22 abc	0.190 AB	70B	60.3D	32.7	21.4	13.6

** Mean separation in columns by Duncan's multiple range test, 5 per cent level (lower case 1 per cent level (upper case), or nonsignificant (no letters).

* Poling and Oberly (1979).

Rymer. In Jonathan, significantly higher TSS was noted on Merton 793 and lower on M.13. Less acid and medium sugar contents in fruits were obtained on M. 799, which increased sugar-acid ratio and resulted in better fruit quality (Table 1.13). The fruits of cv. Nagafu 6 Fuji showed more red colour development on P.22 than on M.9 and M.7. Soluble Solids content of fruit and firmness of fruit were also found higher on P.22 rootstock (Ikjo and Fallahi, 2001).

Table 1.13: Effect of Various Rootstocks on the Chemical Attributes of Apple Cultivars*

Observations	*Mean values for*							
	Cultivars	*Crab C*	*M.13*	*Merton 779*	*Merton 793*	*M.2*	*Local Selection*	
1. Juice (per cent)	Jonathan	69.00	60.33	60.00	63.30	60.00	60.00	NS
	Delicious	58.66	62.00	64.00	55.30	62.00	56.00	NS
	Rymer	59.00	58.00	58.00	62.00	54.00	61.00	NS
2. TSS (Brix)	Jonathan	14.56	14.03	14.30	14.93	14.36	14.23	0.26
	Delicious	10.97	10.90	10.84	10.70	11.21	11.12	NS
	Rymer	13.60	14.60	14.03	13.10	14.70	13.50	NS
3. Acidity (per cent)	Jonathan	0.573	0.682	0.558	0.775	0.620	0.641	0.116
	Delicious	0.960	0.110	0.910	0.139	0.105	0.152	NS
	Rymer	0.673	0.649	0.649	0.576	0.558	0.608	NS
4. Ascorbic acid (mg/100 g)	Jonathan	10.83	9.72	10.27	8.72	8.60	9.44	NS
	Delicious	7.53	5.94	7.13	6.08	6.34	7.53	NS
	Rymer	4.54	4.54	4.79	3.65	4.79	4.41	NS
5. Total sugars (per cent)	Jonathan	9.77	9.70	10.21	10.39	10.35	9.83	NS
	Delicious	8.87	9.03	8.68	9.00	9.72	8.83	NS
	Rymer	10.81	10.83	11.17	10.00	10.57	11.55	NS
6. Sugar- acid-ratio	Jonathan	17.08	14.29	18.61	13.26	16.91	15.36	2.921
	Delicious	9.20	82.09	9.53	64.74	92.57	58.29	NS
	Rymer	16.11	16.81	17.21	17.36	18.94	18.99	NS

* Teaotia *et al.* (1983).

Propagation

Propagation of Nursery Stock

Apples were conventionally propagated by seeds, but the resulting off springs are variable and the economic need to perpetuate desirable features has led to the use of vegetative propagation. The modern apple tree is a highly developed plant embodying among other things varietal fruit characteristics, precocity and size control. The desired range of characteristics does not occur in a single plant and has been achieved by growing a composite plant, comprising a scion cultivar grafted upon a rootstock (Howard, 1971).

Seedling Rootstocks

Crab apple seeds (*M. baccata*) are used extensively for raising rootstocks. Various Asiatic species of *Malus* appear promising as dwarfing or semi-dwarfing rootstocks or intermediate stocks for the cultivated apple. Some of these *M. hupehensis, M. toringoides* and *M.sikkimensis* proved to be apomictic. *M.tschonoskii* produces very uniform progeny, presumably due to absence of cross-pollination with other species, and not due to apomixic (Sax, 1949). Seedling rootstock can also be raised from the seeds of scion cultivar. Seeds of Ben Davis, Delicious, Yellow Newton, Wealthy, Rome Beauty, McIntosh give good germination but seeds from triploid cultivars such as Gravenstein, Baldwin, Stayman, Winesap, Arkansas, *etc.*, are of low viability.

Like the seeds of many other temperate fruit trees, apple seeds do not germinate immediately after fruit harvest. Germination ability is preserved under conditions other than those optimal for germination; hence, these seeds are regarded as being in the state of a deep, innate dormancy (Lewak, 1981). Because of the presence of germination blocks, the seeds require cold and moist stratification to overcome barrier before germinajion takes place (Kar and Bhartiya, 1987). Three consecutive phases are recognized in after-ripening of apples seeds: (i) phase of removal of the primary cause of dormancy, (ii) catabolic phase, and (iii) phase of growth initiation. These phases were distinguished on the basis of changes in hormonal equilibrium and alterations in the intensity of some metabolic activities during after-ripening (Lewak *et al.*, 1975; Lewak and Rudnicki, 1977) as well as on the basis of differential temperature optima (Thevenot, 1980) and differential sensitivity towards some other factors (Sinska and Lewak, 1977). The optimum temperature for stratification of intact seed ranges from 4° to 10°C (Lewak, 1981). Dormant Golden Delicious embryos were found very sensitive to oxygen. They germinated better in hypoxia (2.5 and 10 per cent O_2) than in air or in atmosphere enriched with oxygen. Breaking of dormancy by cold treatments was due to reduction of the noxious effect of high oxygen concentration (Come *et al.*, 1985). Lewak (1981) suggested an integrated scheme which incorporates all the known factors and process involved in the removal of dormancy in apple seeds.

For after-ripening, seeds are first soaked in water for several hours and then kept in a container with moisture retaining material such as damp peat moss. The container is then buried under shelter for the winter months. The media should be kept moist by frequent watering. The after-ripened seeds are sown in the following spring.

The seedlings are allowed to grow for one year in the nursery bed, the laterals are then trimmed and the main shoot headed back to about 30-45 cm in height. The seedlings are then transplanted in nursery rows at a required spacing and are maintained for grafting next year.

Clonal Rootstocks

Clonal rootstocks are multiplied by stooling, trench layering, cutting and by micropropagation.

Stooling

Apple rootstocks are commonly propagated by these techniques. The stools are allowed to remain uncovered during the early part of the growing season, so as not to inhibit their growth and reduce the crop. When they have reached the height of 20 cm or so they are mounded in early spring, they are allowed to grow during the rest of the season and roots form on the new shoots in the covered protions of the stem. In early winter rooted shoots are removed and the cycle repeated. However, the method has a serious disadvantage, namely, the need to grow all shoots at ground level to facilitate earthing up (Howard, 1971). M.27 rootstock gave markedly better rooting in stool beds than M.9 and almost as good rooting as M.26 (Crnko and Marn, 1987b). Ringing of shoots and treatment with IBA at 2500 ppm in lanoline before mounding was reported to increase rooting. Joshi *et al.*(1987) reported higher number of rooting in clones of MM series than in M series (Table 1.14) by using IBA at 2500 ppm. Khatik and Sharma (2013) recorded 84.6 per cent rooting of stool shoots of Merton 793 under open field condition by treating the cuttings with IBA at 2500 ppm. Wounding and application of IBA and bottom heat to unrooted non-basal shoots from stools of M.27 apple rootstock greatly improved their establishment when planted back into the nursery for a second season (Pontikis *et al.*, 1979). From a long term trials in USSR with MM.106, A2, B9, No.54-118 and No.57-490; Bakun *et al.* (1985) concluded that the easy-to-root No.54-118 could be best propagated by stooling and others by cuttings.

Table 1.14: Rooting Performance of different Rootstocks by Stooling*

Rootstocks	*Rooting (Per cent)*	*Rootstocks*	*Rooting (Per cent)*
M.4	40	MM.106	85
M.7	20	MM.109	55
M.9	55	MM.110	60
M.21	45	EMLA.26	50
M.27	50	EMLA.106	20
Crab c	10	Merton 793	85
		Merton 779	60

*Joshi *et al.* (1987).

Trench Layering

In this method, the shoots (suckers) from mother plants are bent down to ground level over trenches on either side of the plant. Notches are made underneath the shoot buds. The upper sides of the shoot portions on both sides of buds are covered with soil, leaving the buds open to facilitate their growth into new shoots and roots appear from the notch underneath (Bhartiya, 1983). The method is, however, more laborious and cumbersome.

Cutting

One approach of propagating apple rootstocks is by cuttings. Although softwood cuttings rooted more easily under mist or in more sophisticated fog house, the difficulty of their subsequent establishment and the need to take two to three years to grow them into useable plants mitigate the use of softwood cuttings except in exceptional circumstances. Cuttings close to the previous year's wood which include the swollen base of the annual shoot lead to better rooting but the inclusion of a 'heel' of older wood should be avoided (Howard, 1971).

Many scion cultivars of apple and dwarfing rootstocks (*e.g.*, M.9) are difficult to propagate by stem cuttings and various treatments were used to alter the physiological or nutritional condition of the stock plants or portions of the plants to increase rooting. Etiolation of a section of the stem by wrapping it with black tape was found to increase rooting of M.9 by 84 per cent over the control (Howard and Schmidtm 1979). Garner (1937) found promotion of rooting in cuttings up to 70 per cent by localized etiolation in McIntosh and 100 per cent rooting was obtained by girdling of the etiolated shoot. Howard *et al.* (1980) found superior rooting in cuttings from the trunk of the apple rootstocks MM.111 and MM.106 compared with cuttings from the crown of the same plants. However, cuttings collected from deheaded hedges gave higher percentage of rooting than those collected from normal hedges of MM.111 and MM.I06 apple stocks. Wounding the bases of M.26 winter cuttings, especially by splitting, increased the rooting percentage as well as the number of roots per rooted cutting. The response was best in situations where rooting was relatively poor normally, such as in non-basal internodal tissue compared with basal cuttings and in cuttings from the hedges compared with those from stool beds. A significant wounding response occurred only in the presence of IBA, (Howard *et al.*, 1984). The percentage of rooting in leafy cuttings of the apple rootstock M.9, increased on an average from 11 per cent to 78 per cent by prior etiolation. A period of exposure to light before taking pre-etiolated shoots as cuttings was found essential and it was not necessary to continue to exclude light from the future-rooting zone. The only limitation to the practical application of the technique is that the black polythene cover used to exclude light from the stock plants favours *Botrytis* infection. Since complete darkness is not essential, this problem can probably be overcome by effective ventilation of the covers (Harrison-Murray, 1982).

Table 1.15: Relative Rooting Ability (per cent) of Apple Rootstocks Propagated from Hardwood Cuttings during Spring*

MM.I06	94.0
Crab c	86.5
M.26	80.5
MM.111	80.0
M.27	76.0
M.7	73.0
M.2	67.5
M.25	67.0
M.9	28.5

* Howard (1971).

A method of propagation, using woody, leafless apple cuttings taken between October and April, was also developed (Howard, 1971). In this method leafless, woody cuttings are prepared from vigorously grown shoots produced on severely pruned hedge plants, treated with a root promoting auxin to stimulate callus formation, and to root by planting them for a few weeks in bins of heated compost. The popular MM. 106, and also M.27, M.26. and MM.111 responded well for successful propagation by winter cutting (Howard, 1981).

Considering the problems of water stress when many leaves are present, summer cuttmgs of smaller size are taken than winter- cuttings, from currently growing leafy shoots in late spring or early summer. These are kept relatively turgid under automatically controlled intermittent fog, mist or at relatively high humidity under polythene covers. It is essential to maintain a correct balance of air and water in the rooting medium and death of both rooted and unrooted cuttings occurred in over-wet conditions, especially at a high degree of bottom heat (Howard, 1971). The level of bottom heat determines the rate at which roots develop (Howard and Garner, 1965) and at a optimum temperature of 70°F (21°C) roots appear on many clones within 2 weeks, and cuttings are fully rooted and ready for transplantation in about one month.

Micropropagation

A less expensive and more efficient method of producing clonal apple rootstock is by micropropagation (Hutchinson, 1984). A considerable progress has been made for the last few years in the micropropagation of apple rootstock. The success of micropropagation of apple is largely based on the findings that the cytokinin BAP can stimulate the growth of shoot tips (Jones, 1967) which subsequently induces shoot proliferation (Pieniazek, 1968) and that phloroglucinol can synergise auxins during rooting (Jones, 1976; Jones and Hatfield, 1976). Investigations with M.7 have indicated that these effects of the phenolic were sustained only during the initial three months of the culture when the capacity for proliferation and rooting was low (Jones *et al.*, 1982). In rootstock A2, the phenolic improved the initiation and growth of roots and also the survival of plantlets in soil was markedly increased when they were derived from adult plants compared to seedlings (Welander and Huntrieser, 1981). At East Malling Research Station, England, Jones and his associates were able to achieve success through *in vitro* propagation of rootstocks M.26, M.27 and M.7 (Jones *et al.*, 1977). The length of the auxin-sensitive phase of root initiation *in vitro* in the apple rootstock M.9 has been determined using the auxin indol-3yl-butyric acid (IBA) at 1.5 x 10^{-5}M in the presence or absence of 10^{-3}M phloroglucinol (PG) (James, 1983a). PG synergised IBA-induced rooting after 4 days' exposure, but contact times exceeding 8 days decreased root number. In contrast, PG consistently synergized IAA-induced rooting in the dark for contact periods up to 13 days with the highest rooting being recorded at 9 days. During the auxin-sensitive phase of root initiation, rates of 3-indolyl-(2-^{14}C) acetic acid (IAA) uptake into the 1 cm bases of shoots of the apple rootstock M.9, *in vitro* were not significantly affected by the presence of 10^{-3} M phloroglucinol using either liquid or agar-solidified media. The use of a liquid medium did, however, reduce the rate of uptake over a 10-day period of auxin application. The distribution of labelled IAA between the 1 cm base and the shoot remainder was, however, not affected by PG (James, 1983b). Aseptic cultures of Northern Spy could be established any time during the year although mid-spring and mid-summer were found to be the best (Hutchinson, 1984). Maximum shoot proliferation could be obtained by using nodal explants from aseptic cultures with 10 $\mu M1^{-1}$ BAP and a light level of 75 $\mu Mm^{-2}S^{-1}$, but the shoots were dwarf and difficult to sub-culture. Better quality shoots, suitable for root initiation, were obtained when BAP concentration was reduced to 5 $\mu M1^{-1}$. Half-strength salts and 1 $\mu M1^{-1}$ IBA resulted in maximum percentage of root initiation and if used in combination with different physical supports, allowed control over the number and length of roots formed. M.7 rootstock multiplication protocol has been standardized by Silveira *et al.* (2001). They obtained best multiplication by using MS media with 100/1 myoinasitol, 30g/1 sucrose, 6g/1 agar and 5 µM BA. Addition of IBA at 1.0µM caused best shoot multiplication.

The procedure for the propagation of an apple rootstock can be initiated at any time of the year with shoot tips of 0.5-1.0 cm, cut shortly after bud break from a previously cold stored plant. Jones (1979) suggested that the tips should be sterilized by washing with dilute solution of sodium hypochlorite before placing on a culture medium with addition of phloroglucinol. Each tip will produce about 20 shoots in three months and subsequent multiplication can be achieved by transferring the portion of culture every three weeks on a fresh medium. For scion cultivars, he suggested to use double amount of auxin for rooting and as the small self-rooted scions are difficult to establish in pots, it was also suggested to harden them for 10 days in the glasshouse in a liquid medium which has the same composition of rooting medium, except that agar, auxin and phloroglucinol should be excluded and the mineral salts and sucrose reduced to half concentration.

Regeneration of shoots from M.27 apple leaf discs was reported by James *et al.* (1983). Success occurred with a high concentration of BAP and a low concentration of 2,4-D. The amount of callus formed at the cut edge of the disc was found variable and organogenesis from the abaxial leaf surface occurred with no or very little callus formation.

Immature embryos (up to 4-week post-anthesis) of Delicious, cultured without testa or endosperm, produced only callus. Embryos 6.5 weeks of age produced multiple shoots and the percentage of embryos producing shoots increased with age until 10.5 weeks old embryos. All surviving embryos produced shoots but no callus (Kouider *et al.*, 1985). Nucellar tissues from 30-day old (post-anthesis) seeds of scion cultivars such as Barmley's, Cox, Greensleeves and Spartan and rootstocks M.9, M.25, M.26 and M.27 formed callus only in the presence of an auxin, 2,4-D or NAA, and a cytokinin, BA. Concentrations of 4.4 x 10^{-6}M and 2.2. x 10^{-5}M were found effective (James *et al.*,1984).

There are many reports on the medium used for shoot culture of apple with varying degrees of success. It is worthwhile to mention that the medium used by Jones and his coworkers (1977) in which it has been possible to produce 60 000 shoots from a single shoot tip of M.26 in eight months. They also reported that these shoots can be readily rooted and grown in pots. The medium used for sterilization consisted of the major and minor elements of Murashige and Skoog (1962) but with the NaEDTA and $FeSO_4$ replaced by FeNaEDTA at 20 mg l^{-1} and with the addition of 3 per cent sucrose and 0.7 per cent agar; the pH was adjusted to 5.2 with 0.1N KOH. The medium for shoot culture used was also of this composition but with the addition of (mg l^{-1}); thiamine hydrochloride 0.4; inositol 100.0; 6-benzylaminopurine (BA) 1.0; gibberellic acid (GA_3) as ammonium salt 0.1; indol-3yl-butyric acid as ammonium salt (IBA) 1.0; phloroglucinol 162.0. In their experiment, all media constituents were sterilized by autoclaving at 103 KNm^{-2} for 5 min except for IBA and phloroglucinol which were passed through millipore filters (0.45 µm dia.).

Tree Production

The traditional method of propagating apple trees consists of budding or grafting of the scion cultivars onto rootstocks which are raised by seed or asexual method. The ideal fruit tree is produced in two years from planting of the rootstock, and depending on the species, it should be about 1.5m tall, with many laterals arising above a height of about 65 cm from the ground (Howard, 1981).

Budding

The common methods of shield or T and chip budding are practised for propagating scions. Budding is usually done in summer when buds are well formed and the bark slips well, *i.e.*, during July to early September. Bud-woods are collected from healthy shoots. Leaves are removed by cutting to a short petiole. In shield budding, a single bud, together with a shield piece of stem, is cut from the scion and inserted beneath the rind of rootstock through a T-shaped incision during periods of active growth (Garner, 1967). The use of an inverted T incision in the rind of the rootstock is sometimes advocated under moist condition because water does not readily enter the incision and cause rotting. In chip budding, a wedge shaped piece of scion-wood with the bud is removed and a similar cut is made in the stock for placement of bud. Cambial contact in chip budding is good and development of the newly placed scion tissue occurs in step with that of the rootstock during the immediate post-budding period. Anatomical studies of unions showed that the essential connection between the cambium of rootstock and scion was not always effected completely in traditionally T-budded plants in which the scion-bud was simply inserted under the rootstock bark. Chip budding, which replaces a piece of rootstock tissue with a scion-bud cut so as to give cambial just a position resulted in quick union formation with rapid and uniform growth in the following spring (Howard, 1981). Larger number of high grade trees result, therefore, from chip budding (Howard, 1977; Table 1.16). In Korea, optimum chip budding time for apple trees appeared to be from mid-August to early September. Untying wrapping tapes after 6-8 weeks from budding led to higher graft success and better shoot growth (Jongseob and TaeMyung, 2001). Karamusel and Kalyoncu (2011) in their experiment with scions 'Red Chief' (dwarf), 'Braeburn' (Semi dwarf) and 'Mondial Gala' (vigrous), grafted by bench grafting and chip budding techniques on MM.106 and M.9 apple rootsstocks. They recorded 82 per cent graft survival inside greenhouse compared with 69 per cent in outdoor. Sapling development was also higher inside the greenhouse. The survival ratio in bench graft was 82 per cent compared with 64 per cent in chip budding. The yield of Nagano Fuji 2 was found higher when it was grafted on *Malus micromalus* and with interstock SH.29 (derived from a cross of Ralls Janet with *Malus henanensis*) compared with M.7 (Dengke and XueMei, 2002).

Table 1.16: Per cent 1st and 2nd Grade Cox Apple Trees by Chip and T Budding* (out of 4. grades)

Rootstock	*Chip*	*T*
M.27	84	36
M.9a	80	72
M.26	88	56
MM.106	80	60

* Howard (1977).

In the UK, chip budding can be done successfully between April to early September when the rootstocks grow actively (Table 1.17), however, chip budding can be done at any time of the year because the bark does not have to slip (Howard. 1977). One inch polythene budding tape is used for tying; light-degraded rubber strips can also be used.

Table 1.17: Per cent Bud-Take of CoxIM.26 Apples Related to Date of Budding (Ties released after 3 weeks)*

Budding Dates	*Bud Take (per cent)*	*Budding Dates*	*Bud Take (per cent)*
1974		*1975*	
18 July	98	27 February	0
8 August	100	10 April	0**
5 September	88	7 May	96
10 October	0		
21 November	0		

** April budding gave 72 per cent take when ties were removed after 6 weeks.

* Howard (1977).

Micropropagation

A wide range of apple scion cultivars has now been rapidly propagated *in vitro* in laboratories throughout the world. Further, at East Malling Research Station in England, the method has been applied by the fruit tree breeders to semi-dwarf cultivar Malling Greensleeves and also to dwarf types derived from cv. Wijick Mclntosh, a dwarf mutant of the cultivar McIntosh. In the field. these Wijick types usually produce single leading shoots and few laterals whereas *in vitro* they produce many axiliary shoots as prolifically as much more vigorous types, such as. cvs. Bramley's or Golden Delicious (Longbottom *et al.*, 1980). Jones *et al.* (1979) achieved success through *in vitro* propagation of scion cvs. Golden Delicious. Cox Orange Pippin, James Grieve, Malling Kent and Malling Suntan. The regeneration capacity from leaf segments of shoots was influenced by wounding, macro-elements, light conditions, polarity and maturity of the leaves (Welander. 1988). Good root initiation from single shoot of cvs. Delicious and York Imperial was obtained following culture for one week in darkness on half-strength MS solid medium with 0.33 mg/l IBA but without activated carbon (Standardi, 1985). The shoots of cv. Fuji grew best on MS medium containing BA at 1.0 mg/l, IBA at 2.0 mg/l and GA_3 at 0.1 mg/l, IBA at 1.0 mg/l promoted root initiation but GA_3 was inhibitory. Phloroglucinol at 180 mg/l promoted rooting of shoot tips under a 16-h light regime but was inhibitory at 80 mg/l in the dark (Yae *et al.*, 1986). Zhang-Xin *et al.* (2014) developed a simple and efficient protocol for obtaining shoots from leaf explants of apple rootstock G.41 (Geneva Series). They obtained the best shoot multiplication index (2.58) from successful subculture medium that was standared MS medium supplemented with 7.5g/l agar, 3.55 micro MN6-benzyladenine, 0.16 micro MIBA and 30 g/l sucrose. Regeneration rates was highest (99 per cent) when MS medium was supplemented with 2.7 micro thidiazuron and 0.9 M 1-NAA and cut wounding explants before placing the abaxial surface in contact with the medium. The best rooting percentage (80 per cent) was obtained on MS medium supplemented with 4.92 micro MIBA. Zheng *et al.* (2004) developed a rapid propagation method per apple cv. Granny Smith. MS medium supplemented with BA at 0.3-0.6 mg/l and NAA at 0.1-0.3 mg/l promoted the growth of vegetative axillary buds and regeneration of lateral buds. Shoots derived from axillary and lateral buds formed roots readily on half-strength MS medium supplemented with IBA at 0.6-1.0 mg/l and NAA at 0.1 mg/l under natural light after one week of culture in the dark. The proportion of rooting was 78 per cent after 30 days. More than 80 per cent survival was recorded when plantlets dipped in NAA at 5mg/l for 1-2 h were transplanted into a substrate containing vermiculite, humus and garden soil at 2:1:1 ratio.

Induction of Feathers in Nursery

The lack of feathers on apple plants in the nursery constitutes a major problem when they are planted in the orchard, since more investment has to be made by the grower in shaping the trees and in bringing them into production (Ramirez *et al.*, 1983). Plants which have produced feathers in the nursery tend to have an increased number of blooms and a better fruit set in the early years of bearing in the orchard and higher yield can be expected (Van Oosten, 1978; Shepherd, 1979). Dwarf rootstocks which are much in use for modern apple trees inhibit branching in the nursery (Quinlan, 1978).

Growth substances are being used to promote feathering in nursery plants (Quinlan. 1978. 1980; Wertheim. 1978a. 1978b; Johann. 1983; Ramirez *et al.*, 1983). The cultivars Gloster and Boskoop develop only few side shoots, if they are not stimulated by growth hormones. Johann (1983) reported that spraying of M and B 25-105 (propyl 3-t-butylphenoxyacetate) and Promalin (GA_{4+7} and 6-benzyladenine) increased the number of feathers in cvs. Gloster and Boskoop. The combined application of growth regulators also induced feathering, especially in cv. Boskoop (Table 1.18). Promalin at 1-2 per cent was also found effective in formation of long feathers at the basal part of tree while the upper shoots were short, thus creating an ideal vertical distribution of feathers in cv. Fuji (Tae Myung *et al.*, 2001).

Table 1.18: Influence of M and B 25-105 and Promalin on the Side Shoot Production of Nursery Trees (cv, Gloster and Boskoop)*

Treatment		*Nursery-I*		*Nursery-2*	
		Gloster	*Boskoop*	*Gloster*	*Boskoop*
M and B 25-105	600 ppm	3.1	5.8	4.5	6.3
	750 ppm	3.2	5.0	4.4	7.6
	1000 ppm	5.0	4.9	4.3	9.7
Promalin	1 per cent	3.8	3.4	6.0	9.2
	2 per cent	4.7	4.8	7.9	9.3
	4 per cent	8.1	4.4	9.1	10.3
M and B 25-105	1000 ppm	7.9	8.6	—	—
	+ Promalin 1 per cent				
Control		0.2	0.4	0.5	5.4

*Johann (1983).

Planting

Planting should be done 2 to 5 cm deeper than what it was in the nursery. Deep planting of apple trees has been recommended as a means of reducing or eliminating such problems as cold injury to the roots (Carlson, 1981), burrknots (Rom and Brown, 1979) and poor anchorage of dwarfing rootstocks (Rogers and Parry, 1968). Some nurseries graft trees on seedling roots 15 cm high and recommend burying the union up to 7.5 cm deep. Many nurseries graft clonal rootstocks 30 cm high and recommend leaving only 2 to 5 cm of rootstock above ground at planting (Lyons *et al.*, 1983). In sandy and dry soil, a deeper planting is advocated. Costante *et al.* (1983) recommended that interstem apple trees be planted with the stem piece at least partially buried. This will tend to decrease the amount of root suckering, lead better trunk development and better cropping efficiency, especially for cvs. Empire and Delicious with M.9 stem-piece and MM.106 or MM.111 rootstocks. Interstem trees are found weak and bear less if the stem-piece is above the soil line (Norton, 1982). Shallow planted trees of Delicious on M.7A and MM.111 rootstocks showed increased frequency of burrknots and were less likely to be loosened in the soil by wind than were deep planted trees (Tables 1.19 and 1.20). It has been suggested that clonal rootstock be grafted not higher than 15 to 20 cm if above ground rootstock portions are to be avoided. By using shorter shanks on the rootstocks and setting most of the shank below the soil surface, the problem of loose trees may be lessened and the problems of excessively deep planting avoided (Lyons *et al.*, 1983).

Planting Plan

In planting trees in the orchard, the prime consideration is planting of sufficient pollinators to ensure effective pollination. Usually one pollinator tree is needed for two-three large trees planted at 10m distance or one row pollinator for two rows main cultivar. In close planted intensive orchards, every sixth tree in the row should be a pollinator.

Table 1.19: Effect of Subsoil Type and Depth of Planting on Growth of Spur 'Delicious'/MM 111*

Treatment	*Depth of First Root (cm)*	*Avg. Leaf Size (cm²)*		*Trunk Diam. (mm)*		*Burrknots (Per cent)*	*Looseness Rating*[y]
		1981	*1982*	*1981*	*1982*		
Loose subsoil 5 cm deep*	1a[w]	13 a	29 a	25 b	39 b	100	1.0b
Loose subsoil 15 cm-deep	1a	10 b	25b	24 a	40 b	20	1.4 ab
Glazed subsoil 5 cm-deep	1a	13 a	30a	28 c	47 a	100	1.0 b
Glazed subsoil 15 cm-deep	1a	11 b	26 b	24 a	37 b	20	1.8 a

1: Total of 5 trees; y: Rated on a scale of 1 to 5 (most severe); X: Planting depth below nursery depth; W: Mean sepraration in columns by Duncan's multiple range test, 5 per cent level.

* Lyons *et al.* (1983).

Table 1.20: Effect of Subsoil Type and Depth of Planting on Growth of Standard 'Delicious'/M 7A*

Treatment	*Depth of First Root (cm)*	*Avg. Leaf Size (cm²)*		*Trunk Diam. (mm)*		*Burrknots (Per cent)*	*Looseness Rating*[y]
		1981	*1982*	*1981*	*1982*		
Loose subsoil 5 cm deep*	1a[w]	12 bc	34 b	27 a	47 a	100	1.2 a
Loose subsoil 25 cm-deep	2.0 a	10 a	32 a	27 a	45 a	0	3.4 b
Glazed subsoil 5 cm-deep	1 a	13 c	36 b	26 a	45 a	100	1.2 a
Glazed subsoil 25 cm-deep	3.2 a	11 ab	30 a	24 a	41 a	20	2.6 ab

1: Total of 5 trees; y: Rated on a scale of 1 to 5 (most severe); X: Planting depth below nursery depth; W: Mean sepraration in columns by Duncan's multiple range test, 5 per cent level.

* Lyons *et al.* (1983).

Square System

This is the most widely used planting system in many of the old apple orchards in different parts of the world. In this system, planting is done in such a manner that four trees on the corner forms a square. Usually one-in-six or one-in-nine plants is used as a pollinator.

Rectangular System

In this system trees are planted close in the row which on maturity will fill the in-row space and give a hedge effect. Cultivation and herbicides are used in between the rows to provide the trees a weed-free strip.

Table 1.21: Planting Density Used by Major Apple growing Countries in the World in 2014

Country	*Planting Density/ha*	*Country*	*Planting Density/ha*
Austria	3000	Canada	980
Belgium	2275	Mexico	250
France	1874	United States	1000
Germany	2361	China	350
Greece	1060	India	220
Italy	2351	Japan	1072
Netherlands	2538	South Korea	1010
Portugal	1591	Turkey	315
Spain	467	Argentina	1063
United Kingdom	1134	Australia	540
Russian Fed.	620	Brazil	875
Hungary	1135	Chile	501
Poland	1221	New Zealand	1120
		South Africa	1204
		Average	1174

Source: World Apple Review 2015 Edition.

Semi-Intensive System

In this system trees are planted at a spacing of 4.5 m X 3m, 5m X 3.5m or 5.5m X 4m. The actual spacing is, however, determined on the basis of the expected tree size. The proportion of pollinators also depends on the spacing, cultivar and growing condition.

Semi-intensive system											
O	O	O	O			O	O	O	O	O	O
O	O	O	O			O	P	O	O	p	O
p	O	P	O			O	O	O	O	O	O
O	O	O	O			O	O	O	O	O	O
O	O	O	O			O	P	O	O	P	O
P	O	P	O			O	O	O	O	O	O

One-in-six plant								One-in-nine			
Intensive system								Solid plant			
O	O	P_1	O	O	P_2			O	O	O	O
O	O	P_1	O	O	P_2			P		P	
O	O	P_1	O	O	P_2			O	O	O	O
O	O	P_1	O	O	P_2			O	O	O	O
O	O	P_1	O	O	P_2			O	O	O	O
									P		P
								O	O	O	O
								O	O	O	O

Intensive System

Planting in this system is usually done at 4m X 2.5m or less. A complete row of pollinators is used in every two rows of main cultivars. This system of planting is more suitable for triploid cultivars.

Solid Planting

In this system, a single cultivar used as pollinator is planted at suitable interval with the main cultivar. The pollinator used as 'solid' plant is selected in such a way that it could pollinate the main cultivar every year. The 'solid' Cox's Orange Pippin or Bramley's Seedlings are used in Golden Delicious orchards, where the pollinators are trained as pillars or vertical cordons.

Contour Method

In hill slopes where erosion and run off are a serious problem, it is advisable to plant on the contour. On the same elevatation the trees are usually planted at the same row in contour. The procedure helps to build up a low terrace at the tree row which minimize the water run off.

High Density Planting

High-density planting (h.d.p) is now in practice in most apple growing countries. Many physical and agronomical factors, such as, climatic condition, soil fertility, rootstock and scion vigour, methods for training and pruning, orchard management, *etc.*, are the major considerations determining distance and system of planting (Sansavini, 1982). A collaborative research project to investigate the relative productivity of the most modern high density planting system has been started in UK, West Germany, Denmark, Poland, Italy and the Netherlands. In East Malling Research Station, the first trials were planted in 1980/81 with Golden Delicious and Gloster 69 at different planting densities on M.9 and M.27 (Jackson *et al.*, 1986). The dwarfing rootstocks M.9, M.4, M.7, J-TE-E, MM.106 and M.26 showed the best promise for h.d.p. in Czechoslovakia (Bartos, 1987). However, higher densities caused a reduction in light penetration into large trees which sometimes reduced bearing capacity of the internal or lower parts of the tree (Cain, 1971; Jackson, 1968). Jackson (1970, 1978) and Verheij and Verwer (1973) defined various parameters to evaluate optimum planting densities, among which are the ratio between tree height and width of the alley (tree height should not be more than 1.5 to 2 times that of the clear alley) and a LAI (leaf area index) not more than 0.50 to 1.00 over the cropping zone of the tree and not more than 2.50 to 3.00 in total, in order to permit the tree to intercept the maximum quantity of light. In Italy, even with the use of dwarfing rootstocks, the medium density planting system (from 800 to 1600-2000 trees/ha) based on a single continuous row (hedgerow) has become very common, with the trees trained as either free palmette or slender-free spindle (Sansavini, 1980). In a 4-year trial at Romania with cv. Jonathan (on MM. 106 or M.9) planted at 833-5555/ha and trained in different ways, showed that trees on MM.106 at 2222/ha in double rows with the crowns trained as dwarf pyramids and limited in height of 2.4 m gave the highest average yield of 68.75 t/ha (Gher Ghi *et al.*, 1984). Urunov (1987) recommended tree spacing of 5.0 x 4.0 or 5.0 m with the total seasonal irrigation at 2500-3000 m^3/ha and at 3500-4000 m^3/ha irrigation, a spacing of 5.0 x 3.0 or 3.5 m is recommended for cv. Starkrimson on *M. sieversii* seedling rootstocks. In Bulgaria on a meadow chernozem soil, the highest cumulative yield for 1980-85 in cv. Yellowspur was noted at 4.0 x 1.2 m spacing (170.9 t/ha) and in cv. Starkrimson at 4.0 x 1.4 m (116.7 t/ha) on MM.106 rootstocks (Stefanov, 1988). Tree growth, orchard management and fruit quality of cvs. Supergolden Delicious, Wellspur and Starkrimson were found best when the trees were grown in single rows at 3.0 x 2.0 or 1.5 m, *i.e.*, 1666 or 2222 trees/ha (Drobota and Drobota, 1986). The cv. Jonagold when grown on a single-row (3.2 x 1.5m) system with 1875 trees/ha yielded 13.44 kg/tree while the double row system (3.2 + 1.7 x 1.5 m) with 2499 trees/ha yielded 12.22 kg/tree. The proportion of fruits with 0, 0-25, 25-50, 50-75 or 75-100 per cent red colouration were 2.8, 10.0, 41.3, 40.3 and 5.6 per cent with the single-row system and 18.9, 17.1, 37.5, 22.3 and 4.2 per cent with the double-row system, respectively(Wirth and Vogt, 1988). Comasescu *et al.* (2012) studied the performance of 14 disease-resistant apple varieties on M.9 rootstock in high density (2500 trees/ha, 4m x 1m) planting system. The trees were grown in spindle crown form. The varieties Ariwa, Inedit and Saturn produced >30t/ha yield while varieties Read, Golden Lasa, Voinicel, Goldrush, Luca, Remar and Iris produced 24.05 to 26.7 t/ha fruits in the same time (Table 1.22). Varieties Golden Lase, Enterprise, Luca, Real, Rebra, Redix, Remar and Saturn produced bigger size (>170g) fruits on average.

The effect of increasing planting density at constant rectangularity on the vegetative growth, light interception tree yield, fruitsize and fruit colour development of apple trees in three training systems (slender spindle, tall spindle and Geneva y-trellis) was assessed in Canada. Five tree densities (1125 to 3226 trees/ha) and 2 cultivars (Royal Gala and Summerland McIntesh) were tested (Hampson *et al.*, 2004 a, b). As tree density increased, tree size decreased, and leaf area index and light interception increased. A planting density between 1800 and 2200 trees/ha (depending on training system) was found needed to achieve at least 50 per cent light interception. Training System though altered tree height and canopy diameter, but not total scion weight. Tall spindles did not intercept more light than slender spindles. 'Royal Gala' and 'Summerland McIntosh' intercepted about the same amount of light. As tree density increased, per-tree yield decreased, but yield per unit area increased. The chief advantage of high density planting was a large increase in early fruit yield. In later years reductions in cumulative yield efficiency and in fruit colour for 'Summerland McIntosh' began to appear at the highest density. The slender and tall spindles were similar in nearly all aspects of performance, including yield. 'Summerland McIntosh' yielded almost 40 per cent less than 'Royal Gala' and seemed more sensitive to the advance effects of high tree density on fruit colour.

Table 1.22: Fruits Production of 14 Apple Varieties in the High Density System (2500 trees/ha)*

	Variety/Graft bearer	*Production Obtained in the Year (t/Ha)*			*Average of the Years 4-5 (t/Ha)*	*Differences ± as to Wt*	*Signifi-cance*
		3 *2009*	*4* *2010*	*5* *2011*			
1	Jonathan/M9 (Mt)	3.5	15.8	24.0	19.90	-	
2	Ariwa/M9	5.5	21.9	42.0	31.95	+12.05	***
3	Golden Lasa/M9	5.5	19.4	29.8	24.60	+4.70	***
4	Goldrush/M9	5.0	20.8	29.5	25.15	+5.25	***
5	Enterprise/M9	1.3	12.5	28.5	20.50	+0.60	N
6	Inedit/M9	6.5	18.5	42.2	30.35	+10.45	***
7	Iris/M9	6.3	20.4	33.0	26.70	+6.80	***
8	Luca/M9	2.8	16.4	35.3	25.85	+5.95	***
9	Real/M9	7.5	18.8	29.3	24.05	+4.15	***
10	Rebra/M9	3.0	10.7	31.0	20.85	+0.95	N
11	Redix/M9	2.8	16.6	23.0	19.80	-0.10	N
12	Remar/M9	3.8	19.8	32.0	25.90	+6.00	***
13	Saturn/M9	5.8	21.6	42.5	32.05	+12.15	***
14	Voinicel/M9	4.3	18.6	29.8	24.20	+4.30	***

DL 5 per cent = 1.80 t/Ha: DL 1 per cent = 2.50 t/Ha: DL 0.1 per cent = 3.47 t/Ha.

* Adopted from Comasescu *et al.* (2012).

Szczygie *et al.* (2000) conducted a study in Brzezna, Poland to determines the effect of planting systems and densities on the growth and cropping of apple cultivars Elster, Gloster, Idared and Lobo on M.9, P.22 and P.60 rootstocks. The planting was done in medium-heavy loam soil as three-, two- and one- row strip systems. Spacing between strips, and spacing in the rows was from 1.00 to 2.25m. Tree were trained as spindle. Biennial bearer cvs. Lobo and Elster showed the highest yield in a single row system and the lowest in three- (Lobo) or two- (Elstar) row systems. Regular bearer cvs. Idared and Gloster showed the highest yield in three-row strips at the highest density. In cv. Elster, the optimum density was suggested as 1904 trees/ha for M.9 and P.22 rootstocks and 1632 trees/ha for P.60 rootstock.

Sansavini *et al.* (1980) studied the yield efficiency of h.d.p. of dwarf apple trees on M.26; 5000/ha in bed system for cvs. Cooper 4, Spur Red Delicious and Yellow

Spur Golden Delicious; trained as free central leader trees until the fifth year and was compared with those of medium density planting (m.d.p.) with 1666 trees/ ha, and palmette hedgerow training, at the Experimental Farm near Bologna, Italy. The result showed that accumulated crop per tree was always greater in the medium density plots. It was around 50 per cent greater than the high density Yellow Spur bed system whereas it was only slightly higher for bed system of cv. Cooper 4 which showed alternate bearing. Yield efficiency of the tree was always greater in the m.d.p. as was fruit size even though there were fewer apples per tree in the h.d.p. plots. More encouraging results were obtained with Yellow Spur than with Cooper 4 (Table 1.23). Fruit colour in respect of both the red and yellow components as well as skin brightness was also better for apples under m.d.p. In addition, in cv. Yellow Spur under h.d.p., the apples from the lower halves of the trees appeared to be greener than those from the upper halves. Fruit quality was lowered under h.d.p.; ripening began earlier and was more uniform on m.d.p., as revealed by skin brightness, lower acidity and firmness (in Cooper 4) and greater sugar content in cv. Yellow Spur.

Training and Pruning

Improved cultural practices which include manuring, irrigation, mulching, *etc.*, make the apple trees to bear for a longer period. However, pruning is one of the most important practices effecting the vigour, productivity and fruit quality and also the susceptibility to diseases and pests.

Pruning is done with a view to giving desired shape to a tree during early growing years and is termed as training. Even after proper training, the yearly pruning of bearing trees is quite important so that it could support a full crop of large and coloured fruits during the subsequent bearing period (Srivastava and Ram Kumar, 1980). Another objective of pruning is to remove the weak-growing shoots from the tree that would never produce fruits of satisfactory size and quality.

In the past, the most common forms of training apple trees were central leader,open centre or modified leader system. Beginning in the mid-1960s, interest in high-density planting developed and spur strains became widely available. It was soon found that pruning practices required revision to obtain early and heavy yields in high-density planting (Roosje, 1970; Wertheim, 1970) and the spur strains produced much less lateral growth than non-spur strains of the same cultivar (Westwood and Zielinski, 1966). In view of its importance, interest had developed in training practices and several new systems of training trees like dwarf bush, cordon, dwarf pyramid, pillar, espailer, fan, palmette, spindlebush, *etc.*, were standardized.

Training

Central Leader

The trees are trained with a main central leader with 6 to 10 strong lateral scaffold. branches, spread in all directions. The central leader is usually tipped throughout the life which results in development of scaffold limbs. This system of training is used to get a large tree, and pruning is done during the dormant season.

Table 1.23: Bearing Parameters of cvs. Cooper 4 and Yellow Spur trees. h.d.p. and m.d.p. from the Fourth to Sixth Years*

Variety, Planting System and Spacing Area	*Cropping Zone (Tree height) cm*	*Fruit/Spurs/Tree [Variation per cent No. (yr 4 = 100)]*			*Fruits/Tree [Variation per cent No. (yr 4 = 100)]*			*Fruit Set (1) per cent*			*Fruit Cluster No.*
		yr 4	*yr 5*	*yr 6*	*yr 4*	*yr 5*	*yr 6*	*yr 4*	*yr 5*	*yr 6*	*yr 6*
A) COOPER 4/M.26											
HDP	<120	133	-99.7	-3.8	23	-95.7	110.8	2.	(2)	6.2	1.02
(I.4m²)	120-240	363	-99.8	-1.2	71	-95.8	106.3	3.	-	6.8	1.03
Total and mean		496	-99.6	-1.9	94	-95.8	107.4	3.	-	6.7	1.02
MDP	<120	45	102.2	-6.7	36	-44.4	43.0	0	3.6	20.2	1.37
(6.0m²)	120-240	132	-38.7	-8.4	82	-30.5	38.4	4	11.7	15.7	1.32
Total and mean		177	-5.1	-8.0	118	-34.8	39.8	8	7.6	16.9	1.34
B) YELLOW SPUR/M.26											
HOP	<120	188	-71.3	-1.6	39	-2.6	102.5	3.4	11.7	7.1	1.21
(I.4m²)	120-240	182	-78.6	-1.7	57	-57.9	100.0	5.2	10.2	10.6	1.15
Total and mean		370	-74.9	-1.7	96	-36.5	101.0	4.3	10.9	8.8	1.18
MDP	<120	91	25.2	-57.2	62	45.7	-46.8	11.3	13.1	14.1	1.23
(6.0m²)	120-240	101	18.8	-57.5	73	42.5	-47.9	12.0	14.4	14.7	1.35
Total and mean		192	21.8	-57.3	135	43.7	-47.4	11.7	13.8	14.4	1.29

(I) Ratio between number of harvested fruits and number of flowers; (2) The low number did not permit statistical analysis.

* Sansavini (1980).

Open Centre

In this type of training, the central leader is removed about 1 m above ground level and three to five well-spaced scaffold branches are retained. Secondary scaffold branches are encouraged to develop the primary scaffold. Both primary and secondary scaffold branches produce fruit bearing laterals. Training by this system keeps the centre of the tree open, permitting entry of adequate sunlight throughout the tree.

Modified Leader

It is also known as delayed open centre. In this system the main central leader is allowed to grow for a few years, until 6 to 8 scaffold branches develop around the central leader. The central leader is then cut to form a side lateral, which in due course will grow as a modified leader. In this type of training, the tree develops well-spaced limbs with strong crotches. The top being open, allows more sunlight to penetrate deep inside the tree.

The formation of a multiple central leader tree and the defruiting of these leaders back into 3-year-old wood to maintain their dominance, are features of a tree form widely used in the past in Hawke's Bay area of New Zealand. The trees frame is formed from three or four closely spaced vertical leaders on which fruiting laterals spread horizontally. The leaders are usually looped together halfway up to tree and tied at the top, thus forming a tall pyramid and one-year shoots on the vertical leaders are shortened each year until tree forming is complete. After development of the tree, lateral pruning is done by the regulated method (Preston, 1954a). Trees in the Hawke's Bay area were found to grow on vigorous rootstock, mainly on Northern Spy and Merton 793, and this pruning method is well suited to upright growing cultivars on these rootstocks, such as Granny Smith with its characteristic habit of producing vigorous primary branches and narrow crotch angles. Other cultivars grown in this manner, include Cox's Orange Pippin, Jonathan and Dougherty (McKenzie, 1964).

Cordon

It consists of a single stem with side shoots or fruiting spurs kept short by summer pruning. In this system, trees are planted 2 to 3 m apart, 60-90 cm apart in rows. The dwarfing M.9, and semi-dwarfing M.26 rootstocks are most suitable for cordons. The trees are usually planted at an angle of about 45° for better exposure to light and secondary growth arising after summer pruning are cut back to their bases in late September. The trees are supported by three rows of wires, stretched tightly between posts. This system offers advantage of early bearing but is not suitable for vigorous or seedling rootstocks.

Dwarf Pyramid

In this system, trees are shaped like a 'Christmas tree,' which are planted 1 to 1.5 m apart in a row and 2-3 m between rows. During planting central leader is headed back to 45 to 50 cm and all the side shoots are trimmed to 10-15 cm. The fruiting spurs or the small branchlets develop from the trimmed side shoots. In the first few years, central leader and branch leaders are shortened during winter to keep the tree strong and upright. Annual spur thinning in winter is desirable with established trees. The dwarfing M.26 and semi-dwarfing MM.106 rootstocks are suitable for dwarf pyramid training on most soils.

Espalier

It consists of a central stem from which horizontal fruiting arms grow at about 45 cm interval. Trees are planted 4 to 6 m apart, according to the vigour of rootstock, cultivar and the soil. About five to six tiers of branches are trianed on wire. The unwanted side shoots arising from the horizontal leaders are trimmed and all the laterals from vertical stem removed. Annual pruning is usually done in summer when growth slows down. Espalier training and pruning method comprising mechanized lowering of tree height and limiting the crown width on one or both sides, supplemented with moderate manual thinning of the crown, generally results in higher yield and good fruit quality (Kudryavets and Mukhanin, 1988).

Free Spindlebush

This type of training was first developed in Germany before 1940 and has been adopted in several countries with modifications. It is basically a dwarf pyramid system, except that the branches are trained horizontally. Trees are planted at a spacing of 1.5 to 2 m (Golden Delicious group) or 2 to 3 m (Cox group) in rows and 4 to 4.5 m between rows. The rootstock M.9 was found most suitable for spindle bush. At planting, unfeathered maiden trees are headed back to a bud 75 to 90 cm above the ground. During the first growing season two or more branches, having natural wide angle, are selected to encourage even growth and train them to the correct angle, which should be about 30 degrees above the horizontal. The shoots are tied down so that they are straight, not arched, or strong vertical shoots will arise from the top of the arch. The shoots which arise on the main branches are retained and trained horizontally in the same manner. Any unwanted branches arising low down on the main stem should be pruned. Branches are allowed to grow from the central leader at regular interval, choosing wide-angled shoots. The higher placed branches are usually kept shorter for penetration of light to the lower branches. The main branches are trained regularly for development of sufficient fruiting laterals.

Slender Spindle

The Groesbeek or slender spindle training system is followed for greater plant density. The main difference between free spindle bush and slender spindle is limiting the lateral extension by partial removal when pruning or cutting them back to the central leader. In Netherlands, slender spindle trees received minimal pruning when young, with limb spreading to avoid stimulation of growth and encourage early fruiting. The trees had a slender conical crown, a central leader, and generally were supported individually with posts about 2 m in length above ground (Lord and Damon, 1983). In this system, in Netherlands, trees are planted at 1 m apart in rows and 3 to 3.5 m between rows for training as slender spindles. Slender spindle is suitable for cultivars like Golden Delicious and James Grieve and other relatively weak growing ones.

Tatura Trellis

The Tatura trellis method of training was developed by Chalmers and van den Eude at Tatura in Australia, aimed to produce a continuous canopy of fruiting wood half a metre deep to maximize yield and at the same time to facilitate mechanized management. Although, this system was originally developed for peach trees, it is now used for training of apple and other pome and stone fruits. By providing a saw-tooth profile, the potential production volume is increased in this method of training (Dunn and Rowe, 1982). Trees are planted 1 to 1.5 m apart and 5.0 or 6.0 m between rows. The rows are oriented north and south. At planting each tree is topped and allowed to develop two branches. These are trained east and west to form the arms in Y shape within a 50° to 60° crotch angle. Tensioned wires on steel frames support the arms to a height of 4 to 5 m. As the name implies, the trees of the Tatura trellis are trellissed and trained intensively until they reach full size, while after the trees reach their permisible size they require summer pruning to regulate canopy depth and shading.

Lincoln Canopy

Lincoln canopy method of training of apple trees was developed in 1970 by Dunn and Stolp at Lincoln College, New Zealand. In this method all branches are trained to a set of six wires on each side of the tree. The fruiting laterals are selected from buds arising from the leaders and the fruiting canopy is established at a height of 1.5 m. All the branches are trained in a single horizontal plane. The small tree form and the easy access that it offers from below and above facilitate mechanical spraying, pruning and harvesting. The trees, planted at about 1000/ha, form a continuous canopy of about 3 m wide (Dunn and Stolp, 1987).

Pruning

Established Spur System

This type of pruning offers permanent fruit spurs for production of fruits. The leader is headed back every year to ensure formation of spurs on the laterals. The strong erect laterals near the central leader are also removed, leaving the wide angle vigorous laterals for formation of spurs.

Regulated System

Regulated pruning has been successfully practised on a wide range of apple cultivars on semi-dwarfing and vigorous rootstocks like MM. 106, MM. 111 and M. 26. At the time of planting trees are headed back to 75 cm, allowing three well-placed feathers to form primary branches. In the earlier years, unwanted shoots are removed. In bearing trees, the crowded, weak and shaded branches are removed leaving leaders and strong laterals to grow.

Renewal System

The system is usually followed for vigorous cultivars. Pruning is practised in such a way as to maintain a continuous supply of new healthy shoots, spurs and branches every year, rather than to develop permanent spurs. A portion of the tree is pruned every year which produced localized shoot growth and may produce fruits in the following year while the unpruned parts will produce fruit buds. In this way the laterals are replaced.

Time of Pruning

The influence of pruning on tree growth and fruiting depends on the time and the method of wood removal; competition between vegetative growth and reproductive organs is particularly severe during flowering in the following 3-4 weeks. Pruning during this period can increase both flowering and fruitset (Quinlan and Preston, 1973). Pruning during the later part of the growing season has little effect on the reproductive processs.

Dormant pruning restricts growth of roots and reduces the trunk circumference increase of a tree but growth is stimulated near the pruning cuts (Elfving and Forshey, 1976). The number of spurs formed is reduced greatly by dormant pruning and the ratio of long shoots to spurs is much higher on dormant pruned trees (Mika, 1975). Dormant pruning also causes a delay in fruiting of young trees and reduces the yield, especially of young trees grafted on vigorous rootstocks (Jonkcrs, 1978; Mika, 1975).

Summer Pruning

In a broad sense, summer pruning means selective removal of shoots to or branches during the growing season (Saure, 1987). Summer pruning can be applied from as early as its beginning of growth season to as late as four weeks before harvest (Morgan *et al.*, 1987) with different severity of thinning cuts, heading, or pinching, based on the vigour of the tree, the variety the training system, Robinson *et al.* (1991) stated that summer pruning is required to obtain good fruit colour for tree types such as slender spindles trees in which canopy has gaps that become filled with shoot growth soon after full bloom. Summer pruning significantly increased light environment in the inner and middle canopies (Li and Lakso, 2004).

Reports on the effect of summer pruning on growth (Marini and Barden. 1982, Perring and Pearson, 1985), flowering (Lord and Greene, 1982) and fruit quality (Uterrnurk, 1977. Lord and Greene, 1982) are conflicting probably because the response to summer pruning is dependent on cultivar, tree vigour, tree age rootstocks *etc.* Greene and Lord (1983) reported that summer pruning of 6-year-old trees of Cortland apple for 3 consecutive years in either early July or early August reduced trunk circumference increase in the year of pruning and increased terminal growth in the following year. When determining the optimum summer pruning date, prime consideration must be given to the activity of leaf growth, which depends not only on the water and nutrient supply to the tree, but also on the varietal patterns of shoot growth the level of fruit-setting in the particular year and for early cultivars, on the date of harvesting (Gonda, 1984). Poniedzilek *et al.* (1987) stated that in Poland earlier pruning (25 June) reduced yield in cv. Spartan while pruning on 10 August increased it. However, earlier pruning increased vitamin C content of fruit. Sosna (2000) in his experiment with different semi dwarf apple cultivars (Sampion, Honeygold, Gloster, Jonagold, Witos and Alwa) on M.26

rootstock at Poland compared dormant pruning in March with summer pruning at the end of August. Yield and growth rate of different apple cultivars were not significantly affected by the time of pruning. However, apples from summer-pruned trees were better coloured. In order to improve light penetration to the crown and to produce high quality fruits, the distance between the branches of the first tier and the second tier should be no less than 1 m, and the distance between the second and third tier should be no less than 50 cm. For winter pruning, thinning out of the dense brances, retaining a flower bud: leaf bud ration of 1 : 5-6 and heading back the weak flowering branches are necessary. Using this system, ShuChang *et al.* (2002) recorded increased fruit weight by more than 32g and improvement of colour in apple cvs. Red Fuji and Starkcrimson planted at 5 x 4m spacing.

Nutrition

Apple trees require all the mineral clements needed by other fruit trees (Shoemaker and Teskey, 1959) for proper growth and bearing. The commercial standards of apple fruits are determined mainly by the colour, size and appearance, storage potential and organoleptic properties of the fruit. These quality parameters are greatly influenced by mineral nutrition, particularly by potassium, calcium and magnesium. Careful and frequent assessment of growth and fruiting characteristics must be combined with leaf and soil analysis for formulation of an economic fertilization programme for the apple orchard.

Role of Nutrients

Nitrogen

Among the different nutrient elements, nitrogen is thought to influence most the growth and yield of apple. However, application of nitrogen alone may affect the nutrient balance of the soil and the tree, and therefore, application of P and K alongwith some other elements in judicial amount and at proper time has been suggested. The deficiency of nitrogen may cause restricted growth, narrowing of top-root ratio, less elongation of branches and reduction in size and colour of fruits. The bark of the tree may be reddish or yellowish brown; leaves become small, thin, tend to be sparse and are pale green in early stages, but later developing tints of yellow (Sharma and Azad, 1983). In acute nitrogen deficiency the fruits may be woody and devoid of flavour. Pant *et al.* (1976) described the N deficiency symptoms as follows: severely resistricted growth and thin shoots with leaves turning pale green to yellow, severity of yellow colour being more pronounced on older leaves than on young leaves, older leaves falling prematurely.Excess nitrogen can severly reduce apple quality and tree hardiness, and increase susceptibility to fireblight. Large dark green leaves that remain on the plant late into the fall may indicate too much N. Apples colour poorly and lose firmness more readily in storage.

The fertilized trees usually show a marked response to nitrogen in respect of darker green leaf colour, greater terminal growth, spur growth and increased trunk circumferance. However, the rates of N fertilizers to be added for obtaining optimum response may vary considerably depending on various factors, such as location, cultivar, rootstock, time of application, *etc.* Weissenborn (1967) recommended application of 60-100 kg N/ha for Cox's Orange Pippin on M.2 rootstock and 80-120 kg N/ha for Golden Delicious on M.2 rootstock. But when these cultivers were grown on M.9 rootstock, the optimum N requirement increased to 200 kg N/ha. Wrona (2004) suggested application of 50 kg N/ha during early spring and another 50 kgN/ha after 5 weeks for 9 years old apple cv. Sampion. Gu Manru *et al.* (1981) found that nitrogen in the form of NO_3 was only present in apple trees when there was ample N available in the soil. NH_4 nitrogen was present in smaller quantities than NO_3 and was highest when the vegetative organs were being formed.

Hikasa *et al.* (1986), in their experiment with cv.Tsugaru apple trees on M.26, applied N at 0 to 24 kg/10 acare for 6 years. They found that high N levels resulted in poor fruit colour development which was attributed to N accumulation by the fruit. Increasing N levels lowered fruit acidity but had no effect on total soluble solids. Based on these results they suggested application of N at 10 kg/10 acare for Tsugaru apples in Japan. In Polese region of the Ukrainian SSR, yields without N fertilization in cv. Jonathan on Antonovka obyknovennaya seedling rootstocks varied between 99.5 and 360.5 and for White Winter Calville from 87.5 to 332.5 h kg/ha while with N (60-120 kg/ha) it ranged from 93.5 to 399.0 and from 86.5 to 369.0 h kg/ha, in the two cultivars, respectively (Slyusarchuk, 1987).

Nitrogen fertilizer placement method has no significant effect on yield (Boon *et al.*, 1973) or root development (Catzeflis, 1970). However, Trivindo (1973) showed that surface application of N fertilizer before irrigation was more effective than injecting N in liquid form to a depth of 40-45 cm. Naumann (1972) suggested application of calcium nitrate via a sprinkler system to Jonathan, Golden Delicious and Cox's Orange Pippin trees, because with sprinkler it was possible to distribute the total fertilizers evenly throughout the whole vegetative period. This not only reduced the occurrence of bitter pits and internal browning, but also enhanced vegetative growth and initiation of flowers.

Many workers have emphasized the usefulness of urea spray for N nutrition of apple trees. Kropp and Ben (1981) found that 0.75 per cent urea applied 6 times at 5-day interval or 3 times at 10-day interval starting from 1 June increased leaf N content by 35-60.8 per cent and yield by 28.8-45.3 per cent in 17-year-old Cox's Orange Pippin trees, the former treatment giving best results in each case. Although urea spray showed some adverse effects on fruit quality and storability, for young apple orchards during the first 5-6 years after planting, it has always been found beneficial (Reckruhm *et al.*, 1983).

Gu *et al.* (1985) found that ^{15}N absorption by the large leaves of the previous year's shoots was highest when applied to the foliage in the autumn. The ^{15}N was translocated mainly to younger leaves and shoot apices, little was translocated basipetally but ^{15}N absorption by large and average-sized leaves on new spring shoots was only 50 per cent that of the large leaves on the previous year's shoots, and the greater part was translocated basipetally. The amount of ^{15}N translocated to young shoots in the spring, when they were growing rapidly, was higher than in the autumn. Atkinson *et al.* (1987) reported that leaf, wood and soil materials taken from an established apple (Cox's Orange Pippin) orchard contain ^{15}N in excess of astmospheric level. The enrichments were highest (0.380 atom per cent) in surface soil and lowest in leaves (0.376 atom per cent). The enrichments varied according to the season, being generally higher in late summer (August) and autumn (November) than in June and July. Angelis *et al.* (2011) stated that different timings and methods of N application modify N dynamics in the plant and could have differential effects on fruit colour. Application of N after fruit hervest resulted in fruits of 'Royal Gala' on M.9 with darker colouration, higher percentage of coverage colour and higher concentrations of anthocyanins. The negative effects of N on fruit colouration can be mitigated by differing N application to the postharvest period.

Phosphorus

Apple has a fairly low requirement for P and its deficiency rarely occurs in apple orchrds. A mature apple tree yielding 25 bushels of fruit removes only 0.5 lb of P per year (Heinicke, 1936). Generally, phosphorus is present in large quantity in tissues during the period of rapid growth, which drops with the age of the tissue. The early deficiency symptoms include stunted growth with short and slender shoots, smaller leaf size with a tendency to become dark green with yellowish or purplish tint on the midrib and larger veins (Davis and Hill, 1941; Pant *et al.*, 1976).

Thomas (1968) found that the applied P promoted both tree growth and crop yield, because it increased the availability of P in the surface soil (Stoilov and Todorov, 1972). Excessive P nutrition was found to increase frost sensitivity of apple trees (Havelka and Brychta, 1967). It was observed that in the Pacific northwest region, certain apple orchards growing in low phosphorus soil (3 ppm available P) showed less tree vigour, produced low yields and responded well to the application by mono-ammonium phosphate (Raese, 1987).

Potassium

Apple trees respond well to potassium fertilization, especially when the trees are growing on soils inherently low in K and on very acid soils. Kopytko and Mikhalevskeya (1984) found the exchangeable K in the soil increased gradually over several years following application of 120 kg K_2O/ha, but the trees responded only after a substantial increase of the potash level in the rooting zone. Dry weather reduces the ability of apple roots to absorb minerals; as a consequence, K deficiency is more conspicuous in dry years than in wet ones. Boynton *et al.* (1944) found that McIntosh apple leaves were significantly higher in K in a year of high rainfall than in a year of low rainfall.

Deficiency of potassium is seen on spurs of old branches, which progressed towards the younger leaves as the season advances. The first evidence of scorch might be loss of normal green colour at the widest part of the leaf margin, accompanied by a water-soaked or stained appearance, which was followed by necrosis. Severely affected trees were mostly stunted in growth with small-sized leaves. The fruits in affected trees do not develop properly, as a result they are small in size and do not develop proper colour (Cain and Shear, 1964, Ram *et al.*, 1988). Although excessive K is not toxic, high soil K levels may inhibit Mg or Ca uptake and so induce deficiencies of these elements.

Heavy doses of commercial KCl or K_2SO_4 proved to be quite efficient in curing leaf scorch. Degtyar and Krolik (1973) found that N and P rates each remaining constant at 100 kg/ha, raising the K rate from 100 to 250 kg/ha increased the total yield over 3 years from 322.2 to 343.3 kg/tree. However, high rates of K application (0.5 g K/kg soil) may cause stunted growth. Application of K as KCl was found superior to K_2S0_4.or $KN0_3$, since it was found to give higher yield and frost resistance (Vlasishin, 1979). Spring application of K fertilizer was found to be better than autumn application, since it increased leaf K levels more appreciably and reduced frost susceptibility (Kul'kov *et al.*, 1971). Boynton and Oberly (1966) concluded that leaf K levels below 1.0 per cent are deficient, while optimum level ranges between 1.25 and 1.50 per cent.

Calcium

The young foliage of calcium-deficient plants showed an upward cupping of the margins and on actively growing terminals the expanding leaves develop a uniform veinal and interveinal chlorosis (Shear, 1971). In older leaves, the margin turns necrotic and they may fall. In fruits, the colour changes due to reduction in chlorophyll II and longitudinal cracks appear as a result of reduced flexibility of the cell-wall. The fruits deficient in calcium showed intangible nuclear membrane rupture, uneven distribution of calcium maculas of the calcium antimonite precipitates, appearance of blank spaces in the nucleoplasm, breaking down of the tonoplast, appearance of calcium maculas in the cytoplasm and disappearance of the compartment phenomenon (JianHui and Wei, 2004). Deficiency of calcium also leads to some physiological diseases such as bitter pit or cork spot (Sharma and Azad, 1983). Bitter pit appears as slight indentations in the skin, usually toward the calyx of the fruit. These areas later turn brown and soft dried pit of collapsed tissues develop (Faust and Shear, 1972). Lidster *et al.* (1975) opined that Ca content below 55 mg kg^{-1} showed more physiological disorder while the limit of Ca content in peel seems to be 500-700 mg kg^{-1} (Chiu and Bould, 1977). The Ca content of fruit was thought to be genetic (Faust *et al.*, 1971) and in general, larger fruits have low Ca content and are usually more susceptible to disorder (Perring and Jackson, 1975).

An increase in the rate of transpiration increases the transport of calcium from roots to other parts of the trees (Stebbins and Dewey, 1972) and redistribution is related to transpiration stream (Bukovac and Wittwer, 1957). The translocation is mainly through xylem while it may occur both in xylem and phloem (Faust and Klein, 1974). In fruits, the first phase of rapid intake was found during cell division to cell expansion (Smith, 1970) while the second phase with an influx/outflux from the fruits (Wilkinson, 1968). Calcium usually accumulates steadily during apple fruitlet development (end of July for Cox apples in the UK) when the uptake usually declines or sometimes ceases. Occasionally, Ca may be withdrawn from the fruit (Perring and Pearson, 1985). Jones *et al.* (1983) reported that calcium uptake continued from flowering to harvest, though there was a tendency for the uptake rate to decline in the last two months of fruit growth. Reducing the transpiration rate of spur leaves by removing some or all of the leaves or by enclosing leaves in polythene bags was found to reduce the Ca uptake by the developing apple fruits, while having only small effects on fruit growth and the accumulation of other minerals. These results support the hypothesis that adequate evaporation from the primary leaves is crucial to maintain high levels of fruit calcium (Jones and Samuelson, 1983).

Calcium is generally applied as a spray solution in humid areas and to soil in the drier region, however, the effect of liming seems to be indirect through sod cover. Spraying of calcium in leaves is not worthwhile (Biddulph *et al.*, 1959) because of poor translocation and, therefore, four late sprayings at 14-day interval on fruit was found beneficial (Schumacher *et al.*, 1976). Spraying is more effective when the humidity is relatively low, because penetration is higher at low humidity (Goor, 1973). In hot and dry weather, regular application of calcium chloride spray throughout fruit development minimize the occurrence of crinkle (Perring *et al.*, 1985).

Magnesium

Evidence so far obtained indicates that Mg deficiency may develop in acidic soils or in soils high in Ca. In either case, very low exchangeable Mg may be found. Replaceable K considerably higher than Mg may be a factor causing the appearance of Mg deficiency symptoms when replaceable Mg is low (Boynton and Oberly, 1966). The symptoms of magnesium deficiency as described by Cain and Shear (1964) and Boynton and Oberly (1966) are as follows: fading between the large veins of the older leaves on some shoots or spurs is the first visible symptom of Mg deficiency in apple trees. The faded areas may turn pale yellow and remain so for sometime or they may scorch rapidly; in either cases, necrotic brown blotches develop. Some of the dead areas between veins may remain as islands surrounded by green tissues, but many of them tend to coalesce as they grow in size, necrosis then follows. The older basal leaves are usually affected first and as the season progresses the symptoms appear on the younger leaves.

The magnesium deficiency was more prevalent in young trees growing on light sand or sandy loam gray-brown podzolic soils (Sadowski *et al.*, 1976). Epsom salts, kieserite, salt water magnesia, magnesium carbonate and dolomite limestone all have brought about improvement in growth of trees suffering from Mg deficiency, while low-Mg limestone is not very effective in curing Mg deficiency symptoms (Kidson *et al.*, 1943). Slowik (1970) found that three sprays of 2 per cent $MgS0_4$, $7H_20$ solution at 2-week interval from petal fall was most effective in alleviating Mg deficiency symptoms in Jonathan and Starking apple while soil applied Mg showed little effect. In general, when the Mg content of mature apple leaves in mid-summer is about 0.40 per cent the trees rarely show Mg deficiency symptoms. The symptoms may, however, appear when the Mg content is between 0.40 and 0.25 per cent, if other conditions are favourable for their appearance.

Zinc

Zinc deficiency of apple is commonly called 'rosette' because of the characteristic symptoms. Rosette usually first appears on the spring growth. If rosetting occurs late in the season, the symptoms are less pronounced, but the affected terminals may die in the following spring after starting a weak growth, leaving behind a number of short lateral spurs. These laterals and terminals are more susceptible to winter injury than normal shoots. Paulechova-Kralikova (1966) described Zn deficiency symptoms as rosettmg, little leaf, chlorosis and occasionally dying of the shoots. Certain cultivars like Jonathan, Delicious and Starking are likely to be more affected than others. Rosetting may develop when the foliar P:Zn ratio is higher than 115 and the foliar Zn level ranging from 16.4 to 21.2 ppm.

Corrective measures of Zn deficiency in apple include (i) use of alfalfa as a cover crop, (ii) application of rather heavy doses of zinc sulphate to the soil under affected trees, (iii) trunk injection of $ZnSO_4$ (iv) dormant spray of $ZnSO_4$, and (v) early summer spray of $ZnSO_4$. Benson (1962) suggested summer application of 3 to 5 Ibs of Zn-EDTA per 100 gallon if there was no dormant application. Tarasov (1966) recommended an early spring (dormant) application of $ZnSO_4$ at a concentration of 2.3 per cent or 5.6 per cent according to the degree of symptom expression. Dev and Kapoor (1972) reported that the foliar Zn level in deficient trees ranged between 17.5 ppm and 26.5 ppm and in normal trees in the range of 24.0-45.5 ppm for Golden Delicious apples, while in Red Delicious, the Zn levels for deficient and normal trees being found in the range between 13.0-28.5 and 24.2-35.0 ppm, respectively. A leaf Zn content of 130-160 ppm was found to be associated with toxicity (Orphanos, 1982).

Iron

The iron deficiency in apple is more common in semi-arid regions where irrigation waters are high in bicarbonates or where the soil is high in pH and low in organic matter. Highly calcareous soils of humid regions also show Fe deficiency symptoms in apple trees. The deficiency symptoms usually first appear on young leaves at the tip of the shoot. It is characterized by interveinal chlorosis of young leaves. In severe cases the new leaves may unfold, but completely devoid of green colour; the veins, however, may turn green later (Cain and Shear, 1964; Boynton and Oberly, 1966). Iron-deficient plants grown at pH 5.5 developed abnormal chloroplasts with dispersed grana and osmophilic bodies (Hi *et al.*, 1984). Fenner and Adamson (1977) suggested spraying mature apple trees with iron sulphate (0.1 per cent), which was found to raise the Fe content of leaves. However, spray application is sometimes found to be hazardous due to toxicity to the foliage. On the other hand, trunk injection was found to be more effective than other methods and its effects are expected to last for several years after application (Boynton and Oberly, 1966). Ferrous sulphate and Fe-sequestrene 330 were more effective than ferric citrate in alleviating chlorosis (Barney *et al.*, 1985).

Copper

Copper deficiency seldom occurs in apple orchards. The symptoms, however, have been recorded in South Africa, western Australia and coastal California, where the soils are mostly acidic. The deficiency symptoms appear as brown spots on the terminal leaves followed by necrosis, the severity of which increases with time, and finally the upper 3 to 12 inches of the shoot die and wither (Dunne, 1946). Dunne (1938) found 1 to 4 ppm of Cu in the uppermost leaves of trees affected with die back and 3 to 12 ppm in leaves of normal trees. Copper deficiency under field condition can be controlled by soil application of copper sulphate or foliar application of Cu containing fungicides like Bordeaux mixture. Tarasov and Kovalenko (1970) recommended a foliar (0.2 per cent) or soil application of copper sulphate to correct Cu deficiency.

Manganese

The manganese deficiency symptoms described by Wallace (1951) and Cain and shear (1964) are as follows: the characteristic chlorosis first appears at the leaf margins and progresses interveinally toward the midrib. The chlorotic areas are sharply defined and bands of normal green colour, varying in width, surround the midrib and main veins. The symptoms may appear on any or all mature leaves. Manganese deficiency in apples can be corrected by soil application, by injection or by foliar spray. Except in acid soils with low fixing power, where soil application may not be satisfactory, leaf sprays seems to be the most effective control measure. Manganese toxicity may cause apple measles which is found to be more prevalent in acid soils and Red Delicious seems to be more susceptible than other cultivars. Epstein and Lilleland (1942) found that apple leaves on an average contained 81 ppm Mn, while leaf sample taken from Mn-deficient trees contained 5 ppm of Mn.

Boron

The most characteristic symptoms of boron deficiency usually occur on the fruit as internal browning and corking or spots on the flesh (internal cork) and sometimes as corky tissue in the skin (external cork). Besides fruit symptoms, in more severe cases, boron deficiency symptoms appear on vegetative growth, but symptoms on fruit usually occur before vegetative symptoms. The first symptom to appear on the foliage is yellowing and red venation of the terminal shoot leaves, followed by marginal and tip necrosis. Death of small areas of bark near the shoot tip is followed by progressive death of the inner bark and cambium from the tip backward. Oberly (1963) obtained highly significant correlation between the per cent of drought spot on fruit and leaf B levels as well as fruit B levels. Trees with 75 to 100 per cent of apples showing drought spot had leaf B contents ranging from 14 to 21 ppm (dry weight) and fruit B contents ranging from 0.4 to 0.7 ppm (fresh weight), whereas trees with less than 5 per cent apples showing drought spot had leaf B content ranging from 18 to 26 ppm and fruit B content ranging from 0.6 to 1.0 ppm. Rom and Morris (1967) controlled B deficiency by applying sodium polyborate, either as a single treatment of 2 Ib per 100 gallon at full bloom or applied 2 or 3 times at the rate of 1 lb per 100 gallon. Excess boron may affect the fruit quality. Normally, fruits suffering from B toxicity mature early and have shorter storage life. Curling of leaf edges and purple colouration appeared to be associated with high boron (Hansen, 1981).

Fertilizer Recommendations

The usefulness of N, P and K in various combinations for better growth, yield and quality of fruits has been demonstrated by several workers. However, no combination was found best suited for all conditions and the relative amount of N, P and K for best results varied widely (Table 1.24). Bugarcic *et al.* (1981) reported that the rate of NPK application had significant effect on yield of Golden Delicious apple on M.2 rootstock. Application in February gave the highest mean annual yield (66140 kg/ha) and application in late April the lowest (55145 kg/ha). Foliar application of NPK fertilizer has been proved beneficial, especially in conditions when soil application is not very effective. In a 3-year trial with Golden Delicious on MM.106 rootstock, Deckers (1977) found that the yield was raised by 23 per cent by foliar application of NPK in July and August, amounting to annual application of 1.5 kg N, 0.7 kg P_2O_5 and 0.5 kg K_2O/tree. Singh *et al.* (1978) obtained best results in terms of yield and fruit quality in Royal Delicious apple with N, P_20_5 and K_20 at 500, 250 and 750 g/tree/year, respectively.

Table 1.24: Optimum Rates of NPK Fertilizers for Apples

Cultivar	*FYM (tonnes/ha)*	*N*	*P*	*K*	*References*
London Pippin	-	30	30	30	Olisaev and Byasov (1971)
Pepin Shafranyi	-	120	120	120	Varkvasov (1971)
King David	30	60	60	60	Luckkov and Ponomareva (1972)
Winter Calville	-	120	60	60	
Renet Simirenko	-	120	120	120	Osipchuk (1972)
Rossoshankoe	20	40-45	40-45	80	Naskidashvili (1973)
Zolotoe					
Jonathan	-	60	60	60	Guz (1974)
Snow Calville	-	120	120	120	
Richard Delicious	-	180	120	90	Kozhukhar (1975)
Silva Pobeditelyam					
Renet Simirenko	-	100	100	200	Degtyar and Lupashko (1977)

Tissue Analysis

Tissue content of NPK for optimum growth have been estimated by many workers (Table 1.25). The values, in many cases, showed wide variation due to many reasons such as sampling procedure, cultivar, soil condition, *etc.* However, correct interpretation of these data may serve a useful purpose in diagnozing nutrient deficiency in an orchard. The nutrient levels in leaves during late July and August, indicating the range from below to excess as suggested by Childers (1966) is presented in Table 1.26.

Table 1.25: Tissue Contents of NPK for Optimum Growth and Yield

Tissue	*Cultivar*	*N*	*P*	*K*	*References*
Shoot and Buds	Aport Alexander	1.3-1.6	0.36-0.39	0.69-0.789	Darkanbaev *et al.* (1975)
Leaf	White Winter Calville	2.35	0.19-0.20	1.59	Kovalev (1974)
Leaf	Golden Delicious	2.4-2.7	0.40-0.70	1.8-2.0	Keremidarsks *et al.* (1977)
Leaf	James Grieve	2.7-3.0	0.18-0.25	1.2-1.4	Dufkova (1977)

Table 1.26: Nutrient Levels in Leaves of Apple during Late July and August*

Nutrient Level	(Percent d. wt.)					(ppm d. wt)				
	N	K	P	Ca	Mg	Mn	Fe	Cu	B	Zn
Below normal	1.5	0.9	0.08	0.2	0.18	20	40	1	30	10
Normal	2.0	1.2	0.12	1.0	0.24	25	50	4	35	18
Above Normal	2.3	3.0	0.30	2.5	1.00	200	400	50	80	100
Excess	3.5	4.0	0.70	3.0	2.00	450	500	100	100	200

* Childers (1966).

The variation in heavy metal content (*e.g.*, Fe, Cu, Zn, Cd, Mn, Co, Ni, Pb) in apple trees was found to be influenced by the location of the orchard (Ligocki *et al.*, 1988a). In an experiment in Poland they (1988b) observed that the cvs. Jonathan, Starkcrimson and Golden Delicious did not differ significantly in leaf Cu, Zn and Cd in Brzezna and Dabrowice. In Swierklaniec, the same lack of differences occurred in Cd and additionally in leaf Co, Ni and Pb. In all experimental stations the concentration of Fe in leaves of Golden Delicious was lower than in other cultivars. In spurs a similar level of heavy metals in each cultivar included Zn, Co and Pb in Brzezna, Fe and Pb in Dabrowice and Mn and Ni in Swierklaniec. In all the experimental orchards there was no significant differences among cultivars in Cd content of fruits. In contrast to the leaves, fruit of Golden Delicious were higher in each station in Fe compared with Jonathan and Stacrkrimson apples, while apple of Golden Delicious was also richer in Cu. The inter relationship between the leaf nutrient contents and fruit quality were studied in five apple cultivars (Generos, Florina, Delicious de vinesti, Janathan and Pionier) by Jivan and Sala (2014). They observed highly significant relationship between N and Fe contents (R_2=0.734; $P < 0.01$) and between Cu and K (R_2=0.702; $P<0.01$). Fruit acidity was found negetively correlated with soluble solids contents in cvs. Generos, Delicios de Vinesti and Jonathan, whereas the respective correlation in 'Pionier' was positive. In cv. Florina no significant correlation was found between acidity and soluble solids content. Among macroelements, nitrogen has a considerable contribution to fruit acidity and this allows to predict this index with high degree of safety (R_2=0.690; RMSEPN= 0.105).

Irrigation

Irrigation to maintain adequate soil moisture usually results in increased yield, decreased incidence of russeting and skin cracking. Severe water deficiency adversely affects flower formation. The terminal shoot growth decreases in water-stressed trees, and stem and leaf water potentials of the stressed trees are markedly reduced (Ebel and Proebsting, 1986). It is necessary to irrigate the plant during the period June/July to September. Irrigation before June/July is also necessary, particularly for newly planted trees, especially those on shallow, light or gravelly soils. Bunea and Stepanescu (1986) recorded water consumption over several years in an orchard in Romania, planted with cv. Jonathan on M.4, and irrigated when the available soil moisture level fell to 30, 50 or 70 per cent of field capacity. The control plants were not irrigated. They found that the water consumption during the growing period rose from 4258 m^3/ha in the control to 5081, 6758 and 6684 m^3/ha in the 3 irrigated variants, respectively. The daily water consumption reached maximum in June and July, and the trunk growth and fruit were found positively correlated with water consumption. The yield of Cox in the UK was recorded significantly higher for the irrigated trees than those receiving natural rainfall. There was also evidence of a carryover effect, with the most severly stressed trees having very little fruit set in the following year (Jones *et al.*, 1986).

There has been widespread adoption of irrigation systems for applying water economically. The modern irrigation systems for apple orchards are mainly sprinkler, drip or trickle irrigation. These systems are considered more economical since these perform functions other than water application (Proebsting *et al.*, 1977); for examples, chemicals, especially fertilizers, can be applied through irrigation system. With overhead sprinklers the temperature of orchards can be modified to alleviate heat stress (Unrath, 1972) and also to prevent freezing and delay bloom (Anderson, 1975).

The drip (called in Israel) or trickle (called in UK and USA) irrigation system is being increasingly used in most of the apple orchards, because of its high water use efficiency. The system also suits the trees on hill slopes. Trees of Gloster 69 on M.27 receiving high water supply from trickle irrigation produced vigorous vegetative growth and highest yield (Freund and Bunemann, 1986). Proebsting *et al.* (1977), while comparing the trickle and sprinkler methods of irrigation, in Redspur Delicious on seedling roots applied water (i) daily by trickle irrigation in amount determined by water use from companion trees grown in Iysimeter, (ii) daily by over-tree sprinkler irrigation in amount determined from daily pan evaporation, and (iii) by over-tree sprinkling every 2 weeks with amount needed to return soil moisture to field capacity. They found that trees grown with trickle irrigation flowered and set fruit in the 3rd and 4th seasons. Most of the sprinkled trees did not flower in the 4th year. The average shoot growth was less, but more buds broke on the trickle irrigated-trees resulting in equivalent total shoot growth on all treatments. The fruits produced under trickle irrigation were at least as large as those from sprinkled trees but had higher soluble solid contents (Tables 1.27 and 1.28). Keipert (1987b), however, found no significant difference in yield from non-irrigated controls in cultivars such as King of Pippins, Cox's Orange Pippin, Roter Boskoop, Jonagold, Golden Delicious and Gloster (all on M.9 rootstock) by providing 9 mm irrigation water per day through drip during the dry period between April and September (1983 to 1986) in West Germany. In a trial at Ukrain with apple cvs. Renet Simirenko and Red Delicious, Avraamov *et al.* (1988) found that drip irrigation at 2 emmitters/tree promoted yield by 15-20 h kg/ha, over irrigation at 1 emitters/tree. Fertigation with 8 kg N/tree from May to July, applied continuously, once or twice a week or once every 2 weeks was reported to increase shoot growth in the first year and yield in the following year in cvs. Cox's Orange Pippin and Jonagold on M.9 rootstock (Bolding, 1987). Trunk diameter and plant transpiration intensity were affected by soil moilture. With increasing content of soil water, trunk diameter and transpiration rate increased. However, excessive irrigation caused shrinking of the trunk. The optimum level of soil water causing the highest increase of trunk diameter ranged from 25-30 per cent of field capacity (Klamkowski and Treder, 2000). Drip irrigation was suggested by Orta *et al.* (2000) under Thrace conditions and water should be applied when moisture reaches 40 per cent of the available water holding capacity.

Table 1.27: Growth Characteristics of 'Delicious' Apple Trees grown with Trickle Irrigation Compared with Over-tree Sprinkler Irrigation*+

	Irrigation System			
	Trickle		*Sprinkler*	
	Year	*Daily*	*Daily*	*2-wk*
Avg shoot length (cm)	1975	54*	78	72
	1976	37	68*	44
Avg no. shoots/tree	1975	95*	68	70
Avg no. shoots/leader	1976	38	21*	30
Total shoot length/tree (cm)	1975	5130	5304	5040
Total shoot length/leader (cm)	1976	1406	1428	1320
Trunk area increase (cm^2)	1976	14.8*	17.7	16.5
Cross-sectional area, trunk (cm^2)	1976	36.4	40.6	34.9

* Different from trees sprinkled at 2-wk interval at 5 per cent level of significance.

\+ Proebsting *et al.* (1977).

Soil Management

Mulching

Mulching with straw, hay, sawdust or other organic matters increases the humus content of the soil and its moisture holding capacity. It increases soil porosity and soil aeration and decreases runoff. Various plastic and polythene mulches are also used. However, all mulching materials influence growth and yield. Foliage colour was found better with hay and seaweed mulches than with sawdust mulch or sod culture (Latimer and Percival, 1947). The different cultivars also behave differently with the mulching materials. The greatest increase in fresh weight, number of shoots and blossom cluster of cv. Macspur (spur-bearing) was found with white plastic, while the least response was with grass; the reverse held for regular-bearing cv. Summerland (Proctor and Crowe, 1983). Moreshet *et al.* (1975) showed that a ground cover of reflective aluminium-coated plastic between headgrows of tall, dense apple trees increased the fruit weight, diameter, colour and sugar content. The cover crops, ploughed in or left as mulch, improved soil's physical properties and N and P_2O_5 contents and increased the shoot growth, especially when used as a mulch. Pea and rape as cover crops were found to increase the fruit size of cv. Borovinka Tashkentskaya on M.9 rootstock (Dzhavakyants and Yusupkhodzhaev, 1985). Stojanowska (1987) found that use of perforated black foil as mulch moderated daily fluctuations of soil temperature and increased soil water content compared with soils kept under a herbicide fallow. McIntosh trees growing in rows mulched with foil were larger and gave 18-24 per cent higher yield than those under the herbicide fallow. Application of foil also ensured better supply of some essential mineral elements to the trees. In orchards of cvs. Boskoop, Golden Delicious and Gloster, uncomposted conifer or oak bark or rape straw were found highly effective in supressing weed growth under the trees. Mulches of partially rotten or nutrient-rich materials usually need herbicide application. Application of bark and rape straw was found to stabilize the highly variable yield of Boskoop; conifer and oak bark markedly improved fruit quality (Niggli *et al.*, 1988). TerAvest *et al.* (2011) reported elevated earthworm activity, total soil carbon, high C:N ratio, better tree growth and yield by using wood chip as in-row ground covers. Hipps *et al.* (2004) reported that straw mulch increased the concentration of K in the surface soil (0-15 cm) and increased shoot growth, yield and fruit weight of apple cv. Bramley on M.9 rootstock. The increase in shoot growth was ascribed to the greater conservation of moisture within the profile during the short extension period.

Table 1.28: Growth Characteristics of 'Delicious' Apple Trees grown with Trickle Irrigation Compared with Over-tree Sprinkler Irrigation*+

	Irrigation System			
	Trickle		*Sprinkler*	
	Year	*Daily*	*Daily*	*2-wk*
Yield (kg/tree)	1976	8.8*	1.0	.4
per cent trees with over 5 flower clusters	1976	87*	15	18
Fruit Size (g/fruit)	1976	2.59	195	204
Soluble solids (per cent)	1976	13.4	11.0	12.0
Leaf N (per cent)	1975	2.52	2.57	2.58
Leaf N (per cent)	1976	2.16	2.29	2.24
Avg total leaf water pot (bars)	1976	-14.2	-11.0*	-14.5

* Different from trees sprinkled at 2-wk interval at 5 per cent level of significance.

+ Proebsting *et al.* (1977).

Weed Control

Complete elimination of weed competition in the tree row is essential, since the young apple tree, although it is a woody perennial, is very vulnerable to competition (White, 1984). The use of herbicides to eliminate weeds and grasses enables tree roots to exploit the surface soil more efficiently than in grassed or cultivated orchards. This results in increased tree vigour and higher crop yields. Other advantages of herbicide management are reduced frost risk and savings on fertilizers, irrigation and time spent on soil management (Robinson, 1983). However, herbicides applied close to the trunk might affect about 10 per cent of the fine roots, if they penetrate the whole root depth. The innate tolerance of young grafted apple trees to simazine was reported by Casely (1960), although this herbicide was found to damage apple seedlings (Dermine, 1960). Controlling weeds significantly increased tree vigour

(Robinson and O'kennedy, 1978). Apple trees (cv. Delicious) treated with dinoseb at 6.7 and 10.1 kg/ha, glyphosate at 1.7 kg/ha, or mechanical weed control showed considerably higher yield than trees on unweeded plots in the first year of production (Rupp and Anderson, 1985). Prica and Coca (1986) in a 4-year trial with cv. Goldspur on MM 106 rootstocks. planted at 4 x 2 m in a chernozem soil with 2.5 per cent humus and 42 per cent clay content and pH 6.6 noted good control of annual and perennial weeds by treatment with caragard combi (terburneton + terbuthylazine) or pitezin (atrazine + simazine + propazine). The best result was, however, obtained with pitezin at 8-10 kg/ha applied after weed emergence in 1.5 m wide bands along the tree rows. Although the advantages of a grass or weed-free environment are most evident in young trees, substantial increase in yield still occurred after almost 20 years (Robinson, 1983). By choosing the best soil maintenance ststem, trees will not comptete with weeds for nutrients and water and these will get constant fruit production. Micu and Iordanescu (2011) suggested use of Roundup 360 SL (3 l/ha) and two manual hoeing on the tree row in 'Jonathan' apple orchards. In apple orchards, Virginia creeper (*Parthenocissus quinquefolia*) could be controlled by application of butoxyethyl ester of trielopyr at 6 kg/ha (Tworkoski *et al.*, 1988). Overall herbicide treatment also reduces the risk of frost injury (O'kennedy, 1974). Morgan and Hennerty (1976) found excessive tree vigour of Golden Delicious and very high levels of leaf N in herbicide treated orchards of Ireland. The reduction or elimination of weed competition by the use of herbicides sometimes reduce the need for N fertilizers in many orchards (Greenham, 1976). Studies over a 3-year period on a clean cultivated orchard soil at East Malling to which no nitrogenous fertilizer had been applied for 30 years, showed that the amount of mineral nitrogen increased up to 50 to 70 kg N/ha in June each year (White and Greenham, 1967). However, herbicide management can result in a significant reduction in population of earthworms (Stringer *et al.*, 1971). This is due to the indirect effect of the herbicides in reducing the availibility of decaying organic matter (Robinson, 1983). A littering of straw between the tree row is desirable to maintain organic matter and earthworm activity (White, 1984).

Flowering

In apple trees the time of flower bud initiation is closely related to the growth cessation of shoots on which they are set (Huang and Cheng, 1984). The period ranging from growth cessation up to the appearance of floral primordia is the period of physiological differentiation of flower buds. During this period, differentiation processes proceed within the buds, increasing the number of nodes, and inducing some physiological changes. When the bud reached a certain stage (a certain number of nodes), it may become a flower bud, or may terminate its differentiation becoming a leaf bud (Huang, 1987). It was found that defoliation promoted the resumption of growth of terminal buds. The earlier the treatment was applied, the higher the percentage of the second flush. Besides, defoliation applied later than 5 weeks after growth cessation not only failed to transform flower buds already formed into vegetative ones, but promoted some of the flower buds into autumn bloom. This indicates that flower bud is irreversible once the floral primordia appear. Under the strong influence of defoliation it develops into flower in the current year (Janick, 1974; Miller, 1982; Huang, 1987). Huang (1987) also stated that some unknown factors are needed to transform vegetative meristem into flower buds. Once a flower bud is formed those factors are no longer necessary for the further development of floral parts. The exact time of flower bud initiation varies with the cultivar, location, season and other factors. Development of floral parts is in the order of calyx, anthers, pistil, ovary cavities and petals.

The fruit buds are usually developed on spurs, laterally on two/three-year-old wood. However, buds may also develop on tips of one-year-old shoots, which are found mostly when dwarfing rootstocks are used. A survey of Gala, Golden Delicious and Fuji orchards in Brazil showed that> 50 per cent of fruits on terminal buds were larger than those on lateral buds (Petri, 1984).

That a change in hormonal balance leads to flower initiation was viewed by Luckwill (1974), while Sach (1977) proposed a redistribution of nutrients inside the apical meristem. Although, the biochemical changes that occur during flower bud initiation is still unknown, the differentiation process was, however, divided into three different stages by Bernier (1970). According to him, in stage (i) RNA and proteins essential for flower initiation are synthesized when the stimulus triggering flower initiation reaches the apical meristem; this is continued, (ii) the stage II constitutes the mitotic cycle when the nuclei divide and (iii) the stage III comprises the morphogenetic cycle when the flower primordia develop. Schmidt (1978), however, did not find any clear relationship between RNA metabolism and flower bud differentiation. When the vegetative buds attain maturity it is transformed to reproductive structure and the induction occurs when a critical number of internodes develop on bud axis. The number of internodes, however, vary according to cultivars; in Cox Orange it is 20 while in Golden Delicious it is only 16 (Abbott, 1970). Rudloff and Lucke (1960) opined that a moderate decrease in shoot growth is advantageous for flower bud formation, but too weak vegetative growth also lower the number of flowers (Feucht, 1961). Schumacher (1963) reported that about 20 per cent of all shoots extended on biennially cropping trees in an 'off' year, whereas fewer than 4 per cent did so in an 'on' year.

A good correlation was found to exist between the leaf area and flower bud formation. In cultivars like Granny Smith and Delicious, no flower bud formation was found at leaf/fruit ratio below 10 which was found to increase progressively by increasing the ratio from 20 to 50 (Bhambota *et al.*, 1969). The sink effect of fruit is important for utilization of photosynthate produced by the leaves (Buban and Faust, 1982). A leaf area of 190 to 230 cm^2 is needed for each Golden Delicious fruit (Hansen, 1969) which may be 400 to 670 cm^2 (25-42 leaves) for other cultivars.

The failure of flower initiation in trees carrying a heavy crop of fruit which, in past, was attributed directly to the effect of the crop in depleting the carbohydrate and nitrogenous reserve of the tree, is now widely recognized to be due to hormonal rather than nutritional causes (Luckwill, 1974). Tumanov and Gareev (1951) noted the inhibiting effects of seeds on flower bud initiation which was later confirmed by Chan and Cain (1967). They observed that 90 per cent of the spurs induced flowers when they bore seedless fruits in the previous year while it was only 13.1 per cent in spurs that produced seeded fruits in the previous year. Grochowska and Karaszewska (1976) reported that the intensity of auxin outflow from seeds of biennial bearing ones was nearly 60 per cent larger than with the annual bearing cultivars which was thought to be an inhibitory effect on flower bud formation. Gibberellins were reported to inhibit the initial stages of flower bud initiation (Luckwill, 1970). Dennis and Nitsch (1966) identified GA_4 and GA_7 in apple, later Dennis (1976) found GA_4, GA_7 and GA_3 in developing seeds and GA_5, GA_6 and GA_3 in the fruit flesh. The activity of gibberellins in the young developing seeds was 3000 times greater (per unit fresh weight) than in the fruit flesh. There is no positive evidence of the effect of cytokinin on flower bud induction (Luckwill, 1974) but Grochowska and Lubinska (1973) noted 2 to 3-fold higher levels of ABA in leaves of fruit-bearing spurs than in those of non-bearing spurs of Landsberger Reinette. Grochowska and Karaszewska (1978) found high levels of cytokinins in shoots of unpruned apple trees at the time of fruit bud differentiation and an increase in fruit bud formation in these trees in the following spring. Treatment with SADH induced differences in catalase activity and isoperoxidase spectrum during growing season. Reduced catalase activity and qualitative changes in isoperoxidase spectrum were associated with reduced vegetative growth and enhanced fruit bud formation (Badescu *et al.*, 1978).

Buban (1981) studied the histological and cytochemical aspects of flower bud formation and opined that preceding the appearance of morphologically distinguishable flower primordia, distinct changes takes place in the zonation of the apex within the bud. Following an increase in mitotic activity all over the apex the position of the central meristem was found directly below the middle part of the subdermatogen. This conversion of the histological structure results in the disappearance of the middle part of the accessory tunica layer. Apices in buds under flower inductive conditions possess a higher relative level of DNA in the cell nuclei of the central meristem and apical axial zone, compared with apices in which flower initiation was inhibited. Spraying 'Anna' apple with hydrogen cyanamide (Dormex) resulted early bud break, reduced flewering period, high percentage of bud break and fruit-set by regulating the contents of nitrogen and polyamines in buds (El-YaZal and Rady, 2012). Studies on the changes of endogenous hormmones in the differentiation period of flower bud in the apical tissues of the flower and leaf bud of apple cv. Red Chief was made by Cao-ShangYin *et al.* (2007). They observed that during flower bud initiation, IAA and GA content decreases while ABA and ZR (zeatin riboside) are maintained at a high level. They concluded that in the course of flower bud differentiation, the low levels of IAA and GA, and high levels of ABA and ZR, are beneficial for promoting flower bud formation.

Pollination

The apple tree has hermaphroditic flowers with a gametophytic type of self incompatibility controlled by a single multiallelic locus. Therefore, at least two different cultivars must be interplanted in the orchard to ensure high levels of cross-pollination in order to achieve adequate fruit set. Frost damage to flowers sometimes may produce some fruits which are, however, abnormal in shape. The cultivars, Jonathan, Golden Delicious, Rome, Grimes were considered as partially self-fruitful. The pollen germination in triploid cultivars is usually poorer than in most diploids, so it is advisable to use diploid pollenizers in triploid orchards. The factors, which affect fruitfulness in apple, are mainly defective pollens, incompatibility, irregular chromosome behaviour, climatic conditions, growth of the plant, *etc.*

Defective Pollen

Some cultivars produced weak, sterile or abortive pollen as most of the triploids such as Suntan, Jupiter, *etc.*

Incompatibility

All most all apple cultivars are self-incompatible to some extent, and some are completely so. Cross-incompatibility have been reported, usually between close relatives. Triploid cultivars behave in much the same way as diploids, varying in the degree of self-incompatibility and producing an increase fruit set when pollinated with diploids, but varying cosiderably when crossed with other triploids. Pollens of some cultivars are incompatible with certain combinations. Golden Delicious is incompatible with Crispin. The pollen of Cox is ineffective for Kent while Worcester, Golden Delicious, Greensleeves and Spartan are suitable pollenizers for Kent. Cox's Orange Pippin is incompatable with Kidds Orange Red. Red Gold and Tydeman's Worcester have been found to be a good pollenizers for McIntosh, Red Delicious, Richared and Royal Delicious. These pollenizers are, however, partially or fully self-fruitful (Chadha, 1978).

Diploid cultivars Malling Greensleeves (James Grieve x Golden Delicious) and Malling Kent (Cox's Orange Pippin x Jonathan) are partially self-compatible, whereas Cox and the triploid Malling Suntan are normally self-incompatible. Suntan and Cox are cross-incompatible and Kent appears to be an unreliable for Suntan. Cross-compatibility tests showed that the following pollenizers are effective: Greensleeves for Kent, Cox for Greensleeves, Golden Delicious and Gloster 69 for Suntan (Spiegel-Roy and Alston, 1982). The site of incompatibility in apple is in the style, seemingly under phenotypic control (Modlibowska, 1945). Cases of self-compatibility in some apple cultivars have been reported (Knight, 1963). In triploids, self-compatibility is more complicated because of some degree of parthenocarpy, low seed content and pollen sterility. The cross incompatibility between triploid and diploids are more than between diploids (Kobel *et al.*, 1939).

Irregular Chromosome Behaviour

Most apple cultivars are diploid (34 chromosomes), although some commercially important cultivars such as Bramley's Seedlings, Jupiter are triploid (54 chromosomes). Although, the diploid cultivars are common and good pollenizers, the pollens of triploids germinate poorly.

Climatic Conditions

The climatic condition during blooming is one of the key factors for pollination. Low temperature (–4° to –2°C) during flowering may kill the pistil. Bee activity is greater in warm conditions, whereas pollen tubes take 216 hours to reach the ovaries at 8°C which is achieved in 48 hours at 15°C (Williams, 1970). The effect of temperature was considered most important because the flowering period is brief and the ovules remain fertile for only a few days (Jackson, 1983). Jackson and his associates (1980,1983) at the East Malling Research Station, observed high correlations in year-to-year variation in temperature in the period immediately after full bloom with those of fruit set and yield in Cox. In general, optimum temperature for pollination and fertilization are between 20° and 25°C with moderate relative: humidity as high humidity prevents release of pollen and low humidity causes poor pollen germination. High temperature immediately after full bloom, resulting in rapid completion of pollen tube growth, was found to increase the yield (Jackson and Hammer, 1980).

Pollinating Insects

The pollination is mainly done by insects as the pollen of apple is heavy, the pollination by wind is negligible. The pollination is mainly carried out by honeybees but variation in climate and extensive use of insecticides can seriously affect the bee population. Bees do not fly well at temperature below 10°C or in rains and strong winds. To increase the population of bees at pollinating time it is advisable to maintain bee hives in the orchard. The hives should be placed in a protected place and in sun. Two to three hives are sufficient for one hectare of plantation, though it depends on cultivars, age of the plant, tree spacing, pollenizer plants. Application of insecticides in soil or spraying on the plant should be avoided after flowering and the hives should be removed before any post-bloosom spray. In years when conditions for bee pollination were favourable, the percentage of fruit set and yield did not depend on the degree of self-fertility of the cultivars, but in seasons unfavourable for bee flight, self-fertile cultivars fruited better (Skripnikova, 1986). Cross-pollination by honey-bees resulted in a good early set, mainly of the control flowers in the inflorescene (Kurennoi and Mantoptin, 1988). Insect pollination not only affects the quantity of production but also have marked impacts on the quality of apples, influencing size, shape and effecting their classification for market. These effects are however, variety specific (Garratt, 2014).

Choice of Pollenizer

Adequate number of pollenizer cultivar/s should be planted at the time of planting, as the fruit set and yield of the main cultivar is largely dependent on selection of pollenizers. A good pollenizer should produce plenty of pollen which are compatible with the main cultivar, if it is a good market cultivar as well; adds to its quality. The bloom period of the pollenizer should coincide with that of choice cultivar and should slightly precede that of main cultivar for optimum pollination. The dehiscence of pollenizer should start before in the main cultivar. The number of pollenizer trees depends on planting density. With Gloster as pollen source, fruitset percentage in cultivars Idared, Spartan, Jonagold, McIntosh, Starkrimson, Jonathan, Gala and Champion were recorded at 15, 33, 49, 36, 10, 6, 2 and 5 respectively (Blasse, 1986). In two seasonal studies, cv. Golden Delicious proved most consistent and effective pollenizer of the seven cultivars tried. *viz.* Golden Delicious, Mclntosh, Duble Red, Red Gold, Red June, Rus Pippin, Tydernan's Early, and Winter Banana. Golden Delicious alongwith Red Gold registered fruit set to a tune of 15.00 and 14.13 per cent, respectively, when the average fruit set of eight main cultivars of

Delicious group was accounted for, Tydeman's Red June, Rus Pippin and Winter Banana were found to be less effective in comparison with Golden Delicious. Red Gold and McInstosh which were suitable pollenizers (Bhartiya *et al.*, 1986). The full bloom date of some apple cultivars in southern England and in Canada (Ontario) are presented in Table 1.29. The appie cv. Dorset Golden was found as the best pollenizer for cv. Anna with fruitset of 43 per cent (Mamouni, 1999).

Table 1.29: Full Bloom Date of some Apple Cultivars

Southern England		*Canada*	
Cultivar	*Date*	*Cultivar*	*Date*
James Grieve	May 8	Quinte	May 15
Greensleeves	May 10	Red Melba	May 15
Idared	May 10	Idared	May 17
Tydernan's Early	May 10	McIntosh	May 17
Discovery	May 12	Tydeman's Early	May 19
Delicious	May 12	Empire	May 18
Jonagold	May 12	Spartan	May 19
Spartan	May 14	Red Delicious	May 20
Red Delicious	May 14	Golden Delicious	May 19
Golden Delicious	May 14	Northern Spy	May 22
Jupiter	May 15	Mutsu	May 21
Kent	May 15		
Gloster 69	May 16		

A good pollenizer should produce plenty of viable pollen. The quantity of effective pollen produced is determined by the number of flowers, the amount of pollen per flower and its viability, and these three factors have all been found to vary among the cultivars (Church and Williams, 1983). Simidchev (1976) recorded cultivar differences in flower number per hectare, and the weight of pollen produced per flower. The number of pollen grains per anther may vary between cultivars (Bhartiya *et al.*, 1983), and also the viability of pollen (Church and Williams, 1983).

Williams and Church (1983) stressed the need for adequate pollination of apple flowers and suggested the use of ornamental *Malus* as pollenizers for apple cv. Cox's Orange Pippin. One important criterion for a pollenizer is its flowering period, as pollen must be available at the right time for cross pollination. Palfitov *et al.* (2011) in their studies with apple cultivars Lobo, Spartan, Orlik, Sinap Orlovskii, Zelenyi Mai, Martovskoe and Bessemyanka were self and cross pollinated under controlled conditions. They observed that with cross pollination, percentage of fruit set in all cultivars increased because content of flavonols in pollen of pollenizers increased as the content of phloridzin in the style bases of recipients reduced. Highest percentage of fruit set (9-30 per cent) was observed in crosses using pollen of Lobo and Spartan, the cultivars with the highest flavonol contents (9-12 per cent on DM basis). A low propensity for self pollination was found associated with a high phloridzin content of style bases and a low flavonol content of pollen. The best pollenizers were that cultivars with pollen flavonols content of > 10 per cent. They also studied eight other apple cultivars and found the highest contents of pollen flavonols (10-12 per cent) in cultivars Kitaika, Melba, Bogatyr and Severnyi Sinap and these produced the hgighest fruit set in the self sterile cultivars. Church *et al.*

Table 1.30. Number of Anthers. Dehiscence Pattern and Petal Fall in Several Dessert and Ornamental Malus in Four Years*

Cultivar				*Days after Opening Unitl*		
	Year	*Number of Anthers per Flower*	*Duration of Dehiscence (Days)*	*Dehiscence of First Anther*	*Peak of Dehiscence*	*100 per cent Petal Fall*
Discovery	1974	20	6	1.1	4	6
	1979	17	7	1.4	6	7
Egremont Russet	1974	16	7	1.6	5	6
	1979	15	7	1.4	6	7
Golden Delicious	1974	17	6	1.0	4	6
	1979	19	7	1.2	5	7
James Grieve	1974	19	6	0	1	6
	1979	11	5	0.8	5	7
King of the Pippins	1975	11	7	0.7	3	9
	1976	17	5	0.6	2	5
Lord Lambourne	1974	19	8	0.2	2	6
	1979	16	7	0	2	7
Worcester Pearmain	1974	20	7	0.8	5	6
	1979	18	7	2.4	6	7
M. cv. Golden Hornet	1974	17	6	0	1	5
	1975	19	6	0.2	1	6
	1976	20	3	0.2	2	5
	1979	19	7	1.2	5	7
M. cv. Hillieri	1974	32	10	1.0	7	12
	1975	21	8	0.9	4	13
	1976	12	7	0.04	1	12
	1979	42	9	0.6	8	10
M. x *purpurea* cv. Aldenha-mensis	1974	40	11	0	1	10
	1975	21	8	0.4	5	11
	1976	40	6	0.1	2	11
	1979	44	9	0.6	5	9
M. cv. Winter Gold	1974	18	9	0.6	2	9
	1976	24	5	0.5	3	7
	1979	19	8	1.0	6	9

* Church *et al.* (1983).

(1983) in the UK recorded flowering dates, pattern of anther dehiscence after flower opening and date of petal fall in seven dessert cultivars and four ornamental *Malus* cultivars over several years (Table 1.30). No single dessert cultivar was found to produce pollen throughout the Cox flowering season every year, but a combination of two or more of the ornamental *Malus* could do. Bhartiya *et al.* (1983) studied the flower characters of 15 commercially important apple cultivars of India and found variations in stamens, anthers and carpel characters among the different cultivars (Table 1.31).

Table 1.31: Flower Characters of Apple Cultivars*

Cultivars	*Flowers per Bud*	*Stamens per Flower*	*Length of Stamen (cm)*	*Weight of Anther (mg)*	*Number of Pistils per Flower*	*Length of Pstil (cm)*	*Weight of Ovary (mg)*
Golden Delicious	4.84	19.50	0.65	1.10	5.00	0.96	28.2
Hardman	4.38	19.50	0.66	1.12	5.00	0.79	32.0
McIntosh	5.58	18.00	0.68	1.10	5.20	0.77	317
Red Delicious	4.95	20.00	0.69	1.12	5.00	0.80	32.0
Red Gold	5.18	20.00	0.68	1.10	5.25	0.63	271
Red June	4.55	19.00	0.73	1.00	5.25	0.66	28.0
Royal Red	4.95	19.25	0.66	1.11	5.25	0.78	31.4
Ruspippin	5. 00	18.25	0.74	1.00	5.20	0.64	27.1
Starking Delicious	4.54	19.00	0.66	1.11	5.25	0.69	31.4
Starkrimson	5.25	18.55	0.88	1.10	5.00	0.79	34.2
Top Red	4.86	18.00	0.65	1.10	5.00	0.69	28.2
Tydernan's	4.75	18.48	0.88	1.00	5.00	0.66	28.2
Vance Delicious	4.85	19.50	0.68	1.10	5.00	0.70	271
Winter Banana	4.96	19.58	0.73	1.00	5.00	0.78	31.4
Average	4.91	19.10	0.71	1.08	5.09	0.75	30.00

*Bhartiya *et al.* (1983).

Fruit Set

An unsatisfactory fruit set may result from a lack of cultivars suitable for cross-pollination, cultivars not having blooming period which overlapped sufficiently, and the presence of inadequate supply of pollinating insects. However, pollination is not allways followed by fertilization of the ovules. The pollen grains may arrive on the stigma too late, or pollen tube growth may be too slow for the male gametes to reach the ovules before they abort (Luckwill, 1980). In some years a proportion of the ovules may be sterile and incapable of being fertilized. This proportion can be as high as 80 per cent in Cox's Orange Pippin (Williams, 1970). In addition to these factors which bring about an unsatisfactory setting of fruits, weather conditions during the blooming season, such as temperature, rainfall and humidity also play important roles in governing the extent of fruit set (Fortes and Petri, 1983). Frost at blossom time can prevent pollination by killing the styles. In India (Himachal

Pradesh), average maximum and minimum temperatures around 20° and 10°C, respectively, during flowering in April, resulted in good fruit set. High relative humidity (>50 per cent) in March delayed flowering without affecting fruit set, but high rainfall and relative humidity during flowering in April adversely affected fruit set, while low rainfall and relative humidity in April improved fruit set and yield (Awasthi *et al.*, 1986). Very low temperature (<4.5°C) reduced fruit set by interfering pollen germination and bee activity. Sprays of aminoethoxyvinylglycine (AVG) inhibit ethylene production and increase fruit set in treated apple flowers, suggesting that endogenous ethylene level limits fruit set. However, sprays of neither aminooxyacetic acid, which also inhibits ethylene synthesis nor silver thiosulphate which blocks ethylene action, improved set. It was, therefore, thought that inhibition of ethylene synthesis is not the mechanism by which AVG enhances apple fruit set (Dennis *et al.*, 1986). Irradiating pollen grains with 5Gy dose of ^{137}Cs γ-rays increased pollen germination by 62 per cent and fruitset by up to 56 per cent (Guo, 1987).

Fruit Drop

In apple, between 20 per cent and 80 per cent of the fruits initially set drop off from the tree during developing. Often several distinct waves of dropping occur, the early post-blossom drop being followed after a short interval by the so-called June drop. As the fruit approaches maturity further loss of fruit occurs through preharvest drop, particularly in early- ripening cultivars. Separation occurs at the base of the fruit stalk where a special zone of weakness, the abscission layer, develops caused by dissolution of the cell walls by enzymes which is under hormonal control (Luckwill, 1980). Although the timing of drop is determined by the auxin flow from the seeds, the extent of the drop is a function of the available food supplies and of the competition for those supplies by alternative sinks. Important competing sinks during the first few weeks after bloom are the shoot tips, particularly the tips of those shoots arising directly from bourse (Abbott, 1959). Fruit set and retention are increased by removing shoots within a 3-4 week period of blossoming (Quinlan and Preston, 1968). Quinlan and Preston (1971) demonstrated that removal of all shoots within 3 weeks of full bloom increased fruit set in apple but resulted in an increased fruit abscission during June drop period in comparison with that occurring in unpruned control trees. They thought that shoot growth competes with fruitlet development for available assimilate resources, particularly in the early part of fruitlet development but bourse shoot leaves are important in contributing photosynthates to fruit at later stages of development and in reducing fruit abscission. The timing of the competition between shoot and fruit meristems is critical and it is clearly important to determine the mechanisms regulating the directional flow of assimilates and relative sink strength (Treharne *et al.*, 1985).

Control of Fruit Drop

Luckwill (1953) observed that when seeds were producing high amount of auxin, the fruit drop was minimum and he (1948) established a definite relationship between the abscission and auxin contents in the developmental stages of fruit. According to him, the auxin first appeared in the seed in larger quantities 30 days after petal fall, its appearance coinciding with the formation of endosperm and cessation of postblossom drop. Following endosperm formation there was a period of rapid embryo growth during which hormone content showed a temporary reduction, rising again to its maximum level 75 days after petal fall when embryo had attained its full size. In the final stage of fruit growth, there was a rapid decline in the hormone content of the seed, which was correlated with the degeneration of the endosperm and occurrance of preharvest drop. With this view in mind that high concentration of auxins inhibit fruit drop (Biggs and Leopold, 1958), extensive investigations have been carried out to control preharvest fruit drop by application of auxins with varying degrees of success.

Aitken (1941) was able to reduce preharvest drop of Gravenstein and McIntosh apples by one spraying of NAA at 4.8 g per 100 gallon water. Application of NAA at 5 and 10 ppm successfully checked preharvest drop of Miller's Seedling (Vyvyan, 1942) and McIntosh (Murphy, 1942) apples. NAA at 10 ppm was advocated by Moretti (1954) for cv. Beurre Gifford and also for Jonathan apples (McKenzie, 1953). Two sprayings of NAA reduced 26 per cent fruit drop and increased fruit size of Jonathan apples. Application of NAA in dust or as a spray was found equally effective (Marth *et al.*, 1946). French (1956) opined that NAA is superior to 2, 4, 5-TCP acetamide in controlling the fruit drop of McIntosh apple while NAA at 10 ppm added to lead arsenate was superior to 2, 4-D or IBA in controlling fruit drop of apples (Samisch and Reich, 1950). NAA can also be effective in controlling drop in 'McIntosh' when ethephon is applied to accelerate colour development (Stover *et al.*, 2003)

Preharvest sprays of 2, 4-D at 10 ppm successfully reduced the fruit drop in Winsap (Batjer and Thompson, 1947) and Stayman (Harley *el al.*, 1947) apples, while 20 ppm 2, 4-D was found effective for controlling fruit drop in Vicar of Winkfield (Moretti, 1954). 2, 4-D is, however, not effective on McIntosh (Batjer and Thompson, 1946) which was thought due to the active decarboxylation of 2, 4-D by McIntosh leaves (Edgerton, 1961).

Marth *et al.* (1950) suggested application of 100 ppm 2,4,5-T in the month of June to control the preharvest fruit drop, while 5 to 20 ppm 2,4,5-T spraying in August or September was suggested by Thompson (1951). Southwick *et al.* (1953) opined that 20 ppm 2,4,5-T was superior to sodium salt of NAA in minimizing the fruit drop of McIntosh apples. 2,4,5-T at 50 ppm also gave excellent control of preharvest drop in Red Delicious (Moretti, 1954). In cultivars, Starking, Stayman and Winsap, preharvest sprays of 2,4,5-TCPA resulted in a significant and economic decrease in fruit drop (Mattus and Moore. 1954). Spraying of 2,4,5-TCPA at 15 ppm three weeks before harvest reduced preharvest drop from 48 to 25 per cent in Jonathan apples (Samisch and Steiner, 1956). It had a longer period of effectiveness than NAA on McIntosh (Edgerton and Hoffman, 1951) and it came into extensive commercial use for several cultivars.

Foliar spray of aminoethoxyvinylglycine (AVG) delayed the preharvest drop of apples (Bangerth, 1978) and foliar spray of AVG at 150 ppm increased the set of self-and open-pollinated Delicious (Edgerton, 1961). AVG is thought to prevent or slow the biosynthesis of ethylene in apple tissues (Baker *et al.*, 1978) and thus helps to increase fruit set while Rahemi and Dennis (1981) concluded that the effects of AVG in increasing set were independent of its effects on ethylene synthesis. Schupp and Greene (2004) in their experiment with apple cv McIntosh reported that treatment with AVG at 225 mg/l was most effective in delaying drop and maturity when applied 8 or 2 weeks before harvest. GA_{4+7}, benzylamino purine (BA), and promalin (GA_{4+7} + BA) were effective in increasing fruit set of apple with or without AVG (Table 1.32). AVG (50 to 200 ppm) increased set of apple by as much as 10-fold when applied between pink and petal fall (Dennis *et al.*, 1983).

Table 1.32: Effects of Branch Sprays of GA_{4+7}, BA and Promalin, with and without AVG, on Fruit Set (fruits/100 flower cluster) in Red Prince Delicious Apple*

Date of Observation	*AVG (ppm)*	*Chemical and Concentration (ppm)*						*Mean*
		None	*GA_{4+7}*		*BA*	*Promalin*		
			25	50		25	50	
June 12	0	6	5	2	3	4	7	5
	200	32	26	25	32	28	30	29
October 10	0	6	3	1	3	2	1	2
	200	26	20	21	13	12	17	18

AVG applied on May 19 (full bloom); all other treatments applied on May 16, 1980.

* Dennis *et al.* (1983).

Rhodofix (a French proprietary product) effectively reduced the preharvest drop of Cox's Orange Pippin (Leib, 1951). In Golden Delicious apples, a mid-September treatment of Obsthormon 24A at 50 ml/100 l proved very effective (Faccioli, 1958). Application of Clingspray, Estone and Fruitone gave excellent control of fruit drop (McKenzie, 1953). Jackson *et al.* (1983) reported that spraying open-pollinated trees of Cox's Orange Pippin on M.9 rootstock at 80 per cent petal-fall with GA_3, DPU and either NAA or NOXA in a 3-component mixture increased both initial and final fruit set. The same treatments applied to Cox trees on M.26, M.7 and MM.106 rootstock, randomized with the trees on M.9 in the same field experiment, showed an increase in initial set but had no effect on final set (Tables 1.33 and 1.34). The differences in Cox's response on different rootstocks appear to be linked to the greater ability of trees on M.9 to retain seedless or few-seeded fruits.

Parthenocarpy

In apples, growth regulators are used to induce parthenocarpic fruits. Flower buds are normally emasculated at the full pink stage by removing the petals, stamens and styles and growth regulators, *viz.*, gibberellins, auxins and cytokinins are applied by brush to the emasculated flowers. Exogenous plant hormone application to unfertilized ovaries might serve as a switch that starts the continuing autonomus development of fruit (Bangerth and Schroder, 1994) and controls the continuity of the flow of assimilates and nutrients required for fruit growth (Treharne *et al.*, 1985). Genetic parthenocorpic cultivars might also have switch-induced autonomous development of the fruit, although the time at which this switch acts in naturally occuring parthenocarpric fruits remains unclear (Watanabe *et al.*, 2008). The type

Table 1.33: Effect of Hormone Spray at 80 per cent Petal-fall on the Initial (I) and Final (F) Fruit Set per 100 Bloom Clusters of Cox's Orange Pippin in a Replicated trial on 4 Rootstocks in 1975*

	Whole Tree		*Split-tree Branch Treatments*							
	Control		*U*		*HP*		*U*		*HS*	
	I	*F*	*I*	*F*	*I*	*F*	*I*	*F*	*I*	*F*
M.9	33	22	26	13	179	68	34	15	84	28
M.26	37	18	33	17	157	68	31	18	77	19
M.7	16	13	21	11	150	74	18	10	44	16
MM. 106	27	21	33	16	174	80	28	15	58	19
Mean	28	19	28	14	165	73	28	15	66	21

U: Utreated; HP: Hand pollinated; HS: Hormone spray (300 ppm DPU + 200 ppm GA_3 +50 ppm NOXA).

*Jackson *et al.* (1983).

Table 1.34: Effect of Fruit-Setting Sprays (200 ppm GA_3 +15 ppm NAA + 100 ppm DPU) and Daminozide (2000 ppm) on Fruit Set per 100 Blossom Cluster: Cox's Orange Pippin, 1978*

Rootstock	*Fruit-setting Spray*		*Control*	*Mean*
	GA_3 + NAA + DPU – daminozide	*GA_3 + NAA + DPU + daminozide*		
M.9	79	69	61	70
MM.106	70	71	65	69
M.26	50	36	55	47
M.7	44	38	40	40
Mean	61	53	55	

*Jackson *et al.* (1983).

or types of growth regulators, *i.e.*, a mixture of two or three compounds to be used usually depends on the cultivar. Occasionally some natural parthenocarpy is also observed in Cox's Orange Pippin, when caged before flowering (Goldwin, 1978), which is, however, very unusual. Parthenocarpic fruits of the cultivar, Cox's Orange Pippin, could be obtained by spraying mixtures of gibberellins, auxin and cytokinin (Goldwin and Schwabe, 1975). Williams and Letharn (1969) recorded parthenocarpy in Red Delicious and Sturmer with a mixture of gibberellin and cytokinin while Golden Delicious did not respond well to gibberellins or cytokinins. Varga (1966) reported that GA_1, GA_2, GA_4 and GA_7 promoted parthenocarpic fruit set in the Lombartscalville but not in Winston or Golden Delicious cultivars. Jonkers (1978) noted parthenocarpic fruit development in Cox's Orange Pippin by application of a mixture of GA_{4+7} at 600 ppm + NAA at 40 ppm + BA at 300 ppm and suggested spraying at 40 per cent flowering. Parthenocarpic fruits of Cox's Orange Pippin was.

however, smaller than seeded control ones (Goldwin and Schwabe, 1974; Jonkers, 1978). Unrath (1974) demonstrated increase in both length/diameter ratio and weight of Delicious fruits by applying sprays of GA_{4+7} plus benzyladenine. Fortes and Petri (1983) obtained parthenocarpic fruits in Golden Delicious by using Wye mixture (NOXA 500 ppm + GA_3 200 ppm + DPU 300 ppm) irrespective of time of application (Table 1.35). The influence of gibberellins, cytokinins and auxins in situation, where the ovules are killed in adverse environmental conditions. is due to substitution by applied hormones for those normally produced by developing seeds and which regulate fruit growth. Since parthenocarpic development does not involve physiological events associated with the full reproductive organ development, it is reasonable to assume the hormone application regulates continuity of flow of assimilates and nutrients required for fruit growth (Treharne *et al.*, 1985). Application gibberellins before flowering trigger parthenocarpic apple fruits (cvs. Fuji and Ohrin). Subsequent fruit growth requires cytokinins and auxin (Watanabe *et al.*, 2008).

Table 1.35: Fruit Set and Parthcnocarpic Fruits as Affected by the Use of Wye Mixture in Golden Delicious on MM. 106*

Time of Treatment	*Mean No. of Floral Buds*	*Number of Fruits*	*Number of Fruits/100 Floral Buds*	*Mean Number of Seeds*	*Percentage of Parthenocarpic Fruits*
50 per cent bloom	121.40	34.20	28.2	0.66	68.0
Full bloom	86.60	40.80	47.1	0.80	48.0
Petal fall	62.80	21.00	33.2	1.36	24.0
Control	81.80	13.20	16.1	1.64	0

Fortes and Petri (1983).

Thinning of Fruits

Thinning is one of the major techniques employed to regulate fruit quality, which tends to deteriorate with too many fruits on the shoots and sometimes even initiate alternate bearing. Thinning of individual fruits manually is costly and tedious, the fruit grower must rely upon chemical thinning.

Chemical fruit thinning agents are impeded by many factors, such as climate, cultivar, internal tree factor, *etc.* One possibility for achieving greater precision of thinning would be the complete removal of fruitlets from certain portion of the tree, provided it could be shown that fruits on the unthinned portion of the tree increased in size and that the amount and distribution of blossom in the following year were adequate (Luckwill and Child, 1978). The commonly used thinning agents like NAA. NAAm and naphthyl N-methylcarbatnate (carbaryl) reduce fruit set in many apple cultivars effectively. Some morphactins, such as, CME72 200P (dichloroflurenol methylester at 62.5 and 100 ppm) and CME 73 130 P (chloroflurenol-NA, 7.5 ppm) and CGA 15281 (bis-benzyloxy-2- chloroethyl-methyl-silan, 400-1500 ppm) gave promising results; CEPA was found effective for thinning of Belle de Boskoop, Cox's

Orange Pippin, Glockenapfel, Jonathan and McIntosh but was ineffective in Golden Delicious and was frequently associated with decreased fruit growth (Link, 1978). Ethrel (2-chloroethylphosphonic acid) and Frufix (NAAm) were found effective in Golden Delicious (Pfammatter, 1978) which also increased the fruit size. The cultivars Cox's Orange Pippin/MM106 and Golden Delicious/M27 responded well to CEPA applied just before June drop, which was proposed as a possible alternative to overall chemical th.inning (Luckwill and Child, 1978).

Luckwill (1953) suggested that the free nuclear stage is the time of maximum sensitivity of fruitlets for NAA and this free nuclear endosperm is the critical stage for chemical thinning sensitivity (Tukey, 1965b). The possible mechanism of NAA action in thinning was suggested by Schneider (1978), which brought about the following responses: (i) less reducing sugar per fruit, (ii) temporarily stopped or slowed fruit growth, (iii) did not affect pedicel capacity to translocate 14_C-IAA or 14_C-sucrose in either direction (iv) reduced 14_C-sucrose movement from foliage to fruit. (v) increased ethylene evolution but not until 24 hours after NAA application, (vi) increased water potential of foliage (vii) shading limbs resulted in reduced ethylene evolution, (viii) shading limbs resulted in reduced fruit retention in the absence of an exogenous ethylene stimulant, and (ix) reduced number of fruits retained by the tree. It has also been suggested that NAA reduced metabolite movement from foliage to fruit which results in the initiation of the fruit abscission process. Anatomical studies were made of seeds from young Golden Delicious fruits collected from trees which received NAA thinning spray of several concentrations and applied at different dates. Fruits from the treated plants were found to contain a greater percentage of seeds with degenerating embryo than from unsprayed trees. The contents of seeds within the same fruit are often affected to different degrees. Some showed degeneration of embryo while in others the embryo was small, arrested in development. The effect of higher concentrations of NAA was more pronounced and expedient in causing embryo degeneration (Hartman and Danho, 1966).

Table 1.36: Effectiveness of Carbaryl as a Fruit Thinner*

Cultivar	*Thinning Method*	*Yield (Kg)*	*Average Fruit Weight (g)*
Cox	Not thinned	173.2	70.9
	Hand thinned	130.7	89.5
	Carbaryl	134.2	91.2
Spartan	Not thinnied	42.3	79.3
	Hand thinned	23.5	103.3
	Carbaryl	33.7	95.2
George Cave	Not thinned	48.9	76.9
	Hand thinned	38.4	99.2
	Carbaryl	37.3	93.8

* Knight and Jackson (1980).

In a number of trials at East Malling, carbaryl at 2.5 kg (85 per cent wettable powder) in 1000 litres of water has been found effective in thinning of fruit of Cox, Spartan, George Cave (Table 1.36). It was also suggested to apply carbaryl at 80 per cent petal fall or when the average fruit diameter is 10-15 mm. Although, the activity of ethrel is strongly intluenced by temperature, it was effective in cultivars like Laxton's Superb when applied at the baloon stage of the axillary blossom (Knight and Jackson, 1980). Carbaryl at 100g/100l showed excellent fruit thinning in apple cv. Ozark Gold when sprayed at 17 days after full bloom. The treatment also produced good sized fruit of (Mahhou and Achachl, 2002).

Knight and Jackson (1980) have forwarded the following points about thinning:

i) Thinning is achieved more readily with young than with old trees.

ii) Fruitlet retention is generally greater on horizontal than on upright branches. A young tree with upright branches is likely to shed its fruits naturally and does not require thinning.

iii) Chemical thinners are least effective if applied in a very hot, dry condition when the spray dries quickly and absorption is decreased.

iv) Fruits on the top of the tree are more likely to be retained than those on shaded side, so spraying should be done thoroughly.

v) The better the conditions for fruit set, *i.e.*, the higher the temperature, the closer the flowering dates of the cultivars, the bolder the blossom and the greater the bee activity, the more likely it is that thinning will be needed, if the trees had light crop in the previous year.

vi) If the axillary fruit set in 1-year-old shoot is high, a separate application of carbaryl may be made.

vii) Effective thinning sometimes enhances the duration of fruit maturity, which should be taken into account for harvesting.

viii) Fruits on thinned plants are generally larger, which often show symptoms of calcium deficiency; thus, adequate attention is to be given for calcium nutrition.

The other chemical which was used since 1940 for blossom thinning is DNOC (sodium 4,6-dinitro-ortho-cresylate) commercially known as elgetol. DNOC at 0.25 per cent completely prevented pollen germination when applied to the stigmas of flowers (Mac-Daniels and Hildebrand, 1940) which also inactivated pollen tubes that had already grown halfway down the style of the pistil (Hildebrand, 1944). DNOC application after pollination and fertilization, *i.e.*, later than early petal fall is ineffective. Thinning is, however, more important for spur type trees and the spraying should be done before 20 days after full bloom. In years of heavy blossoming and fruit set, it is better to start early spraying and if needed a second spraying at the fruitlet thinning stage. In cultivars that crop regularly, pruning of few spurs may be helpful to reduce the need for chemical thinning. A mixture of 750 ppm carbaryl + 5-10 ppm NAA showed considerable promise as chemical thinning agent for apple cv. Summerred (Grauslund, 1981). Kacal and Koyungu (2012) in their experiment at Egirdir, Turkey used 6 years old 'Jerseymac' and 'Jonagold'

Table 1.37: Effect of Thinning Treatments on Physical Fruit Properties in "Jerseymac"*

Treatment	*Fruit Weight (g)*			*Fruit Diameter (mm)*			*Fruit Flesh Firmness (lb)*		
	2006	*2007*	*2008*	*2006*	*2007*	*2008*	*2006*	*2007*	*2008*
Control (no thinning)	126b*	131b	152ab	68.46c	69.74b	73.02^{ns}	13.97ab	14.12	13.65
Hand thinned	159a	142b	167a	74.10a	71.87b	75.58	12.83bc	12.92	13.67
ATS									
1.0 per cent	145ab	132b	149ab	72.64abc	69.13b	73.31	11.60c	12.92	12.90
2.0 per cent	153a	138b	142ab	73.73ab	71.02b	72.01	13.62ab	14.72	11.76
3.0 per cent	142ab	154b	142ab	70.89abc	73.80ab	71.92	12.69bc	12.98	12.78
Dormex									
0.25 per cent	138ab	148b	152ab	71.39abc	73.24ab	73.38	14.74a	13.46	12.32
0.5 per cent	155a	179a	151ab	74.16a	77.51a	73.84	12.70bc	13.20	11.72
0.75 per cent	125b	129b	137b	69.53bc	69.40b	71.18	13.78ab	13.79	11.87

Means in columns followed by the same letters are not significantly different.

ns: Not significant.

*Adopted from Kacal and Koyungu (2012).

trees on M.9 rootstock at 3.0 x 1.5m spacing; evaluated the effect of ammmonium thiosulfate (ATS) and hydrogen cyanamide (Dormex) at different concentration on thinning of flowers. Chemical thinning treatments were applied at full bloom and hand thinning was done after June drop. All chemical thinners increased average fruit weight and size at harvest (Table 1.37). Dormex at 0.5 per cent produced good quality fruit. Hydrogen cyanamide was also reported as an effective blossom thinner for 'Early Spur Rome', 'Law Rome', 'Gala' and 'Red Spur Delicious' apples (Fallahi *et al.*, 1992).

Biennial Bearing

Biennial bearing is one of major problem in apple cultivation. Alternate bearing can be tiggered by many factors. However, the direct cause of alternate bearing was found to be lack of flower buds in the 'off year'. Tumanov and Gareev (1951), Chan and Cain (1967) and Huet (1972) studied the phenomenon of biennial bearing in apples. Luckwill (1970) proposed that apple seeds contain ample amount of gibberellin, which inhibits flowering. Diffusates collected from apple fruitlets during the period prior to flower initiation have revealed that more hormones move out of biennial bearing cultivars than from those, which flower regularly. Within anyone cultivar the amount of gibberellin in diffusates was dependent on the number of seeds in the fruit, but between cultivars no such relationship existed (Head, 1978). Ramirez and Hoad (1978) at Long Ashton, UK, obtained evidence of the promotive effect of zeatin (a natural cytokinin) on flower initiation in apple, and they found that this hormone could induce flower initiation even in completely defoliated spurs. Tentatively, flower initiation in apple is regarded as being controlled by the cytokinin/gibberellin balance in the bourse around the time when flower initiation would normally occur (Luckwill, 1980).

Werzilov *et al.* (1978) recorded a sharp fall in auxin content in shoots during flower initiation. Hoad (1978) found more auxin activity in fruit diffusates of the biennial cultivar Laxton's Superb than the more regular cultivar Cox's Orange Pippin which support the idea that flower initiation in spurs is enchanced with less auxin in the tissues (Hoad, 1979). Luckwill (1970) concluded that so long as growth progresses the shoot apex produces a growth substance inhibiting flower formation. In Luckwill's (1970) hypothesis on the relationship between growth and flower-bud formation on current year's apple shoots, gibberellin produced in the young expanding leaves plays an important role. The anti-flowering role of gibberellin is well founded in apples (Guttridge, 1962; Marcelle and Sirnoval, 1963; Tromp, 1972). Presence of the fruit induced a lower level of nucleic acid and a higher nucleohistone content. These conditions proved to be inhibiting for flower initiation (Buban and Simon, 1978).

In most of the apple cultivars, young trees bear regularly and older trees biennially while in Laxton's Superb the biennial bearing habit is more conspicuous in early years than in older trees (Healey, 1968). Within a particular tree the branches were even found to behave differently, in a biennial year one branch do bear while the tree as a whole does not (Kobel, 1954; Friedrich, 1970). Sometimes one single tree in a plantation was also found to behave differently (Freidrich, 1970). Dvorak

(1959) observed the influence of rootstock on bearing. Apple cultivar Laxton's Superb bears alternatively on M.I, M.2 and M.11 but a regular bearing was found on M.9 spindle bush. The production of secondary shoots was found to be poor in the 'on' year of a biennial bearing cultivar (Kobel, 1954) which was about 3.5 per cent compared to 20 per cent in the 'off year whereas it was about 8 per cent in regular bearing cultivars (Schumacher, 1962).

Table 1.38: Spray Thinning Strategies for some Apple Cultivars

Cultivar	*Time of spraying*	*Chemicals*	*References*
Golden Delicious	Fruits 9.8 to 14.2 mm diameter	Carbaryl 0.1 or 0.2 per cent	Tiscornia and luchi (1984)[1]
	Petal fall	Dilamid (NAAM) 100-125 ppm	Gaash *et al.* (1985)[2]
Gala	Fruits 9.8 to to 14.2 mm diameter	Carbaryl 0.1 or 0.2 per cent	Ebert and Bender (1984)[3]
	13 days after full bloom	Carbaryl 450 ppm + NAA 7.5 ppm	
Fuji	Fruits 9.8 to 14.2 mm diameter	Carbaryl 0.1 or 0.2 per cent	Tiscornia and luchi (1984)[1]
	14 days after full bloom	Carbaryl 450 ppm+ 7.5 ppm	Ebert and Bender (1984)[1]
		Ethephon 500 ppm or Ethephon + carbaryl (600 X)	Kim *et al.* (1987)[4]
Starking Delicious	Young fruit	Carbaryl 0.3 per cent	Gaash *et al.* (1985)[2]
Red Delicious	Full bloom	Ethephon 100 or 200 ppm	Koen *et al.* (1988)[5]
Discovery	Fruitlet diameter of 12 mm	Carbaryl 2100 ppm	Knight (1983)[6]
Cox, Spartan, George Cave	80 per cent petal fall	Carbaryl 2.5 kg of 85 per cent wettable powder/100l	Knight and Jackson (1980)[6]
Gloster	13 days after full bloom	Citowett at 0.1 per cent (a commercially available surfactants)	Noga (1980)[7]

Country where experiment/s were conducted: 1. Brazil 2. Israel 3. Brazil 4. Korea Republic 5. Australia 6. United Kingdom 7. Fedaral Republic of Germany.

The nutrient content of leaves and plant as a whole has a direct correlation with bearing. The nitrogen content of leaves was found to be 20 per cent higher in 'on' year than in off year (Mason, 1955). The leaves of trees in an 'on' year also contained more Ca and less P and K than those in an 'off year (Lamb *et al.*, 1959). Grochowska (1973) recorded lower starch content in trees in 'on' year than in 'off' year, but no definite C/N ratio for flower induction of each cultivar, however, could be established. Starch content in roots, leaves and shoots of apple trees was found lower in the 'on' year than in the 'off' year, especially in the roots. The 'on' and 'off' cycle was considered one of the major factors causing premature senility (Xia *et al.*, 1983). Proteins and aminoacids are indispensible for flowering in apple trees. There is a significant increase in protein synthesis associated with flower induction. This has been attributed to the requirement for a large amount of proteins during flower development (Marcelle, 1984). In this research, a 5-10 fold increase in the concentration of 17 different kinds of amino acids was detected during flower induction, Subsequently, these amino acids were quickly converted into proteins when flower organ development commenced. Changes in the electrophoretic profiles of proteins of two apple cultivars in two different harvest years were studied by Schaeffer (1987). The qualitative electrophoretic pattern of soluble proteins was found nearly identical in the cultivars, Boskoop and Golden Delicious, in all organs through the year. Quantitatively, however, there were differences in the seasons, dependeing on fruit load and on fluctuations of total N and total soluble proteins, and there were also some differences between the two cultivars. Low molecular weight proteins ≤60000 daltons occurred in the shoot before and during the growth period. In this period, nitrogen absorbed by the roots was stored if metabolized into soluble proteins. Until autumn, proteins were polymerized up to 500000 daltons and more, and the number of highly diffused electrophoretic bands increased. In the 'on' year, this polymerization was less marked owing to a higher N consumption than in the 'off year, if began later, and in autumn medium sized proteins could still be detected. Immediately after the autumnal maximum of N in the 'on' year, high molecular weight proteins were degraded, partly because of exhaustion of N pool due to translocation to sinks in older tissues. In the 'off' year, however, the electrophoretic pattern of very high molecular weight proteins remained nearly constant, though several bands decreased to some extent. Owing to metabolic changes at the end of dormant period and at the beginning of mobilization for leafing out, the concentration of some high molecular weight protein bands declined, beginning in February. At this time, a large bud transitory band with a molecular weight of 300000 daltons was detected in some shoots, especially in Boskoop. Protein break down increased during March and April and low molecular species again appeared. At the beginning of the 'on' year this process was somewhat delayed. In the 2-year-old shoots the extent of mobilization in March was lower at the onset of the 'off year, but this was higher before the 'on' year in accordance with the number of fertilie buds in these shoots. High potassium application enhanced the tendency of biennial bearing by increasing yield in 'on' years (Hansen, 1971).

Cold and early winter can cause flower bud abscission in apple in the following spring, particularly in clusters on one-year-old wood, where flower initiation is late in the season and flower development is incomplete (Rudloff and Feucht, 1957). Water stress at a lower level in summer can increase flower induction (Zeng, 1990). Tromp (1984) documented the increase of flower induction when Cox's Orange Pippin apple trees were treated with air humidity of 40 per cent -50 per cent instead of 80 per cent -100 per cent. Appropriate water stress may increase apple flowering indirectly by regulating the balance between vegetative and reproductive growth. Treatment with GA suppressed flowering on spurs and laterals. GA_{4+7} was found more effective that GA_3 (Marino and Greene, 1981)

Genotype is probably the dominant reason for alternate bearing. For example 'Gala is generally a regular bearing cultivar (Hirst, 1999) while 'Fuji' tend to be very biennial bearing (Li *et al.*, 1995). Atay *et al.* (2013) studied the bearing paltern of apple cultivars on MM.106 rootstock at Egirdir, Turkey (Latitude 37°49′ N, longitude 80°16′ W). Based on the modified alternate bearing index (MABI), cultivars were classified into four relative susceptibility groups: not susceptible (cvs. Braeburn and Jerseymac), susceptble to alternate (Starkspur, Golden Delicious, Fuji, Arlet, Redchief Delicious, Rajka and Golden Delicious), medium alternate (Topaz, Granny Smith, Mondial Gala, Jonagold, Starkcrimson Delicious and Clear Red) and highly alternate (Kasel 37, Golden Reinders and Kasel 41).

However, the different cultivars behave differently in a particular environment and with similar cultural practices. A list of cultivars and their bearing habit are presented in Table 1.39. Trees of Red Fuji on Mark and MM.111 have shown little tendency towards biennial bearing in South Australia, whereas on M.9, M.26 and MM. 106, trees have had marked annual fluctuations in yield (James and Middleton, 2001).

Table 1.39: Apple Cultivars According to Tendency of Bearing

Regular	*Intermediate*	*Biennial*
Jonathan	Delicious	Baldwin
Stayman Winesap	Winesap	York Imperial
Ben Davis	Grimes Golden	Yellow Newton
Rome Beauty	Northern Spy	Wealthy
McIntosh	Golden Delicious	Yellow Transparent
Gravenstein	Wagener	Arkansas
Cortland	Stark	Laxton's Superb
Bramley's Seedling	Winter Banana	Cox's Orange Pippin
George Cave	Discovery	Miller's Seedling
Gloster 69	Granny Smith	Antonovka
Idared	Boskoop	Ralls
James Grieve		Newton
Jonagold		Fuji
Lobo		Gravenstein
Spartan		
Tydernan's Early		
Winston		
Gala		

Control of flowering in apple with the use of growth regulator has been carried out in almost all the apple growing regions of the world with varying degrees of success. Ethephon was found to induce flowering without affecting tree vigour (Schmidt *et al.*, 1975). Tromp (1972) proposed that the reduction of growth and the promotion of flowering by SADH (succinic acid 2-2-dimethylhydrazide) are independent phenomena while Grasmanis and Edwards (1974) reported that short exposure of potted trees to ammonia ion promoted flowering without influencing growth. Partial defruiting of spur-type Golden Delicious trees 5 to 21 days after full bloom (DAFB) and/SADH applied 14 DAFB created a gradient in the tendency for 'on' year in the biennial trees to flower again the next season (Looney *et al.*, 1980). Treatment with daminozide promoted flowering by decreasing the level of GA (Hoad and Monselise 1976) and application of daminozide 2 weeks after full bloom increased the level of cytokinins and decreased auxin level in seeds but it did not alter the GA or ABA content of seeds or GA level in shoot tips. It was thought that increased zeatin brought about increased amount of return bloom (Ramirez and Hoad, 1981). Treatment with GA suppressed flowering on spurs and laterals. GA_{4+7} was found more effective than GA_3 (Marino and Greene, 1981). Treatment with alar restricted shoot growth (Hansen and Grauslund, 1980), but resulted in increased in fruit bud set (Seth and Singh,1977). Action of alar on fruiting depended on cultivar. It increased the yield of Starkspur but reduced that of Starkrimson and Jonathan (Stan *et al.*, 1977). The effect of Alar depended on temperature and time of treatment (Mantinger and Vigl, 1978). Ethrel at 144 mg/l, applied to trees of Spartan 5 weeks after full bloom and again a week later, significantly increased yield in the following year (Koupil and Plasilova, 1986). Flower bud development may be correlatd with ethylene content of wood or buds. Robitaille (1975) showed that wood of horizontal apple shoots produced more ethylene than in the vertical ones. Lorette-type pruning may induce ethylene production through wounding (Klein and Faust, 1978). Ethephon (200mg/l) applied at 3, 3+6, and 3+6+9 weeks after full bloom in 'on years' on spur-type 'Delicious' apple increased flowering in 'off years' by 33 per cent and reduced flowering in 'on years' by 17 per cent compared to control (NTC). The mean yield per tree for ethephon- treated trees over three 'on years' and three 'off years' was almost identical to that of the controls (82 vs 80 kg/tree) (Table 1.40). However, the distribution of yield between 'on' and 'off' years was changed, 24 per cent greater in 'off years' and 10 per cent less in 'on' years (Bukovac and Sabbatini, 2006).

Table 1.40: Mean Effect of Ethephone (200mg l^{-1}) Applied at 3+6 Weeks after Full Bloom in Three Consecutive 'on years' (1998, 2000, 2002) on Yield over Six Seasons (Three 'on' and three 'off') of Redclief Delicious Apple Tree*

Parameter	*Yield (kg/ tree)*		
	'On years'	*'Off years'*	*Mean*
NTC	102	58	80
Ethephon	92	72	82
Percentage of NTC	90	124	103

* Adopted from Bukovac and Sabbatini (2006).

Growth retardant, cultar, (paclobutrazol) has been demonstrated significantly to control shoot extension growth and decreased apical dominance which resulted in the production of a greater number of spurs (Buban *et al.*, 1987). Paclobutrazol was thought to interfere with GA synthesis; increase in organic acids (including succinic, malic, citric and quinic acids) and also in total phenols in wood was noted following treatment with paclobutrazol (Wang and Steffens, 1987). Paclobutrazol at 2.8 g ai/tree, applied in soil or trunk was found to reduce the shoot growth but increase fruit bud size in a 12-year-old Fuji apple tree. The treatment also increased the carbohydrate content of the shoot, leaves and wood, but decreased that of the fruit bud (Byun and Chang, 1986). Trials in USA, Italy, Germany, and the UK have demonstrated that programmed sequential foliar spray of paclobutrazol can control vigour and increased yield of apples. The product has given benefits on several cultivars including Red Delicious, McIntosh. Granny Smith, Bramley, Cox's Orange Pippin and Gloster. Varietal differences coupled with choice of rootstock, planting density, soil fertility and growing environment give rise to a range of differing programmed sprays of 125 to 500 g ai/ha applied sequentially as low or high volume spray through the growing season. The degree of vigour reduction was highest with sprays at early tight cluster or two weeks after petal fall, although sprays too close to flowering can adversely affect fruit development. Late season application is necessary to maintain continued growth inhibition and provide a carryover effect for the following season (Shearing and Palmer, 1986).

Thinning of fruit in one year often increases the return bloom in the following year. In a heavy cropping year, thinning is particularly essential for return bloom as the tree become exhausted due to heavy crop. Thinning should be done with in 40 days after petal fall. The number of seeds within develoiping fruit seems to inhibit flower bud initiation and most of the inhibition takes place in the first few weeks after pollination. This is the reason for early thinning. Thinning of Golden Delicious with carbaryl in a heavy cropping year was found to reduce the yield by 30 per cent but overcome the tendency of biennial bearing in the following year (Gaash *et al.*, 1985). Zika (1987) reported that Bercerna NMC 50 (carbaryl) at 500 to 2000 mg/1 was best for regulating alternate bearing. though thinning did not significantly affect 2-year yield, it merely changed the cropping pattern. Thinning with Promonit R-10 (NAA) at 30 ml/l in combination with Flordimex 420SL (ethephone) at 60 ml/l improved flower bud formation and resulted in abundant blooming in the next year of apple cvs. Lobo, Spartan, Szampion, Jonica and Elster (Tyburski and Wazbinska, 2000).

Nitrogen application early in the 'off' year has a slight tendency to promote annual-bearing in trees having a low-nitrogen level (Shoemaker and Teskey, 1959). However, studies with apple cv. Fuji in Japan showed that tree growth and fruit yield were not at first affected by N application (up to 20 kg/10a annually), but after the second year the increases were apparent especially at higher N rates, and from the fifth year onwards alternate bearing occurred with the highest N rates (Kondo *et al.*, 1987). The form of nitrogen applied is also an important factor influencing flower induction. Ammonia (NH_4+) favours flower induction while nitrate (No_3-) does not. Application of nitrogen in the form of ammonia leads to a higher levels of amino acids in the floral buds, especially basic amino acids like Arginine (Zeng *et al.*, 1987). The function of NH_4+ is not only nutritional, but it can also trigger the activity of plant hormones, Gao *et al.* (1992) detected an increase of cytokinin activity in apple xylem soon after NH_4+ application, cytokinins in turn are thought to promote flower initiation.

It is generally accepted that pruning should be severe in the spring of the 'on' year and light in the spring of the 'off year (Jonkers, 1979). Pruning once in two years reinforces the biennial bearing pattern (Vittrup, 1970). Heavy pruning during the 'on' year reduces crop load and improves the nutrient status of the tree in the 'off year to initiate flower buds. Link (1983) reported that summer pruning of cvs. Boskoop and Cox's Orange Pippin, which are very susceptible to alternate bearing, gave a more regular and richer bloom than did winter pruning (Table 1.41).

Table 1.41: Influence of Pruning Time on Yield in Cox's Orange Pippin (kg/tree)*

Treatment	*Experiment A*		*Experiment B*
	1st Year	*2nd Year*	*4th Year*
Winter pruning	4.6	24.7	6.5
Summer pruning	6.4	29.0	27.5

* Link (1983).

Fruit Growth and Development

During the development of the fruit, there is a continued enlargement of the receptacle. The throat of the receptacle becomes narrow through which the style protrudes. The connection between the receptacle and carpel tissues becomes so close one that the receptacle makes up the greater portion of the flesh of the apple. A rapid phase of cell division occurs in the first few weeks after pollination which ceases abruptly 30-40 days after full blossom in Cox's Orange Pippin (Denne, 1960) and after 4 weeks in Granny Smith (Bain and Robertson, 1951). The subsequent fruit growth occurs mainly due to cell expansion. The fruit growth pattern follows a smooth sigmoid curve. Fruit growth in apple is dependent on the presence of growing seeds up to the end of the June drop (Abbott, 1959). After this stage, removal of the seeds no longer results in cessation of fruit growth. Hebdas *et al.* (1982) compared the pedicel structure of larger fruitlets, which are normally retained on the tree, with that of smaller fruitlets which tend to June drop. It was found that in the pedicels of larger fruitlets, development of the xylem especially secondary xylem, was more intensive, and this difference increased with time. The lignification of cortex sclereids, phloem fibres, xylem parenchyma cells and pith cells was also more advanced in the larger fruitlets. Goffinet (1986) studided the fruit growth pattern of Delicious from bloom to October harvest. Fruit length was found to increase at the rate of 82 per cent of fruit expansion, and after 2 months, diameter surpassed the length. The rate of fruit volume increase was 110 per cent of the rate of fruit weight over the season as carpellary space increased and cell packing loosened. The cortical cells enlarged 5 per cent faster than cells of other regions. Intercellular space increased at the rate of 18 per cent of cell enlargement in pith and cortex, but at 14 per cent in the two lobe regions. Pith and cortex converted cell volume and intercellular space into fruit volume 3-4 per cent faster than did lobes. Changes in the structure of peripheral and inner chloroplasts and the pigments during fruit ontogeny reveal that frequencies of 600-6000 stomata per fruit or 2-20 per mm^2 are maximum at petal fall, subsequently decreasing with expansion of the fruit surface area. The pattern of decreasing CO_2 exchange with fruit development is attributed to insensitivity of fruit stomata and decreasing cuticular permeability, and may be affected by temperature, light, CO_2: O_2 ratio, assimilate supply, epicuticular wax composition and structure. CO_2 is captured within the fruit in the intercellular space and accumulates to maximum values of up to 8-10 per cent at the peak of climacteric (Blanke and Lenz, 1988),

Depending on the stage of development; temperature and light (intensity) are the most important environmental factors determining fruit growth and final fruit quality. Final fruit size is mainly controlled by temperature during the first weeks after full bloom, *i.e.* during the period of cell division (Lurie, 2002; Tromp and Wertheim, 2005). Cold spring temperatures lead to an extended period of cell division and presumably larger fruit sizes (Lurie, 2002). Excessive sun light (high light intensity) may result in sunburn or sunscald, whereas insufficient leads to smaller fruit with unsatisfactory quality (Kays, 1999). Other factors such as available soil moisture, nutrient availability in addition to adverse weather effects such as frost, hail and wind also influence fruit development and final fruit quality. Fruit size of cv. Golden Delicious was found to increase with increasing effective temperature sum between full bloom and harvest, and to decrease with increasing cumulative precipitation (-0.5 g/10 mm). Date of full bloom, number of days between full bloom and harvest, and cumulative sunshine hours, however, showed no significant effect on fruit size (Reichel and Schmidt, 1986).

Depending on the cultivars and stage of maturity, the water content in apple fruits varies from about 75 to 90 per cent. The water content increases with the increase in fruit size with a slight decrease towards the end of maturity. The increase in water content of fruit at different stages of development, however, is influenced by several climatic factors. Pruning and thinning are mainly used to determine the crop load and thereby fruit size and quality. Pruning enhances the light penetration into the canopy (Beverly *et al.*, 1993) whereas thinning increases the leaf to fruit ratio which leads to higher quality of individual fruit (Tromp, 2005). Fruit from light cropping trees have a better flesh firmness, higher percentage of soluble solids and a larger fruit size when compound with high-cropping trees (Wunsche *et al.*, 2005).

The bulk of the carbohydrate in an apple fruit is sugar which is mainly synthesized in leaves and transported to the fruits although, a very little sugar is also synthesized by the fruit. Since, most of the sugars are synthesized in leaves, the leaf-fruit ratio was found to affect the sugar content of fruit to a certain limit. Fructose, glucose and sucrose are the three principle forms of sugars found in the apple flesh. The reducing sugars (fructose and glucose) and sucrose content of the fruit increases with the increase in fresh weight. The amount of different forms of sugars in different stages of fruit development and at harvest varied with cultivars, climate, cultural practices, *etc.* Light and temperature during the first four months after fruit set have a marked effect on sugar content of fruit. The type and amount of different nutrients, use of fruit thinning chemicals, herbicides, pesticides and fungicides also have a direct influence on the sugar content of fruit.

In apple, starch accumulates at very early stage of its development. With the advancement of maturity, the accumulated starch is hydrolyzed into sugars. The starch disappearance is higher at the later stages of fruit development (Joshi and Diwakar, 1985). The conversion of starch into sugars in maturing and ripening apples generally starts in the core area and gradually progresses outwards (Kingston, 1992, Tromp, 2005). Different starch degradation characteristics during the growth and maturation processes were observed in two Japanese apple cultivars Tsugaru and Fuji (Thammong and Arakawa, 2009). The loss of starch was earlier in Tsugaru than in Fuji. In Tsugaru amylase (AM) and amylopetin (AP) degraded rapidly between 95 to 110 days after full bloom (DAFB) and almost all starch were lost rapidly at 125 DAFB with simultaneous increase in the rate of respiration and ethylene production. However, in Fuji starch degraded gradually throughout growth and maturation process and was clearly degraded at 170 DAFB with a low level of ethylene production and decreased respiration. In both the cultivars, content of AM and AP were highest in the outer cortex and lowest in the inner cortex. High amounts of auxin in the seed compared with the fruit cortex were observed in Royal Gala apples, with amounts increasing through fruit development. Injection of exogenous auxin into developing apples at the start of cell expansion caused an increase in cell size, suggesting that endogenous auxin concentration are atleast one of the limiting factors controlling fruit size. The expression analysis of ARF 106 linked to a strong QTL for fruit weight suggests that the auxin signal regulating the fruit size could partially be modulated through the function of these gene. One class of Gene (GH_3) remove free auxin by conjugation to amino acids. The lower expression of these GH_3 genes during rapid fruit expansion is consistent with the apple maximizing auxin concentrations at this point (Devoghalaere *et al.*, 2012). The hemicellulose and dextrin contents are higher at the early stages of development, which decline gradually with the advancement of maturity.

The titratable acidity of the fruit steadily decreases as the fruit matures but the actual or absolute amount of acid present in the fruit increases until just before harvest when it decreases slightly. The levels of free polyamines is high only during the first few weeks after full bloom and then decreases gradually. The largest variation in polyamine levels in fruits occurs very early in the season when dry weight and protein contents also change rapidly and fruit diameter is 10-20 mm (Biasi *et al.*, 1988).

Bagging of Fruits

The practice of preharvest bagging of fruits to improve the commercial value of fruit, namely, improving fruit colouration reducing splitting, mechanical damage, sunburn of the skin, reducing pesticides residues in the fruit, minimizing insect, and disease are used in many countries. Preharvest bagging are used widely in apple production in China, especially for Red Fuji apple, which is a major apple cultivar with a bagging rate of over 75 per cent in main production areas of both Shandong and Shaanxi (Zhai *et al.*, 2006). Some countries importing apple from China, such as Mexico, Chile, Argentina, requires bagged fruits only. The fruits are usually bagged after 35-40 days of full bloom and removed the bags before 30 days of anticipated harvest. (Quin *et al.*, 2013).

Russeting

Russeted surface apples are not liked by the consumers. Russeting of apples in a humid environment is a major concern of the fruit growers (Faust and Shear, 1972). Russeting is caused by damage to epidermal cells during their intensive division shortly after petal fall. It depends on clonal susceptibility, physiological stage at fruit set and environmental condition (Vogl, 1985) but there are several thoughts about the location of russeting, it may originate in the epidermis (Zzchokke, 1987) or tissues subjacent to small cuticular cracks (Sorauer, 1972).

The apple cultivars which have thin cuticle (Tukey, 1960) are more susceptible to russet than those with cuticles arranged in free platelets (Faust and Shear, 1972). However, some cultivars with oily and thin cuticles were found resistant to russet (Hochapfel, 1963).

The water balance is considered critical in the formation of russet as it develops under conditions of high humidity (Kofler, 1961), occasional rain (Chandler and Mason, 1942) or dew (Winter, 1964). Lalatta *et al.* (1968) opined that russeting is more likely to develop in areas where the night temperature is high (15-20°C). High turgor pressure in epidermal cells due to diffusion of rain or dew through the cuticular cracks on fruits exposed in the sun, may cause bursting of the skin, and thus russeting is more commonly noticed on exposed fruits than on fruits remaining in shade (Kremer, 1963). Frost during blossom or at the early fruitlet stage may also cause russeting (Hadorn, 1966). Frost damage is caused by ice forming within plant tissues; which may form between the cortex and epidermis of the fruitlet, leading to rupture of the skin and development of primary cracks which in later stages become inelastic, predisposing to the development of secondary cracks closely associated with russeting (Modibowaska, 1961). Khanal *et al.* (2013) stated that russeting is a disorder of the fruit skin that results from microscopic cracks in the cuticle and the subsequent formation of a periderm. Some workers believe that there is a strong correlation between nitrogen nutrition and russeting (Moorre, 1936; Uitterlinden, 1964). It is apparent from the literature that Ca (Simons, 1960), Mg (Eggert and Mitchell, 1966) and K (Greenham, 1965) nutrition sometimes may be responsible for russeting.

Application of high dose of potassium was reported to be the cause of russeting in Cox (Greenham, 1965), particularly in the absence of nitrogen, while Dalbro (1958) recorded russeting under potassium deficiency.

Rootstock vigour was also found to influence russeting. In Belgium, Cox fruits on M.9 and M. 7 showed 32 per cent more russet than those on MM.106 and MM.104 (Monin, 1961). Early mildew infection on blossoms and fruitlets cause russeting (Zorbrist, 1962). Sprays which are toxic to cells are also known to cause russet (Aerts, 1971; Hochapfel, 1963; Winter, 1967).

The fruits of the cultivars that have a platelet type of wax are apparently more resistant to cracking (Faust and Shear, 1972). Reduction in russeting may be achieved by selecting less susceptible clones, *e.g.*, clone A (Danish), clone B (Swiss) and Smoothee (which is however, less productive) of Golden Delicious. Clone/Rootstock relations, freedom from viruses, optimal orchard management (water relations, manuring, judicious control of pests and diseases) all may contribute to a reduction in fruit russeting. In special cases sprays are recommended at petal fall and every 8-10 days thereafter until the fruitlets hang downwards, with 0.7 per cent wettable sulphur +0.1 per cent borax +0.2 per cent kaolin at the rate of 1500 1/ha. Specific russet control sprays include Antirost, Golclair and wettable sulphur (Vogl, 1985). Treatment with GA_{4+7} reduces russeting (Eccher and Boffelli, 1981).

Spraying with 6, 12 or 24 ppm of GA_{4+7} prevented russet without reduction of flower buds (Eccher, 1983; Table 1.42).

Table 1.42: Control of Russeting of Golden Delicious Apple by GA_{4+7} in Two Locations of Sondrio Province, Italy*

GA_{4+7} (ppm)	*Ponte*				*Villa*	
	RST	*RSC*	*RI per cent*	*RST*	*RSC*	*RI per cent*
6	1.02	1.52	69.4	3.00	6.00	50.0
12	0.77	1.59	45.0	2.08	5.02	41.4
24	0.93	1.39	56.3	2.52	5.85	43.0

RST: Mean russet score of treated fruits; RSC: Mean russet score of control fruits; RI: Russet index.

* Eccher (1983).

Sunburn

Sunburn is caused by over-heating of the fruit surface due to excessive levels of incoming solar radiation in combination with high ambient air temperatures. It has been hypothesized that sunburn may be an expression of plant defense mechanisms involved in response of oxidative stress. Schrader *et al.* (2003) identified two types of sunburn in apples. One type (sunburn necrosis) is caused by thermal death of epidermal and subepidermal cells when the peel reaches 52±1°C. A necrotic spot appears on the sun-exposed side of the fruit. The second type (sunburn browning) is sublethal and results in a yellow, bronze, or brown spot on the sun exposed site of the fruit. The fruit surface temperature (FST) at which sunburn occurred varied between 46°C and 49°C depending on the cultivar. The sensitivity of apple cultivars to sunburn may be partly related to the physico chemical properties of the fruit skin. The homogeneity thickness and composition of the epicuticular wax layer, the amount of hair on the skin surface, and the concentration and quantity of skin pigments could offer some protection against sunburn by increased reflection of visible and particularly infrared light accompanied by decreased fruit temperature. Chlorophyll a and b concentrations in the skin decreased considerably from healthy to sun tinted skin tissue, probably resulting from photobleaching of the chlorophyll (Wunsche *et al.*, 2004). Pigments with substantial radical scavenging ability such as beta-carotene and chlorogenic acid concentrations increased substantially with greater sunburn damage. Northern blot analysis of apple tissue RNA for heat exposure treatments has shown that hsp 17 and hsp 70 gene expression was associated with high fruit temperature. To avoid sudden exposure of fruit to intense heat and solar radiation, overhead sprinkling and coating the tree with a reflective kaolin particle film are effective orchard practices for reducing incidence and severity of sunburn.

Maturity

The maturity of apple fruits cannot be judged by firmness of flesh at all times like many other fruits. The gustatory qualities of apples are inferior if the fruits are picked too early. If the apples are picked after the optimal harvesting date, seldom the gustatory quality and the storage ability is reduced (Henze, 1983). Fruits picked at immature stage or insufficiently developed are likely to be inferior in quality (Stoll, 1978). On the other hand, late picked apples have a good internal quality. The optimum date of harvest must be a reasonable compromise between internal fruit quality and storage life (Streif, 1983). The 'optimum' harvest date is dependent on the cultivar, the planned use of the fruit, the given storage technology and the marketing strategy (Tromp, 2005). Several cultivars such as 'Jonagold' and 'Elstar' have a wide within tree maturity variability which is due to extended period of flowering and fruit set as well as differences in fruit position within the canopy. In order to harvest all fruits at the right time, *i.e.*, at the optimum of maturity, sequential picking is recommended (Kingston, 1992; de Jager, 1994).

The days from full bloom to harvest is considered as a fairly good index of maturity; however, climatic factors immediately after petal fall are very important. For most of the English apple cultivars, Kidd and West (1926) recorded a relationship between optimum harvest date and the beginning of the respiratory climacteric rise. The best harvest date for McIntosh seems to be just at or just following the beginning of the climacteric rise in respiration (Smock, 1946). The proposed theory of Kid and West (1926) is, however, not applicable universally because Blanpied (1964) in his experiment with Mclntosh, Cortland, Delicious and Golden Delicious at Cornell University, USA, did not find a relationship between optimum harvest date and a specific point on the climacteric curve. The optimum harvest date for cv. Cox's Orange Pippin apples, required for long-term storage is normally taken when the crop is about to commence ripening as evident by a sudden increase in ethylene production by a small portion of fruits on the tree. The internal ethylene concentration is not a reliable index of maturity for harvest for immediate sale in the fresh market (Blankenship and Unrath, 1987, 1988; Chu, 1988). The rate of increase in internal ethylene concentration varies from year to year. The current methods for monitoring changes in fruit maturity as the harvest date is approached, include measurements of fruit firmness, respiration rate, ethylene production, starch hydrolysis, soluble solids and titratable acidity, and also the colour of the skin and cortical tissue (Fidler, 1973). In 'Cripps Pink' apple delayed harvest and extended cold storage period increased the internal ethylene production and reduced fruit firmness, soluble solids concentration, titratable acidity. Fruits harvested 2 weeks after commercial maturity can be cold-stored for 90 days with acceptable fruit quality. If the fruit is intended to be sold fresh immediately in local markets, the harvest can be delayed up to 6 weeks after commercial maturity to harness substantially improved fruit colour with acceptable fruit quality (Shafiq *et al.*, 2011). Blanpied (1964) and Fidler (1973) concluded that the most reliable harvest index for several apple cultivars was a standard calendar date. However, although this may be adequate in most years, it is desirable to forecast the optimum harvest date every season including those in which the standard date is inapplicable. Luton and Hammer (1983) in predicting the optimum harvest dates for Cox's Orange Pippin in UK, studied meteorological variables between 1959 and 1980 and were able to derive an equation which enabled them to predict the harvest date on 1st September.

$$HD = 68.1 + 0.243\ FB - 3.516\ \overline{Tmean}\ JJA$$

where HD is the harvest date, expressed as the number of days from 31 August, FB is the date of full bloom, expressed as the number of days from 31st March and $\overline{Tmean}$ JJA is the mean of the mean temperatures in June, July and August. The optimum harvest dates for some of the apple cultivars are presented in Table 43.

Table 1.43: Optimum Harvest Date for some Apple Cultivars

Horticultural Experiment Station, Simcoe, Canada+*				*Country*
Cultivar	*Harvest Date*	*Cultivar*	*Harvest Date*	
Quinte	August 1	Jonagold	October 10	British[1] Columbia
Red Melba	August 13	Starking	September 24	Turkey[2]
Idared	October 11	Golden Delicious	September 28	Turkey[2]
McIntosh	September 18	Delicious	October 13	New York[3]
Tydeman Early	August 17			
Empire	October 1			
Red Delicious	October 18			
Spartan	September 26			
Golden Delicious	October 15			
Northern Spy	October 18			
Mutsu	October 18			

* 10 year Mean: 1970-1979; + Elfving and Pierce (1988).

1: Lau (1983); 2: Gunel and Karacalli (1985); 3: Blanpied (1969).

Colour Development

In general, buyers prefer red or yellow apples than to green ones. Nearly all red colours in apples are due to soluble anthocyanin pigments, these arise from a product known as chromogen (Walter, 1966).

Changes in anthocyanin accumulation may be useful for predicting maturity in Jonathan and other early to mid-season, red apple cultivars. While the level of anthocyanin at a particular stage of maturity may be related to environmental factors, the capacity to accumulate anthocyanin is a function of maturity and independent of environmental influence (Chalmers *et al.*, 1973). From annual studies of anthocyanin accumulation (Faust, 1964, 1965), it is clear that there are large difference in seasonal potential for anthocyanin synthesis, although the time course of anthocyanin accumulalion is qualitatively similar from year to year. The tranistion from the immature to mature stages with respect to anthocyanin accumulation occurs rapidly and precedes the normal harvest date by about 2-3 weeks, suggesting that this attribute might be used in the prediction of maturity of suitable red apple cultivars (Chalmers *et al.*, 1973).Development of attractive yellow colouration is a problem with Golden Delicious and Grimes Golden in some areas (Workman, 1963a). The pigments responsible for Golden Delicious skin colour are chlorophyll (green) as well as carotenoids and flavanoids (yellow). During ripening chlorophyll disappears, making the yellow pigments visible (Workman, 1963b; Rhodes and Wooltorton, 1967). Flavanoids may also play a role in the development of yellow colour (Siegelman, 1955). Gorski and Creasy (1977) stated that the minimum amount of carotenoids necessary for yellow colour in Golden Delicious apple was found to be 0.3-0.4µg/cm^2 of skin. Yellow colouration was not visible unless chlorophyll concentration was less than 0.15-0.20 µg/cm^2. Flavanoids play a secondary role in yellow colouration. 'Fuji' apple has gained popularity in many countries. However, the red colouration of Fuji showed variations among its strains under certain climatic condition. In South West Idaho, USA situation, fruits of 'Autumn Rose Fuji', 'Myra Fuji' and 'Top Export Fuji' always had less but 'September Wonder Fuji', 'Dessert Rose Fuji' had more red colour (Fallahi *et al.*, 2014).

Factors Affecting Colour Development

Light

Ultraviolet and blue light affect most the red colouration of apple, but these photo-spectra are readily absorbed by dust and moisture in the air. The ultraviolet rays are found in higher concentration at a high elevation and thus, at higher elevations fruits were found to develop better red colour (Magness, 1928, Damast, 1949). However, that light is indispensible for colour development in apple is proved by the fact that when fruits of cv. Gladstone is kept in bags, they fail to develop any red colour, and remain green (Rogers, 1926). Arakawa *et al.* (1985) studied the relative effectiveness and interaction of ultraviolet-B, red and blue light in anthocyanin synthesis in apple fruits. It was found that high fluence rates of white light provided from a xenon lamp were unable to produce large amount of anthocyanin in cv. Jonathan, and anthocyanin production became saturated at about 30 Wm-^{2}. When UV-B light, provided by a fluorescent lamp, which had an emission peak at 312 nm, was combined with white light, anthocyanin production was synergistically stimulated and increased up to the highest fluence rates of white light tested (44 Wm^{-2}). This UV -B light was more effective than red and blue light provided from fluorescent lamps, but anthocyanin production became saturated at about 1.7 Wm^{-2}

Humidity and Temperature

High night temperatures result in relatively small fruits and many cultivars fail to develop the characteristic red colour at harvest (Jackson, 1983). This effect appears to be due to depletion, by respiration, of the carbohydrate required for red pigment synthesis. The most rapid anthocyanin synthesis was, however, observed at average temperature of 46.5° and 60.5°F (8.1° and 15.8°C) during the night and day, respectively (Uota, 1952). Both cool day and night temperatures or high night and cool day temperatures were also favourable for red colour development but high day temperature was found to inhibit colour development, even when nights were cool (Walter, 1967). Red colour development was found earlier and more intense at cool (11°C) night temperature (Blankenship, 1987). The average absolute humidity favours a greater amount of red colouration.

Leaf Area

The presence of adequate number of leaves increases the sugar content of fruit and helps in development of colour. Delicious and Winesap fruit with 10 leaves (under perfect light exposure) had only 23 per cent of its surface showing solid red colour; whereas fruits with 75 leaves showed 58 per cent red colour development and an increase in sugar content by about 10-15 per cent (Magness, 1928). Preston (1954b) suggested drastic thinning, *i.e.*, leaving one fruit of Duchess Favourite for 30 leaves which was found to increase fruit size, colour development and also to induce regular bearing. In cv. Jonagold, well-coloured fruits were found on trees with more and larger leaf area/fruit. Leaving 1 fruit/cluster at thinning was associated with 49 fruits/100 flower cluster at harvest and 45 leaves/fruit (Wertheim, 1987).

Rootstock

Monin (1961) in his experiment with Cox on different rootstocks at France, concluded that colour development of fruits is inversely proportional to the vigour of the rootstock. The rate of fruit colouration was greater on M.9 at East Malling (Rogers, 1927), and on M.6 and M.9 in Germany (Hilkenbaumer, 1959) which suggests that improvement in fruit colour in dwarfed rootstock is due to greater exposure of the fruits to the sun.

Nutrition

In general, level of nitrogen which increases fruit yield usually inhibit proper colour development (Weeks, 1952, 1954), specially when applied 2 to 3 months before harvest (Reuther *et al.*, 1958). Phosphorus nutrition, however, has no such consistent effect on fruit colour but high potassium rate was found to improve the red colour development in McIntosh (Southwick, 1961) and Red Delicious (Fisher, 1961). The nitrogen content of the leaf ranging between 2.2 and 2.4 per cent (on dry weight basis) at mid-summer was found optimum for fruit colour in Red Delicious and Golden Delicious in the USA (Van Doren, 1961), while the best fruit colour in McIntosh was found with the leaf potassium level of 1.5 per cent or more (O'Grady, 1959).

Position of Fruit on the Tree

Fruits with a better appearance often occur on the top and other exposed parts of the tree canopy (Farhoomand *et al.*, 1977). There is also a variation in the maturation rate between full coloured and less coloured, interior and exterior fruits, and small and large ones on the same bunch but on separate stalks. Ripened bottom apples show poor colour but good eating quality, whereas top fruits have a better colour and poor aroma and taste (Table 1.44). For better storage and organoleptic quality, it is better to harvest apples from the bottom half of the tree earlier than from the top with a gap of 7-10 days between harvests (Krishnaprakash *et al.*, 1983).

Table 1.44: Effect of Position on Quality (Sensory and instrumental mean scores of cv. Royal Delicious)*

Quality Attributes	*Fruit Position*			*SEM (df)*	*Physico-chemical Parameters*	*Fruit Position*			*SEM (df)*
	Top	*Middle*	*Bottom*			*Top*	*Middle*	*Bottom*	
Colour and appearance	4.2c	3.8b	3.2a	±0.15					
Texture	4.2a	4.4a	4.5a	±0.13	MT value (kg) at harvest	4.6a	4.5a	4.2a	±0.12
					MT value (kg) after transit	2.9a	2.83	2.7a	±0.09
Juiciness	3.2a	4.2b	4.4b	±0.11	OTMS yield				
					point (kg)	15.0a	14.3a	14.la	±0.41
Aroma	2.8a	3.6b	4.2c	±0.13					
Taste	3.2a	3.8b	4.4c	±0.14	^{0}Brix	11.5a	12.6ab	13.46b	±0.39
					Acidity (per cent) malic acid (w/v)	0.40a	0.36a	0.35a	±0.004
Overall quality	2.6a	3.2b	3.5b	±0.14					

SEM (df) = Standard error of mean (degrees of freedom).

Means carrying different superscripts in rows are significantly different (P≤0.05).

* Krishnaprakash *et al.* (1983).

Hamadziripi *et al.* (2014) stated that consumers preferred the under outer canopy to the less red inner canopy 'Starking' fruit. The redness of Starking fruit in their studies has been seen as a true signal of eating quality *i.e.,* the redder the fruit the better the eating quality. It was also observed that the outer canopy fruits of cvs. Starking, Golden Delicious and Granny Smith had higher antioxidant capacity, soluble solids concentrations, dry matter, lower titrable acidity and were generally sweeter than inner canopy fruit.

Irrigation

The colur development in fruits of apple is closely related to available soil moisture at different times of the year, irrespective of fertilizer treatments (Kimball. 1933). Chandler (1951) stated that trees which received a continuous moderate water supply generally produced good coloured fruits than those grown under stress. A high precipitation immediately after full bloom (Mowry, 1960-61), or irrigation during the fruit swelling period (Packer *et al.,* 1963) and during ripening (Magness *et al.,* 1935) was found associated with a high intensity of red colour.

Growth Substances and other Chemicals

The different growth substances used for improving colour of fruits are 2,4,5-TP (2,4,5-trichlorophenoxypropionic acid), 2,4,5- T (2,4,5-trichlorophenoxyacetic acid), NAA (naphthaleneacetic acid), SADH (N-dimethylaminosuccinamic acid), ethrel [(2-chloroethyl) phosphonic acid] *etc.*

Preharvest spray of 2,4,5-TP was found effective in colour development in McIntosh (Teskey and Francis, 1954), Red Duchess, Wealthy (Lott and Rice, 1955), Worcester (Whitwell, 1956), Delicious (White, 1953). Treatment with 2,4,5-TP, however, often increased the softness of fruits and sometimes it may advance maturity. 2,4,5-T improved colour of Golden Delicious (Vorijhof, 1952) and James Grieve (Glas, 1958). The effective concentration of 2,4,5-TP varies between 10-25 ppm while for 2,4,5-T, several cultivars showed marked response to an application to 50 ppm, three to four weeks before harvest. Spraying with NAA markedly improved fruit colour to Black Ben Davis (Faccioli, 1958). SADH used 70 days before harvest, has only a slight effect on the colour development, but prolongs storage life, while ethrel stimulates colour, but shortens storage period. In Jonathan apple, the best homogeneous red colour development was found with ethrel at 500 ppm applied 6 weeks before harvest (Gvozdenovic *et al.,* 1978).

Spraying with calcium carbonate at 12 to 100 ppm (Smock, 1964), sodium diethyldithiocarbarnate at 1 or 5 per cent (Uota, 1950) greatly improved the colour of McIntosh apples. From Italy, Carlone (1958a,b) reported that spraying trees of Delicious and Golden Delicious with captan at 0.25 per cent twice in August and September increased red colouration. Fruits of cv. Tsugaru treated with fenoprop matured 2-3 weeks earlier but had a shorter shelf life. Addition of calcium acetate to fenoprop did not alter the increase in anthocyanins following fenoprop treatment (Ryou *et al.,* 1986). Although, several chemicals gave promising results of fruit colouration in laboratory studies (Smock, 1969), but the response to orchard sprays was frequently disappointing (Smock, 1971). This was interpreted as a lack of penetration of the chemical to the cells in the hypodermis which make the red pigment (Creasy, 1976). Smock (1971) showed that wetting agents and other adjuncts could improve the effect of some chemicals. The antitranspirant di-l-p-menthene (Wilt-Purf) increased red colour development in McIntosh apples when used alone or in combination with diuron or $CaCO_3$, (Creasy, 1976).

Firmness of Flesh

Texture is one of the most important characteristic of quality in apples (Arthey, 1975). As apples ripen, either 'on' or 'off' the tree, the insoluble cell-wall cementing material, protopectin, is hydrolyzed to soluble pectin which results in fruit softening (Smock and Neubert, 1950; Arthey, 1975). The firmness varies at different degrees of ripeness in many of the cultivars (Haller, 1941). The firmness range of McIntosh apples at prime eating condition was reported to be 3.2 to 5.0 kg (Magness and Taylor, 1925). Customers of USA rated 76 per cent of the McIntosh apples with a pressure test value of 5.9 kg (13 Ib) as crisp, neither too hard nor too soft. As the pressure test values decreased linearly from 5.9 to 3.6 kg (13 to 8Ib), the percentage of fruits rated as crisp decreased, but the percentage rated as too soft increased linearly. The proportion of fruits rated as too soft was 66 per cent for those with a pressure test value of 3.6 kg (8 lbs) and 92 per cent for those with a value of 3.2 kg (7 lbs) (Liu and King, 1978). Bangerth *et al.* (1972) showed a significant decrease in softening of Jonathan apples when treated with $CaCl_2$ solution prior to storage. Golden Delicious apples individually sprayed with $CaCl_2$ solutions over a period of 9 weeks immediately prior to harvest were 1.0 kg (2.1Ib) firmer than untreated apples at harvest time (Riley and Kolattukudy, 1976). However, the variability of pressure test readings from tree influence fruit firmness and maturity (Shoemaker and Teskey, 1959). Fruit size has an influence on firmness. Large fruits are usually but not always softer than small fruits (Smock and Boynton, 1944). Flesh firmness is an indicator of maturity and also an important parameter for judging quality throughout the distribution chain (Tromp, 2005)

Seed Colour

The unripe or immature apple fruit has white or whitish brown seeds which on maturity turn brown. However, seeds may become brown even when the fruits are not mature enough and hence, this is an unreliable index of maturity. The combination of firmness decrease, soluble solids increase and conversion of starch has been more closely tied to the perception of maturity (Blankenship and Unrath. 1988). Tripathi *et al.* (1986) stated that the best index of maturity is the number of days from full bloom to harvest. Firmness of the flesh also provides a fairly dependable index of maturity. The optimum time of harvest can not be defined by chlorophyll measurement or chemical test. The regression of ground colour, starch and acidity, the appearance of colouration and sugar synthesis are interdependent but do not develop in parallel. Ground colour and colouration are the most important factors, together with meteorological conditions (Siegrist, 1987).

Physiological Disorders

Bitter Pit

The symptoms appear as small, dry, brown pockets usually spherical in shape below the skin and also in cortex. Cultivars like Northern Spy and Delicious are more prone to this disorder, although York, Rome, Yellow Newton, Cox, Spartan were also found to develop bitter pit. Electron-microscope studies revealed that the middle lamella of cells in the bitter-pit area differentiated normally but the finely laminated suberin layer could not be found. In places, the walls of the neighbouring cells were separated and islands of pectin could be seen. Walls of healthy cells, on the other hand, were homogeneous (Buchloch *et al.,* 1961). There was a marked increase in the silicon content of the cuticle in the tissue where bitter pit lesions were forming (Simons and Chu, 1980).

Early picked (at immature stage) and large fruits are more susceptible to this disorder. Several orchard factors were found to increase the disorder. Unbalanced nitrogen supply or high dose of nitrogen promote the incidence of bitter pit (Gorini and Lalatta, 1964). Van Der Boon (1974) suggested that cultivars which are sensitive to bitter pit should receive not more than 0-40 kg N/ha during the first year and about 80 kg N/ha in the bearing stage. Application of N in the form of NH_4 showed significantly highest N content in fruits and leaves of cv. Fuji than application in the form of NO_3. Nitrogen application in the form of NH_4 markedly increased bitter pit incidence compared with NO_3 form of application. High N content in fruit decreased Ca content in fruits and increased the incidence of bitter pit (MongSup and Kwangchool, 2004). Calcium deficiency is certainly the primary cause of bitter pit (Bangerth. 1974). Van Der Boon (1974) concluded that Ca spray is the most effective measure against the disorder and he sugested that spraying with calcium chloride (0.4 per cent) should be started after blossom and continued until harvest. Dipping of Cox apples immediately after harvest in 2 per cent, calcium chloride solution was found to reduce the bitter pit incidence from 50 per cent in water-treated control to 25 per cent, but the effectiveness of calcium chloride decreased markedly when dipped after 16 days from harvest (Anon., 1977). Spraying of fruits (cv. Jonagold) with calcium chloride both in summer (9, 7, 5 and 3 weeks before fruit harvest) at 3 kg Ca/ha and in autumn (one week before fruit harvest) at 2.4 Kg Ca/ha had higher firmness and titratable acidity after storage and were more resistant to bitter pit and internal breakdown (Wojcik, 2001). In sandy soils or soil with high potassium, further addition of potassium was found to aggravate the incidence. Spraying with zinc (Metzner and Hubner, 1971) or boron (Metzner, 1969) increased the calcium content of the fruit and lowered the incidence of bitter pit. Sinks for clacium was observed in auxin-treated plants which increased the auxin content of seeds. Johnson and Wickenden (1982) noted a slight increase of bitter pit incidence in air-stored fruits by GA/auxin treatment.

Cox apples stored in air at 3.5°C developed 20 per cent bitter pit compared with only 6 per cent stored in 2 per cent O_2 (< 1 per cent CO_2). Although subjecting the CA stored fruit to a further 18°C increased bitter pit to 10 per cent, the maximum incidence remained lower than in air-stored samples (22 per cent). A similar result was obtained with Barmley, in which 11 per cent incidence was recorded when stored at 3.5°C in 8 per cent CO_2 and 13 per cent O_2 compared with 20 per cent in air at the same temperature (Johnson, 1977). Calcium appears to have a curative effect on the incipient pit lesions. Bootsma (1984) stated that spraying with calcium chloride 13 times reduced the incidence of bitter pit in Gloster apple from 2 per cent to 0.2 per cent and the fruits which were harvested in the middle or end of September and stored until 16th March showed no incidence of internal bitter pit. Dipping of Golden Delicious fruits in 1-4 per cent $CaCl_2$ for 30 seconds before storage at room temperature (for 30 days) or in cold storage (up to 6 months) reduced the incidence of this malady (Fortes *et al.*, 1984). The apples cvs. Cox's Orange Pippin, Sturmer Pippin, Golden Delicious, Delicious, Cleopatra, *etc.*, are susceptible to bitter pit. Scott *et al.* (1985) suggested dipping of fruits in calcium chloride solution (1-3 per cent) followed by a water rinse after I or 2 days for reducing bitter pit incidence in several cultivars. Controlled or modified atmosphere storage can reduce bitter pit to commercially acceptable levels. Perforated poly bags in apple cartoons resulted in modified atmosphere (3-5 per cent CO_2 and 15-18 per cent O_2 at 0°C) leading to reduced bitter pit, and improve fruit quality in Cox's Orange Pippin (Hewett and Thompson, 1986). Application of calcite lime (15 tonnes/ha to a soil with a pH of 4.2 in 0.01 M $CaCl_2$ and calcium saturation of ± 50 per cent) was found to increase leaf calcium and to reduce the bitter pit incidence to Golden Delicious and Starking apple (Table 1.45). Bitter pit seems to be related to irregular water supply and moisture stress during warm and dry summer when the leaf-canopy transpiration is very high (Treccani and Stanchina, 1982). In such conditions, not only the calcium uptake is impaired, but calcium is transferred from fruits to the leaves (Tromp, 1979).

Table 1.45: Effect of Liming on the Leaf Calcium and Bitter Pit Incidence*

Cultivar	*No Lime*		*Calcite Lime (15 t/ha)*	
	Leaf Ca (per cent)	*Bitter Pit (per cent)*	*Leaf Ca (per cent)*	*Bitter Pit (per cent)*
Golden Delicious	0.92	30.7	1.02	20.3
Starking	0.85	10.4	0.93	4.6

* Terblanche *et al.* (1980).

Cork Spot

This term is used to designate a type of disorder that develops during the growing season on a number of cultivars of apple. The first symptom is usually the appearance of a small blushed area on the skin of the fruit above the affected brown spot which may be anywhere in the cortex between the skin and core (Faust and Shear, 1968). Cork spot on York Imperial is similar in appearance to bitter pit, but the spots are much in the flesh and appear much earlier in the growing season (Faust *et al.*, 1968). The affected tissue is usually much harder than the healthy tissue. Although the exact cause of cork spot is still not fully understood, boron and calcium deficiency are occasionally found responsible for development of cork spot. Smith *et al.* (1986) found that in Yorking apple on seedling rootstock, less severe corking was highly correlated with higher levels of leaf Ca, B and N and peel Ca.

Injecting calcite lime number of spots could be reduced. Similarly, calcium chloride spray was found to reduce the cork spot (Greene and Smith, 1980). Nevertheless, the ratio of N/Ca in the fruit is more closely related with cork spot than calcium alone (Shear, 1974).

Jonathan Spot

Jonathan spot is very common in cv. Jonathan. The spots may be dark brown, black or moderately light brown, depressed on the fruits. High levels of boron in the fruit may cause earlier maturation and increased incidence of Jonathan spot (Bramlage and Thompson, 1963). Spraying of calcium chloride (0.5 or 0.7 per cent) solution, 2,4 or 6 weeks before harvest decreased the incidence of this disorder (Ben, 1986).

Scald

Scald symptom is the browning of the epidermal and hypodermal cells of the fruit (Kang and Lee, 1987). Scald seems to be associated with an oxidation product coming from α-farnesene formed in the waxy coating of the fruit. More α-farnesene was found in earlier picked apples, which scald more severely, and about twice as much in the scald-sensitive Granny Smith as in the scald-resistent Crofton cultivar (Huelin and Coggiola, 1966). Early harvested Delicious apples showed a high incidence of scald after 7 months in 1 per cent O_2 + 0.05 per cent CO_2 atmosphere storage while those harvested at commercial maturity, developed only slight or minimum storage scald after 7 months in same low-oxygen atmosphere storage (Chen *et al.*, 1985). Calcium spray was found to increase the incidence of scald in cv. Fuji (Fukuda, 1984). Meir and Bramlage (1988) studied the antioxidant activity in Cortland apple peel after storage and found that α-farnesene increased during the first two months of storage and then decreased, while conjugated trines increased progressively with storage time. Conjugated trines were found more positively correlated with scald than was α-farnesene.

Scald is controlled, almost everywhere, with antioxidant products such as ethoxyquin and diphenylamine applied after harvest. However, some countries do not allow the use of those products for toxicological reasons (Treccani and Stanchina, 1982). Nenadovic- Mratinic (1986) found the best control of scald in cv. Grany Smith by application of diphenylamine (2000 ppm) and $CaCl_2$ (2.3 per cent) two weeks before harvest.

Water Core

Water core is characterized by water-soaked regions in the flesh of the fruit. Affected fruit may have a dull skin and the fruit is spongy. Severely affected fruit may smell and taste fermented (Smock, 1971). The cvs. Delicious, Stayman, Jonathan, Winesap, Cox's Orange Pippin, Miller's Seedling, James Grieve, Bramley's Seedling, Worcester, Suntan *etc.*, are susceptible to water core. Excessive levels of boron in fruit can cause increased incidence of water core (Bramlage and Thompson, 1963). The symptoms of water core develop rapidly when the fruits are on the tree and decrease gradually at storage. A low Ca content of fruit was sometimes found associated with water core development (Perring, 1968).

Core Flush or Brown Core

Brown core of apples constitutes a major source of waste in storage in certain cultivars. It is also called core browning or core flush. The main characteristic of this disorder is that the flesh turns brown near the core. Its first appearance is noted as a slight discolouration between the seed cavities. Later, considerable portions of the flesh around the core become rather brown. There is no sharp line of demarcation between healthy and affected portion of flesh (Smock, 1946). This is a low temperature storage problem developing in fruits kept at -1°C to 2°C (Smock, 1971). The risk of core flush incidence increases as the level of fruit K increases (Waller, 1980). Core flush in Belle de Boskoop apple was highest in CA storage with 2.4-3.1 per cent O_2 (Schouten, 1984). Johnson and Ertan (1983) found increased incidence of core flush in Idared apples by lowering the storage temperature and was particularly severe at 0°C (Table 1.46). The average incidence of the disorder in 1 per cent O_2 was less than in fruits kept in 2 per cent O_2 or in air.

Table 1.46: Effect of Storage Condition on Incidence of Core Flush in Idared Apples*

	Core Flush (Per cent)
4°C	0.8
2°C	2.5
0°C	59.2
SED	-
21 per cent O_2	21.7
2 per cent O_2	24.2
1 per cent O_2	16.7
SED	± 1.18

* Johnson and Ertan (1983).

Harvesting

In earlier days, when the apple orchards were mostly with large trees, planted 8 to 12 m apart, the use of shaking machinary and large catching frame for shake-and-catch harvesting was no problem. With the introduction of size-controlling rootstocks, the trees are now planted in narrow rows and even with narrower inter-tree-spaces, made difficult for access of existing shake-and-catch equipment (Dunn and Rowe, 1982).

However, shake-and-catch system generally has not been satisfactory for fresh market apple because of their high susceptiblity to bruising (Kader, 1983). Use of this system for processing has been limited. Bruising which caused by impact such as falling onto a surface or being struck by another object injures both the peel and pulp of apple fruit and influences the consumers preference and acceptance (Hyde and Ingle, 1968). Tennes *et al.* (1969) found 0 per cent to 18 per cent bruished fruits in case of hand picking compared with 11 per cent to 40 per cent in mechanical harvesting.

However, for processing, most of the orchardists use mechanical harvester and different modifications have been made to suit a particular tree form. Peterson and Kornecki (1987) in West Virginia University, USA, developed a rod-press harvester for the Lincoln Canopy T-trellis. The harvest mechanism uses direct contact to push the fruit from the canopy. An average of 96 per cent of the apples is removed with low bruise percentage and other mechanical damage. Two catching frames designed to reduce damage caused by apple-on-apple impact during mechanical harvesting were developed at Denmark (Bennedsen, 1986). The equipment, in laboratory tests, showed 8-19 per cent bruised apples compared with 95 per cent using only a padded conveyor. Robotic harvesting of apples has also been tried in France. The robot consists of a telescopic arm with 3 degrees of freedom, a line-scan camera and a micro-computer. A photoelectric sensor at the end of the arm has reached the fruit (Grand D'Esnon, 1985). A pneumatic picker, located at the end of the hydrostatic manipulator arm, detaches the fruit and transfers it to a bin. In suitable light condition, more than 50 per cent of the apples could be picked by the robot with relatively little damage (75 per cent with stem) at a rate of one fruit every 4 seconds (Grand D'Esnon *et al.*,1987).

However, in the past few years, to make the tree more adaptable to mechanical harvesting, there has been a shift in the emphasis from machine design to tree design (Dunn and Row, 1982). This includes shape and size control methods of trees (genetical and cultural) which can help in reducing impact injuries (Kader, 1983).

Yield

Table 1.47: Per Hectare Yield of Apple during 2014 in Major Apple Growing Countries of the World

Country	*Yield/ha (mt)*	*Country*	*Yield/ha (mt)*
Austria	32.4	Canada	22.2
Belgium	36.2	Mexico	10.7
France	34.8	United States	35.2
Germany	33.4	China	20.0
Greece	17.6	India	6.3
Italy	41.0	Japan	20.9
Netherland	36.3	South Korea	21.0
Portugal	11.7	Turkey	17.4
Spain	16.1	Argentina	35.7
United Kingdom	22.4	Australia	15.3
Russian Fed.	3.4	Brazil	34.1
Hungary	24.2	Chile	51.2
Poland	20.3	New Zealand	62.2
South Africa	42.3		
		Average	**24.6**

Source: World Apple Review 2015 Edition.

Ripening

The increase in respiration in the climacteric phase by synthesis of ethylene which causes unripe apples to ripen is now well established. The rise in ethylene production which, for apples, is coincident with the rise in respiration (Reid *et al.*, 1973, Wills *et al.*, 2007) has become experimentally a more convenient indicator of the onset of the climacteric phase (Knee *et al.*, 1983). Apples attached to the tree often contain high levels of ethylene in the pre-climacteric phase (Reid *et al.*, 1973) and the rise in concentration is more gradual than for detached fruit (Knee, 1973); this leads to inaccuracy in estimating the date when a particular concentration is reached. The sensitivity of apples to ethylene increases during development (Harkett *et al.*, 1971). Knee *et al.* (1983) defined the onset of climacteric as the date when the internal ethylene concentration (lEC) exceeded 0.1μl/l. In general, early season varieties have high respiration and ethylene production rates and subsequently show short storability. Late season varieties with low respiration ripen slowly and can be stored long periods (Wertheim, 2005).

The ripening of fruit is associated with the change in colour, texture, quality and the development on characteristic flavour which are dependent on the cultivar and with a number of environmental and agronomical factors during the growing period. Hulme *et al.* (1968) opined that ethylene synthesis triggers many enzymes which influence ripening. Ethylene production in fruit is regulated naturally in at least three ways: (i) 1-amino cyclo-propane-1-carboxylic acid (ACC) synthesis, (ii) conversion of ACC to ethylene, and (iii) conjugation of ACC to form the malonyl derivative (Knee, 1985). It is suggested that abscisic acid (ABA) may be involved in the regulation of ethylene production in ripening apples (Plich and Nowacki, 1987), and that endogenous ethylene is not regulatory prior to the onset of rapid ethylene synthesis (Knee, 1988). Schaffer *et al.* (2013) investigated the relationship between auxin concentrations and the onset of ethylene related ripening in apple at both hormone and transcription level. Transgenic apple suppressed for the SEPALLATA ½ (SEP ½) class of gene (MADS 8/9) that showed severely reduced ripening were compared with untransformed control apples. They found that mature MADS 8/9 – suppressed apples had a higher concentration of free IAA. This was associated with increased expression of the auxin biosynthetic genes in the indole-3-acetamide pathway. It was concluded that delay in ripening observed in MADS 8/9 – suppressed apples may be partly due to high auxin concentrations. Schaffer *et al.* (2013) propose that to achieve low auxin associated with fruit maturation, the auxin honeostasis is controlled in two pronged manner; (i) by the reduction in biosyntaesis and (ii) by an increase auxin conjugation.

Chemical Changes during Ripening

Chloroplasts is the main constituents of apple peel although it also contains chlorophyll. The chlorophyll content of peel and pulp break drown with the advancement of ripening. Workman (1963a) observed that 75 per cent of total chlorophyll content fell during ripening of Golden Delicious apple with five-fold increase in xanthophyll content. The carotene concentration also decreased at this stage but there was no change in the concentration of quercetin glycosides which had been thought to be associated with the yellow colour of the fruit. The red pigment of apple is idaein (cyanidin-3-galactoside), as reported by Sando (1937). Later on, Bonnie and Francis (1967) recorded cyanidin-7-arabinoside and cyanidin-3-arabinoside as minor anothocyanin components in 74 apple cultivars. Siegelman and Hendricks ((1958) reported that idaein synthesis in isolated apple skin is dependent upon two light-requiring phases, the first of which is an induction phase in which no pigment is formed, followed by a production phase, and the light of wave lengths between 600 and 750 mμ is necessary for the formation of anthocyanins in apple peel.

In apple during the process of ripening, there is an increase in the aroma released by the fruit. The volatile compounds responsible for the characteristic aroma in apple fruit are mainly esters, alcohols, acids, ketones and aldehydes (Kader, 2002, Tromp, 2005, Wills *et al.*, 2007). Drawert *et al.* (1968) have detected as many as 120 compounds in different apple cultivars which are involved in developing aroma at different stages of fruit maturity. These are mainly identified as alcohols, aldehydes, ketones and esters of formic, acetic, propionic, butyric, valeric and caproic acid (Pailliard, 1967a,b).Flath *et al.*(1967) suggested that the main component responsible for the characteristic apple flavour are 2-hexanol and hexanol of aldehydes and ethyl-2-methyl butyrate. However, an important compound in this aroma is the ester with acetate as acyl portion (Goodenough, 1983). In the cultivar Cox's Orange Pippin, the development of aroma near the climacteric of apples relies heavily on n-butylacetate, n-hexylacetate and 2,3-methyl butyl acetate (Williams *et al.*, 1977). Schceier *et al.* (1978) also recorded n-hexylacetate, isopentyl acetate and butyl acetate in Jonathan apple. Bachmann (1983) studied the effects of harvesting date and after-ripening onditions at 18°C and 4°C on the formation of volatile compounds in different cultivars of apple (James Grieve, Cox Orange, Golden Delicious). She observed that the number and amount of volatiles increased with delayed harvest. Butyl acetate and hexyl acetate were the main components. After-ripening conditions at 18°C caused an additional rise in the content of volatile compounds. It was proposed that 500 ppm butyl acetate was sufficient to indicate aroma and thus apples from the optimal and late harvest had more flavour than apples picked earlier. The esters, butyl and hexyl acetate, are the principal aroma compounds synthesized by Cox apples during ripening. The precursors of butanol and hexanol for ester synthesis are thought to derive from the oxidation of long chain fatty acids through a number of cycles of β-oxidation pathway (Bartley, 1986). Both post harvest and preharvest treatments can have significant impacts on aroma formation in fresh-marketed as well as in stored fruit (Bangerth *et al.*, 2012). The earlier the apples are harvested, the poorer their ability to produce volatiles and again this was coupled with lower respiration and depressed fatty acid concentrations compared with late harvested fruit (Jong and Bangerth, 2003). All factors affecting fruit development, maturation, ripening and senescence affect the development and metabolism of flavour. Yahia *et al.* (1990) studied the evolution of 22 odour active volatiles during the maturation and ripening of McIntose and Cortland apples on the tree. Only 12 volatiles were produced on the tree and showed four patterns (Table 1.48). Most of the important odour active volatile were not produced until later stages of ripening and their production followed the autocatalytic evolution of ethylene.

Table 1.48: Patterns of some Odour-active Volatiles during the Maturation and Ripening of McIntosh and Cortland Apples on the Tree*

General Behavior	*Volatiles*
High in immature apples and decreased gradually during ripening	Hexyl acetate, methyl butanoate
Continuous increase during maturation and ripening	(E)-z-hexenal in McIntose
Produced at low level before ripening and suddenly increased after ripening	Hexanal, ethyl-methyl butanoate
Not detected or very low before ripening then suddenly increased after ripening	Ethyl butanoate, propyl butanoate, methyl hexanoate butyl hexanoate, hexyl hexanote.
	Ethyl 2-methylbutanoate, hexyl 2-methylbutanoate.

* Adopted from Yahia *et al.* (1990).

Although the major organic acids in apple fruit are malate and citrate, Hulme (1958) identified presence of quinic, glycolic, succinic, lactic, galacturonic, citramalic and mucic acids in whole fruit or juice of whole fruit; malic, citric, quinic, shikimic, succinic, glyceric, α-oxoglutaric, pyruvic, oxalacetic, glyoxylic and isocitric in pulp, while malic, citric, quinic, shikimic, citramalic, glyceric, α-oxoglutaric, pyruvic and oxalacetic in peel of apple fruit. Flood *et al.* (1970) thought marked increase in malic acid utilization in peel as well as in pulp initiates during the period of climacteric. Hulme *et al.* (1968) showed that there was an increase in malic enzyme and pyruvate decarboxylase during the climactric period.

Detailed studies of lipid metabolism in apples by Bartley (1985a) indicated that the fruits ripening in air at 12°C show an enhanced turnover of lipid and it is thought that the long chain fatty acids liberated are the precursor for the synthesis of esters. The principal lipids of apple cortical tissue are sitosterol, phosphatidylcholine (PC) and phosphatidylethanolamine (PE). When apples were allowed to ripen in air at 12°C, the sitosterol content showed little change whilst phospholipids increased by 8 per cent due mainly to an increased PC. Analysis of lipids following incubation of cortical discs of apple with (1-^{14}C) acetate at different stages of ripening showed that incorporation in sterol changed very little but the incorporation into PC increased 27-fold and into PE 8-fold. A very large increase in the rate of phospholipid synthesis and small increase in phospholipid content indicated that phospholipid degradation increased substantially during ripening (Bartley, 1985b).

Regulation of Ripening

Increasing the level of ethylene in fruits by spraying the trees with CEPA hastened ripening (Edgerton and Blanpied, 1968), but spraying with SADH, which inhibits ethylene biosynthesis (Looney, 1968), caused delay in ripening (Looney, 1971). Plich and Hegazi (1978) determined the ethylene production and activity of acid phosphatase during pre- and postharvest ripening of McIntosh fruits treated with SADH, 3 weeks after blooming and/or CEPA and NAA, 2 weeks before harvest. In fruits treated with CEPA, and CEPA in combination with NAA and/or SADH two peaks of ethylene content were observed; one when the fruits were attached to the tree, a few days after spraying, and another, shortly after spraying as with the untreated fruits, but the content of C_2H_2 was slightly lower in fruits treated additionally with SADH. The activity of acid phosphatase in control fruits remained constant before and after harvest and only increased towards the postharvest period while in fruits treated with CEPA + NAA +.SADH, maturity was hastened and an increase in acid phosphatase activity took place just before harvesting and continued during postharvest ripening. Rudnicki and Pieniazek (1970) reported an increase in ABA levels during ripening. The influence of growth substances in enhancing (CEPA, NAA) or delaying (SADH) fruit ripening and ABA content during maturation of cv. McIntosh was studied by Hegazi and Plich (1978). They found that control fruits attached to the tree contained little ABA during the last two weeks before harvest while those treated with CEPA at 250 ppm alone or combined with NAA at 20 ppm or with NAA at 20 ppm and SADH at 1000 ppm contained much more ABA in comparison with control. The ABA content of fruits treated with SADH at 1000 ppm was also increased.

A mixture of 480 mg/l of ethephon and 50 mg/l of NAA, 14 days before harvest accelerated ripening of early apple cultivar Close (Basak *et al.*, 1978). Treatment with ethrel, however, caused some reduction in fruit size (Child, 1973). Methyl jasmonate stimulated ethylene production and ethylene forming enzyme activity in preclimacteric fruits in cvs. McIntosh and Idared. On the contrary, it inhibited ethylene production in climacteric and postclimacteric McIntosh fruit with no effect on ethylene forming enzyme activity. Methyl jasmonate also stimulated chlorophyll disappearance and inhibited anthocyanin accumulation in the skin of apples (Saniewski *et al.*, 1988). In Jonathan apples, methyl jasmonate was found to increase the ACC concentration in preclimacteric stage while it inhibited ethylene production and decreased the ACC level in postclimacteric apples (Saniewski *et al.*, 1986).

Storage

Considerable improvements have been made in growing and storage techniques which have enabled the growers to keep their fruits for much longer periods. These advances are due to a combination of factors including an awareness of the importance of regular cropping, moderate tree vigour and balanced nutrition, and to the continuous development of controlled atmosphere storage conditions which delay senescence changes and development of storage disorders (Sharples, 1974). The apples are usually harvested over a restricted period for 6 to 8 weeks, and it is, therefore, necessary to provide both refrigerated and controlled atmosphere storage to regulate marketing and provide high quality produce to fresh and processing outlets on a year round basis (Sharples, 1983).

Orchard Factors Influencing Storage

Effect of Rootstock

The rootstocks influence the various forms of storage disorders in apples. Fruits of Cox's Orange Pippin were found to develop more bitter pit when grown on M.4 than on M.9 rootstock (Hilkenbaumer and Reinken, 1959). The susceptibility to break down and core flush of Bramley's Seedlings fruits is more on M.1 than on M.5 (Wallace, 1953) which is thought to be due to lower ability of potassium uptake by M.5 rootstock. Sharples (1972) recorded more mature fruits of Cox on M.26 at picking than those on MM.106 and M.7, but it did not show any influence on storage disorders.

Age of the Plant

Fruits harvested in the early years of cropping are, in general, large and are more prone to different storage disorders while on mature trees, the decrease in vegetative growth associated with heavier cropping was found to reduce the susceptibility to disorder.

Pruning

Severe winter pruning is often found to cause bitter pit in apples (Hilkenbaumer and Reinken, 1959). Severe winter pruning encourages vigorous shoot growth which competes with fruits for uptake of water and often causes various disorders in fruits during storage.

Nutrition

Nitrogen

Additional or higher level of nitrogen dressing causes a decline in red colouration and fruit firmness. High nitrogen dressing was also responsible for bitter pit in Cox's Orange Pippin (Nyhlen, 1960) and Jonathan (Yamazaki *et al.*, 1962). Pratella (1961) reported soft scald of Golden Delicious with high nitrogen fertilization; extra nitrogen, however, was found to reduce the incidence of core flush (Anon.,1965). Application of nitrogen in the form of urea is suggested as ammonium sulphate and ammonium nitrate sometimes cause incidence of bitter pit (Martin *et al.*, 1970).

Potassium

The red colour, proper flavour and good fruit quality are mostly dependent on optimum potash application to the tree but excess application of potassium may cause rotting during storage (Sharples, 1968). Van Der Boon *et al.* (1968) noticed bitter pit development in Cox's Orange Pippin by foliar sprays of potassium nitrate while soil application of potassium was found to decrease the incidence of low temperature injury (Montgomery and Wilkinson, 1962). Higher levels of potassium, sometimes decreased the uptake of calcium and magnesium, which was reflected in the fruit composition (Weeks *et al.*, 1958).

Phosphorus

Adequate application of phosphorus reduced the incidence of breakdown. particularly in soil with low phosphorus content (Martin *et al.*, 1965); however, high levels of phosphorus was found to reduce the volatile production in McIntosh apples (Forsyth and Webster, 1971).

Calcium

Application of calcium in the form of gypsum was found beneficial in reducing bitter pit incidence (Van Der Boon *et al.*, 1968). A continuous application of calcium increased the leaf calcium levels and decreased the critical potassium:calcium ratio. Mineral analysis of Suntan apples suggests that water core disorder is unlikely to occur in fruits with a concentration of calcium in excess of 4 mg/100 g. Bitter pit is more prevalent in large fruits of low calcium status stored in air at 3°C, but the threshold concentration of calcium appears to be about 5 mg/100 g which is similar to that found for Cox fruit (Perring and Johnson, 1978).

The multiple effect of calcium is expressed only if the calcium content in the tissue is sufficient *i.e.*, >5 mg/100 g. Studies made by Dierend and Ricken (2007) showed that the calcium content in apples varied between 3 and 15 mg/100g. Shapatava *et al.* (2014) observed that in Reinette du Canada with 3.8 mg/100 g calcium content, 14 per cent fruits were affected by this physiological disaorders, while in fruits with 5.6 mg/100 g calcium content, the percentage of loss did not exceed 3 per cent. They also found an inverse correleation between apple fruit weight and calcium content of Reinttte du Canada.

Table 1.49: Calcium Content in Fruit of Reinette du Canada by Fruit Mass*

Fruit Diameter (mm)	*Weight (g)*	*Calcium Content (mg/100g)*
>75	≥18	3.8
70-75	170-180	4.2
65-70	150-170	5.3
60-65	≤150	5.6

* Adopted from Shapatava *et al.* (2014).

Application of calcium in the form of nitrate and chloride was found more effective in controlling corking disorder than in the form of lactate, acetate, phosphate or hydroxide.

Magnesium

It is believed that application of magnesium in soil or irrigation water increases its susceptibility to bitter pit. Soil or foliar application of magnesium was found to depress the leaf calcium level, and high level of magnesium in the fruit relative to calcium had a tendency to promote bitter pit (Garman and Mathis, 1956).

Boron

Excessive soil dressing with boron reduces the storage life (Holbeche, 1946) by reducing fruit firmness or enhancing the maturity; boron deficiency on the other hand may cause internal cork in the fruit of some cultivars. An increase in storage decay of Jonathan apples was noted by 6 consecutive sprays of boric acid (0.1 per cent) in the season (Bramlage and Thompson, 1963).

Fungicides

Spraying against apple scab with dimethyldithiocarbamate compounds like thiram, ziram was found to increase the incidence of bitter pit in some cultivars (Kirby *et al.,* 1968). The incidence of scald in Golden Reinette apples was higher with organic mercury compounds than with sulphur (Keijer and Dijksterhuis, 1956).

Growth Substances

Auxins

Variable results were reported from different apple growing regions of the world with regard to different preharvest auxin treatments and storage quality of apples. Delicious apple in Australia showed increased incidence of scald with NAA spray (Stevenson and Dodd, 1962). It also increased the core flush and scald of Granny Smith apples (Padfield. 1949), but had no adverse effect on Cox's Orange Pippin (Vyvyan, 1949). 2,4,5-T and 2,4.5-TP, in general, decreased fruit firmness at harvest and induced higher rate of respiration.

Gibberllins

It was observed that calcium concentration in fruits of Bramley's apples from GA-sprayed trees was higher than in the untreated ones and thus the fruits were less susceptible to bitter pit (Johnson and Wickenden, 1982), but the relation between applied gibberellin and calcium contents in apple is not well established. Application of GA, however, reduced the respiration rate and chlorophyll breakdown and thus rotting (Clijsters, 1971).

Kinetin

Very little information is available about effect of kinetin on the storage life of apple. A light reduction in respiration was noted by Smock *et al.* (1962) following the application of BA one week before harvest.

Aminozide

Aminozide (succinic-2, 2-dimethylhydrazide) greatly reduced the rate of respiration (Looney, 1968) and ethylene production (Sharples, 1972). Early application of aminozide markedly reduced the fruit size and fruit nitrogen content but increased the calcium content of fruit (Ashby and Looney, 1968) and thus they were less susceptible to bitter pit. However, there are enough evidence that aminozide treatment may increase the core flush disorder in storage. Johnson and Wickenden (1982) reported that application of Alar delayed the onset of the climacteric in Cox's Orange Pippin and enhanced red colour development; but they did not find any consistent effect on fruit firmness or on the mineral composition of fruits.

Ethephon

Unlike aminozide, ethephon accelerates respiration, increases the rate of chlorophyll breakdown, enhances red colour development and promotes the onset of climacteric when applied at the preclimacteric stage (Clijsters, 1971).

Postharvest Chemical Treatments

Researches have elucidated the mechanism of ethylene biosynthesis and suggested measures for regulating it. Preharvest treatment with ethylene biosynthesis inhibitor, aminoethoxyvinylglycine (AVG) inhibits production of autocatalytic ethylene and the fruits showed a lower respiration rate, better firmness and higher acidity throughout the storage (Halder-Doll and Bangerth, 1987). However, AVG-treated fruits, in general, showed a decreased aroma production as storage duration increased (Bangerth and Streif, 1987).

The addition of calcium chloride to the postharvest dip or drench increases fruit Ca level (Mason, 1979) and helps to control bitter pit in Cox and breakdown in Spartan, McIntosh and Jonathan apples (Sharples, 1984a). Gala apples dipped in 1,2,3, or 4 per cent calcium chloride for 30 seconds and kept at ambient temperature for 30 days or in cold storage for 3 months showed a reduction in bitter pit incidence (Fortes *et al.*, 1984). Drenching with 3 per cent calcium chloride reduces bitter pit incidence of Newton apples, while causing little injury if any (Mielke and Facteau, 1986). Dipping the fruits of Red Delicious at ambient temperature in chilled (2.5°C) calcium chloride solution at 2 per cent + Uphar (a surfactant) at 0.03 per cent for 10 minutes resulted in high Ca content (72.7 mg/100 g d.w.), reduced respiration rate (13 mg CO_2/kg/h), improved fruit quality and extended storage life at 0°C ± 1 (Chopra and Naqash, 1986). In New Zealand, vacuum infiltration of calcium chloride is used commercially to control bitter pit (Scott and Wills, 1979). After 3 months in 0°C normal atmosphere storage, Golden Delicious apples pressure- infiltrated with calcium chloride were 7-15 per cent firmer and had 20-35 per cent less decay than untreated fruits. When stored in 3 per cent O_2/2 per cent CO_2 or in 1 per cent O_2/0 per cent CO_2 controlled atmosphere, infiltrated fruits were 26 per cent and 43 per cent firmer and had 51 per cent and 58 per cent less decay, respectively, than the control ones (Sams and Conway, 1987).

Infections of fruit by fungi, including *Gloeosporium, Botrytis* and brown rot lead to high loss during storage. Their control may be achieved by the use of any of the currently available group on benzimidazole and related fungicides. These fungicides can normally be used in combination with ethoxyquin or diphenylamine for control of superficial scald (Johnson, 1979).

Storage Methods

Cold Storage

The most simple and effective means of controlling ethylene production and reducing the metabolic activity after harvest is by storage at low temperature (Watkins, 2002). The enzymes, ACC-S and ACC-O involved in ethylene biosynthetic patway are sensitive to low temperature (Larrigaudiere *et al.*, 2004). Low temperature also reduced vitamin-C loss, fungal infection and decay during storage.

In cold storage, apples are usually stored at -1°C to 0°C at 85-90 per cent relative humidity. To reduce the temperature of fruits, they are precooled in a separate chamber, which slows down the life process, thereby prolonging the storage life. Different liners and antitranspirants are also used to reduce weight loss in storage.

Royal Delicious apples stored at 1°C and 85-90 per cent relative humidity, remained firm for 135 days (Maini *et al.*, 1985). Moisture removal is considered essential in cold storage, as respiration of apples in cold storage may raise the relative humidity in storage chamber to 95-98 per cent; the wooden boxes containing the fruits may absorb up to 1.8 per cent of the moisture of the fruits. In cultivars like Belle de Boskoop and Cox's Orange Pippin, which are susceptible to flesh browning, a moisture loss of 0.3-0.4 per cent per month from the chamber is considered normal whereas cultvars such as Lombarts Calvillie can be stored well with a moisture loss of <0.1 per cent per month. If moisture extraction from the chamber is inadequate, the incidence of flesh browning is increased, while excessive moisture extraction results in fruit shrivelling (Bliek, 1987). Coseteng and Lee (1987) reported that activity of polyphenol oxidase and polyphenol concentration remained relatively constant during cold storage (0°C, 90 per cent RH) which were related to the flesh browning of apples. Cantillano (1984) in a 3-year trial on cold storage (0°C and 90-95 per cent RH) with Golden Delicious apples, found the best result with polyethylene film liners which reduced weight loss to <1 per cent after 8 months. The thickness or pattern of liner did not show any difference in the effectiveness to reduce the loss in weight. The cvs. Starking, Royal Red Delicious, Jonathan, Renet Simirenko and Renet Shampanskii were less affected by rots in cold storage at 2-4°C with 85-90 per cent relative humidity. Storage quality was found to improve by treatment with calcium chloride (4-8 per cent) solutions for 1 min. (Fataliev, 1985). The glucose and fructose contents in cvs. Spartan, Idared, Melrose, Kidd's Orange, Golden Delicious and Gloster were found to increase during storage at 5°C and 80 per cent RH for 5 months, whereas sucrose decreased or remained largely unchanged (Subrtova, 1987).

Controlled Atmosphere (CA) Storage

Controlled atmosphere (CA) storage has long been established in most of the world's apple growing areas as an effective method to prolong the marketing period of apples. Controlled atmosphere storage is considered as the predominant method to preserve fresh fruits. Modified atmosphere has been used to supplement temperature control and extend the postharvest shelf life of fruits by reducing respiration and delaying senescence. Since, O_2 is a critical substrate in the respiratory process, respiration can be reduced by restricted availability of O_2. Elevated concentration of CO_2 lead to a lesser extent to reduced respiration rates (Mir and Beaudry, 2002) Apples are normally stored at CO_2 concentrations up to 3 per cent, however CO_2 sensitive varieties such as Braeburn, Granny Smith and Fuji are stored at CO_2 concentrations below 1 per cent. The CA rooms are usually filled in 7-10 days and the atmosphere (2-3 per cent oxygen, with 2-3 per cent carbon dioxide) is established by the respiration of the apples over a period of 2-3 weeks. The temperature is maintained at 0°-3.5°C, depending upon the cultivar and the region where the apple is grown (Blanpied, 1982).

Earlier, the commercial recommendation for CA storage of apples in North America suggest 2 per cent to 3 per cent O_2 (Blanpied, 1977) despite demonstrations that Turley (Workman, 1963b) and Delicious (Anderson, 1967) apples were firmer when stored at 1 per cent to 1.5 per cent O_2 than when stored at recommended O_2 concentration. However, modern approach is mostly on use of ultra-low O_2 (1-.5

per cent) (Blanpicd, 1982). Most storage disorders, especially browning, scald and core flush are markedly low in ultra-low O_2 (1 per cent). In ultra-low O_2 storage, cv.Boskoop showed longer shelf life than at 3.1 per cent O_2 (Schouten, 1988), the fruits of cvs. Granny Smith, Top Red and Golden Delicious at 1.5 per cent O_2 had the highest acidity, best taste quality, least colour change and highest firmness (Chapon, 1988).Spartan apples kept in 1.0 per cent O_2 + 2.0 per cent CO_2 at 0°C for 6-9 months (Lau, 1983) and Golden Delicious and Delicious apples stored in 1.0-1.5 per cent O_2 +1.0-1.9 per cent CO_2 at 0° to 0.5°C and 90-92 per cent relative humidity for 6-10 months (Lau, 1988) maintained a higher firmness and had more acid than fruits stored in standard commercial atmosphere at 2.5 per cent O_2 + 1.5-2 per cent CO_2, Storing McIntosh apples initially in 1.0 or l.5 per cent CO_2 and l.0 per cent O_2 at 3°C was more effective in retarding firmness loss than initial storage in 4.5 per cent or 5.0 per cent CO_2 and 3.0 per cent O_2 at 3°C (Lidster *et al.*, 1987). Work has shown the feasibility of using low levels of O_2 for the storage of Cox's Orange Pippin (Bartley, 1986), McIntosh (Lidster *et al.*, 1980, 1981) and Spartan (Lougheed *et al.*, 1980; Lau, 1983). However, a safety margin was deemed necessary to avoid anaerobiosis and injury to the fruit (Lau, 1983). Ethyl alcohol and ethyl acetate may accumulate in apple flesh due to anaerobic respiration (Sharples and Stow, 1982). Best response with regard to ultra-low O_2 storage was recorded by harvesting at or near the respiratory climacteric (Lidster *et al.*, 1981); delaying the harvest may increase internal browning (Lidster, 1982) and internal corking disorders (Sharples, 1982). Fruits of Delicious harvested at commercial maturity, were found to develop only slight or minimum storage scald symptoms after 7 months in 1 per cent and 0.5 per cent O_2 + 0.05 per cent CO_2. However, a high incidence of low O_2 injury (*i.e.*, ribbon-like, depressed skin browning) occurred in fruit stored for 9 months in 0.5 per cent O_2 (with or without CO_2) and in 1.0 per cent -1.5 per cent O_2 with <0.03 per cent CO_2 (Chen *et al.*, 1985). Delicious fruits stored for 14 weeks in 0 to 1.5 per cent O_2 were significantly firmer, lower in soluble solids, and had 10 times more ethanol than fruits stored in 0.5 per cent to l.5 per cent O_2 (Nichols and Patterson, 1987). Weber *et al.* (2011) stated that for Royal Gala apple, the best Ultra Low Oxygen (ULO) condition was 1.0 kPa O_2 + 2.0 kPa CO_2 at 1°C.

At the beginning of CA storage the fruits are first exposed to high CO_2 (15-17 per cent) which reduces loss of firmness and green colour, and incidence of storage scald and flesh disorders (Blanpied, 1982). The period of CO_2 treatment varies between 4 to 10 days (Peggie *et al.*, 1979; Martus. 1981; Blanpied, 1982) after which the room is flushed with air or the CO_2 is reduced to below 2 per cent by scrubbing the atmosphere. A high CO_2 concentration (20 per cent) at 20°C rapidly and strongly inhibited the development climacteric ethylene burst in apple (cv. Granny Smith) but had no significant effect on the l-aminocyclopropane-l-carboxylic acid (ACC) content. Treatment with 20 per cent CO_2 markedly decreased ACC-dependent ethylene biosynthesis at 20°C in climacteric pericarp tissue and it was, therefore, suggested that high CO_2 level inhibited conversion of ACC to ethylene (Cheverry *et al.*, 1988).

Scrubbing of undesirable levels of ethylene is considered necessary to control storage scald (Liu, 1977) and retardation of flesh softening (Knee *et al.*, 1974; Liu,

1977. 1978). To retard flesh softening in a 2-3 per cent O_2 atmosphere, the ethylene concentration inside the apples must be kept below 1 ppm (Blanpied, 1982). A large number of reagents and techniques have been tested over the years (Blanpied *et al.*, 1982) for ethylene scrubbing from CA storage, but only potassium permanganate is common in commercial use (Sherman, 1985). To be effective. $KMnO_4$ must be adsorbed on a suitable carrier with a large surface area. Celite, vermiculite, silica gel, alumina pellets (Ables. 1973), perlite (Saltveit, 1980) and expanded glass (Lidster and Lawrence, 1983) have all been successfully used as carriers. Catalytic converter system was used to remove ethylene from CA store at East Malling, (Dover. 1983). However, in some cultivars (*e.g.*, Boskoop) investment in ethylene-scrubbing apparatus was found not justified (Schaik and Boerrigter, 1987).

Reduction in the production of volatiles largely resembled the general inhibition of ripening under various CA conditions. With increasing CO_2 and decreasing O_2 concentrations, aroma volatile production was reduced, especially if both gas treatments were applied simultaneously. Decreasing the O_2 concentration to 3 per cent had little effect on volatile production, but a further decrease to 1 per cent significantly reduced it. At O_2 concentration above 3 per cent, higher CO_2 concentration became decisive in diminishing volatile production of Golden Delicious. After an extended storage, a clear residual CA effect on volatile production was observed and after 7 or 9 months' storage a complete recovery from this CA effect was not possible. A 3-week treatment at 1°C and ambient atmosphere, after the CA storage and before a shelf-life period, greatly improved production of volatiles. After 9 month storage, however, even this treatment failed to allow a complete recovery (Streif and Bangerth, 1988). Off-flavours in apple may develop by storing apples at very high levels of CO_2 and/or, especially with very low levels of O_2. Apple cvs, Granny Smith and Yellow Newton when stored in an atmosphere containing 0.02-0.25 per cent O_2 at 0, 4 or 10°C for 3 – 35 days developed alcoholic off-flavour (Ke *et al.*, 1991). Delicious apples stored in 0.05 per cent O_2 and 0.2 per cent CO_2 at 1°C for 30 days produced large concentration of ethanol and acetaldelyde (Mathes *et al.*, 1991). It is probable that the precursors of the aromatic esters responsible for fruit aroma are derived from the lipid components of the cell membrane and work has shown that some of the typical Cox flavour can be recovered, without necessarily losing the benefits of improved texture, by raising the oxygen concentration to 2 per cent during the last 4 to 6 weeks of the storage period (Sharples, 1983). Bartley (1986) advocated two approaches which might be used to enhance aroma production, namely, (i) to supply the fruits with precursors of the aroma compounds or (ii) to adjust the cellular metabolism of the fruits so as to generate precursor for synthesis of aroma compounds. Sharples and Stow (1986) recommended CA storage conditions for several apple cultivars which are presented in Tables 1.50 and 1.51.

1 – MCP

1-Methylcylopropene (1-MCP) is thought to act as a competitive substance to ethylene, occupying the ethylene receptor site so that ethylene can not bind to trigger its action (Blankenship and Dole, 2003).1- MCP, a synthetic unsaturated cyclic olefin, which is structurally related to ethylene (C_4H_6 vs C_2H_4) is considered

Table 1.50: Recommended Storage Conditions for the Main Cultivars of Apple**
(Unless otherwise stated the terminal date for each set of conditions allows for a further 2 weeks at ambient temperatures to cover grading and distribution)

Cultivar	*Air Storage*			*Controlled Atmosphere (CA)*							
	Temperature		*Terminate (Month)*	*Temperature*		*No Scrubber*		*Using a Scrubber*			*Note*
						Per cent CO_2	*Terminate (Month)*	*Per cent CO_2*	*Per cent O_2*	*Terminate (Month)*	
	°C	*°F*		*°C*	*°F*						
Bramley's Seeding				4.0-4.5	39-40	8-10	June	6	2	June	1 (a.b.c). 2b
Cox's Orange Pippin	3.0-4.0	37-39	January	3.5-4.0	38-39	5	Early Jan	5	3	Mid Feb	2a
	3.0-3.5	37-38	Mid Dee	3.5-4.0	38-39			<1	2	Late March	2b
				3.5-4.0·	38-39*			<1	1.25	Late April	3 (a.b.c.d.e)
				3.5-4.0·	38-39*			<1	1.0	Early May	3 (a.c.d.e). 4
											(a.b)
				3.5-4.0	38-39	8	May	-	-	-	la. 5
				3.0-3.5	37-38	8	Mid Sept.	<1	2	Mid Sept	2b. 6 (a.b)
Crispin	1.5-2.0	35-36	January	3.0-3.5	37-38	6-8	January	5	3	Early March	7
Discovery	3.0-3.5	37-38	End Aug	3.5-4.5	38-40	8	April	5	3	May	8a
Egremont Russet	3.0-3.5	37-38	December	3.5-4.5	38-40	-	-	<1	1.25	May	8 (a.b)
Idared	3.5-4.5	38-40	March	3.5-4.5	38-40	6-8	March	7-8	3	Late March	9
Laxton's Superb	1.5-3.5	35-38	February	1.5-2.0	35-36	-	-	<1	2	March	2b. 10

Cultivar	Air Storage			Controlled Atmosphere (CA)							Note
	Temperature		Terminate (Month)	Temperature		No Scrubber		Using a Scrubber			
						Per cent CO_2	Terminate (Month)	Per cent CO_2	Per cent O_2	Terminate (Month)	
	^0C	^0F		^0C	^0F						
Spartan	0-0.5	32-23	January	1.5-2.0	35-36	-	-	6	2	June	2b. 10
Worcester Pearmain	0-1.0	32-34	December	0.5-1.0	33-34	7.8	February	5	3	Mid March	7

1. (a) Use postharvest chemical treatment to control superficial scald
   (b) Slight risk of low temperature injury in fruit stored beyond May, especially if picked late or fruit phosphorus is below 9 mg/100 g
   (c) Use scrubbed storage for Bramley's where there is a suspected risk of bitter pit
2. (a) Core flush risk after cool season or after prolonged storage in 5 per cent CO_2 regimes. Use 2 per cent O_2 or 1.25 per cent O_2 (1 per cent CO_2) for storage beyond mid-February
   (b) Ensure O_2 level does not fall below 1.7 per cent when controlling manually at 2 per cent O_2, Aim to keep a mean of 2 per cent
3. (a) Carry out regular independent checks of O_2 levels using portable analyzer and drawing atmosphere sample direct from store
   (b) For operating at 1.25 per cent O_2 or below, use automatic equipment to control O_2 within range 1.1-1.4 per cent
   (c) Operate store at 2 per cent O_2 for at least 1 week before lowering to 1.25 per cent O_2 or below
   (d) Restrict 1.25 per cent O_2 or 1.0 per cent O_2 to Cox of good storage potential only, *viz.*, Ca 5 mg/100 g (min) K 150mg/100g (max). Avoid late picking.
   (e) After cool summers, maintain *minimum* temperature of 4°C (39°F) to reduce risk of low temperature injury
4. (a) Use 1 per cent O_2 for Cox stored beyond March in years when texture is suspect
   (b) Maintain O_2 between 0.9 per cent and 1.1 per cent with automatic equipment: make regular weekly checks for traces of alcohol for first 4 weeks and thereafter at monthly interval.
5. Risk of core flush after April in 8 per cent CO_2.
6. (a) Cool quickly, preferably using forced air pre-cooler. Establish high CO_2 by adding dry ice to store and low oxygen by nitrogen injection.
   (b) Maximum poststorage life of 4 to 5 days only.
7. Risk of shrivel: reduce by loosely covering the top layer of fruit in each bin with polythene liners.
8. (a) Risk of core flush during distribution through market.
   (b) Risk of Jonathan Spot in 1.25 per cent O_2 storage after hot summers.
9. Risk of core flush in long-term CA storage after cool growing seasons.
10. Risk of breakdown if Ca <4 mg 100 g: apply calcium sprays and/or treat postharvest with calcium chloride.

* Sharples and Stow (1986).

Table 1.51: Provisional Storage Recommendations for Less Widely Grown Cultivars of Apple*

Cultivar	Temperature		Terminate (month)	Temperature		Per cent CO_2	Terminate	Per cent CO_2	Per cent O_2	Terminate (month)	Note
	°C	°F		°C	°F						
Charles Ross	3.0-3.5	37-38	November	3.5-4.0	38-39	5	December	5	3	December	
Edward VII	3.0-4.0	37-38	January	3.5-4.5	38-40	8	May				1
Ellison's Orange	4.0-4.5	39-40	December	4.0-4.5	39-40	-		5	3	January	2
								<1	2	January	
Elstar	0-1.5	32-35	January	1.0-1.5	34-35	-		<1	2	March	
Fiesta	1.0-1.5	34-35	January	1.0-1.5	34-35	-		<1	2	June	
Gala	0-1.5	32-35	January	3.5-4.0	38-39	-		<1	2	May	3
Gloster 69	0-0.5	32-33	January	1.5-2.0	35-36	-		<1	2	May	
Golden Delicious	1.5-2.0	35-36	January	1.5-3.5	35-38	8	April	5	3	April	14
Grenadier	0-1.0	32-34	MidNov								
Howgate Wonder	3.0-4.0	37-39	January	3.5-4.0	38-39	8-10	April				
James Grieve	3.0-3.5	37-38	October	3.5-4.0	38-39	7-8	November	6	5	November	
								<1	3	November	
Jonagold	0-0.5	32-33	January	1.5-2.0	35-36	8	April	<1	2	April	
								<	1.25	May	
Jonathan	3.0-3.5	37-38	January	4.0-4.5	39-40	-		6	3	January	
Jupiter	3.0-3.5	37-38	November	3.5-4.0	38-39	-		<1	2	May	5
Katy (Katja)	3.0-3.5	37-38	End Sept	3.5-4.0	38-39	-		<1	2	Mid Oct	6
Kent	3.5-4.0	38-39	March	3.5-4.0	38-39	-		5	3	May	
								<1	2	May	
Laxton's Fortune	3.0-3.5	37-38	November	3.5-4.0	38-39	7-8	December	5	3	Mid Jan	
								<1	2	MidJan	

Cultivar	Temperature		Terminate (month)	Temperature		Per cent CO_2	Terminate	Per cent CO_2	Per cent O_2	Terminate (month)	Note
	°C	°F		°C	°F						
Lord Derby	3.0-4.0	37-38	December	3.5-4.0	38-39	8-10	March				
Lord Lambourne	1.5-2.0	35-36	January	3.5-4.0	38-39	8	May				
McIntosh	3.5-4.0	38-39	Mid Dec	3.5-4.0	38-39			<1	2	Mid March	7
McIrose	0-0.5	32-33	January								
Red Delicious	0-1.0	32-34	January	0-1.0	32-34	6-8	March	5	3	March	
								<1	3	March	
Sunset	1.5-2.0	35-36	January	3.0-3.5	37-38	8	May				
Suntan	3.5-4.0	38-39	January	3.5-4.0	38-39	-		5	3	January	8
								<1	2	January	8
Tydeman's Late Orage	3.0-3.5	37-38	February	3.5-4.0	38-39	5	March	5	3	April	9
								<1	2	April	

1. Use appropriate postharvest chemical treatment to control superficial scald; 2. Susceptible to soft scald (a form of low temperature injury); 3. Scald risk during marketing following storage in 2 per cent O_2; 4. Risk of core flush and breakdown in some years at store; 5. Risk of scald in air storage and during marketing following prolonged storage in 2 per cent O_2: use ethoxyquin to reduce risk; 6. Poststorage life limited to 4-5 days; 7. Pick Mid-September; 8. Risk of low temperature breakdown after January; 9. Risk of brown heart in 5 per cent $C0_2$ + 3 per cent O_2.

* Sharples and Stow (1986).

a safe product. It is sold as a vapour release formulations under the trade name Smart Fresh™ (Agrofresh, inc.).

The efficacy of 1-MCP is dependent on various factors such as cultivar, stage of fruit maturity at the time of treatment, storage atmosphere and duration as well as time between harvest and 1-MCP treatment (Watkins *et al.*, 2000; Blankenship and Dole, 2003, Tatsuki *et al.*, 2007). Ethylene-dependent ripening processes such as softening, yellowing of the background colour and loss of titratable acidity were significantly more reduced in 1-MCP (625 nl/l 1-MCP for 24 hours at 1°C) treated Jonagold apples stored in ultra-low oxygen (ULO, 3 per cent CO_2 and 1 per cent O_2) at 1°C temperature and 92 per cent relative humidity. (Wunsche and Heyn, 2015). 1-MCP acts via a preferential interaction with ethylene receptors, thereby inhibiting the effects of both endogenous and exogenous ethylene and thus slowing down the ripening process and the production of ethylene in fruit (Blankenship and Dole, 2003; Watkins, 2006). 1-MCP not only maintains fruit quality during storage, but also during the shelf life and the marketing period of the fruit (Hohn *et al.*, 2007; Wunsche and Heyn, 2015). Rebeaud and Gasser (2015) reported that quality of Golden Delicious apple treated 6 days after harvest with 1-MCP (625 ppb at 0.5°C for 24 hours) was better maintained in terms of firmness, acidity and TSS during 29 weeks of storage under regular atmosphere (RA) at 0.5-1.0°C and 92 per cent RH conditions than that of fruit treated 12 and 20 days after harvest (Table 1.52).

Table 1.52: Firmness, Total Soluble Solids (TSS) and Acidity of Golden Delicious Apples Harvested in 2007, Treated (Treatment performed 6, 12 and 20 days after harvest) and Non-treated (Control) with 1-MCP and Stored under RA Conditions for 29 weeks. Means with the same letter are not significantly different at $p<0.05$ in Duncan's multiple range test*.

	Firmness (kg cm²)	*TSS (°Brix)*	*Acidity (g malic acid l⁻¹)*
		Optimum harvest	
MCP 6 days	5.9 A	14.1 A	3.7 A
MCP 12 days	5.2 B	13.3 B	3.2 B
MCP 20 days	4.8 C	13.2 B	3.0 B
Control	4.6 C	12.9 B	2.2 C
		Late harvest	
MCP 6 days	5.7 A	14.4 A	3.7 A
MCP 12 days	5.1 B	13.4 BC	3.0 B
MCP 20 days	5.0 B	13.7 B	2.9 B
Control	4.6 C	13.0 C	1.7 C

* Rebeaud and Gasser (2015).

Low Pressure or Hypobaric Storage

In low pressure or hypobaric storage system, the commodities are stored at below atmospheric pressure, usually 0.05-0.10 atmosphere (Lougheed *et al.*, 1978). The process of ripening and senescence are greatly retarded by decreasing respiration and evacuation of ethylene given out by the produce (Dilley, 1978). Commercial applications of this system have not been widespread (Sherman, 1985) due to the high cost for construction and refrigeration of hypobaric chambers and frequent lack of varietal flavour in apples stored under hypobaric conditions (Blanpied, 1982).

Modified atmosphere has been used to supplement temperature control and extend the postharvest self life of fruits by reducing respiration and delaying senescence. A recent approach has been the use of a coating that is edible and semi-permeable to CO_2 and O_2. Sucrose polyester coating (SPE) was found to suppress CO_2 and ethylene production in several apple cultivars (Elson *et al.*, 1982). Chu (1985) determined that a poststorage application of SPE reduced the softening of low O_2 stored McIntosh, and CA stored Delicious apples during a 21-day shelf life period but did not influence the firmness of CA stored McIntosh apples. Smith and Stow (1984) reported that SPE applied after storage reduced yellow colour, loss of firmness and increased internal CO_2 level of Cox's Orange Pippin, during 21 days of a simulated marketing period. Drake *et al.* (1987), working with Golden Delicious apples, found that fruits treated with Semperfresh (a sucrose-fatty acid polyster) showed delayed colour development and higher acidity than control fruits. An internal accumulation of CO_2 and a reduction of ethylene was observed (Table 1.53).

Table 1.53: Quality Attributes of Golden Delicious Apples Influenced by Coating with SPE, Wax or SPE+ Wax*

Treatment	*Skin Colour (Agtron) (E-5W)*	*Firmness (Newtons)*	*Titratable Acidity (Per cent malic)*
Control*	51.6a[Y]	56.2ab	0.32bc
SPE	50.9b	55.2b	0.34a
Wax	51.7a	55.5b	0.31c
SPE+Wax	51.7a	57.6a	0.33ab

Z: Apples were removed from CA storage, coated and stored for 60 days at O°C; y: Means in a column followed by different letters are significantly different (P <0.05)

* Drake *et at.* (1987).

Pests

Codling Moth (*Cydia pomonella*)

The adult moths lay eggs on developing fruitlets and on leaves. The invading larvae start feeding from fruit skin and burrow down to the core. The larvae, after feeding, escape, leaving a small, red-ringed entry hole. The damaged fruit drops off prematurely.

Two to three sprays are necessary because of the extended period of egg laying and hatching. The parasites *Trichogramma embryophagum* and *T. cacoeciae pallidum* showed good parasitism (Pawar *et al.*, 1984). The bacterial preparation dipel (containing *Bacillus thuringiensis*) together with the pesticide adjuvant Elo (containing light oils) at 0.25 per cent when applied to the eggs during mass oviposition, caused the Iarva inside to become infected and it died either before or after hatching. Young larvae became infected when they crawled on the treated surface of apple leaves in search of food. Some larvae became infected when boring into the fruits also died on the surface or after penetrating the fruit (Videnova and Ismail, 1985). Effective sprays are carbaryl, chlorpyrifos, deltamethrin, fenitrothion, malathion, phosalone, *etc.* The growth regulator, alsystin (triflumuron) has an ovicidal, larvicidal and sterilant effect on *Cydia pomonella.* The best time to apply the compound was about 3-4 days after the capture of males in pheromone traps, *i.e.*: against day-old females and newly laid eggs (Velcheva, 1986). NeemAzal-T/S (azadirachtin) at 3.5 l/ha along with Madex (granulosis virus) at 0.25/1 showed effectiveness in control of codling moth in organic apple orchards (Badowska-Czubik *et al.*, 2011).

Apple Clearwing Moth (*Synanthedon mvopaejormis*)

The larvae make tunnels within or just beneath the bark. The bark peels off easily, encouraging entry of pathogens. The pest, however, usually attacks old and sickly trees.

Winter spraying (when the larvae start feeding) as well as summer sypraying (when the adults appear and the first eggs of the following generation hatch) has been suggested to control the pest. Chlorpyrifos at 0.15 per cent is recommended for application 3 times at interval of 20 days, beginning in the week following the appearance of the first larvae (Iren *et al.*, 1984). Golden Delicious on M. 9 is considered particularly susceptible to this pest, because a partial incompatibility between the stock and scion results in weakened trees with swellings, cracks and cavities in the bark at the collar region, where the infestation begins. Traps baited with a synthetic sex pheromone composed of 50 per cent (3Z, 13Z)-3, 13-octadecadienyl acetate placed at the rate of 10 traps/ha, were used for the mass traping of males and provided a useful supplement to the autumn treatment against the larvae (Lozzia and Daolio, 1984). Monnet (1987) suggested autumn and spring applications of dichlorvos as sprays at 200 g/hl to the base of the tree and in summer at least 2 applications of deltarnethrin at 1.75 or endosulfan at 60 g/hl to control the clearwing moth.

Tortrix Moths

Caterpillars of tortrix species which regularly feed on the fruits include the fruitlet mining tortrix (*Pammene rhediella*), the summer fruit tortrix (*Adoxophyes orana*) and the fruit tree tortrix (*Archips podana*). Among these, the fruit tree tortrix which attacks fruit at the later stages of development is most common. The caterpillars usually spin web around several leaves or in between two fruits. They feed on areas of the fruit skin and sometimes in the calyx cavity. *Teleutea striata, Meteorus ictericus* and members of the Tachinidae are the predators for summer fruit tortrix (Galli, 1984). Organophosphorus insecticides (fenitrothion, chlorpyrifos, phosalone, endosulfan) were found effective. Galli (1984) reported that treatment with Insegar (fenoxycarb) markedly reduced the damage of fruits of Golden Delicious and Boskoop by summer fruit tortrix.

Wooly Apple Aphid (*Eriosoma lanigerum*)

This is a small, brown and greyish purple sucking aphid which attacks bark and roots forming a white waxy material, resembling little tufts of wool. They feed on new wood and roots to form galls. Cold and wet condition favour the development of woolly aphid. Over and above the weakening of the apple trees, infestations with woolly apple aphids lead to production of smaller sized fruit. The aphids produce long cottony-white wax filaments, providing a protective coat to the colonies and given then a distinctive woolly appearance. In addition to colonization of the aerial parts of the tree, woolly apple aphids may be found year round on roots of mature trees where they often go unnoticed (Beers *et al.*, 2007). Its main natural enemy, the parasitoid *Aphelinus mali*, is an important control agent. Spraying of vamidothion is particularly useful because of its lasting effectiveness (Staubli and Chapuis, 1987); measures should be adopted when the aphids become established on the twigs and about 10 per cent of the buds bear colonies (Geoffrion, 1985). Spraying of plictran at 0.1 per cent, twice in May and June was also effective to control woolly aphid (Muller, 1987). Imidacloprid 17.1 per cent and 19.6 per cent both at 50 ml/hl and thiametoxam at 0.54 kg/ha gave excellent control of wooly aphid when sprayed after flowering (Pasquatini *et al.*, 2010). Woolly apple aphid control in Belgium is currently a combination of chemical and biological control. Since 2010, two crop protection products are registered for use against *E. lanigerum* in apple; Movento 100 SC (a.i. spirotetramat) and Pirimor 50 WG (a.i. pirimicarb). The main natural enemies are the parasitoid wasp *Aphelinus mali* and the common earwig *Forticula auricularia* (Belien *et al.*, 2011).

Apple Aphids

In addition to woolly apple aphid, the other aphids, such as rosy apple aphid (*Dysaphis plantaginea*), rosy leaf curling aphid (*D. deveeta*), green apple aphid (*Aphis pomi*) and the apple grass aphid (*Rhopalosiphum insertum*) also cause serious damage resulting in curling and distortion of leaves and reduced extension growth. To control these aphids, destroying of eggs before hatching should be done by spraying tar oil, DNOC or DNOC-petroleum oil, thoroughly wetting the bark of the tree. Spraying of organophosphorus insecticides (malathion, phosphamidon, dimethoate, vamidothion) at green cluster stage was found effective. Helsen (2001) reported good control by spraying Imidacloprid at 100 g/ha in the month of October in Apple cvs. Elstar and Boskoop.

Blossom Thrips

There are several species of thrips of which *Taeniothrips* spp, *Thrips flavus, Thrips carthami, Haplothrips ceylonicus, etc.*, have been found to affect apple blossom. The attacked flowers show withering symptoms, as a result either the fruits do not set or may fall off in the early stage of development. Heavily infested bloom produces distorted flowers that open on one side. Excreta are often deposited near the feeding site, which provides a suitable site for fungal infection (Bhardwaj and Bhardwaj, 1985). *Chrysopa* sp. and lady-bird bettle (*Coccinella septumpunctata*) are the most voracious predators of thrips. Prebloom spray at green tip stage with fenitrothion (0.05 per cent) followed by isofenphos (0.1 per cent), methamidophos

(0.05 per cent) or fenvalerate (0.01 per cent) were found effective against the thrips, without any long-lasting effect on pollen viability (Mishra *et al.*, 1987). Singh *et al.* (1986) in their experiment with fenitrothion (0.05 per cent), methyldemeton (0.025 per cent), quinalphos (0.02 per cent), phosalone (0.035 per cent), dimethoate (0.03 per cent), endosulfan (0.035 per cent), reported that fenitrothion was most effective in reducing population of thrips mainly, *Thrips flavus* and *Haplothrips ceylonicus*.

Red Spider Mites (*Panoychus ulmi*)

The adult lay reddish eggs underneath the leaves, and on the spurs. The mites suck sap from the leaves, which may turn bronze. In severe infestation, the growth of the plant is markedly reduced. The key component of integrated management of red spider mites is the regulation of predators *Typhlodromus pyri* and *Amblyseius andersoni* (Bower and Thwaite, 1986). Vigl *et al.* (1985) studided the effect of 12 insecticides and acaricides for control of red spider mites of Jonathan on M.9 rootstock. The insecticides, tumar (chlorpyriphos-methyl), orthene (acephate) and hostathion (triazophos), killed 30-50 per cent of predators but their population again recovered quickly. The use of acaricides, apollo (clofentezine), plictran L (cyhexatin) and cropotex (flubenzimine) was not recommended because these products proved more harmful to the predators than to mites. Folimat (omethoate) at 150 ml, omite (propargite) at 100 ml and neothrin (fenpropathrin) at 100 ml per 100 gal water, gave very good control with 2 applications in cv. Fuji in Brazil. Mitac (amitraz) at 300 ml per 100 gal water gave even better control with 1 application (Kovaleski and Salles, 1984). The fungicide thiozol (wettable sulfur) was also found effective in reducing the *P. ulmi* population (Kolev, 1986).

San Jose Scale (*Quadraspidiotus perniciosus*)

This is one of the most troublesome insect pest in all the warmer regions of the world. In India, it is found in Himachal Pradesh, Kashmir and other apple growing regions. They live on the bark of trunks and branches and feed by sucking sap of the tree, forming greyish specks on the bark surface. It also affects fruits causing reddish spots similar to pear scale (*Q. pyri*). In India, attempts have been made to establish biological control with the parasite *Prospaltella perniciosi*, but trials in Kulu valley have not been encouraging possibly due to unfavourable climatic conditions, extensive use of pesticides in the area, and the cumbersome methods of parasite breeding, release and maintenance. Two summer sprays (early June and a month later) of contact insecticides like phosalone, naled or fenitrothion or systemic insecticide like phosphamidon, vamidothion or demeton-S-methyl gave satisfactory control (Sud *et al.*, 1975). Spraying of fenitrothion to run-off at a concentration of 0.05 per cent when the fruits were 3-4 cm in diameter gave good protection against San Jose scale of apple cv. Red Delicious (Thakur and Kashyap, 1986). Singh *et al.* (2001) recorded maximum control (97.33 per cent) by two spraying of Koranda 28 EC (fenvalerate 3 per cent + acephate 2.5 per cent EC) at 0.05 per cent in the month of September and October. Methyl demeton at 0.05 per cent and Carbosulfan at 0.05 per cent also caused 94.66 and 83.33 per cent control, respectively.

Leaf Miners

The leaf miners *Stigmella malella, Lithocolletis blancardella* (*Phyllonorycter blancardella*) and *Leucoptera scitella* (*L. malifoliella*) are reported in the apple orchards of Yugoslavia (Stamenkovic and Stamenkovic, 1987). The *Phyllonorycter blancardella* is also found in Switzerland (Baumgartner and Severini, 1987). Leaf damage by apple leafminer (*Lyonetia prunifoliella*) was reported from Korea (Kyung Hee *et al.*, 2004). In this study, the leaf damage in cv. Fuji was found much higher when grown on M.9 rootstock compared with Fuji grown on seedling rootstock. The researcher believes that increase in M.9 rootstock is responsible for the outbreak of leaf miner in Korea.

The catterpillars usually feed on young leaves. They also roll the leaves which may fall prematurely. *Leucoptera scitella* and *Phyllonorycter blancardella* could be controlled by treatment with diflubenzuron followed by an application of deltamethrin, except the spring generation of *P.blancardella* (Injac *et al.*, 1987).

Apple Sawlfy (*Hoplocampa testudinea*)

The caterpillars damage apples by tunnelling in the fruit. The 1arvae produce ribbon-like scars which are found on both young and mature apples. Infested fruits usually drop off from the tree, resulting in severe loss of fruits. The cultivars Charles Ross, Ellison's Orange, James Grieve and Worcester Pearmain are very susceptible to sawfly. Spraying of organophosphorus insecticide (dimethoate, vamidothion, deltamethrin) at petal-fall stage was found effective to control sawfly.

Diseases

Apple Scalo (*Venturia inaequalis*)

Apple scab is reported from almost all the apple growing regions of the world and is considered as a most dreaded disease of apples. Apple cultivars differ greatly with regard to their resistance level to scab. In Europe and New Zealand over 50 scab-resistant cultivars have been introduced based on apple breeding program *e.g.* Prima, Redfree, Liberty, Adore, Divine, Inored, Moonlight *etc.*

The disease affects both leaves and fruits. Symptoms usually appear first on the undersurface of leaves of fruit-spurs, the side exposed when the fruit buds first open in the spring. Once the entire leaf has unfolded, both the sides may be infected. Velvety brown to olive-green powdery lesions (spots) whcih turn mousy black with age appear on the leaves. At first the margins of the lesions may cover a large portion of the leaf, a condition known as sheet-scab, and may lead to premature yellowing of leaves, defoliation and fruit drop. Early in the season, the scab spots often develop around blossom end (calyx end) of the fruit, but later in the season they develop anywhere on the fruit surface. Scab spots on fruit resemble leaf spots but with age becoming dark brown and corky. Severe early infection results in misshapen and knotty fruits. Fissures or cracks often develop in the scabbed areas which allow the entry of other pathogens, causing rot of fruit either in the field or in storage (Gupta, 1987a). Apparently healthy fruits carrying infection in late summer develop small, rough black, circular lesions on their skin in storage (Kaul and Gupta, 1987).

Being a dreaded disease all over the world, the strategy for apple-scab management is being vigorously investigated. Several effective fungicides to control scab and spray schedule have been suggested. Some of the findings in different countries to control apple scab are presented below:

Australia

Campbell *et al.* (1988) found good control of *V. inaequalis* by high volume spray (with high volume of air at low air speed) of bitertanol.

Brazil

Thiabendazole (1.4 kg/ha) + dithianon (2.24 kg/ha) gave good control of *V.inaequalis* and reduced premature leaf drop in apple cv. Fuji (Bleicher, 1984).

Denmark

Topas 50 and captan 83, applied as a spray throughout the season, were the most effective treatments for control of scab in apple cv.Rogers McIntosh (Gronborg, 1986).

India

The first severe epidemic of apple scab occurred in 1973 in the Kashmir valley. In Himachal Pradesh, the disease was reported in 1977 and in 1982 and 1983 severe epidemic occurred in all the apple growing areas. It is also prevalent in Naini Tal and Almora districts of Uttar Pradesh, Arunachal Pradesh and Sikkim (Kaul and Gupta, 1987). Gupta (1983) reported that in on season, sprays of protectant fungicides (dodine, captafol, bitertanol and fenapronil) provided better escape for leaves and fruits from primary scab than the systemic ones. Control of secondary scab was achieved by spray of delan (dithianon), benzimidazoles, dithane Z-78 (zineb), wettasul, cuman-L and ziron. Timely application of bavistin (carbendazim). mancozeb, captafol and curnan-L in 6 sprays provided almost complete control (Table 1.54). Gupta (1987b) further stated that two postharvest sprays of carbendazim or dodine in September and October, or a single pre-leaf fall spray of urea. carbendazirn, thiophanate-methyl, carbendazim + mancozeb, bitertanol, dodine or fenapanil completely suppressed ascorpore discharge by *V. inaequalis*. Spray of 5 per cent urea enhanced the yield of treated trees. A spray schedule for scab control and increasing yield as proposed by Gupta (1987a) is presented in Table 1.55.

Israel

Excellent scab control in Israel was achieved by Shabi and Elisha (1987) with 6 spray applications (at early pink, early bloom, full bloom, petal fall, fruit set and fruit development stage) of fungicides DPX H6573 [flusilazol] (punch 40EC, 40 µg/ml a.i.), PP 523 [hexaconazole] (anavil 5SC, 20, µg/ml a.i.) RH 3866 [myclobutanil] (systhane 12EC, 60 µg/ml a.i.), penconazole (topaz 10 EC, 25, µg/ml a.i.), pyrifenox (ACR-3567 or RO 15-1297 48EC, 50 µg/ml a.i.) and diniconazole (S-3308L 12.5 W 50 µg/ml a.i.). Scab causes heavy losses in Starking and other Red Delicious type apple cultivars which predominate in Golan Heights region. Jonathan was found most susceptible to scab, although Golden Delicious and Granny Smith were also affected (Shabi and Elisha, 1988).

Table 1.54: Efficacy of Fungicides in Controlling Primary and Secondary Scab of Apple at Katrain*

Fingicide	*Dose (per cent)*		*Disease Control (per cent)*			
	1st Spray	*2nd and Onward*	*Primary Scab*		*Secondary Scab*	
			Leaf	*Fruit*	*Leaf*	*Fruit*
Syllit	0.15	0.10	92.3	100.0	97.4	98.7
Difolatan	0.30	0.15	87.6	100.0	97.6	99.8
Baycor	0.15	0.10	84.0	100.0	99.4	99.6
Sisthane	0.20	0.15	78.4	100.0	96.0	96.8
MBC	0.05	0.05	62.2	96.2	97.8	98.8
Bavistin	0.05	0.05	55.3	92.8	97.2	99.8
Carbendazim	0.05	0.05	46.1	82.1	97.1	99.6
Daconil	0.30	0.20	35.2	96.2	64.6	97.3
Dithane M-45 (Flowable)	0.60	0.50	23.0	92.8	94.6	96.8
Dithane Z-78	0.40	0.30	8.4	73.2	92.9	95.7
Cuman-L	0.40	0.30	0.0	61.5	84.8	91.6
Ziron	0.40	0.30	0.0	55.3	80.0	90.5
Delan	0.075	0.05	26.1	81.0	94.1	99.0
Wettasul	0.40	0.30	15.3	55.3	92.2	92.7

* Gupta (1983).

Table 1.55: Suggested Spray Schedule for the Control of Apple Scab*

Spray No.	*Tree Stage*	*Fungicide/100 1 of water*
1.	Silver tip to green tip	Captafol (300 g)/Dithianon (75 to 100 g/Mancozeb (400 g) Captan 50 WP (400 g).
2.	Pink-bud stage	Dithianon (50g)/Mancozeb (300 g)
3.	Fruit Set (Pea stage)	Dithianon (50 g)/ Captan 50 WP (300 g)/Mancozeb (300 g)
4.	Petal fall	Carbendazim (50 g)/Thiophanate methyl (50 g)
5.	Fruit set (Pea stage)	Captan 50 WP (300 g)/Dithianon (50 g)/Mancozeb (300 g)/ Mixture of Carbendazim
6.	Repeat fungicides of 5th spray after 14 days	
7.	Preharvest (20-25 days before fruit harvest)	Captafol (150 g)/Mancozeb (30 g)

* Gupta (1987a).

Poland

In field trials against apple scab with rubigen plus (4 per cent fenarimol + 35 per cent dodine), a mixture of 1.2 per cent nuarimol + 75 per cent thiram, and home-made mixtures of rubigan 12 EC with syllit (dodine); captan or sadoplan 75 (thiram), Nowacka and Cimanowski (1987) found that both commercial and home-made mixtures were more effective than the systemic fungicides used alone when rainfall and disease severity were high. Trifloxystrobin at 0.012 or 0.015 per cent alone or in combination with propineb effectively reduced apple scab damage when it was applied up to 72 h after infection (Niemiec and Bielenin, 2003).

Yugoslavia

In Yugoslavia, application of 0.05 per cent baycor (triadimenol), 0.05 per cent rubigen (fenarimol) and 0.15 per cent saprol (triforine) to apple leaves and fruits 48 h after infection showed good results. When concentrations increased (0.1 per cent and 0.2 per cent, respectively) the fungicides were effective even when applied 96 h after infection (Mikec and Ciglar, 1984). Vojvodic and Dulic (1987) reported that triazole-based systemic fungicides and sterol inhibitors gave the best control of *Venturi inaequalis*.

Fire Blight (*Erwinia amlylovora*)

Fire blight has been known in North and Central America for more than 200 years. In New Zealand, it first appeared in 1919 while in Europe, the bacterium was first discovered in pear orchards of Kent. UK, in 1957. Later on, the bacterium was found to spread in Poland, Netherlands (1966), Denmark (1968), West Germany (1971). East Germany, France and Belgium (1972). The bacterium were also found in Egypt (1962).

Paulin (1982) stated that the severity of this desease is mainly due to three factors:

i) The potential host plants (Pomoideae) are widely spread in the temperate countries.

ii) The bacterium is able under certain conditions to kill a tree (even a several-years-old tree) within a few months and does not limit its damage only to the crop of one year.

iii) The disease is very easily transmitted from one place to another. This is due to the presence of ooze on the surface of the disease parts which contains a large number of bacterial cells. This exudates is very specific of fire blight. It plays a role both by placing the bacteria outside the plant, which makes it easily accessible to a number of vectors, and by protecting the very fragile bacterial cells from desiccation.

The symptoms first appear as blossom blight and spread later to shoot. In shoots, the disease causes wilting of shoot tips which drop without browning. Golden droplets of bacterial ooze are occasionally found on the stem. The leaves and shoots gradually turn brown (Billing, 1983). In fruits, symptoms appear as necrotic spots and oozing lesions on the fruit surface.

The spread of the disease can be minimized by restricting the import of host plants (*e.g.*, service berry, sureberry, red chokeberry, flowering Japanese quince, hawthorn, quince, loquat, medlar, firethorn, pear, mountain ash, white beam) in a healthy zone (Paulin, 1982). The affected trees and host plants should be grubbed and burnt as soon as the incidence of fire blight is seen. Sprays of streptomycin brought about reasonable control of infection in spring blossom of apple. Tests on methods of pruning of infected shoots showed cutting back to the base with scissors soaked in the disinfectant, $KMnO_4$, C4 concentrate or Wofasteril before closing the wound give best results (Ficke *et al.*, 1986). Some cultivars are more susceptible (*e.g.* Rome Beauty, Jonathan, Granny Smith) than others and the cultivars Delicious, McIntosh and Golden Delicious are considered as resistant (Moore *et al.*, 1991; Khan *et al.*, 2006).

Powdery Mildew (*Podosphaera leucotricha*)

Powdery mildew is a common disease in all the apple growing regions of the world. The disease appears both in nursery and orchards, and attacks twigs, foliage, blossoms and fruits. The cultivars, such as Golden Delicious, Jonathan, Granny Smith, Cox's Orange Pippin, Jonagold, Starking, Crispin are more susceptible to powdery mildew.

On infection, whitish powdery growth develops on both sides of leaves and twigs. The affected leaves are distorted in shape and smaller in size; become hard and brittle. The powdery coating on twigs finally disappears and a brown, felt-like covering with black fruiting bodies is seen (Pathak, 1980). In case of severe infection, leaf fall and pre-mature fruit drop may occur. The young infected fruit show signs of russeting.

Integrated protection against mildew essentially involves introduction of resistant cuitivars, use of natural resistance, destruction of overwintering fungus structures, forecasting early spring sprays, reduction of inoculum by pruning shoots and spraying, determining the threshold of damage by infection and proper use of spraying equipment (Cimanowski *et al.*, 1986). Spraying of pyrazophos (0.021 per cent), sulphur 0.48 per cent), triadimephon (0.05 per cent) and triforine (0.024 per cent) gave good control of powdery mildew (Bernardi, 1984). Mantinger and Vigl (1985) effectively controlled mildew by spraying of tapas (penconazole + captan). In Israel, Shabi and Elisha (1987) found excellent control of mildew in Anna apple with 8 spray applications of pyrifenox (ACR-3567 or RO 15-1297 48EC, 40 or 50 µg/ml a.i.). Pyrifenox and triflumizole were efficient in controlling powdery mildew even if the spray treatment was started late in the season and the trees were heavily infested.

For young seedlings (2 years), spraying of bayleton (triadimefon) at 500 ppm for 5-7 times was suggested by Verma and Gupta (1986). However, occasional rains during the growing period may affect the effectiveness of the fungicides. Model trials show that preparations, such as, nimrod (bupirimate), morestan (quinomethionate). fundazol (benomyl) and mancozeb are highly resistant to the effects of rain. The effectiveness of sickosul (sulphur) was also little affected by rain because of its high dose. The effectiveness of bayleton spezial (triadimefon), rubigen (fenarimol), afugan (pyrazophos), saprol (triforine) and captan decreased with increasing amount of rain (Jahn and Tathke, 1987).

Crown Galls (*Agrobacterium tumefaciens*)

Globular, elongated or irregular tumours form at or near the graft union mainly on the nursery plants (Kaul and Gupta, 1987). Kim *et al.* (1986) in an experiment with Fuji and Smoothee Golden, grafted with an M.26 interstock onto seedling rootstock, found that the different cultural methods, such as, cutting out infected tissues, treatment with lime or antibiotics or soil disinfection with formalin before planting, had no significant effect on control of crown gall. They suggested deep planting at 30-35 cm to avoid infection by the crown gall bacterium. Planting of rootstock (M.9) after dipping in copac E (ammoniacal cupric sulphate) or with copper oxychloride in clay slurries or mixed with glue showed good control of crown gall (Grimm, 1987).

Breeding and Improvement

The aim of apple breeding is to produce cultivars that have compact, well-branched trees which start bearing in the first or second year after planting. It should be a regular bearer, producing high yield with good fruit size and quality, with improved resistance to pests and diseases and the fruit should stay long in storage. Since, apple is a composite tree consisting of a rootstock and scion, there is also need to develop rootstocks with the aim to manupulate growth, suitable for a specific location as well as for good reproductive ability.

Rootstock Breeding

Although rootstock breeding was initiated in the beginning of nineteenth century, serious interest with considerable range of objectives for developing rootstock was started in the mid-nineteenth century. The first apple rootstock breeding programme at the East Malling Research Station, UK, was started in 1917. Initially the main objective was to develop woolly apple aphid resistant rootstock using Northern Spy as a source of resistance. In 1952, M.25 and MM.101 to MM.115 were selected from 3758 seedlings and were considered as woolly apple aphid resistant. Later on, several rootstocks of M and MM series were bred in UK, with specific objectives, most of which are now well adopted all over the world. Rootstock breeding programme in UK was mainly focussed on productiveness and tree size control with considerable emphasis on adaptation to soil and pest problems (Preston, 1953). The range of rootstocks produced by East Malling in the past, embracing M.27, M.9, M.26, MM. 106, MM. 111 and M. 25 in increasing vigour potential, is considered by many to be adequate for the needs of the UK apple producer. However, limitations in the performance of these rootstocks have been recognized. M 27 produces poor fruit size on many sites and shows poor tolerance. It is too dwarfing in most situations, poorly anchored and is sensitive to woolly apple aphid. M. 9 is difficult to propagate and poorly anchored with brittle roots. In many parts of the world it exhibits sensitivity to winter cold and to fire blight. Like M. 27 it is sensitive to wooly apple aphid (Webster *et al.*, 2000). A rootstock with vigour intermediate between M.9 and M.27 was envisaged for high density production system where vigorous scion varieties could be grown. M.26 produces abundant burknots and, in some situation, it takes up calcium poorly from the soil. It is also sensitive to fire blight and wooly apple aphid. MM. 106 is sensitive to collar rot. Breeding of apple rootstocks was continuing at East Malling and several promising rootstocks have emerged.

AR86-1-25

This is a selection from and M.27 x MM.106 cross. It produces trees with similar vigour, yield and fruit size to MM. 106 but has the great advantage compared with MM. 106 in its strong resistance to collar rot. This selection has performed particularly well on collar rot infected soils in New Zealand. It is also resistant to woolly apple aphid.

AR 86-1-20

This is a sibling from the same cross as AR 86–1–25. It induces trees of similar vigour and cropping performance to those on AR 86 6-1-25. This is a semi vigorous stock, resistance to soil borne pests and diseases and good anchorage. Several other rootstocks from HRI-East Malling program with vigour similar to M.27 or intermediate between M.27 and M.9 (Table 1.56) and with similar vigour to M.9 (Table 1.57) have been developed and are now being evaluated.

Table 1.56: Rootstock Selection from the HRI-East Malling Apple Rootstock Breeding Program with Vigour Similar to M.27 or Intermediate between M.27 and M.9

Rootstock Clone No.	*Origin*
AR 10-2-5	M.27 x MM.106
AR 69-7	AR 10-2-6 open pollinated
AR 360-19	M.9 x M.27
AR 628-2	Ottawa 3 x MM.106
AR 672-1	M.9 x 3426
AR 669-1	M.7 x M.27
AR 682-6	M.26 x M.793

Table 1.57: Rootstock Selections from HRI-East Malling Apple Rootstock Breeding Programme with Vigour Similar to M.9

Rootstock Clone No.	*Origin*
AR 120-242	M.27 x MM.106
AR 295-6	Robusta 5 x Ottawa 3
AR 486-1	Ottawa 3 x M.27
AR 680-2	M.26 x M.27
AR 852-3	AR 363-16 (M.27 x Ottawa 3) open pollinated

The rootstock breeding programme was also taken up in Germany, Sweden, Canada, USSR, Poland, Czechoslovakia, USA and in Japan. The rootstock Jork 9 (J. 9) was developed from open-pollinated seedlings of M. 9 in Germany (Tiemann and Dammann, 1981). Bemali, a hybrid of Mank's Codlin x M.4 developed at Sweden, is said to be winter-hardy and woolly apple aphid resistant (Olden, 1953). Ottawa 3 (0.3) was selected from Robin x M.9 progeny which showed good promise as an interstock (Heeney, 1977). In Poland, the rootstock breeding programme resulted in the development of P-l, P-2, P-16 and P-22 from M.4 x Common Antonovka. All these rootstocks showed considerable resistance to crown rot and are more winter-hardy (Cummins and Aldwinckle. 1982). The Michurin College of Horticulture, Moscow, introduced the Budagovsky series (Bud.9, Bud.490, Bud.491) with the primary objectives centred on adaptation to the severe climate of central Russia (Budagovsky, 1962). At Michigan State University. MAC-9 (Mark) was selected from open pollinated seedlings of M.9 (Carlson, 1981) which is as dwarf as M.9 or M-25. The other more vigorous rootstock in this series include MAC-24, MAC-39 and MAC-46. Romanian researchers have produced 4 new, rootstocks for apple and the dwarfing clonal stock G-21 was introduced (Parnia *et al.*, 1986). In Poland, rootstock research comprises (i) development of new, dwarfing, winter-hardy, efficiently cropping clones, using cross breeding techniques, and (ii) evaluation of developed dwarfing rootstocks of Czechslovakia and USSR origin (Zagaja *et al.*, 1986). The present breeding programme is with primary focus to study the genetics on heritability of vigour, rooting, feathering, earliness of bud break and efficiency of cropping. Evaluation research is mainly aimed to test the efficiency of propagation, influence on the growth of a cultivar, precocity of bearing, cropping efficiency and resistance to pests and diseases.

Breeding of Scion Cultivar

Modern marketing system has emphasized the importance of fruit size and appearance, besides the increasing demand for heavy yield and regular cropping. At the same time, eating quality must be maintained or even improved. The plant breeders are busy in solving these problems. Over the last five decades several new cultivars have been developed and released all over the world, bringing together factors for quality, yield and pest and disease resistance from a wide range of parents. Thomas Andrew Knight (1759-1838) is considered as the father of controlled breeding in apples. In the first half of twentieth century the important US cultivars derived from controlled breeding were Cortland (McIntosh x Ben Davis) and Idared (Jonathan x Wagener). Later on several important cultivars were developed which includes Elstar, Gala, Jonagold, Mutsu, Pink Lady, Empire, Fuji *etc.*

Winter Chilling Requirements

Apple tree requires a certain amount of chilling during dormant season for successful flowering in the following spring. There are some local cultivars which requires very little chilling. However, these apples are usually poor in quality. In Israel, Oppenheimer and Solor (1968) developed good quality low chilling requirement cultivars Anna, Ein Shemer and Schlor by using some of the local cultivars. These cultivars were further used to develop low chilling requirement cultivars. Hauagge and Cummins (1991) stated that low chilling requirement of Anna is controlled by one major gene and several minor genes. The cvs. Adina, Earli Del, Goldina, Princessa, Primicia and Summer Del were subsequently developed as low chill cultivars.

Season of Ripening

Ripening is a highly heritable polygenic character with predominantly additive genetic components of variance responsible for genetic variation. The triploid cv. Sirius (Golden Delicious x Topaz) ripens 10 days after Golden Delicious, Semeralda is a green apple ripens 30 days before, Granny Smith, Galmac (Gala x Jerseymac) developed in Switzerland ripens in late July and are targeted for sale to celebrate Switzerlands National Day on August 1. (Brown and Maloney, 2013).

Fruit Appearance

The present-day consumers' preference is for fruits that have attractive good colour, a regular shape and a skin free from russet. Northern Spy is considered as a valuable parent because of its production of a high proportion of seedlings with a good skin finish in crosses with Cox. The presence of anthocyanin stripes in apple skins is controlled by a single dominant gene. The cultivars Jonagold and Elstar with a yellow background, have good consumers' response. Bicolour (or tricolor apples such as Cox's Orange Pippin) are popular in Europe. Several new selections and hybrids have been developed in recent past such as Autumn Glory (Fuji x Golden Delicious) - Fruits are red on a yellow background; Fuji sport: (Aztec Fuji - very high colouring strain; Jonagold Sport; Jonastar-nice red stripe over a red background; Royal Crimsom Topaz- is a highly coloured sport of the selection Crimson Topaz.

Fruit Size

Fruit size, shape, and productivity are under polygenic control, which means that when two cultivars are intercrossed there will be a wide and continuous range of expression of all these characters in the seedlings. They will not segregate into discrete categories (Janick *et al.*, 1996). Fruit size of about 65 mm diameter is generally accepted, but it is the aim for regular fruit size of 65-75 mm range. Selection of diploid cultivar with large fruit, and natural triploids of diploid x diploid, has been proved successful. The triploids like Crispin and Holstein, produce much higher proportion of large fruits than their parents Golden Delicious and Cox. The triploid, Jonagold, also has large fruits than the parents Golden Delicious and Jonathan.

Eating Quality

In Europe, surveys revealaed that most people prefer cultivars which have medium acid (pH 3.4 to 3.6) and high sugar (14 per cent to 16 per cent) category (Cox). The cvs. Jonagold and Elster have less acid content than Cox but had similar sugar content and have been well accepted in European markets. The cv. Suntan, though has superior aroma to Cox but the fruits are more acidic. The fruits of Greensleeves are crisp, juicy, slightly acidic with attractive flavour. Acidity in apples is controlled by a single dominant gene. Most successful North American apples such as Golden Delicious, Red Delicious and Mclntosh belong to low acid flow sugar category and good culinary apples (*e.g.* Bramley's Seedling) in high acid/medium sugar category.

Yield

A number of heavy cropping selections were made in last few years in different countries. The cvs. Greensleeves, Jupiter and Jester showed high yield in the UK; Fucui in China, Romus-2 in Romania, Himekami in Japan, also showed good yield.

Pest and Disease Resistance

New apple cultivars have been released in many countries which show high resistance to apple scab imparted from a gene derived from ornamental crab. *Malus floribunda* 821 has indicated the potential of interspecific crosses to introduce new genes (Crosby *et al.*, 1992), Breeding resistant cultivars is a rational strategy that has been attempted on a large scale for many years in many countries; new cultivars that are virtually field immune are now available (Janick, and Moore, 1996). Many scab resistant cultivars have been developed of which Enterprise, Gold Rush, Jonafree, Liberty, Pristine, Judiline *etc.* are growing in different countries and more recently Adore, Crimson Crisp, Inored, Kalei were developed (Brown and Maloney, 2013). A very high level of resistance to milew is being transferred to the cultivated apple from *M. robusta* and *M. zumi*. The mildew resistant breeding lines of *M. robusta* also carry resistance to the rosy apple aphid. This resistance was transferred to commercial cultivars in a collaborative projects involving North American and European breeders.

Compact Columnar Cultivars

Breeding for development of columnar apple cultivars at East Malling was started in early 1970s. It was based on a mutant of the Canadian cultivar Mclntosh, known initially as Strain A, that was found in an orchard in British Columbia in 1964 (Fisher, 1969). This mutant, now known as Wijcik, grows rather like a sturdy natural cordon. The internodes are short and axillary buds usually grow into spurs rather than extension shoots, so the flowers and fruits are borne up the single main stem (Tobutt, 1984). When Wijcik was crossed with a normal cultivar, about half the seedlings showed growth similar to Wijcik and it appeared to be heterozygous for a dominant gene of compact or columnar habit (Lapins, 1969; Lapins and Watkins, 1973). Cultivars which have been selected at East Malling, can be grown as natural cordons without pruning, at very close planting distance, less than one metre apart. Experimental crop from this type of orchard has given yield in the second year almost equal to that normally expected in eight years after planting a conventional orchard (Alston and Watkins, 1983).

Mutation Breeding

In Canada and USA, isolation of spontaneous somatic mutations in apple has played a significant role in the production of new apple cultivars. Stadler (1930) first suggested the possibility of using X-rays for inducing bud mutations in apple. Later, the fruit breeders in Sweden and in Canada established the potential usefulness of X-rays in developing new apple strains. Bishop (1954) advocated that X-ray could be used as an artificial means for the induction of colour gene mutation in apples. The mutant cv. Red Elstar was found in 1981 in a Dutch plantation of Elstar which was more coloured than the normal fruits of mother tree (Visser and Verhaegh, 1987).

Important Varieties Developed in Selected Apple Growing Countries in the last 60 Years

- Australia: Cripps Pink (Pink Lady®), Cripps Red (Sundowner™), Applethorpe Earlidel, Applethorpe Summerdel, Kalei, Crimson snow.
- Belgium: Merlijn (King Jonagold x Liberty)
- Brazil: Condessa, Princesa, Eva, Anabela, Caricia, Monalisa (Galax Malus 4)
- Canada: Summerred, Creston, Chinook (Splendour x Gala), Silken (Honegod x Sunrise), Blair, Rouville, Richelieu, Belmac, Primevere, Salish, Aurora Golden Gala.
- China: Huaguan (Golden Delicious x Fuji), Huashuai (Fuji x Starkcrimson), Xinshuai, Hanfu (Dongguang x Fuji), Quinguan (Golden Delicious x Jiguan), Taishan Gala, Qiufuhong.
- Czech Republic: Angold, Julia, Klara, Nabella, Jarka, Resista, Selena, Zuzana, Goldlane, Mercury (Topaz x Rajka), Moonlight.
- England: Kent (Cox x Jonathon), Fiesta (Cox x Idared), Saturn (PRI 1235 x Starkspur Golden Delicious), Maypole, Telamon, Tuscan, Trajan, Charlotte, Obelisk.
- France: Jurella, Judaine, Judor, Initial (Gala x Redfree), Tentation (Golden Delicious x Blushing Golden), Autumn Blush, Gradiyel (Christmas Rose x Gradigold), Adore, Divine, Antares (Elstar x 3191), Choupette.
- Germany: Ahrina, Ahrista, Ahra, Piros, Pilot, Pingo, Pinova (Clivia x Golden Delicious), Remo, Rewena, Reanda, Bella.
- Italy: Prime Red (Prima x Summered), Giongo, Golden Mira, Nova, Summerfree, Golden Orange, Gaia, Gemini, Renoir.
- Japan: Megumi, Mutsu, Sekaiichi, Hokuto, Mellow, Akita Gold (Golden Delicious x Fuji), Hacnine (Fuji x Tsugaru), Kio (Orin x Hatsuaki), Akane, Alps-Otome, Kitaro (Fuji x Hatsuaki).
- Korea (South): Hwahong (Fuji x Sekaiichi), Hongro (Spur Earliblaze x Spur Golden Delicious), Chukwang (Fuji x Mollie's Delight), Seokwang, Hongan (Fuji x Jonathan).
- New Zealand: Joy, Festival, Pacific Rose, Southern Snap, Sci Early, Sci Red, Envy (Braeburn x Royal Gala), Aztec Fuji, Premier Star, Genesis, Southfield.
- Poland: Fantazja (McIntosh x linda), Alwa (Macoun open-pollinated), Lodel (Lobo x Red Delicious), Sawa (Fantazja x Primula), Witos, Alka.
- South Africa: Drakenstein, Gold Gift.
- Sweden: Alice, Katja, Aroma, Sylvia, Kim, Amorosa.
- Switzerland: Maigold, Goro, Arlet, Iduna, Marina, Ariwa, Galmac (Gala x Jerseymac), Mairac (Gala x Maigold), Ladina (Topaz x Fuji)

☆ United States of America: Stellar, Arkcharm, Regent, Sweet Sixteen, Honey Crisp™, Zester, Liberty, Freedom, Goldrush, Enterprise, Pristine, Jerseymac, Mollies Delicious, Vista Bella, Summer Treat, DS-22, Ever Crisp (Fuji x Honey Crisp), Royal Red Honeycrisp, Crimson Crisp.

References

Abbott, D.L. (1959). Rep. Long Ashton Res. Stn. for 1958, pp. 52-56.

Abbott, D.L. (1970). Physiology of Tree Crops. (eds. L. C. Luckwill and C.V. Cutting). Academic Press, New York, pp. 65-80.

Abeles, FA. (1973). Ethylene in Plant Biology. Academic Press. New York.

Aerts, R. (1971). Fruitteelt, 70 : 102-04.

Aitken, H.C. (1941). 79th A.R. Nova Scotia Fruit Grs. Association, 1940, pp. 120-24.

Alekseeva, S.A. (1983). Zashchita plodevo-yagod. Kul'tur i vinograda ot Vreditelei i boleznei v Zone Sev. Kavkaza, pp 42 - 44.

Alston, F.H. and Watkins, R. (1983). Rept. Royal Hort. Soc. Conf. London, pp. 97-105.

Ananthakrishna, S.M. Dhanraj, S. and Govindarajan, V.S. (1983). J. Hort. Sci., 58: 31-36.

Anderson, J.L. (1975). J. Amer. Soc. Hort. Sci., 100: 229-3l.

Anderson, R.E. (1967). Proc. Amer. Soc. Hort. Sci., 91: 810-20.

Angelis, V.de., Sanchez, E. and Tognetti, J. (2011). Int. J. Fruit Sci., 11: 364-75.

Anomymous (1987a). Agronomico, 39: 109-12.

Anonymous (1987b). Productia Vegetala Horticultura, 36: 7-8.

Anonymous (1965). Rept. East Malling Res. Stn. for 1964, pp. 42-43.

Anonymous (1977). Rept. East Malling Res. Stn. for 1976, p.145.

Anonymous (1984-85). Schweizerische Zeitschrift fur obst-und Weinbau, 120: 121 (27: 2). 770-76/38-44.

Arakawa, O., Hori, Y. and Ogata, R. (1985). Physiol. Plantarum, 64: 323-27.

Archbold, D.O., Brown, G.R. and Cornelius, P.L. (1987). J. Amer. Soc. Hart. Sci., 112: 219-22.

Argles, G.K. (1937). Imp. Bur. Fruit Production Bull., 9: 7-115.

Arthey, V.O. (1975). Quality of Horticultural Products. Chapter 6. Wiley, New York.

Ashby, D.L. and Looney, N.E. (1968). Canad. J. Pl. Sci., 48: 422-27.

Atay, A. N., Koyuncu, F. and Atay, E. (2013). J. Biol. Environ. Sci., 7: 81-86.

Atkinson, D., Johnson, M.G., Crisp, C.M., Mande, Gi and Mercer, E.R. (1987). J. Sci. Fd. Agric., 38: 19-22.

Atkinson, D., Jackson, J.E., Sharples R.O. and Waller, W.M. eds.(1980). Mineral Nutrition of Fruit Trees.Butterworths. UK. pp. 71-82.

Avraarnov, G.N., Petrash, A.D., Yaroshenko, S.V., Tomashek, G.I. and Sidorenko, A.M. (1988). Sadovodstvo i Vinogradarstvo, 6: 11-13.

Awad, M.M. and Kenworthy, A.L. (1963). Proc. Amer. Soc. Hort. Sci., 83: 68-73.

Awasthi, R.P., Thakur, G.C. and Chauhan, P.S. (1986). Proc. Nat. Symp. Temperate Fruits, pp. 11-17.

Bachmann, U. (1983). Acta Hort., 138: 77-82.

Badescu, A.P., Sitiriu, D., Sian, S. and Cotorobai, M. (1978). Acta Hort., 80: 467-70.

Badowska-Czubik T., Rozpara, E., Danelski, W. and Kowalska, J. (2011). J. Res. App. Ag. Eng., 56: 20-22.

Bain, J.M. and Robertson, R.N. (1951). Aust. J. Sci. Res., 4: 75-78.

Baker, H.A. (1983). Rept. Royal Hort. Soc. Conf., pp. 79-83.

Baker, J.E., Lieberman, M. and Anderson, J.D. (1978). Plant Physiol., 61: 886-88.

Bakun, V.K., Tarasenko, M.T., Zagurskii, S.G. and Sedov, A.E. (1985). Izvestiya Tirniryazevskoi.Sel'skokhozyaistvennoi Akademii,1 No.1, pp 122-24.

Bangerth, F. (1974). Acta Hort., 45: 9-16.

Bangerth, F. (1978). J. Amer. Soc. Hort. Sci., 103: 401-404.

Bangerth, F. and Schroder, M. (1994). Plant Growth Reg., 15: 293-302.

Bangerth, F. and Streif, J. (1987). J. Sci. Fd. Agric., 41: 351-60.

Bangerth, F., Song, J. and Streif, J. (2012). HortSci., 47: 4-10.

Barney, D.L., Waber, R.H., Davis, T.D. and Williams, C.F. (1985). HortSci., 20: 236-38.

Bartley, I.M. (1985a). Phytochem., 24: 2857-59.

Bartley, I.M. (1985b). Rept. East. Malling Res. Stn. for 1984, p.193.

Bartley, I.M. (1986). Rept. East Malling Res.Stn. for 1985, pp. 179-80.

Bartos, V. (1987). Proc. Symp. 60 yrs. of Hort. Res. in Czeckolsovakia, pp. 124-125.

Basak, A., Soczek, Z., Golic, Z. and Niczborata, B. (1978). Acta Hort., 80: 373-76.

Batjer, L.P. and Thompson, A.H. (1946). Proc. Amer. Soc. Hort. Sci., 47: 35-38.

Batjer, L.P. and Thompson, A.H. (1947). Proc. Amer. Soc. Hort. Sci., 49 : 45-48.

Baumgartner, J. and Severini, M. (1987). Proc. 2nd Int. Cesena Agri. Conf'., Cesena, pp. 225-43.

Beech, F.W. (1983). Rept. Royal Hort. Soc.Conf., pp. 70-78.

Belien, T., Bangels, E., Peusens, G., Goossens, D. Berkvens, N. and Viaenc, N. (2011). Acta Hort., 917: 15-22.

Ben, J. (1986). Acta Agraria et Silvestria, Agraria, 25: 119-29.

Bennedsen, B.S. (1986). Meddelelse,Jordbruksteknisk Institut, KongeligeVeterinaero Landbohojskole,53: 1-73.

Benson, N.R. (1962). Proc. Wash. St. Hort. Assoc., 62: 19-22.

Bernardi, J. (1984). Empresa Catarinense de Pesquisa Agropecuaria. S.A., 3: 782-86.

Bernier, G. (1970). Canad. J. Bot., 49 : 803-19.

Beverly, R.B., Latimer, J.G. and Smittle, D.A. (1993). In: Postharvest. Handling: a system approach (Eds: R.L. Shewfelf and S.E.Prussia,). Academic Press, Inc pp. 73-98.

Bhambota, J.R., Shrestha, A.B. and Uppal, O.K. (1969). Punjab Hort. J., 9: 166-98.

Bhardwaj, S.P. and Bhardwaj, S. (1985). Indian Hort., 40: 25-27.

Bhartiya, S.P. (1983). Indian Hort., 28: 23- 26.

Bhartiya, S.P. Thakur, D.R. and Kar, P.L. (1986). Progressive Hort., 18: 207-12.

Bhartiya, S.P., Thakur, D.R., Kar, P.L. and Chadha, T.R. (1983). Progressive Hort., 15: 200-203.

Biasi, R., Bagni, N. and Costa, G. (1988). Physiol. Plantarurm, 73: 201-205.

Biddulph, O., Cory, R. and Biddulph, S. (1959). Plant Physiol., 34: 512-19.

Biggs, R.H. and Leopold, A.C. (1958). Amer.J. Bot., 45: 547-51.

Billing, E. (1983). The Garden, 108: 206-10.

Bishop, C.J. (1954). J. Hered., 45: 99-104.

Blanke, M. and Lenz, F. (1988). Erwerbsobstbau, 30 : 44-47.

Blankenship, S.M. (1987). Scientia Hort., 33: 205-12.

Blankenship, S.M. and Dole, J.M. (2003). Postharvest Biol. Technol., 28: 1-25.

Blankenship, S.M. and Unrath, C.R. (1988). J. Amer. Soc. Hort. Sci., 113: 88-91.

Blanpied, G.D. (1964). Proc. Amer. Soc. Hort. Sci., 84: 72-81.

Blanpied, G.D. (1965). Proc. Amer. Soc. Hort. Sci., 87: 85-92.

Blanpied, G.D. (1969). J. Amer. Soc. Hort. Sci., 94: 177-79.

Blanpied, G.D. (1977). Proc. 2nd Nat.C.A.Res.Conf., Michigan, pp. 225-30.

Blanpied, G.D. (1982). Proc. Int. Hort. Congress, Hamburg, 1: 248-56.

Blanpied, G.D., Turk, J.R. and Douglas, J.B. (1982). Proc. 3rd Nat. C.A. Res. Conf., pp. 337-44.

Blasse, W. (1960). Arch. Gartenbau, 8: 441-52.

Blasse, W. (1986). Gartenbau, 33: 307-308.

Blazek, J. (2011). Vedecke-Prace-Ovocnarshe, 22: 15-28.

Bleicher, J. (1984). Empresa Catarinense de Pesquisa Agropecuaria S.A., 3: 787-800.

Bleik, W. Van Ocr. (1987). Fruitteelt. 77: 16-17.

Bolding, P.J. (1987). Fruitteelt, 77: 12-14.

Bonnie, H-Sun and Francis. F.J. (1967). J. Fd. Sci., 32: 647-52.

Boon. J. Van Der., Das, A. and Pouwer, A. (1973). Stikstof, 65: 173-84.

Bootsma, J.H. (1984). Fruitteelt, 74: 1083.

Borie, Y. (1985). Japanese J. Breed., 35: 87.

Badowska-Czubik T., Rozpara, E., Danelski, W. and Kowalska, J. (2011). J. Res. App. Ag. Eng., 56: 20-22.

Bower, C.C. and Thwaite, W.G. (1986). Agfacts, Dept. Agric. NSW. 2nd.ed., p.4.

Boyer, J. and Liu, R.H. (2004). Nutrition J., 3: 13-19.

Boynton, D. and Oberly, G.H. (1966). In: Fruit Nutrion (ed. N.F. Childers). Rutgers State Univ. NJ.

Bramlage, W.J. and Thompson, A.H. (1963). Maryland Agric. Exp. Stn. Bull., A-129.

Boynton, D., Cain, J.C. and Compton. O.C. (1944). Proc. Amer. Soc. Hort. Sci., 44: 15-24.

Brown, A.G. (1975). In: Advance in Fruit Breeding. (eds. J. Janick and J. Moore)., Purdue University Press. West Lafayette, Ind., pp. 3-37.

Brown, S.K. and Maloney K.E. (2013). NY Fruit Quart, 21: 3-10.

Brown, S.K. and Maloney, K.E. (2005). *Malus domestica* Apple. In: Litz, R.E. (Ed.). Biotechnology of Fruit and Nut crops. CABI Publishing, UK. pp. 475-11.

Buban, T. (1981). Acta Hort., 120: 113-17.

Buban, T. and Faust, M. (1982). In: Horticultural Reviews. (ed. J. Janick). Vol. 4, AVI Pub. Co. Westport.

Buban, T. and Simon, I. (1978). Acta Hort.,80: 191-98.

Buban, T., Vasarhelyi, E. and Olasz, Z. (1987). Proc. Symp. 60 yrs. of Hort. Res. in Czechoslovakia, pp. 110-13.

Buchloch, G., Baxter. P. and Newbeller, J. (1961). Angew Bot., 35: 259-62.

Budagovsky, V.1. (1962). Proc. 16th. Int. Hort. Congress, Brussels, 3: 132-35.

Bugarcic, V., Jankovic. R. and Gavrilovic, J. (1981). Jugoslov. Vocarstvo, 15: 311-18.

Bukovac, M. J. and Sabbatini, P. (2006). HortSci., 41: 1606-11.

Bukovac, N.J. and Wittwer, S.H. (1957). Plant Physiol., 32: 428-35.

Bunca, A. and Stcpanescu, E. (1986). Lucrarile Stiinlifice ale Inst. de Cerecetare Si Productie Pentru Pimicultura Pitestii, 11: 203-12.

Bangerth, F., Dilley, D.R. and Deway, D.H. (1972). J. Amer. Soc. Hort. Sci., 97: 679-82.

Byun, J.K. and Chang, K.H. (1986). J. Korean Soc. Hort. Sci., 27: 331-37.

Cain, J.C (1971). J. Amer. Soc. Hort. Sci., 96: 664-67.

Cain, J.C. and Shear, C.B. (1964). In : Hunger Signs in Crops (ed. H.B. Sprague)., David Mckay Co. New York.

Cambell, M.M., Loveless, R.T. and Evans, P.T. (1988). Crop Protection, 7: 112-17.

Cantillano, R.F.F. (1984). Empresa Catarinense de Pesquisa Agropecuaria S.A., 3: 818-35.

Cao-Shangyin, Zhang-JinChang and Wei-Littua, (2000). J. Fruit Sci., 17: 244-48.

Carlone, R. (1958). Not. Mal. Piante, Nas., 43/44: 302-303.

Carlone, R. (1958a). Frutticoltura, 20: 43-52.

Carlson, R.E. (1970). Dwarf Fruit Tree Assoc., Michigan State Univ., 3: 95-96.

Carlson, R.F. (1981). Virginia Fruit, Vol. LXIX(3)., pp. 55-58.

Carlson, R.G. (1981). Fruit Var. J., 35: 78-79.

Carlson. R. and Robitaile, H.A. (1970). Proc. 18th Int. Hort. Congress, Israel, 1: 173.

Carlson, R.F. (1975). Compact Fruit Tree, 8: 26-31.

Casely, J. (1960). Rept. Long Ashton Res. Stn. for 1959, pp. 59-63.

Catzeflis, J. (1970). Rev. Suisse Vitic. Arboric., 2: 45-46.

Chadha, K.L. (1978). Indian Hort., 23: 13-19.

Chalmers. O.J. Faragher, J.D. and Raff, J.W. (1973). J. Hort. Sci., 48: 387-92.

Chan, B.G. and Cain, J.C. (1967). Proc. Amer. Soc. Hort. Sci., 91: 63-68.

Chandler, F.B. and Mason, I.C. (1942). Proc. Amer. Soc. Hort. Sci., 40: 120-22.

Chandler, V.H. (1951). Deciduous Orchards. 2nd ed., Henry Kimpton, London.

Chapen. J.F. (1988). Infos, Centre Technique Interprofessionnel des Fruits et Legumes, France. 41: 13-18.

Chen, P.M., Olsen, K.L. and Meheriuk, M. (1985). J. Amer. Soc. Hort. Sci., 110: 16-20.

Cheverry, J.L., Sy, M.O., Pouliqueen, J. and Marcellin, P. (1988). Physiologia Plantarum, 72 : 535-40.

Child, R.D. (1973). Acta Hort., 120: 77-82.

Chiu, F.E. and Bould, C. (1977). J. Hort. Sci., 52: 19-28.

Chopra, S.K. and Naqash, G.S. (1986). Abst. 22nd. Int. Hort. Congress, California. No. 1292.

Chu, C.L. (1985). HortSci., 21: 267-68.

Chu, C.L. (1988). J. Amer. Soc. Hort. Sci., 113: 226-29.

Church, R.M. and Williams, R.R. (1983). J. Hort. Sci., 58: 327-36.

Church, R.M., Williams, R.R. and Andrews, L. (1983). J. Hort. Sci., 58: 349-53.

Cimanowski, J., Nowacka, H. and Goszczynski, W. (1986). In. Materialy XXV Sesji Naukowej Instytutu Ochrony Roslin (ed. W. Wegorek). Ponznan, Poland. pp. 85-94.

Clijsters, H. (1971). Fruit Belge, 354: 69-77.

Comancscu, D., Ptre, G. and Petre, V. (2012). Scientific Papers, Series B., Horticulture, Vol. LVI. pp. 63-68.

Come. D., Perino, C. and Ralarnbosoa, J. (1985). Israel, J. Bot., 34: 17-23.

Coseteng, F. and Lee, C.Y. (1987). J. Fd. Sci., 52: 985-88.

Costante. J.F., Lord, W.J., Howard, D. and Connington, L. (1983). HortSci., 18: 913-15.

Creasy, L.L. (1976). HortSci., 11: 251-52.

Crnko, J. and Marn, M. (19R7a). Zbornik Biotehniskc Fakultete Univerze Edvarda Kardelja u Ljubljani, Krnetijstvo, 47: 49-62.

Crnko, J. and Marn, M. (1987b). Proc. Symp. Fruit Growing: 60 yrs. Hort. Res. in Czechoslovakia, pp. 260-62.

Crosby, J.A., Janicu, J., Pecknold, P.C., Korban, S.S., O'Connor, P.A., Ries, S. M., Goffreda, J. and Voordeckers, A. (1992). Fruit. Var. J., 46: 145-66.

Cummins, J.N. and Aldwinckle, H.S. (1982). Fruit Var. J., 36: 66-73.

Cummins, J.N. and Norton, R. (1974). N.Y.State Agr. Expt. Sta. Food and life Sci. Bul., p.4l.

Cummins, J.N. and Aldwinckle, H.S. (1983). In: Plant Breeding Reviews. (ed. J. Janick). AVI Pub. Westport, pp. 294-394.

Cutting, C.V. and Montgomery, H.B.S. (1973). Long Ashton and East Malling Research Station Bull. p.30.

Czynczyk, A. and Holubowicz, T. (1984). Compact Fruit Tree, 17: 19-31.

Damast, J.Z. (1949). Palestine J. Bot., Rehovot, Ser., 7: 103-12.

Darkanbaev, T.B. Demidenko, R.N. and Lysenko, M.K. (1975). Izvestiya AN Kaz SSR Biologicheskaya, 22: 1-18.

Davis, M.B. and Hill, H. (1941). Dom. Canadian Dept. Agr. Pub. 714, Tech. Bull., 32.

Deckers, Je. (1977). Boer en de Tunder. 83: 13.

Degtyar, L.A. and Krolik, M.A. (1973). Sadovodstvo Vinogradarstvo i Vinodelie Moldavii, 6: 15-18.

Dejager, A. (1994). Proc. 16th Int. Symp. European Concerted Action Program COST 94, Oosterbeek, The Netherlands, pp 267-79.

Dengke, Li, and XueMei, Du (2002). China Fruits, 1: 9-11.

Denne, M.P. (1960). Ann. Bot., 24: 397-402.

Denne, M.P. (1963). N.Z.J.Bot., 1: 295-300.

Dennis, F.G.Jr., Archbold, D.O. and Vecino, C.O. (1983). J. Amer. Soc. Hort. Sci., 108: 570-73.

Dennis, F.G.Jr. (1976). J. Amer Soc. Hort. Sci., 101: 629-33.

Dennis, F.G.Jr. and Nitsch, J.P. (1966). Nature, 211: 781-82.

Dennis, F.G.Jr., Ozga, J.A. and Xia, R.X. (1986). Abst. 22nd Int. Hort. Congress, California, No.775.

Derrnine, E. (1960). Rapport General, pp. 7-26.

Dev, G. and Kapoor, M.L. (1972). Sci. and Cult., 38: 447-49.

Devoghalaere, F., Doucen, T., Gruitton, B., Keeling, J., Payne., W. Lingh, T.J. Rose J.J., Hallett, I.C., Gunaseelan, K., Ga, D., Diack, R., Breen, K.C., Tustin, D.S.; Costes, E., Chagne, D., Schaffer, R.J. and David, K.M. (2012)., BMC. PL. Biol., 12: 64-81.

Dierend, W. and Rieken, S. (2007). Erwerbs-obstbau, 49: 51-56.

Dilley, D.R. (1978). J. Fd. Biochem., 2: 235-42.

Dover, C. J. (1983). Proc. 16th Int. Congress of Refrigeration, Paris, pp. 165-70.

Dozie, W.A.Jr., Latham, A.J., Kouskolekas, C.A. and Mayton, E.L. (1974). HortSci., 9: 35-36.

Drake, S.R., Fellman, J.K.and Nelson, J.W. (1987). J. Fd. Sci., 52: 1283-85.

Drawert, F., Heimann, W., Emberger, R. and Tressl, R. (1968). Phytochem., 7: 881-84.

Drobota, G. and Drobota, M.A. (1986). Lucrari Stiintifice, Institutul Agronomic Horticultura, 30: 59-60.

Dufkova, V. (1977). Vedecke Prace Ovocnarske, 6: 247-66.

Dulbro, K. (1958). Tikdsskr. Planteavl, pp. 112-47.

Dunn, J.S. and Rowe, R.N. (1982). Proc. 21st Int. Hort. Congress, Hamburg, 1: 168-78.

Dunn, J.S. and Stolp, P. (1987). HortSci., 22: 568-72.

Dunne, T.C. (1946). J. West. Australian Dept. Agr. Ser., 23: 124-27.

Dunne, T.C. (1938). J. West. Australian Dept. Agr. Ser., 15: 120-26.

Dvorak, A. (1959). Acta Univ. Agric. Brno. Fac. Silvic., Ser. A. Special Nurnero, pp. 95-99.

Dvorak, A. (1988). Acta Hort., 224: 325-29.

Dzhavakyants, Z.H.L. and Yusuphkodzhaev, M. (1985). Sadovodstvo, 3: 10-11.

Ebel, R. and Proebsting, E. (1986). Abst. 22nd int. Hort. Congress. California, No. 97.

Ebert, A. and Bender, R.J. (1984). Empresa Catarinense de Pesquisa Agropecuaria SA, 3: 891-902.

Eccher, T. (1983). Acta Hort., 137: 375-82.

Eccher, T. and Boffelli, G. (1981). Scientia Hort., 14: 307-14.

Edgerton, L.J. (1961). Proc. Amer. Soc. Hort. Sci., 77: 22-28.

Edgerton, L.J. and BIanpied, G.D. (1968). Nature, 219: 1064-65.

Edgerton, L.J. and Hoffman, M.B. (1951). Proc. Amer. Soc. Hort. Sci., 57: 120-24.

Eggert, D.A. and Mitchell, A.E. (1966). Proc. Amer. Soc. Hort. Sci., 90: 1-8.

Elfving, D.C. and Forshey, C.G. (1976). J. Amer. Soc. Hort. Sci., 101: 290-93.

Elfving, D.C. and Pierce, W.F. (1988). Ann. Rep. Hort. Exp. Stn. Simcoe, Ontario. No.19, p. 33.

Elson, C.M., Hayes, E.R. and Lidstar, P.D. (1982). Proc. 4th Nat. C.A. Res. Conf., North Carolina, pp. 162-68.

EL-YaZal M.A.S. and Rady, M.M. (2012). Scientia Hort., 136: 75-80.

Epstein, E. and Lilleland, O. (1942). Proc. Amer. Soc. Hort. Sci., 41: 11-18.

Evequoz, M. (1985). Revue Suisse de viticulture, d' Arboriculture et d' Horticulture, 17: 145-49.

Faby, R., Clever, M. and Tiemann, K.H. (1986). Mitteilungen des obstbauversuchsringes des Alten Landes, 41: 392-399.

Faccioli, F. (1958). Frutticoltura, 20: 375-76.

Fallahi, E., Simons, B.R., Fellman, J.K. and Colt, W.M. (1992). Plant Growth Regulation, 11: 435-39.

Farhoomand, M.B., Patterson, M.E. and Chu, C.L. (1977). J. Amer. Soc. Hort. Sci., 102: 771-79.

Fataliev, A.T. (1985). Vestnik Sel 'skokhozyaistvennoi Nauki, Baku, Azerbaijan SSR. No.5. 51-52.

Faust, M., Shear, C.B. and Smith, C.B. (1968). Proc. Amer. Soc. Hort. Sci., 92: 82-88.

Faust, M. (1964). Proc. Amer. Soc. Hort. Sci., 85: 85-90.

Faust, M. (1965). Proc. Amer. Soc. Hort. Sci., 87: 1-9.

Faust, M. and Klein, J.D. (1974). Proc. Amer. Soc. Hort. Sci., 99: 93-94.

Faust, M. and Shear, C.B. (1968). Bot. Rev., 34: 441-64.

Faust, M. and Shear, C.B. (1972). J. Amer. Soc. Hort. Sci., 97: 437-39.

Faust, M., Shear, C.B., Oberle, G.B. and Carpenter, G.T. (1971). HortSci., 6: 542-43.

Felius. P.M. and Toorenaar, G. (1959). Jversl. Proefst. Fruit., Whilhelminadorp. p. 19.

Fenner, R.J. and Adamson, E. (1977). Hortus, 24: 36-38.

Fernandez, R.T., Perry, R.L. and Flore, J.A. (1997). J. Amer, Soc. Hort. Sci., 122: 14-19.

Ferree, D.C. and Carlson, R.F. (1987). In: Rootstocks for Fruit Crops. (eds. Roye E.Rom and R.F. Carlson). John Willey and Sons Inc. New York, pp 107-43.

Ferree, D.C. and Schmid, J.C. (2001). Acta Hort., 557: 111-15.

Ferree, D.C. and Schemid, J.C. (1987). Fruit Var. J., 41: 150-52.

Feucht, W. (1961). Gartenbau., 26 : 256-60.

Ficke, W., Schaefer, H.J. and Nachtigali, M. (1986). Nachrichtenblatt fur den Pflanzenschutz in der DDR, 40 : 247-51.

Fidler, J. C. (1973). In: The Biology of Apple and Pear Storage. C.A.B., Farnham Royal, U.K.

Fischer, C. and Patzold, G. (1986). Gartenbau, 33: 276-77.

Fischer, M. (1997). Acta Hort., 451: 89-94.

Fischer, M. (1999). Proc. Int. Sem. On Apple Rootstocks for Intensive Orchards, Warsaw, Poland, 18-21 August, 1999. pp 27-28.

Fisher, V.J. (1961). New Jersy. SI. Univ., pp. 16-21.

Fisher. D.V. (1969). British Columbia Fruit Grower's Asso. Quart Rept., 14: 3-10.

Flath, R.A., Black, O.R., Guadagni, D.G. McFadden, W.H. and Schultz, T.H. (1967). J. Agric. Fd. Chem., 15 : 29-33.

Flood, A.E., Hulme, A.C. and Wooltorton, L.S.C. (1970). J Exp. Bot., 21: 301-10.

Folate, K.M. and Gardiner, S.E. (2009). Plant Genetics and Genomics Vol 6. Springer Science Business Media, pp 162.

Forsline, P.L. (1995). New York Fruit Quart., 3: 3-6.

Fortes, G.R.de L. and Petri, J.C. (1983). Acta Hort., 137: 335-41.

Fortes, G.R.De L., Ducroquet, J.P.H.J. and Ribeiro, P. De. A. (1984). Empresa Catarinese de Pesquisa Agropecuaria S.A. 3: 931-37.

French, A.P. (1956). A.R.Massm. agric. Exp. Stn. for 1955-56, pp. 54-56.

Freund, H. and Bunemann, G. (1986). Abst. 22nd Int. Hort. Congress, California, No. 95.

Friedrich, G. (1970). Obstbau, 10: 131-32.

Fukuda, H. (1984). J. Jap. Soc. Hort. Sci., 53 : 298-302.

Gaash, D., Ziv, Em, Sorkin, B., Cylinder, N. and Benami, Y. (1985). Alon Hanotea, 39: 673-81.

Galli, P. (1984). Obstabau Weinbau, 21: 289-92.

Garman. P and Mathis, W.T. (1956). Bull. Conn. Agric. Exp. Stat. p. 601.

Garner, F.E. (1937). Proc. Amer. Soc. Hort. Sci., 34: 323-29.

GAmer. R.J. (1967). The Grafter's Handbook. 3rd. Ed. Faber and Faber Ltd.

Garratt, M.P.D., Breeze, T.D., Jener, N. Polce, C., Bicomeijer, J.C. and Potts, S.G. (2014). Agr. Ecosystem and Environ., 184: 34-40.

Gautam, D.R. and Chauhan. J.S. (1986). Proc. Nat. Symp. Trop. Fruits. HPAU. India, pp. 7-10.

Geoffrion, R. (1985). Phytoma, 369: 25-27.

Gernyseva, G.P. (1964). Sadovodstvo, No.2. p. 2.

Gher Ghi, A., Millim, K., Suta, A., Burzo, I. and Panit, E. (1984). Lucrari Stiintifice, 15: 103-108.

Giovannoni, J. (2010). Nature Genetics, 42: 882-23.

Glas, A. (1958). Fruitteelt, 48: 1084.

Goddrie, P.O. (1987a). Fruitteelt, 77: 16-17.

Goddrie, P.O. (1987b). Fruitteelt, 77: 15.

Goddrie, P.O. (1987c). Fruitteelt, 77: 20-22.

Goddrie, P.O. (1987d). Fruitteelt, 77: 10-11.

Goffinet, M. (1986). Abst. 22nd Int. Hort. Congress, California, No. 420.

Goldwin, G.K. (1978). Acta Hort., 80: 115-21.

Goldwin, G.K. and Schwabe, W.W. (1974). Proc. 12th British Weed Control Conference, pp. 131-36.

Goldwin, G.K. and Schwabe, W.W. (1975). J. Hort. Sci., 50: 178-82.

Gonda, I. (1984). Kertgazdasag, 16: 17-28.

Goode, J.E. (1975). In: Climate and the Orchard, (ed. H.C. Pereira). CAB. Slough, England, pp. 51-62.

Goodenough, P.W. (1983). Acta Hort., 138: 83-92.

Goor, B.J. Van (1973). J. Hort. Sci., 46: 261-70.

Gorini, F.L and Lalatta, F. (1964). Progr. Agric. Bologna, 10: 995-1004.

Gorski, P.M. and Creasy, L.L. (1977). J. Amer. Soc. Hort. Sci., 102: 73-75.

Grand D' Esnon, A. (1985). Proc. Agri-Mation 1. Conf. Chicago, pp. 210-14.

Grand D' Esnon, A., Rabatel, G., Pelleng, R., Journeau, A. and Alden, M.J. (1987). Amer. Soc. Agri Engineers, No.87. pp. 1-12.

Grasmanis. V.O. and Edwards, G.R. (1974). Aust. J. Plant Physiol., 1: 99.

Grauslund, J. (1981). Acta Hort., 120: 77-82.

Greene, D.W. and Lord. W.J. (1983). J. Amer. Soc. Hort. Sci. 108: 590-95.

Greene, G.M. and Smith, C.B. (1980). In: Mineral Nutrition of Fruit Trees. (eds. D. Atkinson, J.E., Jackson. R.O. Sharples and W.M. Walter) Butterworth,UK. pp 316-17.

Greenhan, D.W.P. (1965). J. Hort. Sci., 40: 213-35.

Greenharn, D.W.P. (1976). Proc. Fertil. Soc. Lond., 157: 32.

Grimm. R. (1987). EPPO Bull., 17: 269-72.

Grochowska, M. and Karaszewska, A. (1976). Fruit Sci. Rept., 3: 5-16.

Grochowska, M.J. (1973). J. Hort. Sci., 48: 347-56.

Grochowska, M.J. and Karaszewska, A. (1978). Acta Hort., 80: 457-64.

Grochowska, M.J. and Lubinska, A. (1973). Physiol. Plant., 29: 425-29.

Gronborg, H. (1986). Frugtavleren, 15: 119-21.

Grubb, N.H. (1939). J.Pom. Hort. Sci., 17 : 1-19.

Gu, M., Shu, H. and Zhou, H. (1985). Acta Hort. Sinica, 12: 89-94.

Gul'Ko, I.P. and Gul'Ko, V.I. (1985). Sadovodstvo, 1: 9-11.

Gunel, T. and Karacali, I. (1985). Ege Univ. Ziraat Fakultesi Dergisi, 22: 1-19.

Guo, D. (1987). J. Shenyang Aric. Univ., 18: 20-26.

Gupta, G.K. (1987a). Indian Hort., 32: 48-52.

Gupta, G.K. (1987b). Int. J. Trop. Pl. Disease, 5: 93-97.

Gupta, G.K. (1983). Int. J. Trop. Pl. Disease, 1: 181-85.

Gur, A., Mizrahi, Y. and Samish, R.M. (1976). J. Hort. Sci., 51: 195-202.

Guttridge, C.G. (1962). Nature, 196: 1008.

Guz, M.L. (1974). Nauchnyc Trudy Ukrainskoi Sel'skokhozyaistvennoi Akad. 138: 3-16.

Gvozdenovic, D., Oulic. K. and Slavic, K. (1978). Acta Hort., 80: 377-80.

Habdas, H., Jankiewicz, L.S. and Borkowska, B. (1982). Acta Agrobotanica, 35: 11-23.

Hadorn, C. (1966). Phytoma, 191: 9-36.

Halder-Doll, H. and Bangerth, F. (1987). Scientia Hort., 33 : 87-96.

Haller, M.H. (1941). US. Dept. Agr. Cir. No.579.

Hamadziripe, E.T., Theron, K.I., Muller, M. and Steyn, W.J. (2014). HortSci., 48: 384-92.

Hammond, M.W. and Seeley, S.D. (1978). J. Amer. Soc. Hort. Sci., 103: 655-57.

Hampson, C.R., Quamme, H.A., Kappel, F. and Brownlee, R.T. (2004a). HortSci., 39: 501-506.

Hampson, C.R., Quamme, H.A., Kappel, F. and Brownlee, R.T. (2004b). HortSci., 39: 507-11.

Hansen, P. (1981). Tidsskr. Planteavl, 85: 405-10.

Hansen, P. and Grauslund, J. (1980). Tidsskr. Plantevl, 84: 215-27.

Hansen, P. (1969). Physiol. Plantarum, 22: 186-98.

Hansen, P. (1971). Tidsskr. Planteavl, 75: 615-25.

Hari, Om and Pathak, R.K. (1983). Indian J. Hort., 40: 168-72.

Harkett, P.J., Hulme. A.C, Rhodes, M.G.G. and Wooltorton, L.S.C (1971). J. Food Technol., 6: 39-45.

Harley, C.P., Mann, H.H. and Regcimbal, La. (1947). Proc. Amer. Soc. Hort. Sci., 50: 38-44.

Harris, Stephen, A., Robinson, J.P. and Juniper, B.E. (2002). TRENDS in Genetics, 18: 426-30.

Harrison-Murray, R.S. (1982). Proc. Int. PI. Propagator Soc. for 1981, 31: 386-91.

Hartman, F.O. and Danho, C.W.Jr. (1966). Proc. Int. Hort. Congr. Maryland U.S.A., Abstr. No.45.

Haugge, R. and Cummins, J.N. (1991). J. Amer. Joc. Hort. Sci., 116: 121-26.

Havelka, B. and Brychta, J. (1967). Zuhradlisty, 60: 330-37.

Healey, A.G. (1968). Commer. Grower., 3765: 451-52.

Heeney, H.B. (1977). Res. Rept. Smithfield Expt. Farm, pp. 33-35.

Hegazi, E.S. and Plich, H. (1978). Acta Hort., 80: 419-23.

Heinicke, A.J. (1936). Proc. Amer. Soc. Hort. Sci., 33: 164-65.

Helsen, H. (2001). Fruitteelt Den Haag, 91: 12-13.

Henze, J. (1983). Acta Hort., 138: 69-76.

Hewett, E.W. and Thompson, C.J. (1986). Abst. 22nd Int. Hort. Congress, California, No. 121.

Hi, Z.H., Korcak, RF., Wergin, W.P., Fan, F. and Faust. M.(1984). J. PI. Nutrition, 7: 911-28.

Hikasa, Y., Muramatsu, H. and Minegishi, T. (1986). Bull. Hokkaido Perf. Agroc. Exp. Stn., 55: 23-31.

Hildebrand, E.M. (1944). Proc. Amer. Soc. Hort. Sci., 45: 53-58.

Hilkenbaumer, F. (1959). Erwerbsobstbau., 1: 147-51.

Hilkenbaumer, F. (1960). Erwerbsobstbau, 2: 3-7.

Hilkenbaumer, F. and Reinken, G. (1959). Erwerbsobstbau, 1: 47-50 and 67-70.

Hipps, N.A., Davies, M.J. and Johnson, D.S. (2004). J. Hort. Sci. Biotech., 79: 610-18.

Hirst, P.M. (1999). Ann Rept. State Hort. Soc. of Michigan pp. 118-120.

Hoad, G.V. (1978). Acta Hort., 80: 105-14.

Hoad, G.V. (1979). Rept. Long Ashton Res. Stn. for 1978, pp. 199-206.

Hoad, G.V. and Monnelise, S.P. (1976). Scientia Hort., 4: 41-47.

Hochapfel, H. (1963). Bad. Obstu Gartenb., 56: 493-94.

Hohn, E., Baumgartner, D., Crespo, P. and Gasser, F. (2007). Agarforschung, 14: 188-93.

Holbeche, J.A. (1946). Agric. Gaz. N.S.W., 57: 75-80.

Holubowicz, T., Cummins, J.N. and Forsline, P.L. (1982). J. Amer. Soc. Hort. Sci., 107: 492-96.

Howard, B.H. (1971). Scientific Hort., 23: 116-26.

Howard, B.H. (1977). Proc. Int. PI. Propagator Soc., 27: 357-64.

Howard, B.H. (1981). J. Royal Agric. Soc. England, 142: 110-28.

Howard, B.H. and Garner, R.J. (1965). Rept. East. Malling Res. Stn. for 1964, pp. 83-87.

Howard, B.H. and Schmidt, G. (1979). Rept. East. Malling Res. Stn. for 1978, pp. 73-74.

Howard, B.H., Hanson-Murray, R.S. and Mackenzie, K.A.D.(1984). J. Hort. Sci., 59: 131-39.

Howard, B.H., Harrisson-Murray, R.S. and Babnik, M. (1980). Rept. East Malling Res. Stn. for 1979, pp. 73-74.

Huang, H. (1987). Fruit Sci. Rep., 14: 1-8.

Huang, H. and Cheng, J.S. (1984). Fruit Sci. Rep., 11 : 45-54.

Huelin, F.E. and Coggiola, I.M. (1966). J.Sci. Fd. Agric., 19: 297-301.

Huet, J. (1972). Physiol. Veg., 10: 529-45.

Hulme, A.C. (1958). Adv. Fd. Res., 8: 297.

Hulme, A.C., Rhodes, M.J.C., Galliard, T. and Wooltorton, L.S.G. (1968). PI. Physiol., Lancaster, 43 : 1154-56.

Hutchinson, A. (1974). Fruit Var. J., 28: 38-41.

Hutchinson, A. (1984). Proc. Int. Pl.Prop. Soc., 34: 38-48.

Hyde, J.F. and Ingle, M. (1968). Proc. Amer. Soc. Hort. Sci., 92: 733-38.

Ikjo, Chun and Fallahi, F. (2001). J. Korean Soc. Hort. Sci, 42: 621-24.

Injac, M., Vravl, S. and Dulic, K. (1987). An Zeiger fur Schadlin. Umweltschutz, 60: 115-20.

Iren, Z., Okul, A., Soylu, O.Z., Bulut, H. and Zeki, C. (1984). Bitki Koruma Bulteni, 24: 65·74.

Jackson, J.E. (1983). In: Apples and Pears. Royal Hort. Soc., UK. pp. 11-18.

Jackson, J.E., Blasco, A.B. and EI-Senfaz, S.M. (1983). Acta Hort., 137: 261-67.

Jackson, J.E. (1968). Rept. East Malling Res. Stn. for 1967, pp. 69-73.

Jackson, J.E. (1970). J. Appl. Ecol., 7: 207-16.

Jackson, J.E. (1978). Acta Hort., 65: 61-70.

Jackson, J.E. and Hammer, P.J.C. (1980). J. Hort. Sci., 35: 149·56.

Jackson, J.E., Hammer, P.J.C. and Wickenden, M.F. (1983). Acta Hort., 139: 75-82.

Jackson, J.E., Palmer, J.W., Bunemann, G., Christensen, V., Mika, A., Sansavini, S., Wertheim, B. and Winter, F. (1986). Abst. 22nd Int. Hort. Congress, California, No.308.

Jahn, M. and Tathke, S. (1987). Nachrichtenblatt fur den Pflanzenschutz in der DDR, 41: 246-51.

James, D.J., Passey, A.J. and Malhotra, S.B. (1983). In: "Genetic manupulation Crop Plants-Five Years on" University of Cambridge/AFRC, 68.

James, D.J., Passey, A.J. and Deeming, D.C. (1984). J. PI. Physiol., 115: 217-29.

James. D.J. (1983a). Physiol. Plantarum., 57: 149-53.

James, D.J. (l983b). Physiol. Plantarum, 57: 154-58.

James, P.A. and Midleton, S.G. (2001). Acta Hort., 557: 69-76.

Janick, J. (1974). HortSci., 9: 13-15.

Janick., J. and Moore, J.N. (1996). Fruit breeding, Tree and Tropical Fruits. John Wiley and Sons, Inc. Oxford, UK pp. 77.

JianHui, Chen and Wei, Zhou (2004). Ag. Sci. China, 3: 349-55.

Jivam, C. and Sala, F. (2014). HortSci., 4: 1-9.

Johann, G. (1983). Acta Hort., 137: 87-93.

Johnson, D. (1977). Rept. East Malling Res. Stn. for 1976, p. 148.

Johnson, D.S. (1979). Scientia Hort., 11: 303-17.

Johnson, D.S. (2000). J. Hort. Sci. Biotech., 75: 697-700.

Johnson, D.S. and Ertan, V. (1983). J. Hort. Sci., 58: 527-33.

Johnson, D.S. and Wickenden, M.F. (1982). Rept. East Malling Res. Stn. for 1981, p. 125.

Jones, O.P. (1983). In: Plant Biotechnology, (eds. S.H. Mantell and H. Smith). Cambridge Univ. Press., pp. 139-62·.

Jones, H.G. and Samuelson, T.J. (1983). J. Hort. Sci., 58: 183-90.

Jones, H.G., Higgs, K.H. and Samuelson, T.J. (1983). J. Hort. Sci., 58: 173-82.

Jones, H.G., Higgs, K.H., Cozens, P. and Wilson, A.B. (1986). Rept. East Malling Res.Stn. for 1985, pp. 81-82.

Jones, O.P. (1967). Nature, 215: 1514-15.

Jones, O.P. (1976). Nature, 262: 392-93.

Jones, O.P. (1979). Scientific Hort., 30: 44-48.

Jones, O.P. and Hatfield, S.O.S. (1976). J. Hort., Sci., 51: 495-99.

Jones, O.P., Hopgood, M.E. and O. Farrell, D. (1977). J. Hort. Sci., 52: 235-38.

Jones, O.P., Pontikis, C.A. and Hopgood, M.E. (1979). J. Hort. Sci., 54: 155-58.

Jones, O.P., Marks, T.R. and Waller, B.l. (1982). Rep. East Malling Res. Stn. for 1981, p.159.

Jongseob, L. and Taemyung, Y. (2001). Korean J. Hort. Sci. Tech., 19: 335-57.

Jonkers, H. (1978). Acta Hort., 80: 123-27.

Jonkers, H. (1979). Scientia Hort., 11: 303-17.

Joshi, K.R., Misra, R.S. and Seth, J.N. (1987). Progressive Hort., 19: 143-44.

Joshi, S.M. and Diwakar, B.L. (1985). Progressive Hort., 17: 304-308.

Juniper, B.E., Watkins., R. and Harris, S.A. (1999). Acta Hort, 484: 27-33.

Kacal, E. and Koyumeu, F. (2012). Bulg. J. Agri. Sci., 18: 898-904.

Kader, A.A. (2002). In: Fruit Quality and its Biological Basis, (ed. M. Knee). Sheffield Academic Press Ltd. Pp. 1-6.

Kader, A.A. (1983). HortSci., 18: 409-11.

Kal'gina, R.S. (1988). Sadovodstvo i Vinogradarstvo, 4: 13-15.

Kang, S.D. and Lee, S.K. (1987). Jr.Korean Soc. Hort. Sci., 28: 343-51.

Kar, P.L. and Bhartiya, S.P. (1987). Progressive Hort., 19: 105-108.

Karamursel, O.F. and Kalyoncu, I. H. (2011). African J. Biotech., 10: 19375-84.

Kaul, J.L. and Gupta, V.K. (1987). Indian Hort., 32: 43-47.

Kays,S.J. (1999). Postharvest Biol. Technol., 15: 233-47.

Keijer, E.J. and Dijksterhuis, H.P. (1956). Meded. Dir. Tuinb., 19: 396-400.

Keipert, K. (1987a). Obstbau, 12: 547-54.

Keipert, K. (1987b). Obstbau, 12: 455-57.

Kellerhals, M. and Hohn, E. (1987). Erwerbsobstbau, 29: 241-42.

Kerernidarska, S., Diochev, K., Milanov, B. and Lekhora, E. (1977). Ovoshcharstvo, 56: 27-32.

Khan, M.A., Duffy, B., Gessler, C. and Patocchi, A. (2006). Molecular Breeding, 17: 299-306.

Khanal, B.P., Grimm, E. and Knoche, M. (2013). AOB-Plants, 2013: pls 048.

Khatik, P.C. and Sharma, D.D. (2013). Progressive Hort., 45: 388-91.

Kid, F. and West, C. (1926). Dept. Sci. and Ind. Res. Food Inv. Bd. Rept., 1925, pp. 37-41.

Kidson, E.B, Askew. H.O. and Chittenden, E. (1943). N.Z.J. Sci. Tech., 25: 31-42.

Kim, K.Y., Cho, M.D., Kim, J.K., Kim, S.B. and Moon, B.W. (1987). J. Korean Soc. Hort. Sci., 28: 30-36.

Kim, S.B., Jang, H.L., Yiem, M.S. and Kim, J.H. (1986). J. Korean Soc. Hort. Sci., 27: 249-60.

Kimball, D.A. (1933). Scientia Agric., 13: 566-75.

Kingston, C.M. (1992). Hort. Rev., 13: 407-32.

Kirby, A.H.M., Bennet, M. and Sharples, R.O. (1968). Rept. East Malling Res. Stn. for 1967,pp.I77-81.

Klein, J.D. and Faust, M. (1978). HortSci., 13: 164-66.

Klamkowski, K. and Treder, W. (2000). Zeszyty Naukowe Instytutu Sadownictwa I Kwiaciarstwa-W-Skierniewicach, 8: 141-47.

Knee, M. (1985). Abst. Spring Conf. Growth Regulators in Horticulture, pp. 10-11.

Knee, M. (1988). J. Plant Growth Regulation, 7: 111-19.

Knee, M., Cockburn, J.T. and Watt, J.B. (1974). Rept. East Malling Res. Stn. for 1973, p.79.

Knee, M., Smith, S.M. and Johnson, D.S. (1983). J. Hort, Sci., 58: 521-26.

Knee, M. (1973). Phytochem., 12: 1543-49.

Knight, J.N. (1983). Acta Hort., 137: 71-76.

Knight, J.N. and Jackson, J.E. (1980). Rept. East Malling Res. Stn. for 1979, pp. 227-30.

Knight, R.L. (1963a). Tech. Commun. Commonw. Bur. Hort. Plantn. Crops, 29: 1-535.

Knight, R.L. (1963b). Abst. Bibliography of Fruit Breeding and Genetics to 1960, *Malus* and *Pyrus*, CAB.

Kobel, F., Steinegger, P. and Anliker, J. (1939). Landwirtschaftichen Jahrbuch der Schweiz, 53: 160-91.

Kobel, F. (1954). Lehrbuch des Obstaus auf Physiologiscacr Grundlage, Springer, Berlin, 2. Auflage, pp. 270-79.

Kochanov, E.M. and Konovalov, V.T. (1986). Bull. Nauchnoi Informatsii Tsentral noi Geneticheskoi Laboratorii, 41: l0-14.

Koen, T.B., Jones. K.M. and Longley, S.B. (1988). J. Hort. Sci., 63: 31-35.

Kolev, K. (1986). Agrokhimiya i Rastitelna Zashchita, 21: 75-81.

Kondo, S., Mizuno, N. and Takahashi, Y. (1987). Bull. Akita Fruit Tree Exp. Stn., 18: 23-33.

Koner, A. (1961). Erwerbsobstbau, 3: 11-12.

Korban, S.S. and Skirvin, R.M. (1994). HortSci., 19: 177-80.

Kouider, M., Korban, S.S., Skirvin, R.M. and Joung, H. (1985). J. Amer. Soc. Hort. Sci.,110: 93-96.

Koupil, S. and Plasilova, H. (1986). Agrochemia, 26: 274-75.

Kovaleski, A., and Salles, L.A.B.De (1984). Empresa Catarinense de Pesquisa Agropecuaria S.A., 3: 725-30.

Kovalev, V.V. (1974). Referativnyi Zhurnal, 1.55.722.

Kozhukhar, S. V. (1975). Trudy Kishincvskogo SeJ'skokhozyaestvennogo Instituta. 135: 112-15.

Kppytko, P.G. and Mikhalevskya, N.I. (1984). Agrokhimiya, 6: 61-68.

Kremer, Fr. W. (1963). Pflanzenschutz. Nachrichten Bayer, 16: 17-22.

Krishnaprakash, M.S., Aravindaprasad, B., Krishnaprasad, C.A., Narasimham, P., Anathakrishna, S.M., Dhanraj, S. and Govinadarajan, V.S. (1993). J. Hort. Sci., 58: 31-36.

Kronenberg, H.G. (1979). Netharland J Agric. Sci., 27: 131-35.

Kropp, K. and Ben, J. (1981). Hugona Kollatozaw Krakwie, Ogrodnictwo, 8: 119-32.

Kudryavets, R.P. and Mukhanin, L.V. (1988). Sadovodstvo i Vinogradarstvo, 2: 16-17.

Kul'kov, O.P., Koreisha, Z.I. and Sattarov. M. (1971). Referativnyi Zhurnal. 8.55.697.

Kurennoi, N.M. and Mantoptin, A.l. (1988). Sadovodstvo i Vinogradarstvo, 5: 14-16.

Kwon-Sooni, L., Kim-Jungin, Kim-Mokjong, Paek-Pongnyeol, Shin-Yonguk, Huang-Junghwan, Kang-Sangjo, Kim-Daeil, Choi-Cheol and Kang-Inkyu (2011). Korean J. Hort. Sci. Tech., 29.: 655-58.

KyungHee, C., Soonwon, L., DongHyuk, L., DongA, K. and Sangje, S. (2004). J. Sci. Food Agric., 84: 1871-77.

Lalatta, F.B., Nicotra, A. and Cornai, M. (1968). 2nd Conf. Nat. Fruttic. Montana.

Lamb, G.D., Golden, J.D. and Power, M. (1959). J. Hort. Sci., 34: 193-98.

Lamb, R.C., Aldwinckle, H.S. and Terry, D.E. (1985). HortSci., 20: 774-75.

Lane, W.O. and Mac-Donald, R.A. (1987). HortSci., 22: 515-16.

Lapins, K.O. (1969). Canad. J. PI. Sci., 49: 765-68.

Lapins, K.O. (1962). Proc. Amer. Soc. Hort. Sci., 41: 26-34.

Lapins, K.O. and Watkins, R. (1973). Rept. East Malling Res. Stn. for 1972, p.136.

Larrigaudiere, C.R., Vilaplana, R., Soria, Y. and Recasens, I. (2004). J. Sci. Fd. Agric., 84: 1871-77.

Latimer, L.P. and Percival, G.P. (1947). Proc. Amer. Soc. Hort. Sci., 50: 23-30.

Lau, O.L. (1983). HortSci., 18: 953-57.

Lau, O.L. (1985). J. Amer.Soc. Hort. Sci., 110: 541-47.

Lau, O.L. (1988). J. Amer. Soc. Hort. Sci., 113: 564-69.

Lewak, St. (1981). Acta Hort., 120: 149-59.

Lewak, St. and Rudnicki, R.M. (1977). In: The Physiology and Biochemistry of Seed Dormancy and Germination. (ed, A.D. Khan). Elsevier, North Holland, pp. 193-17.

Lewak, St., Rychter, A. and Zarska-Maciejewska, B. (1975). Physiol. Veg., 13: 13-22.

Li, S.H., Meng, Z.Q., Li, T.H., Liu, H.Z. and Tu, Y.C.(1995). Gartenbau. 60: 240-45.

Lidster, P.O. (1982). Proc. Nat. C.A. Res. Conf. Oregon, USA. pp. 109-120.

Lidster, P.O. and Lawrence, R.A. (1983). HortSci., 8: 615.

Lidster, P.O., Forsyth, F.R. and Lightfoot, H.J. (1980). Canad. J. Pl. Sci., 60: 299-301.

Lidster, P.O., Porritt, S.W., Easton, G.W. and Mason, J. (1975). Canad.J.Pl. Sci., 55: 443-46.

Lidster, P.O., McRae, K.B. and Sanford, K.A. (1981). J. Amer. Soc. Hort. Sci., 106: 159-62.

Lidster, P.D., Lougheed, E.C. and McRae, K.B. (1987). J. Amer. Soc. Hort. Sci., 112: 787-93.

Ligocki, P., Olszewski, T. and Slowik, K. (1988a). Fruit Sci. Rep., 5: 35-41.

Ligocki, P., Olszewski, T. and Slowik, K. (1988b). Fruit Sci. Rep., 15: 43-52.

Li-LinGuang, Yang-JianMing, Li-HuiFeng, Wang-HaiBo, Liu-JiaFen. and He-Ping. (2012). J. Fruit. Sci., 29: 145-46.

Link, H. (1978). Acta Hort., 80: 283-86.

Link, H. (1974). Acta Hort., 45: 53-60.

Link, H. (1983). Acta Hort., 137: 269-75.

Liu, F.W. (1978). J. Amer. Soc. Hart. Sci., 103: 388-92.

Liu, F.W. (1977). J. Amer Soc. Hart. Sci., 102: 93-95.

Liu, F.W. and King, M.M. (1978). HortSci., 13: 162-63.

Longbottom, H., Tobutt, K.R. and Jones, O.P. (1980). Rept. East Malling Res. Stn. for 1979, p.187.

Looney, N.E. (1968). Plant Physiol, 43: 1133-37.

Looney, N.E. (1969). Canad, J. PI. Sci., 49: 489-94.

Looney, N.E., Karnicnska, A., Leege, R.L. and Pharis, R.P. (1980). Acta Hort., 80: 105-l4.

Looney, N.F. (1971). J. Amer. Soc. Hort. Sci., 96: 350-53.

Lord, W.J., Greene, D.W., Damon, R.A.Jr. and Baker, J.H. (1985). J. Amer. Soc. Hort. Sci., 110: 422-25.

Lord, W.J., Greene, D.W. and Damon, R.A.Jr. (1983). Compact Fruit Tree, 16: 65-69.

Lord, W.J. and Damon, R.A. J. (1983). J. Amer. Soc. Hort. Sci., 108: 867-71.

Lord, W.J. and Greene, D.W. (1982). HortSci., 17: 372-73.

Lott, R.V. and Rice, R.R. (1955). Bull. III. Agric. Exp. Stn., p.31.

Lougheed, E.C., Proctor, J.T.A. and Miller. S.R. (1980). HortSci., 15: 424.

Lougheed, E.C., Murr, D.P. and Luce Berard (1978). HortSci., 13: 21-27.

Lovelidge, B. (1987). Erwerbsobstbau, 29: 139-41.

Lozzia, G.E. and Daolio, E. (1984). Informatore Fitopatologico, 34: 34-38.

Luchkov, P.G. and Ponomareva, G.A. (1972). Sadovodstvo Vinogradarstvo i Vinodelie Moldavii, 10: 14-15.

Luckwill, L.C. and Child, R.D. (1978). Acta Hort., 80: 271-74.

Luckwill, L.C. (1980). Scientific Hort., 31: 60-68.

Luckwill, L.C. (1970). In: Physiology of Tree Crops. (eds L.C. Luckwill and C.V. Cutting). p.237.

Luckwill, L.C. (1948). J. Hort. Sci., 24: 32-44.

Luckwill. L.C. (1953). J. Hort. Sci., 28: 14-24.

Luckwill. L.C. (1974). Proc. 19th Int. Hort. Congr. (Warsaw)., 3: 237-45.

Lurie, S. (2002). In: Fruit Quality and its Biological Basis. (ed. M. Knee). Sheffield Academic Press Ltd., pp. 107-21.

Luton, M.T. and Hamer, P.J.C. (1983). J. Hort. Sci., 58: 37-44.

Lyons. C.G.Jr., Byers, R.E. and Yoder, K.S. (1983). HortSci., 18: 913-14.

Mac-Daniels, L.H. and Hildebrand, E.M. (1940). Proc. Amer. Soc. Hort. Sci. 37: 137-40.

Magness, J.R. (1928). Proc. Amer. Soc. Hort. Sci., 25: 289-92.

Magness, J.R. (1941). USDA Farmers' Bull., 1883.

Magness, J.R. and Taylor, G.F. (1925). US Dept. Agr. Cir. No.350.

Magness, J.R., Degrnan, E.S. and Farr, J.R. (1935). Tech. Bull. US Dep. Agric., p.491.

Mahhou, A. and Achachl, K. (2002). Fruits, Paris, 57: 115-27.

Maini, S.B., Diwan, B., Lal, B.B. and Anand, J.C. (1985). Indian J. Agric. Sci., 55: 60-61.

Mamouni, A. (1999). Al Awamia, 100: 85-93.

Mantingcr, H. and Vigl, J. (1978). Obstbau Weinbau, 15: 81-82.

Mantinger, H. (1987). Obstbau Weinbau, 24: 35-38.

Mantinger, H. and Vigl, J. (1985). Obstbau Weinbau, 22: 140-42.

Mantinger, H., Stainer, R. and Nardin, C. (1986). Obstbau Weinbau, 23: 298-301.

Marcelle, R. (1984). Acta Hort., 14: 65-69.

Marcelle, R. and Sironval, C. (1963). Nature, 197: 405.

Marini, R.P. and Barden, J.AA. (1982). J. Amer. Soc. Hort. Sci., 107: 34-39.

Marino, F. and Greene, D.W. (1981). J. Amer. Soc. Hort. Sci., 106: 593-96.

Marjanovic, D. (1983). Jugoslovensko Vocarstvo, 17: 50.

Marth, P.C., Batjer, L.P. and Mann, H.H. (1946). Proc. Amer. Soc. Hort. Sci., 20: 109-12.

Marth, P.C., Harley, C.I. and Havis, A.L. (1950). Science, 111: 331-32.

Martin, D., Lewis, T.L., Cerny, J. and Grassia, A. (1970). Tech. Pap. CSIRO. Div, Plant Ind., p.29.

Martin, D., Stenhonse, N.S., Lewis, T.L. and Cerny, J. (1965). Aust. J. Agric. Res., 16: 617-25.

Martus, G.E. and Moore, R.C. (1954). Proc. Amer. Soc. Hort. Sci., 64: 199-208.

Mason, A.C. (1955). Rept. East Malling Res. Stn. for 1954, pp. 193-98.

Mason, J.L. (1979). Comm. Soil Sci. and Plant Anal., 10: 349-71.

Matsui, I., Sasaki, T., Murai, Y. and Sasaki, M. (1984a). Bull. Akita Fruit Tree Exp. Sta. Japan. 15: .I-15.

Matsui, I., Sasaki, T. Murai, Y. and Sasaki, M. (1984b). Bull. Akita Fruit Tree Exp. Sta. Japan. 15: 17-25.

Mattheis, J.P., Buchanan, D.A., and Fellman, J.K. (1991). J. Agr. Food Chem., 39: 1902-906.

Mattus, G. E. (1981). Rep. Virginia Polytechnic Inst. Contribution, No.103.

McKenzie, D.W. (1953). N.Z.J. Sci. Tech. Sec. A., 35: 45-52.

McKenzie, D.W. (1964). J. Hort. Sci., 39: 69-77.

McKenzie, D.W. (1985). J. Hort. Sci., 60: 7-11.

Mehlenbacher, S.A., Hough, L.F., Bailey, C.H. and Frecon, J.L. (1985). Fruit Var. J., 39: 32.

Meir, S. and Bramlage, W.J. (1988). J. Amer. Soc. Hort. Sci., 113: 412-18.

Metzner, R. (1969). Obst Und Garten, 44: 215-16.

Metzner, R. and Hubner, H. (1971). Obst Und Garten, 40: 190-91.

Micu, R.E. and Iordanescu, O. (2011). J. Hort. Forestry Biotech., 15: 167-72.

Mielke, E.A. and Facteau, T.J. (1986). Abst. 22nd Int. Hort. Congress, California, No.119.

Mika, A. (1975). Fruit Sci. Rep., 2: 31-42.

Mikec, I. and Ciglar, I. (1984). Zastita Bilja, 35: 83-89.

Miller, J.M., Wickenden, M.M. and Jackson, J.E. (1986). Abst. 22nd Int. Hort. Congress, California. No.l681.

Miller, S.S. (1982). J. Amer. Soc. Hort. Sci., 107: 975-77.

Mir, N. and Beaudry, R. (2002). Fruit Quality and its Biological Basis (Ed. M. Knee). Sheffield Academic Press Ltd., UK, pp. 122-56.

Mishra, R.C., Verma, A.K. and Gupta, P.R. (1987). Indian J. Hort., 44: 115-20.

Modibowaska, I. (1945). J. Pomol., 21: 57-89.

Modlibowska, I. (1961). Bull. Soc. Phys. Veg., 7: 123-34.

MongSup, K. and Kwangchool, Ko (2004). Korean J. Hort. Sci. Technol., 22.: 200-205.

Monin, A. (1961). Revue Agric. Brux., 14: 631-39.

Monin, A. (1961). Fruit Bleige. 28: 113-22.

Monnet, Y. (1987). Phytoma, 393: 31-33.

Montgomery, H.B.S. and Wilkinson, B.G. (1962). J. Hort. Sci., 37: 150-58.

Moore, M.H. (1936). J. Pomol., 14: 77-96.

Moreshet, S., Stanhill, G. and Olmo, H.P. (1975). HortSci., 10: 391.

Moretti, K. (1954). Attl. I. Conv. Naz. Fruttic. Montana. St. Vincent for 1953: 328-48.

Morgan, D.C., Stanley, C.J., Volz., R. and Warrington, I.G. (1984). J. Amer. Soc. Hort. Sci., 109: 637-42.

Morgan, M.A. and Hennerry, M.J. (1976). Ir. J. Agric. Res., 15: 75-83.

Mowry, J.B. (1960-61). Fruit Var. Hort. Dig., 14: 53-57.

Mrino, F. and Greene, D.W. (1981). J. Amer. Soc. Hort. Sci., 106: 593-96.

Muller, M. (1987). Obstbau, 12: 504-506.

Murashige, T. and Skoog, F. (1962). Physiol. Plantarum, 15: 173-97.

Murphy, L.M. (1942). Proc. Amer. Soc. Hort. Sci., 40: 42-44.

Nachtigall, G.R. (2004). Frutas do Brasil, 37.

Naschitz, S., Naor, A., Wolf, S. and Godschmidt, E.E. (2014). Trees: Structure and Function, 28: 879-90.

Naskidashvili, I. K. (1973). Referativnyi Zhurnal, 7.55. 686.

Naumann, W.O. (1972). Erwerbsobstbau, 14 : 38-42.

Nenadovic -Martinic, E. (1986). Abst. 22nd Int. Hort. Congress. California. No. 1300.

Nestcrov, Y.S. (1986). Sbornik Nauchnykh Trudov po Prikladnoi Botanike, 101: 67-74.

Nichols, W.C. and Patterson, M.E. (1987). HortSci., 22 : 89-92.

Niemiec, B.A. and Bielenin, A (2003). Zeszyty Naukowe Instytatu Sadownictwa-i-Kwiaciarstwa-W-Sikierniewicach, 11: 121-25.

Niggli,U., Weibel, F.P. and Potter, C.A. (1988). Zeitschrift fur Ptlanzenkrankheiten and Pflanzenschutz. 11: 367-65.

Nightingale, G.T. (1935). Bot. Gaz., 96: 581-37.

Njuguna, J.K., Leonard, S.W. and Teddy, E.M. (2004). HortSci., 39: 841.

Noga, G. (1986). Abst. 22nd Int. Hort. Congress. California. No.109l.

Norton, R.1. (1982). Proc. Massachusetts Fruits Grow. Assn. Inc., 88: 70-78.

Nowacka, H. and Cimanowski, J. (1987). Fruit Sci. Rep., 14: 35-43.

Nyhlen, A. (1960). Sver. Pomol. Foren. Arsskr., 60: 91-110.

Nyhlen, A. and Johannsen, J. (1964). K. Skogs - Lantbr Akad. Tidskr, 103: 307.

Oberly, G.H. (1963). Proc. N.Y. St. Hort. Soc., pp 149-51.

O'Grady, L.J. (1959). Rept. Pomol. Fruit Grow. Soc. Queb., p.79.

O'Kennedy, N.D. (1974). Res. Rept.Foras Taluntais, Dublin. pp. 54-55.

Olden, E.J. (1953). Sver. Pomol. Foren. Arsskr., Sci., pp. 40-57.

Oleszck, W., Lee, C.Y., Jaworski, A.W. and Price, K.R. (1988). J. Agric. Fd. Chem., 36: 430-32.

Olien, W.C. (1987). J. Amer. Soc. Hort. Sci., 112: 209-14.

Olisaev, A.A. and Byasov, K.Kh. (1971). Sadovodstvo, 11: 19-20.

Oppenheimer, C. and Slor, E. (1968). Theor. Appl. Genet., 38: 97-102.

Orphanos, P.1. (1982). J. Hort. Sci., 57: 259-66.

Orta, A.H., Akcay, M.E. and Erdem, T. (2000). J. Applied Hort., 2: 108-10.

Osipchuk, V.V. (1972). Sadovidstvo, Vinogradarstvo i Vinodelie Moldavii, 11: 16-17.

Packer, W.J. Chalmers, D.J. and Baxter, P. (1963). J. Agric. Viet., 61: 453-60.

Padfield, C.A.S. (1949). Orchard. NZ. p.22.

Paillard, N. (1967a). Physiol. Veg., 5: 85-87.

Paillard, N. (l967b). Fruits, 22: 141-43.

Palfitov, V.F., Savelev, N.I., Kozlov, N.E. and Molodtsov, M.A. (2011). Sadovodstvo i Vinogradarotvo, 4: 29-32.

Pant, P.C., Singh. R. and Divakar. B.L. (1976). Progressive Hort., 8 : 93-99.

Parnia, P., Mladin, G., Dutes, I. and Stanciurn, N. (1986). Abst. 22nd Int. Hort. Congress, California. No. 1344.

Parry, M.S. and Rogers, W.S. (1972). J. Hort. Sci., 47: 97-105.

Parnia, P., Dutu, L., Vladianu, D., Movileanu, M., Mizilu, C. Bragau A. and Stefan, D. (1997). 30 de Ani de Activitate (1967-1997). Inst. De Cercetare, Mracineni, pp. 45-52.

Pasqualini, E., Civolani, S. and Pollini, A (2010). Informatore Agrario, 66 : 67-72.

Pathak, V.N. (1980). Diseases of Fruit Crops. Oxford and IBH Pub. Co. New Delhi. India. pp. 188-230.

Patzold, G. (1986). Gartenbau, 33: 241-43.

Paul, A.A. and Southgate, D.A.T. (1978). The Composition of Food. (eds. McAnce and Widdowson.). HMSO., London.

Paulcchova-Kralikova, K. (1966). Biologia Bratislava, 28: 356-64.

Paulin, J.P. (1982). Proc. 21st Int. Hort. Congress, Hamburg. 1: 138-49.

Pavlenko, L.V. (1986). Sbornik Nauchnykh Trudov. Vsesoyuznyi Nauhno-lssledovatel'skii Institut Sadovod- stva imeni I. V. Michurina. No.44, pp. 44-48.

Pawar, A.D., Tuhan, N.C. and Parry, M. (1984). Pesticides, 18: 39-42.

Peggie, I., Minnis, D., Little, C. and Brohier, L. (1979). Bull. Victoria Dept. Agric., Hort. Res. Inst. Knoxfield.

Perring, M.A. (1968). J. Sci. Fd. Agric., 19: 186-92.

Pereira-Lorenzo, S., Ramos-Cabrer, A.M. and Fischer, M. (2009). Breeding Plantation Tree Crops: Temperate Species, pp. 33-81.

Perring, M.A., Samuelson, T.J. Caverly, D.J. and Marks, M.J. (1985). Rept. East Malling Res. Stn. for 1984. p.223.

Perring, M.A. (I968). J. Sci. Fd. Agric., 19: 640-45.

Perring, M.A. and Jackson, C.H. (1975). J. Sci. Fd. Agric., 26: 1493-502.

Perring, M.A. and Johnson, D.S. (197R). Rept. East Malling Res. Stn. for 1977. p.143.

Perring. M.A. and Pearson. K. (1985). Rept. East Malling Res. Stn. for 1984. for 1984. p.110.

Peterson, D.L. and Kornecki. T.S. (1987). Transactions Amer. Soc. Agril. Engineers. 30: 596-600.

Petri. J. L. (1984). In: Annis do VII Congresso Brasileiro de Fruticultura, Brazil, 3: 749-59.

Pfurnrnattcr, W. (1978). Acta Hort., 80: .279-82.

Pieniazek, J. (I968). Bull. Acad Pelon. Sci., CIV. 16: 179-83.

Piolanti, G. (1964). Frutticoltura. 26: 199-206.

Plich, H. and Nowacki, J. (1987). Fruit Sci. Rep., 14: 9-19.

Plich, H. and Hcgazi, E.S. (1978). Acta Hort., 80: 409-18.

Poling, E.B. and Oberly, G.H. (1979). J. Amer. Soc. Hort. Sci., 104: 799-801.

Poniedzilek, W., Nosal, K., Porebski, S. and Sobolewsk, A.A. (1987). Zeszyty Naukowe Akademii Rolmiczej W Krakowic, Ogrodnictwo, 220: 35-44.

Pontikis, C.A., Mackenzie, K.A.D. and Howard. B.H. (1979). J. Hort. Sci., 54: 79-85.

Powcl, L. E. (197R). Acta Hort., 80: 27-38.

Pratella, G.C. (1961). Inform. Fitopat., 11: 137-38.

Pratt, C., Joh, S.P., Way, R.D. and Einset, J. (1978). J. Amer. Soc. Hort. Sci., 103: 690-93.

Preston, A.P. (1966). J. Hort. Sci., 41: 349-60.

Preston, A.P. (1953). Rept. East Malling Res. Stn. for 1952. pp. 169-70.

Preston, A.P. (1954a). J. Hort. Sci., 29: 269-77.

Preston, A.P. (1954b). Rept. East Malling Res. Stn. for 1953. pp. 200-201.

Preston, A.P. (1955). J. Hort. Sci., 30: 25-32.

Preston, A.P. (1959). J. Hort. Sci., 34: 32-38.

Preston, A.P. (1958). J. Hort. Sci., 33: 29-38.

Preston, A.P. (I978). Acta Hort., 65: 229-35.

Priea, D. and Coca, N. (1986). Lucrarile Stiin tifice ale Institutului de Cercetare Si Productie pentru Pomicultura Pitesti. 11: 309-17.

Prive, J.P. and Embree, Ch. (1997). Acta Hort., 451: 179-86.

Proctor, J.T.A. and Crowe, A.D. (1983). HortSci., 18: 470-72.

Proebsting, E.L., Middleton, J.E. and Roberts, S. (1977). HortSci., 12: 349-50.

Provorchenko, A.V. (1986). Trudy. Kubanskii Sel'skokhozyaistvennyi Institut., No.267/295. pp. 41-45.

Qin, S.J., Li, F.D., Lv, D.G. and Gao, W.S. (2013). Word Applied Sci. J., 28: 1742-47.

Quinlan, J.D. (1980). Rept. East Malling Res. Stn. for 1979. p.36.

Quinlan, J.D. (1978). Acta Hort., 65: 129-38.

Quinlan, J.D. and Preston, A.P. (1968). J. Hort. Sci., 43: 373-81.

Quinlan, J.D. and Preston, A.P. (1973). Rep. East Malling Res. Stn. for 1972, p.91.

Quinlan, J.D. and Preston, A.P. (1971). J. Hort. Sci., 46: 525-34.

Raese, J.T. (1987). J. PI. Nutrition. 10: 2007-15.

Raherni, M. and Dennis. Jr.F.G. (1981). Acta Hort., 120: 257.

Ram, R.B., Bhatnagar, V.K. and Joshi. M.C. (1988). Farmers and Parliament, 23: 11-12.

Ramirez, H., Rumayor, A. and Martinez. V. (1983). Acta Hort., 137: 83-85.

Ramirez, H. and Head, G.V. (1981). Acta Hort., 120: 131-36.

Ramirez, H. and Hoad, G.V. (1978). British Plant Growth Regulators Group. Monograph.2: 37-47.

Rebeaud, S. G. and Gasser, F. (2015). Europ. J. Hort. Sci., 80: 18-24.

Reckruhrn, I., Muhe, K. and Kauske, B. (1983). Archiv fur Gartenbau, 31: 81-90.

Redalen, G. (1987). Gartneryrket, 77: 455-56.

Rehder, A. (1954). Manual of Cultivated Trees and Shrubs. New York.

Reichel, M. and Schmidt, G. (1986). Gartenbau, 33: 303-305.

Reid, M.S., Rhodes, M.J.C. and Hulme, A.C. (1973). J. Sci. Food Agric., 24: 971-79.

Reuther, W., Embleton, T.W. and Jones, W.W. (1958). Rev. PI. Physiol., 9: 175-206.

Rhodes, M.J.C. and Wooltorton, L.S.C. (1967). Phytochem., 6: 185-90.

Riesen, W. and Widmer, A. (1987). Schweizerische Zeitschrift fur Obst-Und Weinbau, 123: 338-42.

Riley, R.G. and Kolattukudy, P.E. (1976). HortSci., 11: 249-51.

Roach, W.A. (1947). Rep. East Malling Res. Stn. for 1946, pp. 88-89.

Robinson, T.L., Lasko, A.N. and Ren, Z. (1991). HortSci., 26: 1005-12.

Robinson, D.W. (1983). Scientific Hort., 34: 12-22.

Robinson, D.W. and O'Kennedy, N.D. (1978). Scientia Hort., 9: 127-36.

Robitaille, H.A. (1975). J. Amer. Soc. Hort. Sci., 100: 524-27.

Rogers, W.S. (1926). Dip. Hort. Thesis, University of Cambridge.

Rogers, W.S. (1927). Rept. East Malling Res. Stn. for 1926, 11 Suppl, pp. 16-32.

Rogers, W.S. (1946). J. Pomol., 22: 209-25.

Rogers, W.S. and Parry, M.S. (1968). J. Hort. Sci., 43: 103-106.

Rogers, W.S. and Vyvyan, M.C. (1934). J. Pom. Hort. Sci., 12: 110-50.

Rom, R.C. (1965). Fruit Var. Hort. Dig., 19: 56.

Rom, R.C. and Brown, S.A. (1979). HortSci., 14: 231-32.

Rome, R.C. and Morris, J.R. (1967). Arkans Fm. Res., 16: 14.

Roosje, G.E. (1970). Proc. N.Y.Sta. Hort. Soc., 115: 21-36.

Rubauskis, E., Skrivele, M., Rezgale, Z., and Ikase, L. (2011). Sodininkyste-ir-Darzininkyste, 30: 3-4.

Rudloff, C.R. and Feucht, W. (1957). Gartenbau., 22: 26-42.

Rudloff, C. R. and Lucke, R. (1960). Erwerbosobstbau, 2: 21-24.

Rudnicki, R. and Pieniazek, J. (1970). Bull. Acad. Polon. Sci., 18: 577-80.

Rupp, L.AA. and Anderson, J.L. (1985). HortSci., 20: 727-29.

Ryou, M.S., Lee, J.C. and Hwang, Y.S. (1986). J. Korean Soc. Hort. Sci., 27: 261-68.

Sach, R.M. (1977). HortSci., 12: 220-22.

Sadowski, A., Pekacki, J., Szablowski, W. and Seibisz, K. (1967). Act Agrobotanica, 29: 159-77.

Sako, J. (1986). Abst. 22nd Int. Hort. Congress, California, No.208.

Saltveit, M.E.Jr. (1980). HortSci., 15: 759-60.

Samisch, R.M. and Reich, S. (1950). Katavirn, Vol.1, pp. 293-307.

Samisch, R.M. and Steiner, A. (1956). Hassadeh., 37: 119-22.

Sams, E.E. and Conway, W.S. (1987). Plant Disease, 71 : 1003-1005.

Sando, C.C. (1937). J. BioI. Chem., 117: 45-51.

Saniewski, M., Nowacki, J., Lange, E. and Czapski, J. (1986). Fruit Sci. Rep., 13: 193-200.

Saniewski, M., Nowacki, J., Lange, E. and Czapski, J.(1988). Fruit Sci. Rep., 15: 97-102.

Sansavini, S. (1980). Riv. Ortoflorofr. Ital. 62: 597-602.

Sansavini, S. (1982). Proc. 21st Int. Hort. Congress, Hamburg, 1: 182-97.

Sansavini, S., Bassi, D. and Giunchi, L. (1980). Acta Hort., 114: 114-34.

Saure, M.C. (1987). Sci. Hort., 30: 253-82.

Sax, K. (1949). Proc. Amer. Soc. Hort. Sci., 53: 219-20.

Sax, K. (1953). Proc. Amer. Soc. Hort. Sci., 62: 201-204.

Schaeffer, H. (1987). Angewandte Botanik, 61: 203-15.

Schafter, R.J., Ireland, H.S., Ross, J.J., Ling, T.J. and David, K.M. (2013). AOB-Plants, 2013: pls 047.

Schaik, A.C.R. Van. and Boerrigter, H.A.M. (1987). Fruitteelt, 77: 12·13.

Schceier, P., Drawert, F., Sterger, A. and Mick, W. (1978). J.Fd. Sci., 43: 1797-1800.

Schmidt, S. (1978). Arch. Gartenbau, 26: 177-87.

Schmidt, S., Katzfuss, M. and Buban, T. (1975). BioI. Plant., Prague, 17: 231-34.

Schneider, G. W. (1978). Acta Hort., 80: 225-32.

Schouten, S.P. (1984). Fruitteelt, 74: 1078-79.

Schouten, S.P. (1988). FruitteeIt, 78: 20-22.

Schrader, L., Zhang, J. and Sun, J. (2003). Acta Hort., 618: 397-405.

Schumacher, R. (1963). Schweiz. Z Obst-Weinbau, 72: 253-59.

Schumacher, R. and Fankhauser, F. (1985). Erwerbsobstbau, 27: 108-10.

Schumacher, R., Frankhauser, F. and Stadler, W. (1976). Schweiz, Z. Obst-Weinbau, 112: 3-7.

Schumacher, R. (1962). Landwirtsch. Forsch., 1: 361-69.

Schupp, J.R. and Greene, D.W. (2004). HortSci., 39: 1030-35.

Scott, K.J. and Wills, R.B.H. (1979). Aust. J. Agric. Res., 30: 917-28.

Scott, K.J., O'Loughlin, J., England, B. and Robetts, E.A. (1985). Aust. J. Agric. Res., 36: 305-13.

Seth, J.N. and Singh, B.B. (1977). Progressive Hort., 9: 25-28.

Seth, J.N., Lal, S.D., Solanki, S.S. and Kukoal, R.P. (1981). Fruit Var. J., 35: 77-78.

Shabi, E. and Elisha, S. (1988). Alon Hanotea, 42: 539-50.

Shabi, Z. and Elisha, S. (1987). Alon Hanotea, 41: 1011-14.

Shafiq, M., Singh, Z. and Khan, A.S. (2011). Int. J. Fd. Sci. Tech., 46(12).: 2520-29.

Shapatava, Z., Bobokashvili, Z., Jgenti., M. and Dzeria, K. (2014). Acta Hort., 1032: 265-70.

Sharma, R.P. and Azad, K.C. (1983). Indian Hort., 28: 23-25.

Sharples, G.C. (1961). Inform. Fitopat., 11: 137-38.

Sharples, R.O. (1973). Rept. East Malling Res. Stn. for 1972 p.69.

Sharples, R.O. (1972). Rept. East Malling Res. Stn. for 1971, p.71.

Sharples, R.O. (1974). Rept. East Malling Res. Stn. for 1973, pp. 189-90.

Sharples, R.O. (1975). In: Climate and the Orchard, (ed. H.C. Pereira). CAB, Slough, UK, pp. 113-24.

Sharples, R.O. (1982). Proc. Nat. CA. Res. Conf. Oregon, USA, pp. 131-38.

Sharples, R.O. (1983). Rept. Royal Hort. Soc. Conf., pp. 63-69.

Sharples, R.O. (1984a). Rept. East Malling Res. Stn. for 1983, p.147.

Sharples, R.O. (1984b). Acta Hort., 157: 93-104.

Sharples, R.O. and Stow, J.R. (1982). Rept. East. Malling Res. Stn. for 1981, pp. I99-202.

Sharples, R.O. and Stow, J.R. (1986). Rept. East Malling Res. Stn. for 1985, pp. 165-70.

Shear, C.B. (1974). Proc. 7th Int. Colloq. PI. Analysis Fertil. Probls, Hanover, pp. 427-36.

Shear, G.B. (1971). J. Amer. Soc. Hort. Sci., 96: 415-17.

Shearing, S.J. and Palmer, B. (1986). Abst. 22nd Int. Hort. Congress, California, No.876.

Shepherd, U.M. (1979). Rept. East Malling Res. Stn. for 1978 p.40.

Sherman, M. (1985). HortSci., 20: 57-60.

Shoemaker, J.S. and Teskey, B.J.E. (1959). Tree Fruit Production. John Wiley and Sons. Ivc, pp. l-129.

ShuChang, F., TianFeng, Du., ShuHuan, Ma and YuQiu, L. (2002). China Fruits, 3: 27-28.

Sidnon, A.S. (1986). Instituta Ras Tenievodstva imeni N.J. Vavilova, No.66, pp. 9-12.

Siegelman, H.W. (1955). J. Biol. Chem., 213: 647-54.

Siegelman, H.W. and Hendricks, S.B. (1958). PI. Physiol., Lancaster, 33: 188-92.

Siegrist, J.P. (1987). Revue Suisse de Viticulture, d' Arboriculture et d' Horticulture, 19 : 295-98.

Silveira, C.A.P., Citadin, I. and Fortes, C.R. de. L. (2001). Revista Brasileira de Agrociencia, 7: 107-109.

Simidchev, T. (1976). Proc. Int. Symp. Melliferows Flora Budapest, pp. 85-90.

Simmons, R.K. (1960). Proc. Amer. Soc. Hort. Sci., 76: 41-51.

Simmons, R.K. and Chu, M.C (1980). In: Mineral Nutrition of Fruit Trees (eds. D. Atiknson, J.E. Jackson, R.O. Sharples and W.M. Waller). Butterworths, UK, pp. 57-69.

Singh, M. (1986). Proc. Nat. Symp. on Temperate Fruits. Himachal Pradesh Ag. Univ. Solan. India. pp. 363-66.

Singh, R., Pant, P.C., Singh, M. and Adhikari, K.S. (1978). Progressive Hort., 10: 29-38.

Singh, S.S., Tiwari, H.C. and Rai, K.M. (2001). J. Entomoligical Res., 25: 69-71.

Sinha, M.M., Awasthi, D.N. and Mishra, R.S. (1982). Indian Hort., 27: 23-24.

Sinska, I. and Lewak, St. (1977). Plant Sci. Lett., 9: 163-70.

Skripnikova, M.K. (1986). Sbornik Nauchnykh Trudov, Vsezoyuznyi Nauchno-Issledovatel'skii Institut. Sadorodstva irneni I.V. Michurina No.48. pp. 31-34.

Skrzynski, J. and Poniedziaek, W. (2000). Zeszyty Naukowe Instytutu Sadownictwa-i-Kwiaciarstwa-w-Skierniewicah, 8: 53-58.

Slowik, B. (1970). Proc. Instytute Sadownictwa w. Skerniewicach, 14 : 71-84.

Slyusarchuk, O.E. (1987). Tezisy Dokladov Konferentsii, Zhitomir, pp. 62-63.

Smith, C.B., Greene, G.M. and Demchak, K.T. (1986). Abst. 22nd Int. Hort. Congress, California. No.858.

Smock, R.M. (1946). Proc. Amer. Soc. Hort. Sci., 47: 67-74.

Smith, H.C (1955). Orchardist of N.Z., 28: 16-21.

Smith, S.M. and Stow, J.R. (1984). Ann. Appl. Biol., 10: 383-87.

Smith, W.W. (1970). In: The Biochemistry of Fruits and their Products. (ed. A.C Hulme)., Academic Press. New York, p.456.

Smock, R.M. (1969). J. Amer. Soc. Hort. Sci., 94: 49-51.

Smock, R.M. (1964). Proc. N.Y St. Hort. Soc., 109: 185-88.

Smock, R.M. (1971). J. Amer. Soc. Hort. Sci., 96: 337-71.

Smock, R.M. and Boynton, D. (1944). Proc. Amer. Soc. Hort. Sci., 77: 45-46.

Smock, R.M. and Neubert, A.M. (1950). Apple and Apple Products. Interscience, New York.

Smock, R.M., Martin, D. and Padfield, CA.S. (1962). Proc. Amer. Soc. Hort. Sci., 81: 51-56.

Song, J. and Bangerth, F. (2003). Posth. Biol. Technol, 30: 113-21.

Sorauer, P. (1972). Manual of Plant Diseases. Record Press, p.908.

Sosana, T. (2000). Zeszyty Naukowe Instytutu Sadownictwa-i-Kwiaciarstwa-w-Skierniewicah; 8, 31-37.

Southwick, F.W. (1961). New Jersy St. Univ. 1.

Southwick, F.W., Domoranville, I.E. and Anderson. J.F. (1953). Proc Amer. Soc. Hort. Sci., 61: 155-62.

Spangelo, L.P.S., Fejer, S.O. and Ranger, R.L. (1974). Canad. J. PI. Sci., 54: 601-603.

Speigel-Roy, P. and Alston, F.H. (1982). J. Hort. Sci., 57: 145-50.

Srivastava, R.P. (1975). Indian Hort., 19: 3-8.

Srivastava, R.P. and Ram Kumar (1980). Indian Hort., 25: 8-10.

Stadler, L.J. (1930). J. Hered., 21: 3-19.

Stamenkovic, S. and Stamenkovic, T. (1987). Jugoslovensko Vocarstvo, 21 : 29-35.

Stan, S., Cototobai, M. and Popescu, I. (1977). Lucrarile Stiintifice Ale Institutului De Cercetari Peotru Pomicultura Pitesti, 6: 109-16.

Standardi, A. (1985). Annali delia Facolta di Agraria, 39: 109-120.

Staubli, A. and Chapuis, P. (1987). Revue Suisse de Viticulture, Arboriculture, Horticulture, 19: 339-47.

Stebbins, R.L. and Dewey, D.H. (1972). J. Amer. Soc. Hort. Sci., 97: 471-74.

Stefanov, K. (1988). Rasteniev dni Nauki, 25 : 61-64.

Stevenson, C.D. and Dodd, B.C. (1962). Qd. J. Agric. Sci., 19: 299-303.

Stoilov, G. and Todorov, V. (1972). Gradinarska i Lozarska Nauka, 9 : 3-8.

Stojanowska, J. (1987). Fruit Sci. Rept., 14: 79-84.

Stoll, K. (1978). Schweizer, Z. Obst. Weinbau, 114: 605-609.

Stover, E., Fargione, M. J., Watkins, C.B. and Iungerman, K.A. (2003). HortSci., 38: 1093-99.

Streil, J. (1983). Acta Hort., 138: 69-76.

Streil, J. and Bangerth, F. (1988). J. Hort. Sci., 63: 193-99.

Stringer, A., Lyons. C.H. and Milsom. M. (1971). Rept. Long Ashton Res. Stn. for 1970. pp. 106-108.

Subhadrabandhu, S. and Punsri. P. (1986). Thai. J. Agric. Sci., 19: 141-44.

Subrtova, D., Borek, V. and Hubacek, J. (1987). Sbornik UV TIZ Zahradnictvi, 14: 155-61.

Sud, V.K., Hameed. S.F. and Sharma. L.O. (1975). J. Hort. Sci., 50: 165-68.

Szczygie, A., Kadzik, F. and Mika. A. (2000). Zeszyty Naukowe Instytutu Sadownictwa-i-Kwiaciarstwa-w-Skierniewi-cach, 8: 87-97.

Tabuenca, M.C. (1983). Anales de la Estaction Experimental de Aula Dei. 16: 341-46.

TaeMyung, Y., KyuRae, K., SeakWon, Co., SuGon, H., YoungJae, W., JaeYeol, L. and JongGil, S. (2001). J. Korean Soc. Hort. Sci., 42: 545-48.

Tamai, H., Ono, T., Koike, H. and Shigehara, I. (2002). J. Jap. Soc. Hort. Sci., 71 : 670-74.

Taper. C.D. and Liu, P.S. (1969). Canad J. Pl. Sci., 49: 97-98.

Tarasov, V.M. (1966). Izv. Timiryazev. Sel'hoz Akad., No.1.

Tarasov, V.M. and Kovalenko, V.F. (1970). Izv. Izv. Timiryazev. Sel'hoz Adad. 1: 142-52.

Tatsuki, M., Eudo, A. and Ohkawa, H. (2007). Postharvest Biol. Technol., 43: 28-35.

Teaotia, S.S. Srivastava. R.P. and Misra. R.S. (1983). Progressive Hort. 14: 99-103.

Temple,R.C. (1914). Rept. Hakluyt Society. London.

Tennes, B.R., Levin, J.H. and Wittcnberger, R. (1969). Proc. N.Y. Sta. Hort. Soc., 114: 115-21.

TerAvest, D., Smith, J.L., Carpenter-Boggs, L., Granatsteen, D. Hoagland, L. and Reganold, J.P. (2011). HortSci., 46: 1687-94.

Terblanche, J.H., Gurgen, K.H. and Hescbcck, I. (1980). In: Mineral Nutrion of Fruit Trees. (eds. D. Atkinson, J.E. Jackson, R.O. Sharples and W.M. Waller). Butterworth, UK. pp. 71-82.

Teren'Ko, G.N. and Kal'gina, R.V. (1988). Sadovodstvo i Vinogradarstvo, 2: 14-16.

Teskey, BJ.E. and Francis, F.J. (1954). Proc. Amer. Soc. Hort. Sci., 63: 220-24.

Thakur, A.K. and Kashyap, N.P. (1986). Proc. Nat. Symp. Temperate Fruits, pp. 381-84.

Thammawong, M. and Arakawa, O. (2009). J. Applied Hort., 11: 23-30.

Thevenot, C. (1980). Doctorate Thesis, L'Universite Pierre et Marie Curie, Paris.

Thomas, L.A. (1968). Qld. Agric. Anim. Sci., 25: 211-18.

Thompson, A.H. (1951). Proc. Amer. Soc. Hort. Sci., 58: 57-64,.

Tiemann, K.H. and Dammann, H.J. (1981). Mitt. Obstb. Alten Landes, 36: 49-68.

Tiscornia, J.R. and Iuchi, T. (1984). Empresa Catarinense de Pesquisa Agropecuaria S.N., Vol.3, pp. 878-90.

Tohutt, K.R. (1984). Scientific Hort., 35: 72-77.

Treeccani, C.P. and Stanchina, G.De. (1982). Proc. 21st lnt. Hort. Congress, Hamburg. 1: 310-22.

Treharne, K.J., Quinlan, J.D., Knight, J.N. and Ward. O.A. (1985). Plant Growth Regulation, 3: 125-32.

Tripathi, S.N., Chadha, T.R., Chopra, S.K. and Bhargava, J.N. (1986). Proc. Nat. Symp. Temperate Fruit. Himachal Pradesh Agril. Univ. Solan, India. pp. 301-305.

Trivindo, A.Z. (1973). Vestnik Sel'skokhzyastvennoi Nauki, 16: 102-17.

Tromp, J. and Wertheim, S.J. (2005). In: Fundamentals of Temperate Zone Tree Fruit Production (Eds. J. Tromp, A.D. Webster and S.J. Wertheim)., Backhys Publishers, The Netherlands, pp. 240-66.

Tromp, J. (1972). J. Hort. Sci., 47: 525-33.

Tromp, J. (1979). Comm. Soil Sci. Pl. Anal., pp. 325-33.

Tromp, J. (2005). In: Fundamentals of Temperate Zone Tree Fruit Production (Eds. J. Tromp, A.D. Webster and S.J. Wertheim). Backhuys Publishers, Leiden, The Netherlands, pp. 295-10.

Tromp, J. (1984). Acta Hort., 149: 39-47.

Tukey, L.D. (1965a). Dwarfed Fruit Trees. Macmillan. New York.

Tukey, L.D. (1965b). Trans. Ill. St. Hort. Soc., 99: 68-75.

Tukey, L.D. (1982). Penn. Sta. Hort. Rev., 31: 3-5.

Tukey, L.O. (1960). Farmer, 7: 6.

Tukey, R.B., Langston, R. and Cline, R.A. (1962). Proc. Amer. Soc. Hort. Sci., 80: 73-78.

Tumanov, I.I. and Gareev, E.Z. (1951). Inst. Fiziol. Rast. Timirjazeva, 7: 22-108.

Tworkoski, I.I., Young, R.S. and Sterrett, J.P. (1988). Weed Technol., 2: 31-35.

Tyburski, M. and Wazbinska, J. (2000). Biuletyn Naukowy, 8: 297-60.

Uitterlinden, L. (1964). Fruitteelt, 54: 166-67.

Unrath, C.R. (1974). J. Amer. Soc. Hort. Sci., 99: 381-84.

Unrath, C.R. (1972). J. Amer. Soc. Hort. Sci., 97: 55-58.

Uota, M. (1950). Can. Grow., 73: 9.

Uota, M. (1952). Proc. Amer. Soc. Hort. Sci., 59: 231-37.

Urunov, F. (1987). Vinogradarstva i Ovoshchevodstva, 3: 19-22.

Uterrnark, H. (1977). Compact Fruit Tree, 10: 86-90.

Van der Boon, J. (1974). Acta Hort., 45: 9-16.

Van der Boon, J., Das, A. and Schreven, A.C. (1968). Agric. Res. Rep., PUDOC, Wageningen, p. 711.

Van Doren, A. (1961). New Jersey. St. Univ., pp. 22-26.

Van Oosten, H.J. (1978). Acta Hort., 65: 123-27.

Varga, A. (1966). Proc. Biol. and Med. Sci. (Amsterdam). 59: 641-44.

Varkvasov, V.A. (1971). Khimiya v Sel'skom khozyaistve, 9: 26-27.

Velasco, R., Zharkikh, A., Affourti, J., Dhingra, A., Cestavo, A. (2010). Nat. Genet., 42: 833-39.

Velcheva, N. (1986). Agrokhiyiya i Restielna Zashchta, 21: 121-30.

Vercammen, J. (2004). Fruit Belge, 72: 73-80.

Verheij, E.W.M. and Verwer, F.L.J.A.W. (1973). Scientia Hort.,1: 25-42.

Verma, K.D. and Gupta, G.K. (1986). Proc. Nat. Symp. Temperate Fruits. pp. 407-409.

Videnova, E. and Ismail, K.H. (1985). Agrokhiyiya i Rastielna Zashchita, 20: 90-98.

Vigl, J., Boscheri, S. and Mantinger, H. (1985). Obstbau Weinbau, 22: 108-12.

Villeneuve, F., Ferre, G. and Blanc, R. (1988). Infos, Centre Technique Interprofessionnel des Fruits et Legumes. France, 38: 4-8.

Visser, T. and Verhaegh, J.J. (1987). Fruitteelt, 77: 11.

Vittrup, C. I. (1970). Tidsskr Plantevl., 74: 481-84.

Vlasishin, V.F. (1979). Khimiya v Sel'skorn Khozyaistve, 17 : 29-32.

Vogl, M. (1985). Gartenbau, 32: 84-85.

Vojvodic, D. and Dulic, K. (1987). Zastita Bilja, 38: 33-41.

Vorijhof, B. (1952). Fruitteelt, 42: 546-47.

Vyvyan, M.C. (1949). Ann. Appl. Biol. 36: 553-58.

Vyvyan, M.C. (1942). Rept. East Malling Res. Stn. for 1941, pp. 38-40.

Wallace, T. (1951). The Diagnosis of Mineral Deficiencies in Plants by Visual Symptoms, H.M. Stationery Office, London.

Wallace, T. (1953). Science and Fruit. Univ. Bristol. UK. pp. 140-61.

Waller, W.M. (1980). In: Mineral Nutrition of Fruit Trees. (eds. D. Atkinson. J.E.Jackson, R.O. Sharples and W.M. Waller). Butterworth. UK. pp. 383-94.

Walter, T.E. (1966). Rept. East Malling Res. Stn. for 1965. pp. 83-95.

Walter, T.E. (1967). Rept. East Malling Res. Stn. for 1966, pp. 70-82.

Wang, S.Y. and Steffens, G.L. (1987). J. Pl. Growth Regulation, 6: 209-13.

Warne, L.C.C. and Wallace, T. (1935). J. Pomol., 13: 1-31.

Watanabe, M., Segawa, H., Murakami, M., Sagawa, S. and Komori, S. (2008). J. Jap. Soc. Hort. Sci., 77 : 350-57.

Watkins, C.B. (2002). In: Fruit Quality and its Biological Basis (Ed. M. Knee). Sheffield Academic Press Ltd, UK, pp. 180-224.

Watkins, C.B. (2006). Int. J. Postharvest Technol. Inn., 1: 62-68.

Watkins, C.B., Nock, J.F. and Whitaker, B.D. (2000). Postharvest Biol. Technol., 19: 17-32.

Watkins, R. (1979). In: Evolution of Crop Plants. (ed. N. W. Simmonds). Longman, UK., pp. 247-50.

Weber, A., Brackmann, A., Anese, R-de-O., Booth, V. and Pavanello, E.P. (2011). Pesquisa Agropecuqria Brasileira, 46: 1597-602.

Weeks, W.D., Southwick, F.W., Drake, M. and Steckel, J.E. (1958). Proc. Amer. Soc. Hort. Sci.,66: 11-19.

Weeks, W.D. (1952). Proc. Amer. Soc. Hort. Sci., 60: 11-12.

Weissenborn, K. (1967). Mitt. Obst. Versuchsringes. Jork., 22 : 210-19.

Welander, M. (1988). J. Pl. Physiol., 132: 738-44.

Welander, M. and Huntrieser, I. (1981). Physiol. Plantarum, 53: 301-306.

Wertheim, S.J. (1970). The apple : 1. Systematics. Handbuch der Pflanzenziichtung, 6. Berlin. 2nd ed. pp. 632-45.

Wertheim, S.J. (1978a). Scientia Hort., 9: 337-45.

Wertheim, S.J. (1978b). Acta Hort., 80: 49-54.

Wertheim, S.J. (1987). Fruitteelt, 77: 15-16.

Wertheim, S.J. (2005). In: Fundamentals of Temperate Zone Tree Fruit Production. (Eds: J. Tromp, A.D. Webster and S.J. Wertheim). Backhys Publishers, The Netherlands, pp 311-24.

Werzilov, W.F., Plotnikova, L.V. and Aleyandrova, W.S. (1978). Acta Hort., 80: 175.

Westwood, M.N. (1978). In: Temperate Zone Pornology, Freeman and Co. San Francisco, p.242.

Westwood, M.N. and Zielinski, O.B. (1966). Proc. Amer. Soc. Hort. Sci., 88: 9-13.

White, D.C. (1953). Proc. Amer. Soc. Hort. Sci., 61: 180-84.

White, G. (1984). Fruit Veg. Sci., February, pp. l2-13.

White, G. and Greenham, D.W.P. (1967). J. Hort. Sci., 42: 419-28.

Whitfield, B.A. (1964). Rept. East. Malling Res. Stn. for 1963. pp. 107-10.

Whitwell, J.D. (1956). Comm. Grower, 3170: 705.

Wilcox, AmN. (1962). The apple: I. Systematics. Handbuch der Ptlanzenziichtung. 6., Berlin, 2nd ed. pp. 637-45.

Wildung, D.K., Weiser, C.J. and Pellett, H.M. (1973). Canad. J. PI. Sci., 53: 323-29.

Wilkinson, B.G. (1968). J. Sci. Fd. Agric., 19: 646-47.

Williams, M.W. and Letham, D.S. (1969). HortSci., 4: 215-56.

Williams, R.R. (1970). Pollen-tube growth rate. In: Towards Regulated Cropping (eds. R.R. Williams and D. Wilson). Grower Books. London, p.61.

Williams, R.R. and Chuurch, R.M. (1983). J. Hort. Sci., 58: 337-42.

Williams, A.A. Tucknott, O.G. and Lewis, M.J. (1977). J. Fd. Sci., 43: 1797-800.

Winter, F. (1967). Erwerbsobastabu, 9: 203-207.

Winter, F. (1964). Erwerbsobstbau, 6: 88-92.

Wirth, H. and Vogt, T. (1988). Obstbau, 13: 100-102.

Wojcik, P. (2001). J. Plant Nut., 24: 1925-36.

Workman, M. (1963a). Proc. Amer. Soc. Hort. Sci., 83: 126-34.

Workman, M. (1963b). Proc. Amer. Soc. Hort. Sci.,83: 149-61.

Wrona, D. (2004). Folia Hort., 16.: 55-60.

Wunsche, J.N. and Heyn, C.S. (2015). Europ. J. Hort Sci., 80: 3-10.

Wunsche, J.N., Bowen, J., Ferguson, I., Woolf A and Mc Ghie, T. (2004). Acta Hort., 636: 631-36.

Wunsche, J.N., Greer, D.H., Lian, W.A. and Palmer, J.W. (2005). Tree Physiol., 25: 1253-63.

Xia, C.S., Wang, L.Y. and Zhou, P. (1983). Shanxi Fruit Trees, 4: 11-12.

Xu, K. (2010). NY Fruit Quart., 18: 11-14.

Xu-Guixuan, Wen-Shuying, He-Mingli., Song-Zhe., Shang-ChunBo, Li-HongJian; Wang-ChunHong and Zhang-Delong (2012). J. Fruit Sci., 29: 143-44.

Yae, B.W., Vim, Y.J. and Jo, H.M. (1986). J. Korean Soc. Hort. Sci., 27: 353-58.

Yahia, E. M., Liu, F.W. and Acree, T.E. (1990). Lebens. Wiss. U-Technol 23: 488-93.

Yamazaki, T., Mori, H., Yokomoizo, H. and Fukuda, H. (1962). Bull. Tohoku Agric. Exp. Stn., 23: 153-161.

Yoshida, Y., Haniuda, T., Tsuchiya, S., Sanade, T., Masuda, T., Bessho, H. and Sadamori, S. (1985). Bull. Fr. Trees. Res. Sta. Japan, 12: 1-10.

Zagaja. S.W., Czynczyk, A., Jakubowski, T. and Omiecinska, B. (1986). Abst. 22nd Int. Hort. Congress, California, No. 1345.

Zeng, X. (1990). Physiology of Fruit Trees. Beijing Agriculture Univ. Press, Beijing pp 134-85.

Zeng, X., Meng, Z. Q., Fu, Y.H., Li, G.C. and Wu, X.R. (1987). Sinicia Hort., 14.: 2-38.

Zhai, H.C., Ren, E.M., Li, D.C., Shi, G.Y. Lin and Shu, H.R. (2006). Acta Hort. Sinica, 33: 921-26.

Zhang-Xin, Yuan-Qin, Liang-Dong; Zhou-Yangjun and Ma-FengWang (2014). In Vitro Celluar and Dev. Biol. Plant, 50: 263-70.

Zhaonian Guo, (1986). Abst. 22nd Int. Hort. Congress.California. Abst. No.609.

Zheng, Z., Zhezhi, W. and Zhiqin, Z. (2004). Acta Botanica Boreali Occidentalia Sinica, 24: 621-26.

Zika, J. (1987). Proc. Symp. 60 yrs. of Hort. Res. in Czechoslovakia, pp. 103-109.

Zorbrist, L. (1962). Proc. Br. Conf. Insect. Fung., Brighton, U.K. 1961, 1 : 237-38.

Zzchokke, A (1987). Landw. Schwciz., pp. 153-97.

Chapter 2

Apricots

The apricot is one of the most cultivated stone fruit in the world. The apricot (*Prunus armeniaca* L.) is an attractive, delicious and highly nutritious fruit. It is particularly a rich source of vitamin A and also contains high amount of carbohydrates, protein, phosphorus and niacin than majority of other common fruits (Teskey and Shoemaker, 1972). Apricots are grown with minimum cultural care in the tracts which otherwise unfit for cultivating other temperate fruit crops. It is fairly drought resistant, salt tolerant, hardy plant being rather less susceptible to pests and diseases.

Composition and Uses

The percentages of various ingredients in the edible portion of fresh Indian apricot fruits have been recorded as (g $100g^{-1}$) moisture 85.3, protein 1.0, fat 0.3, fiber 1.1, other carbohydrates 11.6, mineral matter 0.7 per cent, calcium 20 mg, phosphorus 25 mg, iron 2.2 mg, vitamin A 3600 I.U., thiamine 0.04 mg, riboflavin 0.13 mg, nicotinic acid 0.6 mg, ascorbic acid 6 mg and calories 53 (Anon., 1969). Detailed analysis of various stone fruits has been carried out by Wills *et al.* (1983) in Australia. Nutrient composition of fruits of two apricot cultivars, Bavinity and Trevatt, is given in Table 2.1.

Nineteen amino acids have been identified in apricot and predominant ones are glutamic acid, aspartic acid, histidine, arginine, tyrosine, cystine, methionine and hydroxyproline (Hegazi and Salem, 1972). Glucose, fructose, sucrose, sorbitol, mesoinositol and a number of oligosaccharides are the major constituents of the sugars and polyols (Anon., 1969). Reducing sugars of the fruit increase steadily with its development while sucrose appears at the fully ripe stage. Variation in the contents of malic and citric, the main acids present in the fruit, is a varietal characteristic in

Table 2.1: Nutrient Composition of Apricot Fruit (g $100g^{-1}$ edible portion)*

Cultivar	*Sugars*								*Dietary Organics*		
	Water	*Protein*	*Fat*	*Fructose*	*Glucose*	*Sucrose*	*Starch*	*Fibre*	*Acids*	*Ash*	*Energy*** (kJ)*
Bavinity	85.3	0.9	0.1	0.3	1.1	4.4	0	2.0	2.6	0.5	141
Trevatt	85.6	0.7	0.1	0.5	2.7	5.1	0	3.0	1.5	0.5	167

Cultivar	*Organic Acid Composition and Soluble Solids*				
	Organic acids (mg $100g^{-1}$)				
	Malic	*Citric*	*Quinic*	*Titratable Acidity (meq H^+ $100g^{-1}$)*	*Soluble Solids (°Brix)*
Bavinity	440	2130	62	33.7	12.5
Trevatt	770	660	76	17.5	11.0

	Vitamins and Minerals Content (mg $100g^{-1}$ edible portion)										
	Vitamins							*Mineral (mg)*			
	Vitamin C	*Carotene (µg)*	*Thiamine*	*Riboflavin*	*Niacin*	*K*	*Na*	*Ca*	*Mg*	*Fe*	*Zn*
Bavinity	7	617	0.02	0.03	1.4	350	1	16	9	0.3	0.1
Trevatt	16	153	0.03	0.04	1.1	320	3	15	9	0.3	0.2

*Wills *et al.* (1983).

Energy (kJ) = (g protein 17) + (g fat 37) + (g sugar as monosaccharide 16) + (g organic acid 10).

apricot (Souty *et al.*, 1976). The pulp contains about 1 per cent pectic substances as calcium pectate. In apricot, the predominant phenolic compound is catechin, with significant amounts of epigallocatechin, neochlorogenic acid, chlorogenic acid, rutin and epicatechin, and relative minor amounts of cyanidin-3- glucoside, quercetin -3- glucoside and a quercetin derivative. Beta-carotene is the main carotenoid, with other compounds are beta-cryptoxanthin, lutein, zeaxanthin and lycopene (Campbell *et al.*, 2011). The apricot fruit has got a distinct pleasant aroma imparted to it by many volatile constituents. The main volatiles in apricot are aldehydes, alcohols, acetates, esters, terpenes and acids. Among these compounds, Gokbulut and Karabulut (2012) working with eight Malatya apricot cultivars of Turkey noted ethanol, hexanol, hexylacetate, (Z)-3-hexanol acetate (E)-2-hexanol, hexylacetate, (Z)-3-hexand and (E)-2-hexen-1-ol. Geneovese *et al.* (2004) stated that the aroma volatiles of the apricot distillate were characterized by a high concentration of alcohols and by a specific terpene profile that included linalool, ocimenol, alpha-terpineol, nerol, geraniol, cis-and trans-linalool oxide. Wang *et al.* (2009) working with 14 apricot cultivars in China identified 208 aromatic components, including 80 kinds of esters, 25 kinds of aldehydes, 21 kinds of ketones 39 kind of acid substrate and 15 kinds of other substances.

Apricot fruits are used in many ways. A tree-ripen apricot is an excellent dessert fruit. Because of its perishable nature, the fruit is canned, candied, frozen and dried. The fruit is also processed into a number of products, such as, jam, nectar and *papad.* The fruit pulp is thinly spread, dehydrated and pressed to make *papad.* Apricot leather (*kamaradin*) is similarly prepared by drying the pulp after treating it with sulphur dioxide as preservative and mixing the same with sucrose and glucose syrup. Another popular beverage, apricot nectar, is made by converting the ripe fruits into puree and mixing it with sugar syrup and citric acid. Apricots are also consumed as dry fruits.

The apricot seed represents about 15 per cent of the fruit and kernel represents about 34 per cent of the seed. The kernel is a valuable byproduct. Depending upon the cultivar, a kernel is either sweet or bitter. Sweet kernels taste like almond and are used as its substitute in confectionery and can be added to jam. The proximate composition of whole kernel on dry bases were reported as 5.50 per cent moisture, 27.83 per cent crude protein, 50.30 per cent fat, 2.20 per cent ash, 11.32 per cent total sugar and 3.16 per cent crude fibers. The kernels contained 210 mg K, 400 mg Mg, 280 mg P, and 2.8 mg/100g Mn (Azouz *et al.*, 2009)

Apricot kernel oil is an important commercial commodity which is quite similar to almond oil in physical and chemical properties both containing over 90 per cent of their fatty acids as oleic and linoleic acids (Hallobo *et al.*, 1975). The characteristic of apricot kernel oil was reported as 0.916g/cm^3 for specific activity. The defatted apricot kernels cakes were free from HCN and contained 16 amino acids. The major amino acids were identified as glutamic (20.75 per cent), aspartic (14.11 per cent) and glycine (11.86 per cent). The kernel oil is also rich in fatty acids and the major fatty acids were detected as oleic and linoleic acids (Azouz *et al.*, 2009)

Origin and Distribution

It is generally agreed that the cultivated apricot has its origin in western China although it has been included in the three centres of origin by Vavilov (1951), *i.e.* the Chinese centre, the central Asian centre and the Near Eastern centre, the last one being the secondary centre of origin. To be very precise, its centre of origin is northeast China between the town Kan-Techeou and Russian border (Crossa-Raynaud and Audergon 1987) The Northern China group includes forms of *P. mandshurica* and *P. sibirica*, and the Eastern China group includes forms of *P. ansu* (Romero *et al.*, 2003). Tibet eco-geographic group includes forms of *P. armenicaca* var. *holosericea*. The Central Asian and Irano-caucausian including Turkish and Iran cultivars eco-geographical groups show the richest phenotypic variability, while European group including cultivars grown in North America, Australia and South Africa showed the least diversity (Halasz *et al.*, 2010). Kostina (1936) reported the distribution of the cultivated apricot, its wild forms and allied species in temperate zone of Asia between 33° and 70° E longitude and 53° and 30° N latitude. It is believed that apricot was introduced to Anatolia, Turkey by traders from China and central Asia through silk road, and then the Roman soldiers carried apricot to Italy. Apricot was spread to Southern Europe from Eastern Europe (Asma, 2011), to England in 1524 or 1548 from Italy which was later moved to America by the Spanish in 1626 (Faust *et al.*, 1998) On the basis of many explorations and a very detailed study of living collections comprising seven species and 600 forms, Kostina (1969) distinguished the following eco-geographical groups and regional sub-groups among the cultivated apricot.

1. **Central Asian**
   - ☆ Fergana
   - ☆ Upper Zeravshan
   - ☆ Samarkand
   - ☆ Shahrisyabz
   - ☆ Horezm
   - ☆ Kopet-dag
2. **Irano-Caucasian**
   - ☆ Irano-Caucasian
   - ☆ Dagestan
3. **European**
   - ☆ West European
   - ☆ East European
   - ☆ North European
4. **Dzhungar-Zailij**
   - ☆ Dzhungar
   - ☆ Zailij

Apricot Orchard

Brown Rot

Blight

Harcot

Goldbar-150x150

Moorpark

New Castle

Tilton

Wenatchee

Wilt

Bailey and Hough (1975) added two more groups and five sub-groups as under:

5. North Chinese

- ☆ Northeastern Tibet
- ☆ North China
- ☆ Southeastern Manchuria
- ☆ Khingan Mountains

6. East Chinese

- ☆ East China

The cultivars grown in North America, South Africa and Australia are included in European group. The descriptions of the characteristics of the major groups are very useful in breeding programmes for climatic adaptability.

This delectable fruit moved westward from central Asia through Iran and Transcaucasian region and reached Italy during the first century, to England in 13th century and to North America by 1720 (Westwood, 1978). The specific name *armeniaca,* indicates its introduction to Italy and Greece by Armenian traders (Bailey and Hough, 1975).

Area and Production

Apricots are cultivated in temperate climate of all the continents of the world, Asia and Europe being the largest producers' while South America and Oceania being the least. The major apricot producing nations are Turkey, Iran, Uzbekistan, Algeria and Italy (Table 2.2).

Table 2.2: Top Ten Apricot Producing Countries of the World

Name of Country	*Production (Tonnes)*
Turkey	7,95,768
Iran	4,60,000
Uzbekistan	3,65,000
Algeria	2,69,308
Italy	2,47,146
Pakistan	1,92,500
France	1,89,711
Morocco	1,22,405
Spain	1,19,400
Egypt	98,772
Total World Production	**40,38,520**

FAOSTAT, 2013.

In India, the apricot is grown in Jammu and Kashmir, Himachal Pradesh, Uttarakhand and to a limited extent in north eastern hills but it has not been successful in South India (Hayes, 1957). The fresh fruits for the market are produced in mid-hills whereas dry fruits are common at higher elevations in drier situations. In 2012 India produces 18,000 tonnes of apricot (FAOSTAT, 2013)

Taxonomy

Most of the cultivated apricots belong to the species *Prunus armeniaca* which is also sometimes called *Armeniaca vulgaris* (Kostina, 1936; Bailey and Hough, 1975; Watkins, 1976; Okie, 1983)

Several species of apricot are classified either as sub-species or as botanical varieties of *P. armeniaca,* but most of them are accepted as distinct species and are grown for fruits, rootstocks or ornamental purposes.

Table 1.3: Different Species of *Prunus* and their Uses

Species	*Common Name*	*Where Grown*	*Uses*
P. ansu	-	China, Japan	Fruit, rootstock
P. armeniaca	Apricot	Entire world	Fruit, rootstock, oil
P. brigantiaca	Briancon apricot	France	Fruit, oil
P. dasycarpa	Purple apricot	China, USSR	Cold and disease hardy
P. davidiana	-	China, Korea	-
P. mandschurica	Manchurian apricot	China, Korea	Cold-hardy
P. mume	Japanese apricot	Japan, China	Fruit, ornamental
P. sibirica	Siberian apricot	Siberia	Breeding

P. ansu is adapted to humid conditions and is considered by many authors as belonging to *P. armeniaca.* Likewise *P. dasycarpa* is a natural hybrid between *P. armeniaca* and *P. cerasifera* (Kostina and Riabov, 1959). *P. ansu* does not set fruits under Simla, India conditions while the tree thrives well and attains the vigour like any other apricot cultivar. *P. mume* is the earliest apricot to come into bloom in the month of February and if the weather remains good, the fruits set heavily. The fruits are, however, used in pickle and liquor making in Japan where it is mostly grown under warmer and humid conditions (Bailey and Hough, 1975).

P. sibirica and *P. mandschurica* are very cold-hardy species and can withstand as low as - 50° C during dormancy. Hence, these are used for imparting the cold-hardiness trait in apricot breeding programmes. *P. mandschurica* is a bigger tree than *P. sibirica,* which blooms early in the season when grown in warmer regions. The almond-like fruits of *P. sibirica* are not edible.

Cultivars

For commercial cultivation, the selection of cultivars of apricot for a particular site is very crucial because any one cultivar has only limited climatic adaptability and it cannot be grown on a large tract in different situation which have highly variable climatic conditions. Broadly speaking, four climatic zones, *viz.*, low hills, mid hills, high hills and cold dry hills are existing in the hill states. Except the very low and high hills, in other zones, the apricot is cultivated successfully and each zone has its own cultivars. The cultivars New Castle, Shipley Early and Kaisa are grown in mid hills, while Moorpark, Turkey, Royal, St. Ambroise, Nugget and Charmagz in mid and high hills. In cold arid zone comprising Kinnaur, Ladakh and higher ranges of UP hills, of India, Nari, Kaisi, Charmagz, Shakarpara, Halman and Rakcha Karpu cultivars are predominant. In Kashmir Valley, cultivars New Castle, Orgrand and Harcot produced most attractive bigger sized fruits with moderate to higher total soluble solids, ascorbic acid, acidity and sugar content of fruit (Srivastava *et al.*, 2012). The cultivars are also chosen from utilization viewpoint such as fresh market, canning and drying. Drying of apricots is more popular in cold arid zone and the cultivars for this purpose must have high sugar content and sweet kernel. The canning of fruits in syrup is not common in India, therefore, the cultivars producing fruits suitable for fresh market are preferred in low and mid hill regions. In order to avoid the glut of fruits in the market by prolonging the period of fruit harvest, the cultivars ripening at different periods should be planted in an orchard. A brief description of leading commercial apricot cultivars are as follows.

Shipley Early

The fruits ripen in the middle of May making it an early cultivar. The medium-sized fruits are creamy white. Flesh is sweet and juicy but the kernel is bitter. It is a dessert cultivar.

New Castle

It is an early cultivar ripening in the month of May, the exact period depends upon the altitude. The fruits are round, medium-sized and barium-yellow in colour. Flesh is very sweet and juicy. It is a popular dessert cultivar for warmer locations (Bajwa and Misra, 1971a).

Moixent

A hybrid of Goldrich and a Spanish cultivar developed at Valencia, Spain (Martinez Calvo *et al.*, 2011). It is a self compatible, early ripening apricot cultivar with excellent fruit quality and resistance to plum pox virus. It is well adapted to area with warm winter and Mediterranean climate and showed good fruit set and productivity.

Charmagz

The medium-sized fruit is roundish flat. The skin is straw-yellow and the flesh is light yellow, very sweet and highly flavoured. The fruit is suitable for both dessert and drying purposes. The kernel is sweet. This is a self-incompatible cultivar which gives good fruit set when the cultivar Turkey is used as pollenizer (Seth and Kuksal, 1977). It has a high chilling requirement and its cultivation is not increasing because of its shy bearing nature.

Kaisha

The roundish fruits are medium in size having yellowish skin with red blush. The flesh is orange-yellow sweet and tasty. It is a prolific bearing cultivar ripening in early June. The kernel is bitter.

Nugget

In India has been found to be one of the best cultivars for mid hills, in respect of fruit quality, regularity in bearing and yield (Rathore, 1982). It is the earliest cultivar to flower in the' season (Mitreskii and Ristevski, 1985). Fruits are medium to large in size and round in shape. The skin is glabrous, deep orange with vermillion blush. The flesh is deep yellow in colour, firm and juicy with fine texture. Both flesh and kernel are sweet in taste. The fruit is ready to harvest in the last week of Mayor the first week of June.

Royal

Royal is identical to Blenheim and is a popular cultivar of mid hills. The fruits are large in size, yellow in colour, ripening in the middle of June. The flesh is yellow, firm, juicy and sweet in taste. It is a good dessert cultivar and the fruits are suitable for canning. It can be utilized as a parent in breeding for heat resistance (Khalin *et al.*, 1985).

Moorpark

In size and lusciousness, Moorpark is considered as the best apricot cultivar when fully ripe but is shy in bearing. The roundish fruits are large in size. The skin is light orange in colour with a light red blush. The flesh is juicy, sweet and excellent in flavour. The fruit ripens in the first week of June.

St. Ambroise

It is a mid- to late season cultivar ripening from the last week of June to the first week of July. The fruit is oval, large and orange-yellow, having tiny dots of pink colour. The flesh is deep yellow, firm, juicy and very sweet. This cultivar is not suitable for the lower mid hill region.

Nari

Small to medium in size, oblong, depressed at both base and apex ends. The skin is smooth and pale yellow in colour. The flesh is light yellow with slight fibre having sweet pleasant aroma. The kernel is sweet like almond. The fruit is ready to harvest in the last week of May.

Shakarpara

The fruit is similar to the cultivar Nari in size and shape having creamy yellow skin with slight pink blush on the shoulders. The pale yellow flesh is exceedingly sweet with a good aroma. The kernel is bold and sweet. The fruit ripens in the third week of May but it is a shy bearing cultivar and requires high chilling (Misra, 1983).

Halman

Halman is the most important dessert and drying cultivar of Ladakh (Kabu, 1978). The fruit is medium-sized and golden in colour. The flesh is very sweet but not very juicy. It is harvested in early August.

Rakcha Karpu

It is another important cultivar of Ladakh which also fruits well in mid hills. The kernel colour is white in contrast to dark brown colour of the other cultivars. It is dessert cultivar but the kernel is used for oil extraction. It ripens in the last week of May in mid hills while in August in Ladakh.

Wenatchee

Originated as a seedling and was widely grown in the USA and Canada in 1990s as a fresh fruit because of its heavy and regular, cropping. The fruit is large, oblong, irregular, having fair to good flavour. Fruit softens easily on the tree and has yellow skin with green shoulders.

Tilton

It was the most important canning cultivar of the USA as well as Canada. Regular bearing, uniform ripening and greater hardiness of floral buds are its several advantages over Royal. It, however, ripens seven days later than Royal.

Harcot

Layne (1978) has described it as an early high quality apricot which is cold-hardy and resistant to bacterial spot, brown rot and perennial canker. The plant is upright to spreading, vigorous and moderately productive.

Vivagold

It has been evolved from a cross of Veecot x V 49024 (Bradt and Anderson, 1979). The fruit is medium in size varying from 45 to 53 mm in diameter with a good flavour. It ripens a week after Veecot and does not split or crack in wet weather on the tree. The cultivar is resistant to brown rot but not to bacterial spot.

Harogem

It has resulted from the cross Rouge du Roussilon x (Modern 604 x Open). The fruit is very attractive, very firm and of high quality (Layne, 1979). Harogem is a mid-season cultivar suitable for fresh market. The plant is hardy and resistant to brown rot.

Roksana (Rexona)

Roksana is a late ripening cultivar which has originated in Afghanistan (Serafimova and Brisov, 1980). The fruit is of large size (72 g), yellow with red blush. The flesh is crisp, juicy and sweet-acid. The kernel is also sweet. The cultivar is resistant to spring frost and is most suitable for intensive orchards (Ristevski and Mitreski, 1986).

New Cultivars and Hybrids

Daniel

Early season free stone apricot with good productivity, self-fertile and aromatic fruit, developed at Israel. Fruit round and flattened, flesh yellow-orange, texture firm, very good eating quality.

Bora® (B090610010)

A hybrid of Early Blush x PA 7005-2, developed at Italy. Early ripening large fruited apricot, fruit oblong in shape, symmetrical, large, 80g, orange skin with significant blush, melting flesh, free stone, good flavour, self-incompatible, resistant to Dideron and Marcus Plum pox virus isolates.

Boreale® (B08160-4334)

A hybrid of San Castrese x Reale d'Imola from Italy. A self-compatible, skin-cracking tolerant apricot, fruit oblong, symmetrical, large, 80g, skin light yellow with significant blush, flesh firm, yellow, melting.

Brittany Gold

A large sized free stone apricot cultivar developed at California. Fruit globose, light orange ground colour with orange-red blush, free stone, firm textured, slight tendency for fruit cracking before maturity.

Goldensweet

A sweet-fruited self-fertile apricot for late-season fresh and dry market developed at California. Fruit globose, medium, light orange skin with reddish blush, free-stone, strong orange flesh colour, juicy, excellent blend of sugar and acid regular bearer and very productive.

Ninfa

A hybrid of Quardi x Tirynthos from Italy. Early-ripening, fruity-flavoured apricot, fruit round, symmetrical, medium (55g), skin light yellow with blush, flesh yellow, free stone, good flavour, self-compatible, highly productive.

Petra

A hybrid of Goldrich x Pelese di Giovanniello developed at Unviersity of Bologna, Italy. Excellent-flavoured, late ripening apricot, fruit oblong-round, medium (65-70g), orange skin and flesh, melting, free stone, sweet, aromatic, productive, and self-incompatible.

Zhouping Shuixing

This is a semi free stone apricot cultivar. Fruits are very large (105g), orange-yellow skin, flesh yellow-white, juicy, TSS content 16.5 per cent, highly tolerant to drought. (Wang-YuMing *et al.*, 2009)

Zaojinyan

Early maturing, precocious and productive apricot cultivar. Smooth fruit surface,bright golden-yellow in colour, Flesh is juicy, sweet, fragrant, TSS 15.6 per cent (Chen-YuLing *et al.*, 2011)

Xinshiji

It was developed by the combination method of sexual hybridization and embryo culture. Early ripening cultivar, large fruit, good quality, suitable for both open and protected cultivation (Chen-Xue Sen *et al.*, 2001)

Zaomeihong

A hybrid of Chuanzhihong x Xiaomanhuang. Early maturing, free stone cultivar, round in shape, medium (55g), skin orange-yellow with purplish blush, juicy, acid-sweet flavour, good eating quality and productive (Zhao-Xiping and Chai – JuHua, 2000)

Apricot Cultivars Growing in Different Countries

- ✰ **China:** Luotuohuang, Maihuang, Hebao, Shisheng, Huaxiangdajiexing, Shajinhong, Yin Xiangbati, Jidanxing, Chuanzhihong, Wanxing, Badou.
- ✰ **Tuekey:** Aprikoz, Cataloglu, Cologlu, Hacihaliloglu, Hasanbey, Kabasi.
- ✰ **Japan** (*P. mume*): Ryuko Koume, Koshu Koume, Koshu Saisho, Gojiro, Shirokaga, Oshuku, Benisahi, Baigaou, Nanko, Bungo.
- ✰ **Korea:** Choha, Harcot.
- ✰ **Canada:** Harcot, Goldcot, Velvaglo, Goldrich, Veecot, Harglow, Vivagold, Harval, Harlayne, Hargrand, Harojoy.
- ✰ **United States:** Orange Red, Goldbar, Goldstrike, Tomcat, Peterson, Blenheim, Candycot, Nugget, Goldkiss, Royal Rosa.
- ✰ **India:** New Castle, Nugget, Moorpark, St Ambrose, Halman, Nari, Charmagz, Conian Italy, Roundel, Kaisha, Australian Sweet, Quetta.
- ✰ **Pakistan:** Red Flesh Early, Old Cap, Charmaghz, Moorpark, Trevett, Nuri, Shakarpara.
- ✰ **Iran:** Nassiri, Ghermez Shahrood, Dorosht Malaier, Ghorban Maraghe, Kanino.
- ✰ **Afghanistan:** Saqi, Nuri, Travel, Badami, Qaisi, Ameri, Tomcot, Oldcup, Redflesh, Moorpark, Ameri.

Three cultivars, *viz.*, Chaubattia Alankar, Chaubattia Madhu and Chaubattia Kesri, have been evolved at Chaubattia, India which ripen in the last week of May, first week of June and third week of June, respectively (Seth *et al.*, 1984). All these cultivars are of high dessert quality, being very sweet, juicy and aromatic and are improvements over one of the parent cultivars, Charmagz, in bearing. These cultivars also extend the harvest period of the crop and can be grown in mid hills. As a result of regular good breeding programme and varietal evaluation trials being carried out in different countries, many cultivars have been released for cultivation. Some of them are cold-hardy and resistant to bacterial spot, brown rot and perennial canker, besides being more productive and bearing fruits of high quality. The fruit ripening period can also be extended with their cultivation. The cultivars are Flaming Gold, Sundrop, Harlayne, Hungarian Best, Stark Early Orange, Akademik and Jerseycot (Kazimin and Marusich, 1977; Layne, 1978, 1981b; Lane, 1978; Pauriovic *et al.*, 1982; Mehlenbacher *et al.*, 1987). Rival is a superior cultivar to Tilton for canning purposes (Drake *et al.*, 1977).

Climate

The climatic requirements of apricot cultivars are specific making them less adaptable to the variations in the environment. In its natural habitat in northwestern China, apricot grows successfully between 35° and 43° N latitude at an elevation of 700-1500m where annual rainfall is about 50 cm and minimum temperature does not go below -33° C (Baieiy and Hough, 1975). The continental climate without much fluctuating winter temperatures and with warm summer is more favourable for apricot.

Temperature

The temperature effect on apricot is quite intricate as it is directly involved with the cold hardiness, dormancy, frost injury and fruit growth and development. Deep dormancy commencing from the end of October is over by the mid-January to early February (Pulatov, 1978). Termination of deep dormancy is also indicated by a rapid increase in flowering (Martinex, 1979). The sum of effective temperatures in 0°C to 4.5°C range, necessary for completing deep dormancy, vary from 28° to 84°C (Khloptseva, 1977). In a comparison of cyclic and constant temperatures, Guerriero *et al.* (1986a) found that slightly fewer days were needed to get the same percentage of bud break at constant temperature than at the cyclic temperature. They suggested weak chilling effect even at 15°C. The severe freezing temperature during this period do not cause much harm but as soon as induced dormancy begins, the temperatures below -5°C are very injurious to the flower buds,limiting the extension of apricot cultivation in many regions. Induced dormancy lasting from the start of meiosis to flower bud growth is broken by 60° to 100°C sum of effective temperatures (above 2-3°C) and about 90-120°C are required for the start of flowering (Khloptseva, 1977). The temperature requirement though varies with the cultivar and region, yet the calculation at one site gives an approximate idea which can be taken advantage of elsewhere. A study of temperature requirement of apricot showed that the minimum temperature required for peak flowering of all the cultivars was 5-6°C and for fruit ripening varied from 15-19°C. In early cultivars the sum of daily temperatures higher than 10°C needed for fruit ripening was 1155-1245°C and in late cultivars was 1725-2186°C (Ageeva, 1984).

The long cool winter and uninterrupted warm spring are favourable for apricot fruiting. If the temperatures are hot for a long period during the dormancy, the flower bud may drop which is a serious problem in warmer climate. Likewise, once the flower bud has commenced growth in induced dormancy, the occurrence of frost may injure it. Thus the cold hardiness refers to the survival of flower bud at the end of dormant season. The cultivars differ in their ability to withstand the cold in dormancy which depends upon the two stages of floral development. The duration from the initiation of pollen mother cells to meiosis is the first stage when the floral buds are most resistant to cold, but in the second stage from the reduction division to fully formed pollen grains, the buds lose their hardiness following a warm weather. The most hardy cultivars should have long durations of both these stages in order to tolerate fluctuating winters and spring frosts. A considerable amount of work has been done in the USSR on the evaluation of frost resistant germplasm and cold hardy cultivars are available for cultivation (Moskalenko, 1983). The cultivar Royal needs about 85 chilling hours at 5°C threshold temperature to break the dormancy while other cultivars require more or less chilling hours (Teskey and

Shoemaker, 1972). According to Martinex (1979), cultivars Fenornenos, Valenciano-2 and Valenciano-3 have the shortest cold requirement (300 hours) and Bulida-l and Bulida-2 the longest (700 hours). The chilling hour requirements of different Turkish apricot cultivars were reported as 350 hours for cvs. Precoce de Tyrinthe, Feriana and Beliana; 250 hours for cv. Priana and 750 hours for cvs. Precoce de Colomer, and Canino (Kuden *et al.*, 1992, 1995). The effect of climate change on productivity of apricot has been reported (Guerriero *et al.*, 2010). During warm winter of 2006-2007, apricot cultivars native to northern Italy showed unusually low fruit set. On the other hand genotypes from warmer climate (Southern Italy, Northern Africa) produced heavey crop under unusual mild winter conditions.

The use of various plant regulators such as 2, 4, 5-T, ethrel and alar has increased the resistance of apricot to frost injury either by delaying flowering or inducing growth of embryo aborted fruits (Crane, 1954; Buban and Turi, 1979; Hricovsky and Takac, 1981). In the localities where insufficient chilling occurs, Erez and Lavi (1985) were able to avoid deep dormancy by adopting a sequence of treatments, *i.e.*, a period of desiccation followed by manual defoliation, renewed irrigation, and use of rest breaking chemicals [1.5 per cent thiourea and 7 per cent KNO_3), 5 or 7 per cent nerol 5 (DNOC), or 5 or 7 per cent ovamort special (DNOC)]. In order to obtain the best results, the rest breaking chemicals should be applied about six months after the initiation of the former vegetative cycle.

Other Climatic Factors

The effect of other climatic factors, such as, light, photoperiodism, rainfall, humidity and wind on apricot has not been investigated in detail. Apricot trees, however, thrive satisfactorily under low atmospheric humidity. High humid conditions in summer increase the possibility of brown rot (*Sclerotinia* sp.) attack. Late frost severely affect fruit set. Frost-affected fruits developed external symptoms (white scaly spotting) within 3 days of the frost and dropped within 3 weeks (Rodrigo and Herrero, 2000). Guericro and Martelloni (1999) however, observed that cvs. Farmingdale, San Castrese, PI2.29.11 (Farmingdale x Amabile Vecchioni) and PI3.48.7 (Goldrich X 34.6) produced good yield irrespective of spring frosts. An annual precipitation of about 100 cm is sufficient for obtaining a good apricot crop. Because of early flowering, a north eastern side is better for apricot plantation (Teskey and Shoemaker, 1972).

Rain at bloom time and near harvest can cause damage from diseases. Very hot days during fruit development can cause a disorder called 'pit burn' in some cultivars or may cause pre-mature ripening or softening.

Soil

Apricots are quite hardy to grow in most of the soils which are deep and well-drained. The horizontal and vertical root distribution of a ten-year-old tree reached 4.5 and 2.2m, respectively, indicating that the soil for apricot should be about 3 m deep (Vashchenko and Vodyanitskii, 1976). In an experiment, Kitayev (1983) found that the tree growing in wet soil had depressed physiological activity of leaves and the growth of shoots and of the whole plant. In Kinnaur region of Himachal

Pradesh, India where apricots grow in large wild stands, the soils are sandy, well-drained but not very fertile. High lime content of the soil does not depress the tree growth provided the drainage is good (Molchanov, 1976). Loamy-sand, sandy-loam, loam and clay-loam soils have been successfully used in many areas of California for cultivation of apricot. Bacterial canker disease can be problematic in sand and loamy-sands, or old gravelly creek beds.

Rootstocks

Apricot seedlings are mainly used as rootstocks giving satisfactory performance everywhere. The apricot and peach seedlings are suitable for sandy soils and dry conditions while plum rootstocks are good for heavy soils and wet situations. A large number of rootstocks, such as, seedlings of apricot cultivars, the seedlings and clonal rootstocks of peaches and plums and various *Prunus* species have been tried in different countries but without tangible results. Since the cultivars, rootstocks, soils and climatic conditions differ in these trials, it is not possible to arrive at any definite conclusion.

Characteristics of Rootstocks

Apricot

When apricot cultivars Blenheim, Caninos, Damasca, Moorpark, Tilton and Wenatchee were grafted on apricot, almond, peach and *P. cerasifera* seedling rootstocks, the best tree development and productivity were obtained on apricot rootstock (Welkerling *et al.,* 1973). The apricot rootstock is resistant to root-knot nematode and crown rot but it is not tolerant to poor soil drainage conditions. The apricot cultivars Monique, Bulida and Paviot showed slightly higher chilling requirement on apricot seedlings than on Myrobalan and Marianna stocks (Tabuenca, 1976). Seeds of the Royal and other cultivars give rise to equally good rootstock seedlings. Layne and Harrison (1975) recommended use of seeds of apricot 'Haggith' for rootstock purpose which they claim as cold-hardy, vigorous and compatible with a wide range of scion cultivars. INRA Manicot, a wild apricot selection, has been assessed as a very uniform rootstock which is not as vigorous as Haggith (Crossa-Raynaud and Audergon, 1987). The seedlings of North African wild apricot (Mech Mech) showed good performance as rootstocks in the soils having high salt and lime content.

Peach

Peaches are popular rootstocks for apricot mainly adapted to acidic to neutral and sandy dry soils under irrigation. In Italy, seedling peach A/VI rootstock gave excellent performance for the cultivars Canions and San Castrese in comparison to Myrobalan clones 1/2 and 1/10 (Manzo and Salvador, 1978). Stadler and Stassen (1986) reported the highest yield of the cultivar Royal on Okinawa and Nemaguard rootstocks in sandy soils of South Africa. The cultivar Bulida, however, performed best on the Royal seedlings followed by Okinawa. Some other rootstocks such as Montclar, Rubira (France) and Nemared (US) are under observation (Crossa-Raynaud and Audergon, 1987). Peach rootstocks are unsuitable to the poorly drained soils or the soils which have been occupied previously by the peach roots.

Plum

The plum rootstocks for apricot are used mostly on heavier soils or under wet soil conditions but P-8-1-1/16 plum is even promising for dry calcareous soils (Damigella *et al.*, 1973). Nitransky (1978) reported the susceptibility of apricot seedling rootstocks to root asphyxia caused by waterlogging whereas Myrobalan (*P. cerasifera*) was resistant. In a pot experiment also Marianna plum cuttings were found to be the most tolerant to waterlogging followed by Myrobalan, Royal and Nemaguard cuttings. Myrobalan rootstock induced earlier fruit ripening than the seedling apricots and peaches (Costa and Grandi, 1975). Different clonal rootstocks have been recommended in different regions, *i.e.*, Myrobalan 29 C and Marianna 2624 for California, INRA clones GF 31, GF 8-1 and GF 1380 for France and Marianna clone 7/7 for South Africa (Norton *et al.*, 1963; Nicholas, 1979; Stassen and Hurter, 1981). Audubert (1987) has recommended clone MP43 as a dwarfing rootstock which is compatible with different cultivars and more resistant to root asphyxia than Reine-Claude GF 1380. In Poland, apricot cvs. Harlayne, Hargrand, Dariana, Sirena and Kompakta showed significantly higher yield on Myrobalan seedling rootstock (Sosna and Licznar-Malanczuk, 2012) Perez-Romero *et al.* (2014) reported from Spain that apricot cv. Ninfa produced larger sized fruit on Myrobalan 29 C compared with trees on apricot cv. Real Fino seedlings. However, fruit from trees on Real Fino seedlings showed higher soluble solids concentration than on Myrobalan 29 C. In Czech republic rootstocks CF-677 and Myrobalan BN4Kr were recommended in areas of endemic plum pox virus (PPV) infection or in regions where PPV is a threat (Polak and Kominek, 2014)

Two rootstocks 'Penta' (derived from the open pollination of *P. domestica* cv. Imperial and Epineuse) and 'Tetra' (derived from the open pollination of *P. domestic* cv. Regina Claudia Verde) in Italy showed good performance for apricot cvs. San Castrese, Antonio Errani, Bella d' Imola, Palummella, Sungold, Boccuccia Liscia and Vitillo (Nictora and Moser, 1999). Bas and Paydas (2000) studied graft compatibility status of different prunus clonal and seedling rootstocus with some apricot cultivars. They used 2 dried (Kabaas and Hachaliloglu) and 5 table cultivars (Aprikoz, Karacabey, Beliana, Rouge de Roussillon and Joubert Foulon) grafted on 2 seedlings (apricot and Nemaguard) and 3 clonal Prunus rootstocks (Myrobalan CF-31, from the combinations of apricot seedling and Kays Erigi with all apricot cultivars, except Beliana/Kays Erigi, Aprikot/GF-31, Kabaas/Nemguard, Aprikoz/Nemaguard, Karacabey/Nemaguard, Beliana/Kays Erigi and all apricot combinations with Pixy give moderate grafting union. Some incompatibility problems was noted with Joubert Foulong/Nemaguard, Joubert Foulong/GF-31 combination. However, Beliana/Nemaguard, Rouge da Roussillon/Nemaguard, Beliane/GF-31, Rouge du Roussillon/GF-31 and Karacabey/GF-31 combinations were found fully incompatible.

Related Species

The use of other species related to apricot as rootstocks has been attempted with limited success. Both bitter and sweet almond (*P. amygdalus*) gave poor results when used as rootstocks for the cultivars Bulida and Cafona (Damigella *et al.*, 1973).

The use of sand cherry (*P. besseyi*) as rootstock produced dwarf tree (Norton *et al.*, 1963). Tukey (1964) suggested *P. sibirica* and *P. mandschurica* as rootstocks for colder region but *P. sibirica* did not reduce bud damage caused by temperature fluctuations in early spring (Vachua, 1971). It however, significantly retarded the vigour of the scion cultivars (Vachun, 1972). *P. tomentosa* has been found incompatible with apricot cultivars (Fridlund,'1979). Similarly *P. mume* gave poor results with Royal and Bulida (Stadler and Stassen, 1986). The vegetatively propagated Kozlienka (*P. insititia* var. *rubra dulcis*) was assessed as the most promising rootstock against cold damage (Nitransky, 1983). In order to identify a dwarfing rootstock, the orchard performance of Alfred cultivar propagated on 8 clonal and seedling rootstocks was investigated (Cummins 1985). According to him, PI 304921, a triploid hybrid of *P. cerasifera* x *P. spinosa* deserves further testing as a semi-vigorous stock for apricots.

In Romania, Buouruz (*P. insititia*) selected from a local population seedling was found vigorous and compatible with prune cultivars and with most apricots. It is resistant to low winter temperature. Another selection Reine Claude Verde also proved useful as a seed-propagated stock. The seedlings were resistant to plum pox (sharna) and imparted high degree of vigour to the grafted tree (Parnia *et al.*, 1988). Japanese apricot cvs. Xiaoqingmei and Xiyeqing showed better fruit quality on *Amygdalus dividiana* rootstock while trees on *Prunus mume* showed higher yield (Fang-Jin Gui *et al.*, 2000). The performance of apricot cv. Bergeron grafted on 12 rootstocks was evaluated in Germany by Wurm (2007). The highest tree yields were obtained with Ishtara, Torinel, Citation, Julior and 29 C. For a planting distance of less than 2m, Citation, Torinel and Ishtara were more suitable and for distance of more than 2m, Julior and 29 C were highly suitable.

Propagation

Apricots are commercially raised by vegetative methods of propagation, such as grafting and budding (Crossa-Raynaud and Audergon, 1987). The multiplication through cuttings is rarely done. The apricot seedlings are used for rootstock purposes.

Seed Propagation

Apricot seeds require cold stratification for a period of 72 days for 100 per cent germination. The cold treatment at 6°C is better than 4°C (Seely and Damavandy, 1985). The growth inhibitory activity of the seed-coat disappears with continuous cold treatment and accumulation of growth promoters occurs (Fadl *et al.*, 1978). The stratification medium should have proper moisture retaining capacity, good aeration but no toxic substances. The seeds should be kept moist during the period of stratification. The germination of apricot seeds can be hastened by removal of kernel from shell, scarification and gibberellic acid or kinetin treatment (Bogdanova, 1974). The seeds are soaked for 24 hours in 500 ppm GA_3 or 5 ppm kinetin solution before planting. Both the growth regulators also increase the seedling growth (Abohassan *et al.*, 1979). Soaking of seeds in 0.25 or 0.50 per cent Thiourea for 20h showed good germination of seeds of cv. Al Amar (Samaan *et al.*, 2000).

Few seeds begin to germinate at the end of stratification period. These are then ready for sowing in well-prepared planting beds which should be free from weeds and pebbles and supplied with a good amount of compost. The seeds should be sown carefully in 30 cm apart rows. Ample moisture in the nursery bed ensures good germination and seedling growth. It can be achieved by frequent watering and mulching, the latter also helps in controlling weeds and protecting against temperature extremes. After germination the mulch is removed and clean cultivation is practiced in the nursery. Gautam and Chauhan (1982) reported good weed control in the nursery with the application of terbacil at 2-3 kg/ha before weed emergence. If the seedlings are not making good growth, urea may be applied in small quantity. Plant protection measures need also to be adopted against pests and diseases which are very few in apricots.

Vegetative Propagation

Budding and Grafting

June budding, fall budding and tongue grafting are generally adopted for apricot multiplication. The seedlings which attain pencil thickness in the month of June are budded and the other good seedlings are utilized for fall budding in September. T-budding (shield buding) is the usual method followed in apricot (Hartmann and Kester, 1959). The rootstock should be more than 0.6 cm in diameter with a bark separating readily from the wood. The stock may be budded at a height of about 10-25 cm above the ground level in a smooth bark surface. About a week after insertion of a bud, the rootstock can be cut back about 10 cm above the bud. It will help in bud take and after a fortnight of budding operation the rootstock-portion just above the bud is to be removed completely. When the bud starts growing, the union becomes strong enough then the wrapping material should be removed. Some of the June budded plants become ready for permanent planting in the dormant season while fall budded can only be utilized in the dormant season. The graft union is considered better in budding than in grafting. Ijaz *et al.* (2006) reported maximum bud success (100 per cent) by budding on 5 July followed by on 12 July (95 per cent), while minimum bud success (40 per cent was noted on 7 June. Chauhan *et al.* (2004) reported higher bud take in tongue grafting over cleft and crown grafting. Tongue grafting caused 75.0 per cent bud take when grafting was done on 9 February at the mid-hill condition of Himachal Pradesh, India. The time of budding had significant effects on budding success and radial and linear growth of budded plants. Chip budding of cv. Australian in mid-February resulted in highest budtake (84.73 per cent) followed by budding in mid March (Srivastava *et al.*, 2008)

The seedlings which do not make enough growth for budding are usually grafted in the following dormant season by whip or tongue grafting. The grafting in apricot is more successful when it is done late in the season at the beginning of sap flow and the buds become active. A high percentage of success is obtained in apricot both in budding and grafting.

Cutting and Stooling

According to Crossa-Raynaud and Audergon (1987) the grafting (including budding) is adopted because the cultivars cannot be propagated successfully by means of cuttings. Bondok *et al.* (1984) could get about 30 per cent rooting in hardwood cuttings when collected in December and treated with 300 ppm IBA. Avanzato and Nicotra (1986) obtained better rooting in hardwood cuttings in acid media (pH 4 and 5) when adjusted with $Ca(OH)_2$ rather than with $Mg(OH)_2$. Rooting was markedly depressed at pH 7. Japanese apricot was successfully propagated by herbaceous cuttings (Nachtigal *et al.*, 1999) by treatment with IBA at 2000 mg/l and using vermiculite medium under mist. Fifty per cent stool layers rooted when treated with 2500-5000 ppm IBA (Sud and Mishra, 1980). Survival of IBA-treated stools was also significantly better than untreated stools.

Topworking

Topworking of wild plants, seedlings, inferior cultivars and unproductive plants can be done successfully by cleft grafting or budding on butt end of the shoots (Legave *et al.*, 1985; Yanin, 1975). Grafting in April is better than May and the scions should be supported to protect against wind breakage. Yanin (1975), describing budding on butt end, reported better sprouting and callusing in this method than in conventional budding.

Micropropagation

Micropropagation of apricot cultivars and its plum, Myrobalan and Marianna rootstocks has been accomplished successfully (Snir, 1984; Nedelcheva *et al.*, 1985). The excised shoot segments of Canino apricot proliferated on a Lloyd and McCown medium under low light intensity. The segments of the new shoots rooted well on half strength Murashige and Skoog's medium containing 0.5 mg/litre NAA and rooted plants established successfully on transplanting outdoors. The rootstocks can be grafted within a year of transplanting in the field. Apricot has also been regenerated from immature embryo taken 20-30 days after anthesis and cultured on Murashige and Skoog's medium supplemented with BA and 2, 4-D (Lane and Cossio, 1986). Perez-Tornero *et al.* (2000) studied the effect of different basal media and cytokinin concentration on the proliferation of apricot cv. Canino shoot. They obtained optimum proliferation, healthier and greener shoots by using modified woody Plant (WP) medium. BA concentration between 0.5 and 0.6 mg/l produced optimum number of shoots. The best rooting (92.8 ± 2.5 per cent) was induced by NAA at 2 mg/l while the largest number of roots per shoot (5.3 ± 0.5) was obtained with IBA at 6 mg/l. A complete protocol for an Egyptian apricot cv. EL-Hamaway was developed by Soliman (2012). Meristem was isolated from the actively growing plants in the spring. The survival rate of meristem tips was 89.33 per cent, average number of shoots/explant (3.98) and average shoot length/explant (1.85) was obtained on WP medium supplemented with 1.0 mg/l Zeatin, 0.1 mg/l IAA in presence of 100 mg/l ascorbic acid and 150 mg/l citric acid. The highest rooting (91 per cent) of adventitious shoots, mean number of roots (7.64) and mean length of roots (3.43 cm) were obtained in the dark on MS medium with IBA at 2 mg/l and

0.5 mg/l NAA before transfer to a 16h photoperiod. In total, 73 per cent planttets survived which was acclimatized in a greenhouse.

Planting

Undesirable trees and shrubs should be removed from the soil during its initial preparation by digging and ploughing. The natural slope of the soil should be left undisturbed in order to ensure good drainage. The pits having one metre dimension in length, breadth and depth should be dug in winter season and properly filled with a mixture of soil and well decomposed farmyard manure. The pebbles, roots of the plants and stones, if any, found on the sides of pit, are to be removed. On a flat land a regular planting pattern layout, such as, square, rectangular or triangle is followed but on hill slopes contour layout is the general practice. The spacing of plants varies with the soil, climate and vigour of the cultivar to be grown. In a fertile soil and humid climate more distance between plants is provided. Square planting system with a distance of 6 x 6 m is traditionally adopted. There is, however, a recent trend towards rectangular pattern because more plants get accommodated in a row (Westwood, 1978). In comparison to traditional wide spacing (8 x 6 and 10 x 6 m) a closer spacing (7 x 5 m) gave higher yield in apricot. When the planting density was increased from 333 to 1250 trees per hectare the yield rose from 11.3 to 28.3 tonnes per hectare (Bunea *et al.*, 1983). To determine the maximum limit of planting density, Guerriero *et al.* (1986b) planted 1110 to 8880 trees/hectare adopting free spindle, Tatura and open centre training systems. The highest cumulative 4-year yield was obtained with 2500-3333 trees per hectare. According to them, the closest spacing did not always give the highest yield nor did it reduce the tree vigour.

In the absence of a suitable dwarfing rootstock and scion cultivar, the high density planting in apricot is still to be standardized with proper training and pruning systems. Apricots are planted in winter or early spring season. Well-grown plants of about one metre height should be firmly placed in the pits with immediate watering to establish close contact between roots and soil. If summers are dry, the plants should receive adequate irrigation in order to check-the early mortality and maintain good growth.

High Density Planting

In comparison to other temperate fruits, high density plantings in apricot have begun rather late. Gaash (1986) planted the cultivar Canino at densities ranging up to 7200 trees per hectare. He controlled the plant growth by pruning, topping, removing some rows, limbs and hedging. The best result however, were obtained by applying paclobutrazol to the root zone in the autumn at 2 g per tree. Clanet and Salles (1986) sprayed trees of the cvs. Canino and Polonais with paclobutrazol at 250 or 1000 mg/l or applied to the soil at 2 g/tree in April. The shoot growth rate in both cultivars was significantly reduced by treatment with paclobutrazol. Multiple spray of the regulator also improved the fruit set. Ende and Kenz (1986) grew 1670 to 2000 trees per hectare of the cultivar Trevatt employing Tatura trellis as the training method. Annual yield of fully established orchard was about three times more than the conventional planting with 300 trees/ha. According to them,

Tatura trellis was more expensive to establish and maintain than conventional plantings but returns compensated more than the increased costs.

Training and Pruning

Training

Apricot pruning depends upon the soil, cultivar, rootstock, planting density, tree age and harvesting method adpoted. On very fertile soil and with vigorous cultivars, heavier pruning is exercised to check the excessive vegetative growth in order to allow more sunlight and air inside the plant. Similarly, in high density planting, the fruiting area is kept close to trunk. In older trees the amount of pruning is increased so as to stimulate more growth and to increase fruit size. In younger plants, lesser pruning gives higher yield. Pruning resorted to building up of the frame of the plants in early stage is termed as training. Various training systems, such as, modified leader, vase-palmette, Kechement hedge and werder hedge system are adopted for training of apricot plants. Basic principle of the least pruning is applied in pyramid crown, round crown or in free crown systems. Free crown trees begin fruiting earlier and give double yield in the first four cropping years than vase-trained trees (Sansavini *et al.,* 1983). The trees also had the greatest girth increment. The modified leader system is conventionally followed in apricot. Palmette system requires the most severe pruning and produces the smallest trunk girth increment (Bencat and Matuskovic, 1980). Vase system has, however, been recommended for establishment of mechanically harvested orchards because vase-trained trees are more suitable for mechanical harvesting (Adamov, 1984; Sansavini *et al.,* 1983). Evequoz and Evequoz (1983) found palmette to be the best system for mechanization and bush forms under other conditions.

Pruning

Pruning of Young Plants

The main stem of one-year-old plant is headed back at about 70 cm above the ground on planting and only three to five well-spaced side shoots are allowed to grow in all directions which later on turn into main scaffold branches. The pruning becomes more important in the second year because the frame developed in this period will give ultimate shape to the tree. The branches chosen to become scaffold branches are headed back at different heights and all other branches are removed from the base. In the third year, the scaffold and lateral branches on them have much growth. Some shoots may also emerge directly from the trunk. The scaffold branches along with few lateral shoots are only allowed to grow by removing all other shoots. If the leader branch is making excessive growth, it should be headed back to enhance the growth of other main limbs. Thereafter, in the fourth year, the pruning comprises the thinning of overcrowding and criss-cross branches. Shortening of shoots by two-third of their length in the cultivars Hungarian Best and Royal decreased the yield up to 50-60 per cent compared with shortening by one-third only (Cepoiu, 1979).

Pruning of Bearing Trees

The apricot bears on spurs and laterally on one-year-old shoots. The spurs have a short life of 2-3 years and many of them also get broken during the fruit plucking. The production of young growth is, therefore, essential for the initiation of new spurs which generally takes place at the bases of the growing laterals. Thinning of some one-year-old shoots combined with one-third shortening of the other shoots had given the highest yield than the other pruning methods (Tawfik *et al.*, 1985). The shoots should make 50-60 cm growth each year in young trees and 25-35 cm in older ones. In order to check the willowing growth of the long branches, they are headed back creating overcrowding of laterals. This needs thinning out of some laterals as well. The old, unbranched and diseased branches should also be removed to replace with younger ones. Mechanical pruning should be combined with hand thinning to avoid overcrowding in the inner and upper part of the crown (Babayashev, 1983). Apricot researcher from Italy observed that in apricot, growth habit and fruiting behaviour are strongly inter-related. Accordingly, cultivars of apricot can be classified into five groups (Gouerriero and Xiloyannis, 1975a). Since some changes in shoot morphogenetic gradient during the rest period were evident in response to dormancy and chilling (Guerriero and Scalabrelli, 1982), any classification of branch habit should be strictly associated with a specific environment. Depending on the type of bud (floral or vegetative) and shoot (sylleptic, long brindle, brindle, brindle or spur), the need for chilling may be very different and this could be a factor in regulating branch habit and fruiting behaviour (Guerriero and Xilloyannis, 1975b). To decide how to manage pruning, the shoot should be identified by its specific growth rate after bud break. Therefore, its physiological behavior should be constantly monitored. There is some evidence that the emergence of sylleptic shoots is highly probable when a certain threshold of growth rate is exceeded (Zucconi, 2003). This means that in some conditions, apical dominance is not able to stop the growth of lateral meristems, which thus originate sylleptic (or anticipated) shoots rather than buds.

Rejuvenation of old apricot trees can be done successfully by cutting back 3- to 4-year old branches in September than by cutting back 8 to 10 year old branches in March or partial thinning of the crown (Rozsnyay, 1978). All the pruning operations should be carried out in late winter or about a month before bud break because the wounds become resistant to pathogen in the shortest period in spring season (Ramos *et al.*, 1975). The pruned sapwood should be protected against pathogen infection by integrated control employing benomyl and *Fusarium laterititum* (Carter and Price, 1975).

Nutrition

Role of Nutrients

Nitrogen

Among macroelements, nitrogen plays an important role in apricot nutrition as its deficiency adversely affects the leaf dry matter content, fruit set and number of mature fruits (Zvara, 1967). Kecskemeti-Vincze and Nyujto (1986) found a direct correlation between N and trunk girth. But high N has a negative effect on yield and frost-hardiness of flower and it delays fruit maturity which begins early in the cell division phase of fruit development (Szucs, 1986; Albigo *et al.*, 1966). Urea application results into the highest accumulation of N in the shoots (Margarayan, 1976). The effect of N fertilization on its accumulation is more in the spring declining towards the autumn. Weinbaum *et al.* (1980) showed that late summer application of $K^{15}NO_3$ increased N level of the blossoms in the following year in comparison with the dormant season application. They further suggested that summer application of N was effective for early fruit development in the following year to a greater extent than winter application.

Potassium

Potassium application increases the growth, flowering, fruit size and yield (Bohomaz, 1960). Szucs (1986) studied the effect of N fertilization on trees of the cv. Hungarian Best. Leaf analysis showed the effect of NPK application only on the K content in the leaves but not on N or P content. Interestingly, positive correlations were found between leaf K level and foliage conditions, flowering rates, tree growth and yield. In the second part of July, the optimum leaf K concentration was 2.6-3.2 per cent with NK ratio as 0.8-1.2. Kecskemeti-Vincze and Nyujto (1986) also reported a very close correlation between leaf N and K and yield.

Phosphorus

Phosphorus requirements of apricot are low and it responds to P application only when applied with N (Bohornaz, 1960). The deficiency of P causes reduction in fruit set, fruit growth and Ca and Mg contents of leaves (Zvara. 1967).

Sulphur

When sulphur content in immature leaves becomes less than 100 ppm, the deficiency symptoms appear (Benson *et al.*, 1963). Restricted plant growth, pale green leaves devoid of chlorophyll and broom-like plant structure are the deficiency symptoms of this element. Sulphur dioxide at higher level restricts pollen tube growth (Facteau and Rowe, 1981). Most of the mineral nutrients are available to the plants at a soil pH above 5.5 and below 8.0. If soil pH is low, improved soil structure, aeration and root growth are obtained by liming.

Zinc

Zinc deficiency causing 'little leaf' and rosette are common in irrigated orchards (Teskey and Shoemaker, 1972). Soil application and foliar sprays of Zn salts or use of Zn-coated nails have been suggested (Linder and Luce, 1944). Apricots are, however injured by fall sprays of Zn (Westwood, 1978).

Boron

Visual symptoms of boron deficiency in the fruit have been described by Demetriades *et al.* (1971). The affected fruit contained 12 ppm B on a dry weight basis. Internal browning, formation of corky tissue, fruit cracking external browning and fruit shrivelling have been rectified by field applications of 225 g of borax per tree (Bullock and Benson, 1948). Application of boron (soil application or foliar spray) increased boron content in flowers, leaves and short branches, promoted pollen germination, reduced flower and fruit drop and increased yield (Xiaoling and Shi Dan, 1999)

Iron

Yellowing between the veins of younger leaves or entire leaf and marginal necrosis are the main symptoms of lime-induced iron deficiency (Cibes, 1950). Fe deficiency is often associated with the high concentrations of other elements. Bould (1955) recommended soil application and spraying of iron chelates to correct the deficiency.

Manganese

Manganese deficiency is exhibited by interveinal chloroisis of leaves followed by complete absence of chlorophyll in extreme cases. The disorder can be corrected by Mn sprays, injections and soil treatment.

Fertilizer Recommendations

Generally, complete NPK fertilizers are applied. In field trials, the higest yield of cv. Hungarian Best was obtained with 300 kg N, 160 kg P_2O_5 and 200 kg K_2O per hectare (Marinov, 1982). A positive correlation between leaf N content and total length of annual shoot growth was found in pot culture experiments (Marinov, 1983). The optimum leaf contents for high yield of field grown apricots were 2.80-2.85 per cent N, 0.30-0.40 per cent P and 3.90-4.10 per cent K. Maintaining a 0.86-0.92 N: K ratio in the leaves ensured maximum yield in the following season. Bunea (1986) carried out trials with trees of cv. Hungarian Best grafted on Myrobalan, planted at 4.5 m spacing and trained as free palmette. Highest yields were recorded on the trees receiving NPK at 100-200, 80, 100 kg per hectare with leaf NPK contents of 2.5-2.7, 0.17-0.18 and 3.3-3.5 per cent, respectively. Application of organic manure is necessary for good yield on poor sandy soils along with the ferfilizers providing NPK (Kecskemeti and Farago, 1976). Raina *et al.* (2011) suggested fertigation of apricot tree instead of soil application. Fertigation resulted in significantly higher fruit weight and yield compared to soil fertilization. Drip fertigation with water soluble fertilizer found superior than conventional fertilizer and a saving of about 30 per cent fertilizers could be achieved through fertigation.

Tissue Analysis

Leaf nutrient levels in apricots are non-uniform showing seasonal variations (LeeceandVanDenEnde, 1975) and depend upon cultivar (Milosevic *et al.*, 2010), rootstock (Velemis *et al.*, 1999), interstock (Milosevic *et al.*, 2010), fertilization (Szucs, 1986) and soil levels (Liferali *et al.*, 2008). Malik (1965) reported that most suitable period for leaf sampling to determine annual average values is between 20 July and 1 August. The nutrient values are stabilized during this period. Stoilov and Marinov (1979) studied the seasonal changes in N, P, K, Ca and Mg content of the leaves in cv. Hungarian. NPK values decreased during the season but Ca and Mg values increased. Leaf K values declined during fruit growth and ripening because of enhanced utilization. They recommended August as a suitable period for leaf sampling because least changes occurred in the nutrient levels of various elements in that month. Stoilov and Marinov (1983) further carried out both container and field trials by applying different fertilizer treatments of NPK to the trees of cv. Hungarian on Zhlta Dzhanka rootstock. A high positive correlation was established between N application rate and leaf N content in the container trial but a lower level of correlation was found in field trial. A medium high correlation was, however, noted between available P and K in the soil and their contents in the leaves. Boskovic-Rakocevic *et al.* (2012) evaluated nutrient status of five apricot cultivars (Vera, Aleksandar, Biljana, Harcot, Roxana) at 120 days after full bloom (DAFB) grown on acid soil in the region of Cacak, Western Serbia. The results showed no statistically significant variations in the N, K and Ca content of leaves among cultivars, while significant differences was noted in P, Mg, Fe, M, Cu, Zn and B contents.

Tissue analysis coupled with soil analysis serves as a useful guide for fertilizer application. Westwood (1978) is not in favour of using complete NPK fertilizers where only one of the elements is needed. Reviewing the literature on nutritional ranges in deciduous tree fruits and nuts, the ranges given for apricot by Shear and Faust (1980) are given in Table 2.4.

Table 2.4: Nutritional Ranges of Various Elements in Apricot Tissues*

Element	*Time Sampled*	*Tissue Sampled*	*Deficiency Range*	*Normal Range*	*Toxicity Range*
N (per cent)	July	Leaf	< 2.0	2.3-3.0	NE
P (per cent)	July	Leaf	< 0.15	0.15-0.3	NE
K (per cent)	July	Leaf	< 1.75	2.0-3.0	NE
Ca (per cent)		Leaf	NE	-3.0	NE
Mg (per cent)	Summer	(Basal shoot leaves)	NE	0.8-1.5	NE
			< 30	30-	NE
Mn (ppm)					
B (ppm)	July	Leaf	< 20	20-50	>80
B (ppm)	Harvest	Fruit	< 20	30-60	NE
Cu (ppm)	July	Leaf	<5	5-10	NE
Zn (ppm)	July	Leaf	< 12	12-	NE

NE: Not estimated.

* Shear and Faust (1980)

The information on root system of the tree is essential for proper placement of fertilizers so that these are readily available for growth and yield. Root spread and penetration were studied in 1 to 9-year-old trees of cv. Hungarian on apricot rootstocks spaced at 6 x 5 m (Tsolov *et al.*, 1980). They reported that trees of all ages had well-developed root system exceeding the crown projection 1.3 to 2.0 times. Horizontally growing roots were mostly confined close to the trunk but spreading into the interrows upto 4.6 m from the trunk of 9-year old trees. Most roots of l-year-old plants penetrated to a depth of 40-50 cm and those of older trees to 65-70cm. Because of more growth of horizontal roots in older trees, greater concentration of roots was found in the surface soil layers. The fertilizers should, therefore, be applied in the uppermost layer of the soil and well beyond the periphery of trees.

Irrigation

Though apricots are tolerant to dry atmosphere, yet good soil moisture is beneficial because of its shallow roots. In an experiment, heavy irrigations between May and October showed delayed flower bud differentiation but increased the number of flower buds, fruit set and yield per tree (Agabbio, 1973). Water consumption varies with the soil, tree age and irrigation method, it increases until August, then decreases but is higher in October than in September (Ruggiero, 1986). To produce a quintal of the fruit about three times the quantity of water is required under furrow irrigation than with drip irrigation (Vakhidov, 1984). Sprinkler is a better method than furrow irrigation, particularly on steep slopes. Besides irrigation, sprinklers may also be used for frost protection and heat control. Under scarce water conditions, trickle, drip or daily flow irrigation has become important. Westwood (1978) narrates the advantages of trickle irrigation. These are: reduced water consumption, even distribution on steep, rocky hill sides, easier use of equipment and easier weed control. In the rainfed conditions mulching and water harvesting techniques should be practiced but the cultivation of drought resistant cultivars is more desirable. Some drought resistant cultivars have been identified by Kurmanova (1984) which, in decreasing order, are Krasnoschekii, Nantarnyi and Kishinevskii Ranii. The cultivars having the best capacity to restore turgor after water stress and increased water retaining capacity of the leaves under the stress are the most resistant. Lishchuk *et al.* (1985) presented a formula to determine a coefficeint of relative drought resistance on the basis of water movements in the pot plants at different soil moisture levels (90-100 per cent and 50 per cent field capacity) using measurement of rates of water transport through xylem, shoot diameter and transpiration rate.

Bybordi (2013) stated that apricot trees are highly sensitive to drought stress at particular phonological stages such as stage III of fruit growth and during 2 months after harvest. Apricot drought tolerance is mainly based on avoidance mechanisms, such as sotmatal control, epinasty and limitation transpiration by reducing leaf area. Regulated deficit irrigation is the practice of reducing applied water at selected phonological stages less sensitive to water deficit, thus imposing plant water stress in a controlled manner and can be a feasible water saving practice for arid areas with a minimum impact on yield and fruit quality. Abrisqueta *et al.* (2001) reported that availability and uptake of phosphorus and potassium from the root zone of apricot trees happens more easily with drip irrigation than with surface irrigation.

In young apricot plants, stomata closure and epinasty observed in response to water stress represented adaptive mechanisms to drought, allowing the plants to regulate water less more effectively and prevent leaf heating. A substantial reduction in the supplied irrigation water (25 per cent) with a high frequency of application (daily) promoted plant hardening, the plants enduring drought better, due to their greater osmotic adjustments, which prevented severe plant dehydration and leaf abscission. Such a preconditioning treatment may be valuable for young apricot plants in the nursery stage in order to improve their subsequent resistance to drought, (Ruiz-Scanchez *et al.*, 2000)

Micro irrigation can mean using 7-10 times less water in apricot orchards than traditional irrigation methods, provided the number, positioning and through output of the outlets is correct. New plantings initially need one outlet 20-30 cm from the trunk delivering 4-8 l/hour; in years 2 and 3, 2 more outlets at 75 cm either side of the trunk are needed and when production starts a further 2 outlets. Management requires, knowledge of the root spread and of crop evapotranspiration. So as to adjust delivery to plant needs. The aim is to maintain water reserves at root level, irrigating at 1 to 2 day intervals in high summer and 4 to 5 day intervals in spring and autumn was suggested (Xilloyannis *et al.*, 1999)

Soil Management

The orchard management depends upon the type of soil, rainfall and availability of labour and irrigation facilities. In irrigated soil, the plants under green manure had more annual tree growth, leaf surface area, number of fibrous roots and yield than under clean cultivation and sod (Mikaelyan *et al.*, 1976). The weeds are cleared with frequent light hoeing. The use of chemical weedicides has been tried successfully. Pre-emergence application of simazine or diuron each at the rate of 4 to 6 kg per hectare with 15-year-old trees of cv. New Castle gave good control of emerging weeds for 90 days (Dhuria *et al.*, 1980). The cost of applying weedicides is much less than manual weeding. But during the rainy season, growing of a cover crop is preferred to check the soil erosion. When the plants are small, the uncovered land should be utilized by growing leguminous crops.

Flowering

In apricot, usually three buds develop in the axil of a leaf at each node on a shoot and spur, the central one being a vegetative bud, the two side buds are floral. The number of flower buds on a shoot decreases from the base towards the tip. According to Cappellini *et al.* (1978), the time of flowering is correlated with the number of hours below 7°C from first October to 15th February. Khloptseva (1977) reported flowering in 25 cultivars from late March to early April. Under Shimla, India condition, the flowering in apricot cultivars mostly occurs in the month of March. The duration of flowering varies with the cultivar and prevailing weather conditions. High temperature, low humidity and wind shorten the flowering duration by increasing the transpiration.

Japanese apricot (*Prunus mume* Sieb. et Zucc) is widely cultivated in China. Japanese apricot produces two types (perfect and imperfect) flowers. The imperfect flowers are characterized by either pistils below stemens, withered pistils or absence of pistils, leading to falure in fruit set. The percentage of imperfect flowers varies among cultivars, with an average of 35 per cent but extremes as high as 75 per cent (Gao *et al.*, 2006), as a consequence, the yield of some Japanese cultivars are very low. Gao *et al.* (2012) made proteomic analysis of perfect and imperfect flowers in Japanese apricot by 2D-PAGE and mass spectrometry. Out of 35 differentially expressed spots between perfect and imperfect flowers, only 10 spots with most abundant proteins were successfully indentified, which provided biochemical evidence that the development of the flowers in Japanese apricot involved the physiological processes of sugar metabolism, starch metabolism, photosynthesis and photorespiration.

Apricots suffer greater damage due to spring frost because of blooming earlier than other temperate tree fruits. A delay in its flowering is, therefore, very much desired to reduce the frost damage which has been made possible by spraying growth regulators such as ethrel, abscisic acid, alar, maleic hydrazide, gibberellic acid and succinic acid (Buban and Turi, 1979; Koska, 1978; Soni and Yusuf, 1978; Ugolik *et al.*, 1981). The type of growth regulator, concentration and time of spraying are, however, to be standardized for different climatic conditions and cultivars. Spraying of hydrogen cyanamide (49 per cent) at 1to 2 per cent at four weeks before full bloom increased flowering, fruit set and yield of apricot cv. Canino (Hegazi, 2012).

Pollination and Fruit Set

Most of the commercial apricot cultivars have been found self-fruitful in pollination studies and set fruits without pollenizers but open-pollination has always given better fruit set (Cappellini and Limongelli, 1981, Limongelli and Cappellini 1978). The cultivars Charmagz, Perfection and Riland are self-incompatible (Seth and Kuksal, 1977; Teskey and Shoemaker, 1972). Priore and Sannino (1981) conducted a trial to determine the reasons for poor productivity despite good flowering. The results indicated that the failure in fruit set and *Sclerotina laxa* infection were mainly due to persistent rains. The superiority of pollination by honey bees (18.76 per cent) over wind pollination (10.48 per cent) was also confirmed. The cultivar Trevatt benefited more from pollination by honey bees (Longridge and Goodman, 1981). Therefore the provision of honey-bees in an apricot orchard for good fruit set is required. Physiological disturbance due to low temperature during flowering caused a prolonged effect with repercussions in the microspore maturation, regarding the pollen viability and germinative capacity in some apricot cultivars and hybrids (Iordache and Coroianu, 2013). In the ovules of cvs. Moorpark and Trevatt Night which are considered fertile, all reproductive structures were found present including embryo sacs with a complete set of 8 nuclei and the anthers contained mature pollen grains, while in Travel Blue which has low fruit, multiple ovules which were small and retarded in development were observed. The anthers contained degenerated microspores with some failure in tapetal break down (Lillecrapp *et al.*, 1999).

There is generally a good fruit set in apricot cultivars growing in appropriate climatic conditions. Sud and Mishra (1979) reported 45-56 per cent fruit set in New Castle, in India, Royal, Shipley Early and Kaisha in India. Asma (2000) reported that fruit set percentage of apricot range from 31.6 to 46.9 per cent and fruit set percentages in bearing shoots are higher in flowers close to the shoot tip than close to the base. McLaren *et al.* (1996) obtained fruit set values between 0.4 to 46 per cent in Sundrop and between 1 to 54 per cent in Moorpark apricot cultivar in New Zeeland. Alburquerque *et al.* (2004) recorded fruit set percentage of 6.5-28 per cent in Bebeco, 23.2-30.7 per cent in Palstein, 36.5-61.4 per cent in Beliana, 1.1-5.4 per cent in Goldrich, 22.1-66.4 per cent in Piriana and 1.4-8.7 per cent in Pepioto apricot cultivar, depending on the different irrigation treatments in different years. Water stress during flowering caused low fruit set. The highest fruit drop occurs in the second and third week after fruit set and ranges from 78.33 per cent in Kaisha to 61.4 per cent in New Castle (Lal and Sharma, 1979). Water deficit along with high temperature (>25°C) in autumn increase flower bud drop (Medeira, 2000). Gibberellic acid (GA_3) at 10 ppm when sprayed at the start of flowering increased fruit set, yield and to a lesser extent fruit size (Forlani, 1984). A spray of 10 ppm NAA or 50 ppm of 2, 4, 5-T at the beginning of pit hardening also reduced the preharvest drop.

Fruit Thinning

Fruit set in apricot is rather heavy which results into the harvest of under-sized fruits and increases the tendency of biennial bearing. Fruit thinning increases the fruit size, promotes regular bearing, decreases limb breakage due to heavy crop load and maintains the tree vigour (Petu *et al.*, 2011). It should, however, be done within forty days after full bloom. *i.e.*, during the last week of April or first week of May, because this is the effective period influencing fruit bud formation. Both hand and chemical thinning methods are employed. Depending upon the crop load, the fruit may be thinned 6-10 cm apart, more space between the fruits being provided in heavy than in lighter crop. Not more than two fruits be left on a spur. In order to reduce the amount of hand thinning, blossom thinning can be done by spraying 75-100 ppm dinitro-ortho-butylphenol when 60-100 per cent flowers have opened (Looney and Killick, 1979). Suranyi (1979) obtained the best fruit thinning by applying 50 ppm NAA in May and reported ethrel to be phytotoxic to apricots. Suranyi (1986) again reported good thinning on warm days with high humidity when the fruitlets were 8-9 mm long. NAA at 25-75 ppm was best for cv. Hungarian Best and at 50-100 ppm for Rose apricot. Thinning reduced yield (Rettke, 2005), though the increased size, enhanced colour and fruit quality usually compensate the yield reduction (Son, 2004). Coskun and Ozguven (2000) reported effective thinning and increased fruit size by spraying of NAA at 50 ppm at 3 weeks after full bloom in apricot cvs. Priana and Belliana. The effect of different thinning methods on fruit quality of apricot cvs.Tom Cot, Harval and Flavorcot were investigated in France by Jay *et al.* (2000). They suggested that if lower thinning should be carried out on highly productive, self-fertile cultivars, thinning should be best carried out at 25-30 days after flowering having one fruit for every 8-10 cm of shoot. Fruits obtained from thinning treatment were larger and sweeter than on control shoots. Taghipour *et al.* (2001) obtained effective thinning of apricot cv. Gerdi at Brazil by spraying of NAA at 40 mg/l, NAD at 80 mg/l and ethephon at 50-200 mg/l at three weeks after full bloom. However, hand thinning improved fruit quality, pit weight and flesh to pit ratio. The commercial apricot cultivar Canino of Israel, is self-compatible and therefore sets large number of small fruits. Application of unniconazole at 50-100 mg/l at 30 per cent of full bloom, reduced fruit set by about 50 per cent. The remaining fruits were larger in size and there was no impairment in return bloom the following year (Raz *et al.*, 2010)

Fruit Growth and Maturity

The apricot fruits have double sigmoidal growth pattern having a retarded growth period at the time of pit hardening. Fruits develop rapidly and ripen early in summer season. A temperature of about 23°C acclerates the developmental process of the fruits but not at the higher or lower temperatures. Bajwa and Mishra (1971b) working with apricot cultivars in India reported the approximate period of fruit development to be 73, 82, 85, 88 and 96 days in case of Shipley, New Castle, Kaisha, Royal and Shakarpara cultivars, respectively. Arzani *et al.* (2000) studied the vegetative and fruit growth pattern of mature close planted apricot cv. Sundrop trees grown under humid climate of Palmerston North, New Zealand. They observed that fruits as a proportion of new top growth increased on both fresh and dry weight basis, until the time of harvest. Fruits showed a sharp increase in size up to the time of harvest, but this increase was influenced by fruit number. The highest proportion of early season top growth was in shoots with leaves, but from about 90 days after full bloom (dafb) this proportion decreased in favour of fruits until harvest. There was a large increase in fruit size during stage III of fruit growth (from 97 dafb), while shoot growth was more rapid during 60-90 dafb. Valentini and Botta (1999) studied the chemical and physical changes of four apricot cultivars (Aurora, Perla, Tonda di Costigliole and San Castrese) during development and ripening. They reported that fruit growth followed a double sigmoid pattern. Sucrose was the main sugar in mature fruits followed by glucose and fructose. At maturity sugar content ranged between 6.3g/100g fresh weight in Perla and 10.2g/100g in Tonda di Costigliole citric and malic acids were the main organic acids with malic acid predominating in Aurora and Perla and citric acid predominating in Tonda di Costigliole and San Castrese. Titratable acidity ranged from 23.5 meq/ml in Perla to 43.1 meq/ml in Aurora. They concluded that TSS and titratable acidity were good indicators of maturity.

Maturity Indices

Molnar and Turi (1975) found a strong positive correlation between fruit size at the start of pit hardening and at maturity. The ripening period in a particular area can be determined on the basis of computation of mean and maximum diurnal temperatures above 10°C affecting the tree development (Klimenko and Shiman, 1975). Apricot fruits develop maximum flavour and are very delicious when ripen on the tree but such fruits cannot be transported to distant markets because of their perishable nature. Fruits harvested sufficiently in advance of ripening do not attain full taste. The standardization of maturity and harvesting is, therefore, essential for obtaining good return from the marketing of fresh apricot fruits. These standards vary with the cultivar, location and demand of the market. The fruit colour, flesh colour, size, firmness, total soluble solids, titratable acidity, drymatter content, days from bloom and accumulated heat units above 10°C are the main maturity indices. In California, harvest date is determined by changes in skin ground colour from green to yellow. The exact yellowish-green colour depends on cultivar and shipping distance. Apricot should be picked when still firm because of their high bruishing susceptibility when fully ripe and soft. High consumer acceptance is all aimed for fruit with high (>10 per cent) soluble solids content and moderate total acids (0.7 to 1.0 per cent). Apricots with 8.9 to 13.3 N flesh firmness are ready to eat. Sud *et al.* (1979) found 9 May as the best harvesting date 64 days, after full bloom for the cultivar New Castle under mid-hill conditions of India when the fruit attained good size, barium-yellow skin colour and amber-yellow flesh colour. The total soluble soilds and total sugars were 10.03 and 5.48 per cent, respectively. On the basis of studies on optimum picking maturity in cvs. Bulida, Palsteyn Piet Cillii, Peeka and Royal, Visagie (1985) reported that fruit firmness penetrometer reading of 5.5-7.0 kg (using a penetrometer with 11 mm point) is a reliable indicator of the optimum maturity stage while fruit pH and acidity are not dependable indicators as they differ very little between maturity stages. Kock (2011) also stated that flesh firmness is a valuable parameter in determining the optimum harvest maturity of apricots and the use of a penetrometer with a 6-mm plunger accurately determined flesh firmness. Jooste and Taylor (1999) however, indicated that skin colour and flesh firmness as determined by a penetrometer with an 11-mm plunger, gave the most reliable indication of change in maturation. Oe *et al.* (2012) working with Japenese apricot (*P. mume* Sieb. *et* Zucc) cv. Nako observed that fruit located on the inner side of the canopy, the time that the trichome fell out over 30 per cent of their surface is an indicator of harvest maturity. Fruit weight and b and L* values of skin colour were found to increase and fruit firmness decreased with fruit development. However, there were no differences between the fruits of inner and outerside of the canopy when they were compared at the same stage. The contents of citric acid, sorbitol and β-carotene increased as the fruit develops while phenolic content and antioxidant activity decreased with fruit development. These results indicate that delaying the harvesting of the inner canopy fruits makes their fruit weight and some functional components equal to the fruit on the outerside of canopy but causes poorer phenolic content and antioxidant activity than on the outerside of the canopy.

Better-coloured and early-ripened fruits get a premium in the market. Successful attempts were, therefore, made to enhance fruit colour and ripening in apricot fruits by spraying 2, 4, 5-T, 2, 4, 5- TP and ethrel (Martinez *et al.*, 1974; Sharma, 1978; Visagie *et al.*, 1978). Gautam and Dhuria (1979) obtained 2-5 days advance ripening in the fruits of the cultivar Kaisha by spraying 25-150 ppm ethrel two weeks before the expected harvest.

Marketing of high quality fresh apricots is difficult because of their rapid ripening. The respiration of Japanese apricots (*P. mume*) follows a climacteric pattern with much ethylene production (Miyazaki, 1983). Ethylene treated fruits also show the advancement in the onset of the climacteric and in ripening. Mature fruits can be ripened quickly for canning by treating with 100 ppm ethylene for 48 hours at 20°C (Brecht *et al.*, 1982).

Harvesting

Apricot fruits are soft and have a short shelf-life, so they must be picked as close to optimum maturity as possible, in order to offer good quality fruit to consumers.

In view of the period required in marketing of fresh fruits, the fruits are plucked when they change colour from green to yellow, lose flesh firmness slightly depending upon the cultivar. Fully ripe fruits are, however, harvested for freezing, canning and drying.

The optimum stage of fruit ripening for machine harvesting starts when about 10 per cent fruits are still green and ends when more than 10 per cent of the fruits are overripe (Belyakov and Strandzhev, 1979). It is also determined by measuring fruit flesh firmness, dry matter content and refractive index of juice and harvesting time is considered optimum on ripening of 70-85 per cent of the fruits (Szenczi, *et al.*, 1982). During mechanical harvesting, fruit losses of about 15 per cent are quite common (Sansavini *et al.*, 1983).

Yield

An apricot plant starts bearing at the age of four or five years attaining a full bearing age at about eight or ten years when it yields about 50-80 kg fruit. The productive life of a tree is about 30-40 years.

Storage

Apricot fruit is usually stored in cold storage (1°C) and many apricot cultivars can be stored to 4-6 weeks and these cultivars are able to keep an acceptable level of fresh firmness and quality attributes. Fruits start to lose their water content reaching 12 per cent after 28 days at 1°C. This is accompanied with fruit softening, increased fruit acidity and reduction in soluble solids content (Ezzat *et al.*, 2012). Fruits harvested at optimum maturity can be stored for a week at 5°C and 85 per cent relative humidity in normal atmosphere. Respiration in fruits was about 30 per cent lower in 0.05 per cent CO_2 and 5 per cent O_2 atmosphere than in normal atmosphere (Golias, 1974a). A lower consumption of reserve substances in the fruit during the first ten days of CA storage at 3°C was also found (Golias, 1974b). Loss in acidity and deterioration of flavour were more rapid in air storage than in CA storage (Claypool and Pangborn, 1972). Claypool and Pangborn (1972) also observed about equal flavour deterioration of Blenheim in 3 weeks in air, 5 weeks in low O_2 atmosphere with 5 per cent and 10 per cent CO_2, and 7 weeks in low O_2 atmospheres containing 2.5 per cent to 3.5 per cent CO_2. For the storage of Blenheim fruits, the lower concentration of CO_2 (2.5-3.5 per cent) was found better than higher concentrations (5-10 per cent). Apricot cvs. Cegledi biborkajszi, Gonci magyar kajszi and Pannonia could be stored for 28 days in CA storage in an atmosphere with 1.2 per cent O_2, 4 per cent CO_2 at 1°C and 95 per cent relative humidity (Hitka, 2011). Exposure to less than 1 per cent O_2 may result in development of off flavours and higher than 5 per cent CO_2 for more than 2 weeks can cause flesh browning and less of flavour (Turter *et al.*, 1994). However, Drake and Yazdaniha (1999) working with apricot cvs. Perfection and Rival at Israel reported that fruits stored at 2 per cent O_2 and 12-15 per cent CO_2 retained firmness with reduced decay and very little internal breakdown after storage for 60 days. Colour developed much more slowly in fruits stored in 12 or 15 per cent CO_2 that lower CO_2 concentrations.

The cultivars differ in their potential storage life, Blenheim stores much better than Tilton (Claypool and Pangborn, 1972). A spray of 2000 ppm maleic hydrazide six weeks before harvest reduces storage losses of fruits (Suranyi, 1973). Hydrocooling of Japanese apricots prior to storage decreases the chilling injury (Quata and Sakamoto, 1979). Beside the chilling injury, surface pitting and enzymatic browning are the other common physiological disorders. The postharvest diseases infecting the apricot fruits generally are brown rot (*Monilinia laxa, M. fructigena*) whisker's rot (*Rhizopus stolonifer*) black mould rot (*Aspergillus niger*) and green mould rot (*Alternaria alternata*). The infection occurs either in the orchard or during handling, packing and transportation. The management of disease should begin from the orchard. No disease should be allowed to infect the fruit during fruit harvest as it makes fruit susceptible to black mould and green mould rot. Prompt cooling of the fruits to temperatures below 5°C by hydrocooling or aircooling after harvest helps in controlling these diseases. The dipping of fruits in hot water (50°C) for three minutes also reduces the chances of infection when combined with fungicidal treatments. The fungicide bortran (DCNA) at 500 ppm concentration has been found very effective for the control of brown rot and whisker's rot. The fruits are dipped in the fungicide solution for four minutes. Black mould rot is effectively checked by treating the fruits with 1000 ppm, concentration of benlate.

Treatment with I-MCP at I µl/l for 4h at 20°C delayed the onset of ethylene production and respiration rate. The treatment also resulted in less firmness and titratable acidity loss during storage of apricot cv. Perfection when stored at 0 or 20°C. It also delayed production of volatile alcohols and esters during ripening at 20°C. MCP treated fruits also exhibited less colour change and were greener than controls. However, the effect of MCP were less pronounced following treatment of fruits at a more advanced stage of development (Fan *et al.*, 2000). Dong *et al.* (2002) stated that I-MCP has potential to delay ripening, but the cultivar maturity of fruit, and time of application must be chosen carefully. It is suggested that I-MCP is effectively extending the shelf life and improving the quality of 'Canino' apricots directly marketed or after storage.

Ripening

Apricots are climacteric fruits with a high susceptibility to flesh softening and loss of flavour during postharvest storage, and most of the ripening processes are regulated by ethylene, which also has an effect on its own biosynthesis. Ethylene production rate increases with ripening and storage temperature from under 0.1 µl/kg/ha at 0°C to 4-6 µl/hg/ha at 20°C for firm ripe apricots and higher for soft-ripe apricots. The respiratory rate and the activities of ACO and ACS were inhibited when treated with salicylic acid at 0.01 g/l (Zhu-Xuan *et al.*, 2013). Cardarelli *et al.* (1998) working with apricots cv. Boccuccia Spinosa, harvested the fruit at an average penetrometer reading of 62.6 N and soluble solids content (SSC) of 12.6. The fruits were then kept in air at 18°C and 85 per cent RH. They studied physiological and quality measurements at pre-ripe stages 0 and 1 (PRI and PRL) ripe stages 3, 5 and 7 (R_1, R_2 and R_3) and the overripe stage at 10 days (OR). In the first day, a burst of ethylene production and respiration was observed in relation to significant weight loss (1.1 per cent). A second greater ethylene peak was observed at R_3. A good edible quality was achieved at day 5, when SSS was approximately 14 per cent and acidity was between 1.1-1.2 per cent. α-L-arabinofuranosidase, α-D-galactosidase, α-D-manosidase and β-D-glucosidases had activities 10 times greater than glucosidases. The parallel loss of firmness and increase in the activity of some glycosidases, indicates the involvement of glycosidases in fruit softening. Munoz *et al.* (2012) reported that ethylene inhibitors such as I-MCP and AVG were effective in reducing ethylene production, respiration rate and fruit softening. They characterized three ACC genes and found that only the expression of ACS2 was highly reduced by ethylene inhibition, suggesting a key role in ethylene synthesis at ripening. Contrarily, ACS1 and ACS3 showed a higher expression under individually regulated in a specific mode as observed in other climacteric fruits. Changes in ACC oxidase genes did not show a consistent pattern of ethylene modulation.

Pests

As many as thirty insect species have been recorded on apricot but only the following ones are important (Butani, 1979).

Indian Gypsy Moth (*Lymantrio obfuscate*)

This destructive pest has been recorded in all apricot growing regions. The damage is caused by caterpillar which are gregarious and nocturnal in habit. In case of severe attack the trees are completely defoliated as the larvae eat up the entire lamina, leaving the hard veins behind (Pruthi and Batra, 1960). The moth can be controlled by dusting with 10 per cent BHC in March-April. The egg masses having yellowish brown hairs should be located in August and March and destroyed.

Peach Stem Borer (*Sphenoptero lafertei*)

It is a polyphagus pest attacking trunk and branches of apricot but causes more damage to peach and plum. Sometimes the entire tree dries up due to boring of grubs into trunk. Infested inner bark becomes a powdery mass like sawdust which, mixed with frass, oozes out from the splits on the surface. The pest undergoes three generations in a year by overwintering as larva. Heavily infested branches and trees are to be removed and burnt. The adult beetles should be killed by spraying 0.15 per cent BHC in the month of March and it should be repeated after a fortnight again.

Peach Fruit Fly (*Dacus* spp.)

The flies lay eggs on fruits and feed on pulp rendering them unfit for human consumption. Pupation takes place in soil under infected trees. In April-May, on the appearance of flies on leaves the bait comprising 0.1 per cent malathion and 1 per cent sugar (2 ml cythion 50 EC + 10 g sugar in one l of water) should be spayed on leaves and other resting places of the flies. Two sprays, first one month before and second a fortnight before the commencement of fruit harvest, with fenthion (250 ml lebaycid 100 l) or fenitrothion (500 ml sumithion/folithion/accothion 50 EC in 500 I water) should be done.

European Red Mite (*Panonychus ulni*)

Depending upon weather condition and fruit exposure, highest incidence of apricot fruit damage at harvest is associated with this mite. Interestingly, cv. Roxburgh Red is resistant while cvs. Moorpark and New Castle are highly susceptible (McLaren, 1981). The best control is obtained by applying carbophenothion at 58g/100 l in the middle of October when maximum hatching of eggs takes place (McLaren, 1983).

Apricot Chalcid (*Eurytoma samsonavi*)

This is a serious pest of apricot. Eggs are laid inside the tender fruits. On hatching the grub feeds inside the kernel causing drop of premature fruits. Grubs, which are translucent white, feed during April-May and remain quiescent from June to February. Thereafter, they pupate inside the fruits. All the infested and fallen fruits should be collected and destroyed to control this pest (Pruthi and Batra, 1960).

Leaf Roller (*Archips micaceana*)

Leaf roller also damages apricot in various parts of the world. The leaves and petioles are cut and rolled up by caterpillars and then they feed inside on soft tissues (Butani, 1979). After puncturing the ripening fruits, the caterpillars also eat up inner pulp. The affected fruits are not fit for human consumption. The leaf roller can be controlled by spraying 0.05 per cent endosulfan or fenitrothion. The rolled leaves and webbed flowers should be destroyed after collecting.

Diseases

Brown Rot (*Monilinia laxa., M. fructigena, Sclerotinia* spp.)

The disease attacks apricot in all the countries wherever it is grown. It is positively related to the rainfall in the first 7 days after flowering begins (Fideghelli *et al.,* 1978). Both blossoms and fruits are affected. In case of severe infection, blossoms turn brown and die. After infection, tiny brown spot enlarges to a large, soft, brown area within a few days. The pathogens overwinter in fruit mumies, in twig cankers and in rotten fruits. The penetration takes place via cuticle and stomata. Stoyanov (1977) recommended treatment with 1 per cent DNOC in dormant season, 0.3 per cent zineb in growing season and 0.1 per cent benomyl during anthesis for effective control of the disease. Bordeaux sprays cause phytotoxicity (Stoyanov, 1977; Karova, 1972). Twenty cultivars resistant to this disease have been reported from Italy (Simeone, 1982).

Bacterial Canker and Gummosis (*Pseudomonas syringae*)

The disease has a world-wide spread, attacking many other stone fruits besides apricot. Except root, all the other parts of the trees are attacked. The bark and outer sapwood show symptoms of circular to elongated water-soaked lesions. The affected bark later turns brown, gummy and sour-smelling (Anderson, 1956). Sunken black lesions are found on the infected fruits. Sometimes, shoot withering is also noticed as the bacteria invade the older twigs. Bordeaux mixture can be sprayed in the fall as well as spring to prevent infection. Mashobra paste (stearic acid 25.5 g, morpholine

9.0 ml, water 340 ml, lanolin 136 g and non-sterile dihydrostreptomycin sulphate 3.1 g) formulated by Agarwal (1964) has been found effective as a control measure. The infected limbs can also be burnt with a propane burner (Hawkins, 1976). Donmez *et al.* (2010) working with 15 apricot cultivars in Turkey reported cvs. Hasanbey, Cloglu, Soganci and Sekerpare were resistant and cvs. Sam, Tokaloglu (Erzincan) and Erken Agerik were susceptible to bacterial Canker.

Powdery Mildew (*Sphaeratheca pannosa*)

The fungus infects young leaves and shoots while the older ones are fairly resistant. The white, powdery appearance is caused by great number of conidia produced in chains arising from the wefts of hyphae on the leaf surface. The infected young leaves get curled and narrowed as they expand. The disease checks the growth of the affected plant parts. Three sprays of 2 per cent wettable sulphur at different stages, *i.e.*, before flowering, at petal fall and two weeks thereafter, are necessary to control the mildew.

Apoplexy

Sometimes an apricot tree or a part of it begins wilting and dies suddenly due to a syndrome known as apoplexy or permature die back. The problem is very serious in all the apricot growing regions in the world and several causes, environmental and pathological, have been associated with it. Low and sharp fluctuating temperatures are held responsible for bark cracking followed by pathogen infection which produces resin droplets in the xylem, phloem and bark tissues (Babos *et al.*1976; Ambartsumyan and Nersisyan, 1978). The infection increases cell-wall thickness and severe cambium infection inhibits the new tissue development. Badly affected plants die within one or two years but in some cases within ten to fifteen years (Ambartsumyan and Nersisyan, 1978).

Pseudomonas syringae and *Cytospora cinta* are the principal pathogens which have been found associated with apoplexy (Rozsnyay, 1986; Kropis, 1984). Several control measures include pruning in spring instead of winter season, spraying of plants with Bordeaux mixture before and after pruning and removal and burning of all infected wood (Arsenijevic, 1982; Rozsnyay and Klement, 1973) were recommended. To prevent low temperature injury, Ambartsumyan and Nersisyan (1978) recommended a perlite sand and 20 per cent milk of lime mixture spray on the trunk and main branches of the tree. All mechanical injuries to the plant should be avoided. Planting of resistant and tolerant cultivars is also suggested.

Wilt (*Verticillium albo-atrum*)

The disease occurs throughout the temperate zone. Gupta and Saharan (1973) reported the die back of apricot due to wilt. The leaves drop down from the affected limbs in early summer. The whole tree or a part of it may be affected. No external evidence of the disease except general stunting is visible. The infected tree is not always killed, but may remain stunted and unproductive for several years (Anderson, 1956). This is a soil-inhabiting fungus and infects through the roots, obviously its control through protective sprays is not possible. Before new planting, the pit soil should be sterilized with formalin.

Coryneum Blight (*Stigmina carpophylla*)

The blight attacks twigs, leaves, blossoms, fruits as well as unopened buds. Small, circular deep purple spots appearing on the fruit become raised and rough (Pathak, 1980). The raised spots on twigs expand into necrotic cankers. Abscission of the diseased area results in shot hole effect. Diseased buds become darker in colour than normal ones and often show a glistening layer of gum (Anderso. 1956). Copper oxychloride at 300 g/100 l water should be sprayed before leaf fall and at bud swell.

Breeding and Improvement

For any good breeding programme, the availability of wide genetic diversity in the gene pool is a pre-requisite which is abundantely present in apricots. Even then comparatively less breeding work has been done in this crop than other stone fruits. Although apricots are geographically widely distributed, yet a particular genotype of the fruit is ecologically less adapted. The damage due to spring frost is the greatest limiting factor in extending its cultivation. Therefore, the climatic adaptability has been the main objective of apricot breeding throughout the world (Watkins, 1976; Bailey and Hough, 1975). Depending upon the local requirements, the cultivars need to be developed for cold hardiness and for varying temperate climatic conditions, *i.e.*, high and low chilling cultivars.

In an attempt to develop cold resistant material, Kostina (1936) did a commendable work on the collection, characterization and utilization of apricot genetic resources. As a result, the cultivars resistant to fluctuating winter temperatures such as Oranzhevokrasnyi, Zard, Suphany and Tokshion have been identified (Kostina. 1969).

Since the climatic factors vary from region to region and year to year, it is very important to develop cultivars which are least affected by such fluctuations like the Delicious group of apples. For cooler regions the genotypes should not only develop flower buds slowly during winter but also should not respond to warming spring temperatures quickly which is the only escape from spring frost hazard. Good low chilling cultivars are not available in apricots which restrict its extension to warm temperate regions. The breeding of such cultivars is very much desired.

Good pomological fruit qualities such as large size, good skin colour with blush, firm flesh, proper blend of sugar and acidity, good texture and high keeping quality and resistance to pests and diseases are common objectives of all the fruit improvement programmes. Harcot, Harogem Harlayne, Hargrand and Harglow are the apricot cultivars evolved with these aims (Layne, 1978, 1979, 1981a, 1981b, 1984). Italy and Hungary have the richest diversity and number of accessions of apricot germplasm collections (Zanetto *et al.*, 2002). It is commonly accepted that European apricots are more tolerant to fungi than Central Asian and Iran-Caucasion cultivars. The major objective of the European apricot breeding programme are to biotic stresses like Plum Pox Virus, bacterial diseases, chlorotic Leaf Roll Phytoplasma and Apricot decline. Many good cultivars have been bred in last 20 years (Table 2.5). These cultivars along with the others help in extending the ripening period also. The cultivars for processing and drying have also been evolved. Drying apricot must have high sugar content and sweet kernel. Resistance to apoplexy beside adaptation is the principal concern in European countries (Bailey and Hough, 1975).

Table 2.5: Cultivars Developed in different Countries

Cultivar	*Origin*	*Characters*
Boreale	Italy (Milano and Bolgona)	Better fruit quality and extended ripening season
Ardore	- Do -	- Do -
Pieve	- Do -	- Do -
Angela	Italy (Pisa)	Early maturing
Priana	Italy (Pisa)	Good quality
Gheriana	Italy (Pisa)	Late maturing, high-yielding
Soledane	France, INRA	Suitable for Mediterranean coastal areas
Florilege	- Do -	Suitable for lower part of Rhone Valley
Rojo Passion	Spain (Murcia)	PPV resistant with good fruit quality
Selone	- Do -	- Do -
Murciana	- Do -	- Do -
Asli, Atef, Fakher, Meziana, Quafer, Raki	Tunisia	Low-Chilling cultivars with good fruit quality

In India, the breeding of apricot has been undertaken with a view to evolve low chilling and drying types by Seth *et al.* (1984). They have released three cultivars, *viz.*, Chaubattia Alankar (Kaisha x Charmagz), Chaubattia Madhu (Turkey x Charrnagz) and Chaubattia Kesri (St. Ambroise x Charmagz). All the three cultivars are, however, suitable only for fresh market and there is a great need to develop prolific bearing drying types because all the presently grown drying cultivars are shy bearers. Better adaptability, lower chilling, drought resistance and dwarfness should be the main objectives of apricot breeding in India.

Breeding Systems

Selection of seedlings raised either from open-pollinated or controlled-pollinated seeds is the usual system adapted in fruit breeding. Proper selection of desirable parents for crossing and screening of resultant hybrid seedlings are very crucial for the success. Although apricots are highly heterozygous, yet desirable traits are usually inherited in the hybrids resulting in the evolution of many cultivars by breeding. However, specific characters such as disease resistance, cold tolerance and late blooming should be incorporated with other desirabie pomological characters by backcrossing.

In order to make slight improvement in an otherwise good parent, the mutation breeding through chemical mutagens and irradiation treatments is desirable. Lapins (1972) selected 'Early Blenheim' following thermal neutron irradiation as an early ripening cultivar.

Dramatic improvements in the progenies can only be brought about by resorting to interspecific hybrization, though the system is cumbersome, time consuming and with unpredicatable results. Recent advances in biotechnical methods have however, made interspecific hybridization comparatively easier than the past, the embryo culture has been found very helpful. The apricot is nearest to plum than other stone fruits. Occasional natural mutant branches spring up in apricot bearing plum-like leaves. Similar mutants were also obtained following thermal neutron irradiation. Interspecific hybrids between plum and apricot are known as plumcots or cotplums. Hybridization of apricot with *P. sibirica* and *P. mandschurica* should provide cold-hardy progenies. In a varietal evaluation, the most cold-hardy cultivar: were found of specific hybrid origin, notably apricot x Myrobalan plum (*P. cerasifera* hybrids (Moskalenko, 1983). Likewise, the cultivars Aleksandrcherneyi and Amerikanskii Cherneyi, interspecific cherry plum and plum apricot hybrids, have been found highly resistant to *Cytospora,* the principal cause of apoplexy (Kropis, 1984). *P. ansu* and *P. mume* cultivars grown in mild and humid tracts of China and Japan can form the basis for the breeding of the material adapted to such conditions

A process show gametophytic self-incompatibility controlled by a single locus with multiple genes, S-haplotypes (De Netta-ncourt, 2001). The S-haplotype contains a female determinant, (S-RNase) encoding for a ribonuclease enzyme and male determinant, S-haplotype-specific F-box gene (Halasz *et al.*, 2010). 21 S-RNase alleles are known in European apricots, 20 of which (S_1-S_{20}) code for self-incapability and one (Sc) allowing self-compatibility (Halasz *et al.*, 2007) and it was confirmed that Schaplotype is a pollen part mutant of S_8 haplotype. Beside some additional S-alleles have also been identified in Chinese cultivars (Halasz *et al.*, 2010).

Plum box virus (PPV) has spread thousand European countries. Some North Amereican cultivars like Stark Early Orange, Goldrich, Harlayne, Stella and Harcot are known to be resistant to the disease (Martinez-Gomez *et al.*, 2000). These cultivars are mostly used as resistant source of breeding. Asian apricots are most likely the source of resistant genes for PPV. Zhebentyayeva *et al.* (2008) reported *Prunus davidiana* allels in cultivar Stark Early Orange and *Prunus mume* allels in Stark Early Orange and Goldrich which suggest the contribution of these species to PPV resistance.

Various type of DNA markers were used for genetic analysis in apricot. RAPD (Williams *et al.*, 1990), ISSR (Yilmaz, 2008), RFLP (De Vicente *et al.*, 1998), AFLP (Yuan *et al.*, 2007), SSR (Harmaza, 2002) and SRAP has been used in apricot to characterize different cultivars belongs to abundant genetic resources. Yuan *et al.* (2011) analyzed the apricot cultivars of Shandong province using AFLP marker techniques. They divided the 26 cultivars in 4 groups at 58.58 per cent level of similarity co-efficients. Cultivars Yubadan and Yangjiyuan were considered important for broadening genetic diversity of Shandong apricots, breeding new apricot cultivars and creating new germplasm. Janatizadeh *et al.* (2012) studied 39 Iranian apricot cultivars, using RAPD markers to understand the genetic diversity. The highest genetic similarity was observed between cultivars Noori-Dirras and Noori-Pishras (0.95) while the lowest between Tanasgol and Shahrood-48

Prunus mume, the Japanese apricot (mei) originated in China, was divided into two types: fruiting mei and flowering mei. The fruiting mei was divided by their pericarp colour into green mei, red mei, and white mei (Shen-YuYing *et al.*,

2011). Eighty-four cultivars (43 fruiting mei and 41 flowering mei) were studied to identify their genetic relationship by genetic diversity analysis using REMAP and IRAP technologies. The results showed that the observed number of allels in Japanese apricot was 1.9955 and the effective nuclear of allels was 1.4887, Neil gene diversity was 0.2910. Eighty-four Japanese cultivars were divided into 18 groups at the similarity co efficient of 0.743 and 4 groups at the similarity co-efficient of 0.705 with the UPGMA cluster method. Fruiting mei and flowering mei clustered into different groups which showed that the complexity of inheritance and the extent of evolution were different. PopGen 32 analysis suggested that Japanese fruiting mei may have introduced from Zhejiang province in China and Japanese flowering mei from Jiangsu province in China (Shen-YuYing *et al*. 2011).

References

Abohassan, A.A., El-Hamady, A.A.M., and Hamouda, M.A. (1979). Proc. Saudi Biol. Soc., No.3, pp. 1-6.

Abrisqueta, J.M., Ruiz, A. and Franco, J.A. (2001). Agric. Water Management, 50: 211-27.

Adamov, M.G. (1984). Soorruk Nauchnykh Trudov, Vsesoyuznyi Nauchno-Issledovatelskii Institut Sadovod-stva Imeni I.V. Michurina, No 42, pp. 51-4.

Agabbio, M. (1973). Stud; Sassaresi, Ill, 21: 283-286.

Agarwal, R.K. (1904). RFRS, Mashobra Agric. Res., ICAR 4: 139-40.

Ageeva, G. (1984). Byulleten' Gosudarstvennogo Nikistskogo Bot. Sada. 54: ·49·52.

Albburequerque, N., Burgos, L. and Egea, J. (2004). Sci. Hortic., 102: 397-406.

Albigo, L.G., Cloypool, L.L. and Uriu, K. (1966). Proc. Amer. Soc. Hort. Sci., 89: 53-60

Ambartsumyan, M.A. and Nersisyan, T.S. (1978). Sadovodstvo, 5: 24.

Anderson, H.W. (1956). In: Diseases of Fruit Crops. McGraw Hill Book Co. Inc. New York, pp. 189-61.

Anonymous (1969). The Wealth of India. Publications and Information Directorate, CSIR, New Delhi, 13: 256-61.

Arsenijivic, M. (1982). Jugoslovensko Vocarstvo, 16: 43-50.

Arzani, K., Wood, D. and Lawes, G.S. (2000). Acta Hort., 514: 295-300.

Asma, B.M. (2000). Apricot Culture. Evin Pub., Malatya, Turkey, pp. 243.

Asma, B.M. (2011). Her yoniiyle Kaysi. Uyum Ajans, Ankara, Turkey, pp. 367.

Audubert, A. (1987). CTIFL-INRA, 109-122.

Avanzato, D. and Nicotra, A. (1986). Rivista della Ortoflorofrutticoltura Italiana, 70: 13-25.

Azouz, A., Alia, M.E. and Rizk, E.M. (2009). Ann. Agric. Sci. Cairo, 54: 373-83.

Babayashev, E. (1983). Acta Hort., 121: 347-51.

Babos, K., Rozsrnav, Z.D. and Clement, Z. (1976). Acta Phyto. Sci. Hungarica, 11: 71-79.

Bailey, C.H. and Hough, L.F. (1975). In : Advances in Fruit Breeding, (eds. J. Janick and J.N. Moore), Purdue University Press, Indiana, Lafayette. pp. 367-83.

Bajwa, M.S. and Misra, K.A. (1971a). Himachal Hort., 121: 17-18.

Bajwa, M.S. and Misra, K.A. (l97Ib). Progressive Hort., 2: 67-71.

Bas, M. and Paydas, S. (2000). Bahce, 29(1/2): 81-89.

Belyakov, V. and Strandzhev, A. (1979). Gradinarska i Lozarska Nauka, 16: 3-10.

Bencat, P. and Matuskovic, J. (1980). Vedecke Prace Vyzkumneho Ustavu Ovocnych a Okrasych Drevin V Bojnicach, 2: 5-12.

Benson, N.R., Degman, E.S., Chernelir, Le. and Chennault, W.R. (1963). Proc. Amer. Soc. Hort. Sci., 83: 55-62.

Bogdanova, V.M. (1974). Byulleten Glavnogo Botanicheskogo Sada, 92: 67-68.

Bohomaz, K.I. (1960).Visnyk Kyyivs'k Univ. Ser. Biol., 3: 56-66.

Bondok, A.Z., Salama, M.A., Hassan, M.M., Tadros, M.R. and Mohanna, S. (1984). Ann. Agric. Sci. Ain Shams Univ., 29: 433-64.

Boskovic-Rakocevic, L., Milosevic, T., Milivojevic, J. and Paunovic, G. (2012). Acta Sci. Pol, Hortorum Cultus, 11: 227-37.

Bould, C. (1955). Nature, 175: 90-91.

Bradt, O.A. and Anderson, E.T. (1979). HortSci., 14: 82.

Brecht, J.K., Kader, A.A., Heintz, E.M. and Norona, R.E.(1982). J. Fd. Sci., 47: 432-36.

Buban, T. and Turi, L. (1979). Kartgazdasag, 11: 29-36.

Bullock, R.M. and Benson, N.R. (1948). Proc. Amer. Soc. Hort. Sci., 5: 199-204.

Bunea, A. (1986). Acta Hort., 192: I27-36.

Bunea, A., Muler, G., Cociu., V., Burloi, N., Ionescu, P. and Patru, N. (1983). Acta Hort., 121: 323-26.

Butani, D.K. (1979). Insect and Fruits. International Book Distributors, Rajpur Road, Dehradun. pp. 309-15.

Bybordi, A. (2013). Middle. East J. Sci. Res., 14: 423-31

Campbell, O.E., Merwin, I.A. and Padilla Zakour, O.I. (2011). New York Fruit Quart., 19(4): 12-16

Cappellini, P. and Limongelli, F. (1981). Annali dell'Instituto Sperimentale per la Frutticoltura, 12: 47-55.

Cappellini, P., Strada, G.D., Fideghelli, C., Quarta, R. and Simeone, A.M. (1978). Annali dell'Institul Sperimentale per la Frutticoltura, 9: 9-19.

Cardarelli, M.T., Botondi, R. and Mencarelli, F. (1998). Italus Hortus, 5 (5/6): 27-31

Carter, M.V. and Price, T.V. (1975). Aust. J. Agric. Res., 26: 537-43.

Cepoiu. N. (1979). Locrasi Stiintifice: Institut Agronomic "Nicolae Balccscu" Horticultura, 22: 45-48.

Chandlar, W.H. (1987). Bot. Gaz., 98: 624-46.

Chauhan, A., Dwivedi, M.P. and Tomar, C.S. (2004). Scientific Hort., 9: 73-77

Chen-XueSen, Gao.-DongSheng, Li-XianLi, Zhang-Yanmin and Zhang-Lianzhong (2001). Acta Hort. Sinica, 28(5): 475

Chen-Yuling, Feng-YiBin, Chen-ChangWen and Qiao-ShuRui(2011). J. Fruit Sci., 28(66): 1128-29

Childers. N.F. (1983). Hort. Publ., Gainesville, Florida. pp. 381-83.

Cibes. H.R. (1950). M.Sc. Thesis, Rutgers University.

Clanet, H. and Salles, J.E. (1986). Acta Hort., I60: 207-15.

Claypool, L.L. and Pangborn, R.M. (1972). J. Amer. Soc. Hort. Sci., 97: 636-38.

Coskun, M. and Ozguven, A.I. (2000). Turkish J. Agric. Forestry, 24: 309-16

Costa. G. and Grandi, M. (1975). Atti Incontro-Frutticolo su Problemic Prospective delia Coltura dell' Albicocco Sociera Orticola Italiana, Imole, pp. 55-62.

Crane, J.C. (1954). Proc. Amer. Soc. Hort. Sci., 64: 225-31.

Crossa-Raynaud, P. and Audergon, J.M. (1987). In: Rootstocks for Fruit Crops. (eds. R.C. Rom. and R.F.Carlson) John Wiley Sons. Inc. New York, pp. 295-320.

Cummins, J.N. (1985). Fruit Var. J., 39: 41-43.

Damigella, P., Alberghina, O. and Tribulata, E. (1973). Tecnica Agricola. 25: 363-84.

De Vicente, M.C., Truco, M.J., Egea, J., Burgos, L and Arus, P. (1998) PI. Breed., 117:153-58

De.Nettancourt, D. (2001) Incompatibility and incongruity in wild and cultivated plants. 2nd. Ed. Springer-Verlag, New York

Demetriades, S.D., Holvas. C.D. and Dimopovlov-Lekka, D. (1971). Annales del I'lnstitut Phytopothologique Benaki, 10: 226-28.

Dhuria, H.S., Bhutani, V.P. and Dhar, R.I. (1980). Haryana J. Hort. Sci., 9: 118-21.

Dong, L., Lurie, S. and Zhou, H. (2002) Postharvest Biol. Technol., 24: 135-45

Donmez, M. F., Karlidag, H. and Esitken, A. (2010) European J. Pl. Pathol., 126: 241-47

Drake, S.R. and Yazdaniha, A. (1999). J. Fd. Processing and Preservation, 23: 57-70

Drake. S.R., Toyama, T.K. and Carter, G.H. (1977). HortSci., 12: 75-76

Ende, H. Van-den, and Kenz, *I.E.* (1986). Acta Hort., 192: 93-97.

Erez, A. and Lavi, B (1985). Acta Hort., 158: 239-48

Evequoz, M. and Evequoz. N. (1983). Revue Suisse de Viticulture d' Arboriculture et.d.' Horticulture. 15: 333-37.

Ezzat, A., Nyeki, J., Soltesz, M., Amrisko, L., Balazs, G., Mikita, T. and Szabo, Z. (2012). Int. J. Hort. Sci., 18: 35-38

Facteau, T.J. and Rowe, K.E. (1981). J. Amer. Soc. Hort. Sci., 106: 77-79.

Fadl, M.S., Baz, A.G.1.O. and Tayal, S. (1978). Egyptian J. Hort., 5: 105-14.

Fan, X., Argenta, L. and Mattheis, J.P. (2000). Postharvest Biol. Technol., 20: 135-42

Fang-JinGui, Zhang-Zhen, Gai-ZhiHong and Cai-BinHua (2000). China Fruits, 3: 23-24

Faust, M., Suranyi, D. and Nyujto, F. (1998). Hort. Rev., 22: 248-49

Fidcghelli, C. and others. (1978). Rivista della Ortoflorofrutticoltura Italiana, 62: 603-14.

Forlani, M. (1984). Annali della Facolta di Scienze Agrarie delia Universita degli Studi di Napoli Portici., 18: 1-8.

Fridlund, P.R. (1979). Fruit Var. J., 33: 90-91.

Gaash. D. (1986). Acta Hort., I92: 107-10.

Gao, Z.H., Wang, S. Cai, B.H., Fang, J.G., Zhang, Z. and Zhang, Y. M. (2012). Acta Hort., 932: 43-50.

Gao, Z.H., Wu, S.T., Wang, S. and Zhang, Z. (2006). Jiangsu Agri. Sci., 6: 231-33

Gautam, D.R. and Dhuria. H.S. (1979). Punjab Hort. J., 19: 160-62.

Gautam. D.R. and Chauhan. J.S. (1982). Indian J. Agric. Res., 16: 10-14.

Genovese, A., Ugliano, M., Pessina, R., Gambuti, A., Piombino, P. and Moio, L. (2004). Italian J. Fd. Sci., 16: 185-96

Gokbulut, I. and Karabulut, I. (2012). Food Chem., 132: 1098-1102

Golias, J. (1974a). Acta Universitatis Agriculturae Brno, A., 22: 695-700.

Golias, J. (1974b). Acta Universitatis Agriculturae Brno, A., 22: 701-10.

Guerriero, R. and Martelloni, V. (1999). Italus Hortus, 6(3): 55-56

Guerriero, R., Viti, R., Iacona, C. and Bartolini, S. (2010). Acta Hort., 862: 265-72

Guerriero. R., Indiogine, S.E.P. and Scalabrelli. G. (1986a). Acta Hort., 192: 40-41.

Guerriero. R., Scalabrelli. G. and Franceschini. M. (1986b). Acta Hort., 192: 99-100.

Gupta, V.K. and Saharan, G.S. (1973). FAO Plant Prod. Bull., 21: 140-141.

Halasz, J., Pedryc, A., Ercisli, S., Yilmaz, K.U. and Hegedus, A. (2010). J. Amer. Soc. Hort. Sci., 135: 410-17

Halasz, J., Pedrye, A.; Ercisli, S., Yilmaz, K.U. and Hegedus, A. (2010) J. Amer. Soc. Hort. Sci., 135:410-17

Hallobo, S.A.S. EI., Wakeil, F.A. and Morsi, M.K.S. (1975). Egyptian J. Fd. Sci., 3: 1-6.

Hartmann. H.T. and Kester. D.E. (1959). Plant Propagation- Principle and Practices. Prentice-Hall. Inc. Englewood Cliffs. New Jersey. pp. 460-61.

Hawkins. J.E. (1976). Plant Dis. Rep., 60: 60-61.

Hayes. W.B. (1957). Fruit Growing in India. Kitabistan. Allahabad. p.444.

Hegazi, A.A. (2012). World J. Agric. Sci., 8: 169-73

Hegazi. S.M. and Salem. S.A. (1972). J. Sci. Fd. Agric., 23: 497-99.

Hitka, G. (2011). Ph. D. thesis, Corvinus Univ. Budapest, Hungary.

Hormaza, J.I. (2002) Theor Appl. Genet., 104:321-28

Hricovsky, I. and Takac, J. (1931). Vedecke Prace Vyzkummeho Ustavu Ovocnycha Okrasjnch Drevin V Bojruaach, 3: 5-15.

Ijaz, A., Hussain, S.A. and Nawab, A. (2000). Sarhad J. Agric., 16: 163-65

Iordache, M. and Coroianu, L. (2013). Sci Paper Series-B, Horticulture, 57: 273-77

Janatizadeh, A.; Moghaddam, M.F., Zamani, Z.A. and Zeraat-gar, H. (2012) Iranian J. Hort. Sci., 42: 255-65

Jay, M., Lichou, J., Piner, C., Pouget, C. and Michel, S. (2000). Infos Paris, 159: 11-13

Jooste, M.M. and Taylor, M. A. (1999). Dec. Fruit Grower, 49(8): 1-10

Kabu, K.L. (1978). Indian Hort., 23: 24-25.

Karova, V. (1972). Rastitelna Zashchita, 20: 12-14.

Kazirnin, G.T. and Marusich, V.A. (1977). Trudy Dal'nevost, Nil S.K., 22: 142-44.

Kecskemeti - Vincze, M. and Nyujto, F.E. (1986). Acta Hort., 192: 143-47.

Kecskemeti. J. and Farago, M. (1976). Gyumolcstermesztes, 3: 43-57.

Khalin. G.A., Moskalenko, K.M. and Lishchuk, A.I. (1985). Byulleten Gosudarst Vennogo Nikitskogo Botanicheskogo Sada, 58: 80-85.

Khloptseva, I.M. (1977). Trudy Po Prikladnoi Botanika Genetikei Selektsii, 59: 44-54.

Kitaev, C.R. (1983). Sadovnitstvo, 31: .69-71.

Klimenko, S.V. and Shiman, L.M. (1975). Ukrains'k ii Botanichnii Zhurnal. 32: 703-707.

Kock, A.de. (2011). SA. Fruit J., 10(5): 79-81

Koska. N. (1978). Acta Hort., 80: 219-24.

Kostina, K.F. (1936). Bul. Appl. Bot. Gen. Plant Beed. Suppl., 83: 16-48.

Kostina, K.F. (1969). Trud. Nikit. Bot. Sad., 40: 45-63.

Kostina, K.F. and Riabov, I.N. (1959). Trud. Nikit. Bot. Sad., 29: 113-17.

Kropis, E.P. (1984). Sadovodsrvo, Vinogradarstvo i Vinodeiie Moldavii. 11: 30-34.

Kuden, A.B., Kaska, N. and Cebeci, E. (1992). Turk. 1st. Nat. Hort. Congr., pp. 9-12

Kuden, A.B., Kaska, N. and Paydas, S. (1995). Acta Hort., 304: 309-13

Kurmanova, A.V. (1984). In: Geneticheskia Osnovy Selektsii sel Skakhogyaistvennykh rastenii Izhivotnykh. eds. Zhuchenko. A.O., Kishinev, Moldavian SSR; Stiinca p. 134.

Lal, B.B. and Sharma, S.D. (1979). Haryana J. Hort. Sci., 8: 198-99.

Lane, W.D. and Cossio, F. (1986). Canad. J. PI. Sci., 66: 953-59.

Lane, W.O. (1978). Canad. J. PI. Sci., 58: 905-906.

Lapins, K.O. (1972). Canad. Dep. Agr. Publ., 1471, Summerland. BC.

Layne, R. E.E. (1979). HortSci., 14: 758-59.

Layne, R.E.E. (1978). HortSci., 13: 64-65.

Layne, R.E.E. (1981a). HortSci., 16: 97-98.

Layne, R.E.E. (1981b). HortSci., 16: 98-99.

Layne, R.E.E. (1984). HortSci., 19: 136-37.

Layne, R.E.E. and Harrison, T.B. (1975). HortSci., 10: 428.

Leece, D.R. and Van Den Ende, B. (1975). Aust. J. Exp. Agric. Anim Husb., 15: 123-28

Legave, J.M., Ratier, R., Garcia, G. and Fur, G. Le (1985). Arboriculture Fruitiere, 32: 39-41.

Liferdi, R., Poerwanto, R., Susila, D.A., Idris, K. and Mangku, W.I. (2008). Indonesian J. Agric. Sci., 1: 95-102

Lillecrapp, A.M., Wallwork, M.A. and Sedgley, M. (1999.) Scientia Hort., 82: 255-63

Limongelli, F. and Cappellini, P. (1978). Rivista della Ortoflorofrutticol tura Italiana, 62: 623-31.

Linder, R.C. and Lure, W.A. (1944). Proc. Wash. State Hort. Assoc., 40: 154-60.

Lishchuk, A.I., Flnitskii, O.A. and Radchenko, S.S. (1985). Byulleten Gosadarstvennogo Nikitikogo Botanicheskogo Sada, 56: 86-89.

Longridge. D.F. and Goodman, R.D. (1981). Aust. J. Exp. Agric. Anim. Husb., 21: 241-44.

Looney, N.E. and Killick, R.G. (1979). Canad. J. PI. Sci., 59: 741-46.

Malik, T. (1965). Ved. Pr. Vysk. Ustt. Vyroby V Piestanoch, 3: 239-64.

Manzo, P. and Salvador, F.R. De. (1978). Annali dell' Jnstituto Spcrimentale per la Frutticoltura, 9: 33-41.

Margarayan, A.A. (1976). Izvestiya S. Kh Nauka, 1: 62-66.

Marinov, P. (1982). Gradinarska i Lozarska Nauka, 19: 28-35.

Marinov, P. (1983). Gradinarska i Lozarska Nauka, 20: 45-53.

Martinex, C.A. (1979). Anali dell' Instituto National de Ubvestigaciones Agrariae Production Vegetal, 9: 193-202.

Martinez Calvo, J., Llacer, G. and Badenes M.C. (2011) HortSci., 46(4): 655-56

Martinez, Javega, J.M., Reig Feliv, A., Cuquerella Cayuela, J. and Carreres Orius, R. (1974). Annals del Institute National de Investigacious Agrarias Technologis Agraria, 1: 137-45.

Martinez-Gomez, P., Dicenta, F. and Audergon, j.M. (2000) Agronomie, 20:407-22

McLaren, G.F. (1981). Orchardist of NZ., 53: 305.

McLaren, G.F. (1983). NZ. J. Exp. Agric., 11: 53-57.

McLaren, G.F., Frasher, J.A. and Grant, J.E. (1996). NZ. J. Crop. Hort. Sci., 24: 55-63

Medcira, M.C. (2000). Agronomia Lusitana, 48: 75-88

Mehlenbacher, S.A., Hough, L.F. and Voordeckers. A. (1987). HortSci., 22: 680-81.

Mikaelyan, V.M., Petrasyan, Zh. A. and Marutyan, S.A. (1976). Trudy Arm. NIl Vinogradarstva Vinodeliya i Plodovodstva, 13: 167-80.

Milosevic, T., Miklosevic, N., Glisic, I. and Paunovic, G. (2010). PKV Agroekonomik, 16: 31-41

Misra, R.S. (1983). Progressive Hort., 15: 229-31.

Mitreskii, Z. and Ristevski. B. (1985). Jugoslovensko Vocarstvo, 19: 229-35.

Miyazaki. T. (1983). J. Jap. Soc. Hort. Sci., 52: 85-92.

Molchanov, C. F. (1976). Bylleten Gasudarstvennogo Nikitskigo Botanicheskago Sada, 2: 55-58.

Molnar, L. and Turi, I. (1975). Gyurnolestermesztes. 2: 3-13.

Moskalcnko, K.M. (1983). Trudyu pe Priklanoni Botariika Genetikei Selektssi, 78: 58-62.

Munoz, R. P., Rubio, P., Infante, R., Campos, V.R., Manviquez, D., Gon Zalez, A.M. and Defilippi, B.G. (2012). Postharvest Biol. Technol., 63: 85-90

Nachtigal, J.C., Pereira, F.M., Dali-Orto, F.A., Ojima, M. and Martins, F.P. (1999). Rivista Brasileira de Fruiticultura, 21: 226-28

Nedelchcva. S., Geneva. D. and Atanasov, A.I. (1985). Rasteniev dni Nauki, 22: 98-104.

Nicholas. J. (1979). Bull. Tech. Pyrenees Orientales, 90: 35-41.

Nictora, A. and Moser, L. (1999). Italus Hortus, 6: 73

Nitransky, S. (1983). Pol, Nohospodorstvo, 29: 457-71.

Nitransky. S. (1978). Pol. Nohospodorstvo, 24: 105-14.

Norton. R.A., Hansen, C.J., Reilly, H.J.O. and Host, W.H. (1963). Calif. Agri. Exp. Sta. Leaflet, 158: 8

Oe, T., Sakurai, N., Okamura, M., Negoro, K., Tsuchida, Y., Nakanishi, K. and Hosohira, M. (2012). Hort. Res. Japan, 11: 235-40

Okie, W.R. (1983). Fruit Var J., 37: 15-23.

Parma, P., Mladin, Gh., Dutu, I. and Stancie, N. (1988). HortSci., 23: 107-109.

Pathak, V.N. (1980). Diseases of Fruit Crops. Oxford and IBH Publishing Co. New Delhi. pp. 240-41.

Pauriovic,K. and others (1982). Jugoslovensko Vocarstvo, 16: 65-74.

Perez-Romero, L.F., Arroyo, F.T., Santamaria, C., Camacho, M., and Daza, A. (2014). Acta Alimentaria-Budapest, 43: 273-79

Perez-Tornero, O., Lopez, J.M., Egea, J. and Burgos, L. (2000). J. Hort Sci. Biotechnol., 75: 283-86.

Peter, A.R., Sefferou, V., Deaxa, N., Tsantili, E. and Stathis, V. (2011). Sci. Hort., 129: 472-78

Polak, J. and Kominek, P. (2014). Canad. J. Pl. Path., 36: 116-20

Priore, R. and Sannino, C. (1981). Fruits, 36: 785-88.

Pruthi, H.S. and Batra, H.N. (1960). Important Fruit Pests of North West India. ICAR., New Delhi, pp. 4-8.

Pularov, A. P. (1978). Referativnyi Zhurnal, 8: 56: 681.

Quata, V. and Sakamoto, T. (1979). Studies from Inst Hort., Kyoto Univ., 9: 146-150.

Raina, J.N., Sharma, T. and Suman, S. (2011). J. Indian Soc. Soil Sci., 59: 268-77.

Ramos, D.E., Moller, W.J. and English, H. (1975). Phytopath., 65: 1359-74.

Rathore, D.S. (1982). Indian Hort., 26: 15.

Raz, A., Goldway, M. and Stern, R.A. (2010). J. Hort. Sci. Biotech., 85(1): 30-34

Rettke, S.Z. (2005). South Aust. Res. Hort. J., 39: 885-89

Ristevski, B. and Mitreski, Z. (1986). Acta Hort.,2: 369-79.

Rodrigo, J. and Hrrero, M. (2000). ITEA, 21: 123-27

Romero, C., Pedryc, A., Munoz, V., Llacer, G. and Dadenes, M.L. (2003). Genome, 46: 244-52

Rozsnyay, Z. (1978). Koronaijjilon Kiserlet Kajsziparachfak on Kestagazelasag, 10: 1-6.

Rozsnyay, Z. (1986). Kerteszetes Szoleszet., 35: 5.

Rozsnyay, Z. and Klement, Z. (1973). A Kajszigutautes Koroktar.a Hazankban Kiserletugui Kozlemenyek, 66: 61-73

Ruggiero, C. (1986). Rivista delia Ortoflorofrutticoltura Italiana, 70: 35-47.

Ruiz-Sanchez, M.C., Domingo, R., Torrecillas, A. and Perez-Pastor, A. (2000). Pl. Sci. Limerick, 156: 245-51.

Samaan, L.G., Iraqit, M.A., El-Baz, E.E.T. and EL-Dengawy, E.F.A. (2000). Egyptian J. Hort., 27: 171-86

Sansavini, S., Costa, G., Grandi, M. and Lunali, U. (1983). Acta Hort., 121: 365-74.

Seely, S.D. and Damavandy, H. (1985). J. Amer. Soc. Hort Sci., 10: 726-29.

Serafirnova, R. and Brisov, N. (1980). Gradinarska i Lozarska Nauka, 17: 54-58.

Seth, J.N. and Kuksal, R.P. (1977). Progressive Hort., 9: 39-42.

Seth, J.N., Ghildiyal, P.C. and Joshi, R.P. (1984). Fruit Var. J., 38: 51-54.

Sharma. M.P. (1978). Progressive Hort., 10: 23-26.

Shear, C.B. and Faust, M. (1980). Hort. Rev., 2: 142-63.

Shen-YuYing, XiaJun, D., Fei, W., Binttua, C., Zhihong, G. and Zhen, Z. (2011) Scientia Hort., 132:50-58

Simeone, A. (1982). Abst. 21st Int. Hort. Cong., Hamburg, No. 1021.

Snir, I. (1984). HortSci., 2: 229-30.

Soliman, H.I.A. (2012). World Appl. Sci. J., 19: 674-87

Son, L. (2004). NZ.J. Crop Hort. Sci., 32: 331-35

Soni, S.L. and Yuusuf, Y.H. (1978). Indian J. Agric. Sci., 48: 197-200.

Sosna, I. and LicZnar-Malanczuk, M. (2012). Acta Scientiarum Polonorum Hortorum cultus, 11: 27-37

Souty, M., Breuils, L., Reich, M., and Poygi, A. (1976). Fruits, 31: 1775-79.

Srivastava, K.K., Sharma, M. K. and Bhat, S.K. (2012). Indian J. Agric. Sci., 82: 78-80

Srivastava, K.K., Singh, S.R., Sharma, M.K., Sundouri, A.S. and Sheikh, A.S. (2008). Indian J. Ecol., 35: 179-180

Stadler, J.D. and Stassen, P.C. (1986). South Afr. J. PI. Soil, 3: 203-205.

Stassen, PJ.C and Hurter, N. (1981). Dec. Fr. Grower, 31: 102-103.

Stoilov, G. and Marinov, P. (1979). Gradinarska i Lozarska Nauka, 16: 28-37.

Stoilov, G. and Marinov, P. (1983). Gradinarska i Lozarska Nauka, 20: 33-44.

Stoyanov, I. (1977). Ovoschharstovol, 56: 33-35.

Sud, G. and Mishra, K.A. (1979). Haryana J. Hort. Sci., 8: 21-23.

Sud, G. and Mishra, K.A. (1980).Sci. and Cult., 46: 368-69.

Sud, G., Kar, P.L, and Dhuria, H.S. (1979). Indian J. Hort., 36: 262-64.

Suranyi, D. (1973). Botanikai Kozlamenyek, 60: 232-41.

Suranyi, D. (1979). Kertgazdasag, 10: 9-22.

Suranyi, D. (1986). Acta Hort., 192: 73-80.

Szenczi, G., Kemens, M., and Kollor, G. (1982). Kertgazdasag. 14: 51-58.

Szucs, E. (1986). Acta Hort., 192: 137-42

Tabucnca, M.C. (1976). Anals dela Estacion Experimental de Aula Dei., 13: 325-32.

Taghipour, L., Rahemi, M. and Assar, P. (2011). Brazilian J. Physiol., 23: 279-84

Tawfik, M., Zaki, M.M. El. Gebbli, M., and EI Sharif, M. (1985). Agri. Res. Rev., 59: 63-70.

Teskey, B.l.E. and Shoemaker, J.S. (1972). Tree Fruit Production. The AVI Publishing Co. Inc., Westport Connecticut, p. 315.

Tsolov, T., Dirnitrov, I. and Treyakyashki, P. (1980). Nauchni Trudove Virsh Selskostopanski Institut "Vasil Kolarov", 25: 127-34.

Tukey, H.B. (1964). Dwarfed Fruit Trees. Macmillan, New York. p. 562.

Turter, A.B., Combrink, J.C. and Wood von Mollendorff, L.J. (1994). Dec. Fruit Grower, 44: 421-27

Ugolik, M., Ugolik, D., and Garwocka, M. (1981). Akademia Rolnicza, pp. 74-77.

Vachun, Z. (1971). Acta Universitatis Agriculture Brono A., 19: 299-307.

Vachun, Z. (1972). Zahradnictvi, 314: 159-74.

Vakhidov, A. (1984). In: Perspectivy Prakticheskogo Ispol 'zovaniy Rapel' nogo Orsbeniya "Sacovodstvei. Vinograderstve Materialy Sovcshcneniya Izri, Taskent, pp. 61-66.

Valentini, N. and Botta, R. (1999). Italus Hortus, 6(5): 3-8

Vashchenko, I.M. and Vodyanitskii, V.I. (1976). Nauchnye Doklady Vyssh Shkly Biol. Nauka, 12: 89-93.

Vavilov, N.I. (1951). In: The Origin, Variation, Immunity and Breeding of Cultivated Plants, Chronica Botanica.

Velemis, D., Almaliotis, D., Bladenopoulou, S. and Karayiannis, I. (1999). Acta Hort., 488: 489-94

Visagie, T.R. (1985). Dec. Fr. Grower, 35: 401-404.

Visagie, T.R., Eksteen, G.l., Blommaert, K.L.J., and Theoron, l.C.D. (1978). In: Apricot,Pear and Peach Grower's Association Annuals, 12: 47-48.

Wang YuMing, Wangt Ping, Lit ShHua, litHongxia, and HuntJiashan (2009). China Fruits, 5: 56-58

Wang, H., Feng, J., Fan, X. and He, F. (2009). Acta Hort. Sinica, 36: 1043-48

Watkins, R. (1976). 1n: Evolution of Crop plants, (ed. N.W. Simmonds) Longman, London, pp. 242-47.

Weinbaum, S.A., Uriu, K. and Meraoka, T.T. (1980). J.Pl. Nutrition, 2: 699-706.

Welkerling, De, Tacchini, E.M.L. and Azar De Dedoy, E. (1973). Revista Della Faculted de ciencias Agrarias, Universidad Nacional di Cuyo, 19: 97-99.

Westwood, M.N. (1978). Temperate Zone Pomology. W.H. Freeman and Co., San Francisco.

Williams, J.G.K., Kubelik, A.R., Livak, K.J., Rafalski, J.A. and Tingey, S.V. (1990) Nucleic Acids Res., 18:6531-35

Wills, R.B.H., Scriven, F.M. and Greenfield, H. (1983). J. Sci., Fd. Agric., 34: 1381-99.

Wurm, L. (2007). Mit. Klosterneburg, Rebe-und-wein, Obstoou-und-Fruchteverwertung, 57: 51-60.

XiaoLing, Y. and Shidan, B. (1999). Pedosphere, 9: 363-68.

Xiloyannis, C., Dichio, B., Nuzzo, V. and Celano, G. (1999). Italus Hortus, 6: 43-46.

Yanin, G.I. (1975). Sadovodstvo, 7: 38-39.

Yilmaz, K.U. (2008) PhD thesis, Univ of Cukurova, Turkey.

Yuan, Z.H., Yin, Y.L., Feng, L.J., Zhao, X.Q.,Wang, J.Z., An, G.N. and Zhang, C.Y. (2011). Acta Hort., 918: 209-15

Yuan, Z.H.; Chen, X., He, T., Feng, T. and Zhange, C. (2007) J. Genetics and Genomes, 387:1037-47

Zhaot Xiping and Chait JuHua (2000). China Fruits, 4: 26-27.

Zhebentyayeva, T.; Reighavd, G.; Lalli, D., Gorina, V., Krska, B. and Abbott, A. G. (2008) Tree Genet. Genomes, 4:403-17

ZhutXuan, Hout YuanYuan, Wangt Ying and QitHuiNan, (2013). Fd. Res. Dev., 34: 251-53.

Zvara, J. (1967). Gartenbau., 32: 416-22.

Chapter 3

Cherries

Cherries occupy an important position among temperate fruits all over the world. The cultivated cherries are divided into two main groups, *i.e.*, sweet cherries (*Prunus avium* L.) and sour cherries (*Prunus cerasus* L.). The sweet cherry is mainly used for table purposes and sour cherry for processing. Over the last four decades increasing demands in cherry cultivation all over the world have resulted in the development of new size-controlling rootstocks, cultivars suitable for optimal yield. The suitable canopy shapes, training and pruning systems to meet modern requirements such as mechanical harvesting, have also been developed. It is now possible to keep fuits after harvest even for 8-10 weeks in controlled atmosphere storage without loss of any commercial quality.

Composition and Uses

Cherries are rich in protein, sugars, having more calories than apple. The carotene and folic acid contents are also high. The fruit is a rich source of potassium, calcium, magnesium, iron and zinc. The detailed composition of cherry fruits as estimated by Wills *et al.* (1983) is given in Table 3.1.

Methyl anthranilate and methyl salicylate are the flavouring ingredients of cherry juice and the colouring principle of the fruit skin is keracyanin chloride which is a diglucoside of cyanidin (Anon., 1969). Pectic substances are recorded in small quantities. Enzymes invertase and amylase occur at all stages of fruit growth. The major acid in sweet cherry is mallic acid while the major sugar is glucose. Kelebek and Selli (2011) working with four sweet cherry cultivars (Van, Noir de Guben, Larian and 0-900 Ziraat) reported that the total sugar content varies between 103.87 (Larian) to 113.13 g/kg fresh weight (0-900 Ziraat) and that of organic acids from 12.01 (0-900 Ziraat) to 14.17 g/kg fresh weight (Noir de Guben). They also

Table 3.1: Proximate Nutrient Composition of Cherry Fruit (g $100g^{-1}$ edible portion)*

Cultivar	*Water*	*Protein*	*Fat*	*Sugars*			*Starch*	*Dietary Fibre*	*Organic acids*	*Ash*	*Energy (kJ)*
				Fructose	*Glucose*	*Sucrose*					
Eagle	84.8	0.5	0.2	4.9	7.6	0.0	0	1.5	1.2	0.3	228
Lambert	77.9	0.9	0.2	5.7	12.8	0.0	0	1.4	1.0	0.7	329
Ron's	83.7	1.2	0.2	3.7	4.9	0.0	0	1.7	1.1	0.6	168

Cultivar	*Organic Acid Composition and Soluble Solids*				
	Organic Acids (mg $100g^{-1}$)			*Titratable Acidity (meq H + $100g^{-1}$*	*Soluble Solids (^{o}Brix)*
	Malic	*Citric*	*Quinic*		
Eagle	1150	12	5	8.9	15.0
Lambert	990	14	4	13.2	21.5
Ron's	1030	37	7	10.7	13.4

Vitamin and Mineral Content (mg $100g^{-1}$ edible portion)											
Vitamins				*Minerals (mg)*							
Cultivar	*Vitamin C*	*Carotene (µg)*	*Thiamine*	*Riboflavin*	*Niacin*	*K*	*Na*	*Ca*	*Mg*	*Fe*	*Zn*
Eagle	21	190	0.04	0.02	0.4	200	0	14	7	0.2	0.1
Lambert	16	412	0.02	0.03	0.6	240	0	28	7	0.3	0.1
Ron's	18	747	0.03	0.03	0.6	280	1	30	12	0.4	0.1

Energy (kJ) = (g protein 17) + (g fat 37) + (g sugar as monosaccharide 16) + (g organic acid 10).

*Wills *et al.* (1983).

identified eleven phenolic compounds including three hydroxycinnamic acids, five anthocyanis, two flavan-3-ols and one flavonol compounds. Vangdal *et al.* (2006) recorded the anthocyanins cyanidine 3-rutinoside and pelagonidine 3-rutinoside of which cyanidine 3-rutinoside comprised 77-95 per cent of the total anthocyanins. The major phenolic acids detected were neochlorogenic acid and *P*-coumaroylquinic acid. The high level of phenolic compounds and antioxidant activity in sweet cherry fruits implied that they might be source of bioactive compounds that are relevant to human health (Ballistreri *et al.*, 2013). Serradilla *et al.* (2014) working with sweet cherry cv. Pico Colorado identified and quantified four anothocyanins with cyanidin-3-0-rutinoside being most abundant (38.8) ± 0.59 mg/100 g fw; 3 hyroxycinnamic acids, principal of which was neochlorogenic acid (14.18 ± 0.15 mg chlorogenic acid/100 g fw); a flavanol, rutin (4.56 ± 0.10 mg/100 fw); and a flavan-3-01, epicatechin, which was the least abundant of all the phenolic compounds.

The sweet cherry is mainly consumed as dessert and also in canned fruit cocktails, confectionary, ice-cream, bakery, juice making and for various other purposes. The sour cherry is used for cooking, juice making, fruit salads and distilling liquor. The wine made from sour cherry is very popular. Since the fruits are highly perishable, these are largely preserved by canning, freezing and sun drying. Cherries are brined with 1.5 per cent sulphur dixide and 0.9 per cent calcium carbonate or hydroxide (Anon., 1969). Both fresh and frozen fruits are utilized for making cherry juice. By judicious mixing of juice from sweet and sour cultivars a delicious product can be obtained with or without addition of sugar. Cherry syrup is made by concentrating the juice.

Cherry seeds are very rich in fatty acids. Vidrih *et al.* (2014) working with three cherry cultivars (Lala Star, Starking Hardy Giant and Cristalina) grown in Solvenia and Norway recorded the following fatty acids from seeds: palmetic (7.2-8.5 per cent), stearic (2.4-3.7 per cent), oleic (29.2-42.5 per cent), eleostearic (4.45-11.74 per cent) and arachidic (0.84-1.39 per cent).

Origin and Distribution

Three species, *viz.*, *Prunus avium* (sweet cherry) *P. fruticosa* (ground cherry) and *P. cerasus* L. (sour cherry) are involved in the development of present-day cherry froms. *Prunus avium* is the oldest species with a basic chromosome number of x=8. *P. avium* is usually diploid (2n=16) but triploid and tetraploid (2n=24 and 32) froms are also found. *Prunus cerasus* is known to have originated by a cross of unreduced pollen of *P. avium* with *P. fruticosa* (Westwood, 1978; Olden and Nyborn, 1968). Tetraploid hybrids (2n=4x=32) between sour cherry and sweet cherry have been known as Duke cherry; *V gondouini.* Tree and fruit characteristics of Duke cherries are intermediate of sweet and sour cherries (Fogle, 1975). The ground cherry might be a probable parent species of both sweet and sour cherry. The progenies of crosses between ground cherry and either diploid or tetraploid sweet cherry resemble sour cherries (Olden and Nyborn, 1968).

Cherries are considered natives of South Central Europe, Asia Minor and as far east as northern India and China (Westwood and Wann, 1954; Hedrick, 1915). The wild forms of sweet cherry occur throughout the temperate regions of mainland

Europe and Soivet Union with the greatest concentration being between south of the Caspian and Black Sea (Kolesnikova, 1975). Diverse forms of sour cherry exist in eastern Europe and western region of Soviet Union whereas tremendous diversity in ground cherry is found in the Trans-Volga region, in the southern Ural mountains, and in the southern regions of western Siberia (Hilling and Tezzoni, 1988). The geographical distribution of some cherry species is furnished in Table 3.2 (Hooker, 1879; Hedrick, 1915; Bailey, 1924; Cummins, 1979; Randhawa, 1987).

Table 3.2: Geographical Distribution of Cherry Species

Sl.No.	*Distribution*	*Species*
1.	North America	*P. pumila, P. besseyi, P. pennsylvanica, P. virginiana, P. serotina.*
2.	Europe, Western Asia, Caucasia, Siberia.	*P. avium, P. cerasus, P. mahaleb, P. fruticosa, P. laurocerasus*
3.	China, Manchuria, Korea	*P. japonica, P. macradenia, P. tomentosa, P. glandulosa, P. serrulata, P. mugus, P. psuedocerasus, P. canescens, P. concinna.*
4.	Japan, Taiwan	*P. incise, P. subhirtella, P. campanulata, P. sieboldii, P. vedoensis, P. sargentii and P. nipponica*
5.	Eurasia, Himalayas	*P. nepaulensis, P. cornuta, P. cerasoides, P. prostrate and P. tomentosa.*

The earliest records of cherry cultivation show that it was first domesticated in Greece around 300 BC. (Hedrick, 1915). From Greece it spread to Italy where it was established as a fruit crop by 37 BC. Hedrick (1915) assumes that by Christ's time it spread to England, Germany, Belgium and Portugal. Early settlers brought seeds to North America which eventually spread all over the northern parts and southern-most parts of South America.

It is thus apparent that the cherry has been in existance as a cultivated fruit since the dawn of civilization, but little was accomplished in improving it until the sixteenth and seventeenth centuries (Marshall, 1954). The cherry cultivation is mainly confined between 35° N and 55° S latitudes or at some places outside these limits where temperature and other factors are favourable (Westwood and Warm, 1954).

Production

The US maintained a position as the largest producer until the last decade, when it was surpassed by Turkey. Turkey is the largest producer (494325 tons) producing 21.5 per cent of the world's Cherry production, followed by USA (13.1 per cent) and Iran (8.7 per cent). The other important cherry producing countries are Italy, Uzbekistan, Spain, Chile, Ukraine, Russian Federation, Syria, Greece, Poland, Austria, France, Bulgaria, China, Serbia, Germany, Lebanon, Japan, Australia, Albania, Morocco and India. India produced about 13500 tons of Cherries. The world's average yield per hectare increased from 5375 kg in 2010 to 5618 kg in 2012. Major cherry producing countries of the world is presented in Table 3.3 (FAOSTAT, 2016).

Table 3.3: Major Cherry Producing Countries of the World

Name of Country	*Production (tons)*	*Name of Country*	*Production (tons)*
Turkey	494325	Ukraine	81200
United States	301205	Romania	80477
Iran	200000	Russia	78000
Italy	131175	Syria	62372
Uzbekistan	100000	Greece	58200
Spain	97200	Poland	47552
Chile	90703	Austria	41430

FAOSTAT, January, 2016.

Taxonomy

Cherries belong to the family Rosaceae, genus *Prunus* anu subgenus *Cerasus* (Benson, 1957). The subgenus *Cerasus* comprises 150 species (Kolesnikova, 1966), though only a few of them have value for fruit production. However, subsequently sour cherry was referred to as *P. cerasus* and sweet cherry as *P. avium* (Marshall, 1954). *Prunus avium* and *P. cerasus* are the two most important species enjoying commercial importance. However, there is a possibility of utilizing some of the little known species in breeding new rootstocks or in improving the existing cultivars through hybridization. Most of these species are used for rootstocks, roadside planting and as firewood. Sweet cherries are tall, attaining a height up to 20m, and pyramidal form. The branches are usaually erect and leaves are large, thin, variable pubescene beeneath and coarsely serrated. The petioles are long having two or more swollen glands. Sweet cherries are often called as Gean or Mazzard cherries. The flowers are white; with diameter of about 2.5 cm. Flowers are formed on the previous year's growth singly or in groups of five. The fruit is cordate in shape, has a deep cavity, and apex is rounded or pointed. The fruit colour varies from yellow through red to purplish black. The texture of flesh ranges from tender to firm, sweet and may be yellow, red or dark purple in colour (Hedrick, 1915).

Ground cherry has a spreading habit attaining a height of about I m only (Kolesnikova, 1975). The leaves are generally short, thick, glabrous and finely serrated. Two to four swollen glands are found on basal leaf edge (Hilling and Tezzoni, 1988). The fruits, pits and pedicels of ground cherry are smaller than sweet cherry.

Sweet Cherries

On the basis of the texture of flesh, sweet cherries have been divided into two pomological groups, *viz.*, Heart, and Bigarreau. The Heart cherry cultivars have soft and tender flesh and heart-shaped fruits. The fruit colour varies from dark with reddish juice to light coloured with colour less juice. The Bigarreau group of cherries are usually roundish, though some of them may be more or less heart-shaped. The colour of fruit and juice also varies from dark to light red.

Sour Cherries

The trees of sour cherry (*Prunus cerasus*) are usually 6 to 9 m tall, open-headed, round-topped or spreading and often without a central leader. The branches are slender and more or less drooping. The fruit is roundish-oblate or cordate with slightly compressed sides and averages about 2 cm in diameter. The leaves are much smaller, darker green in colour and more numerous than those of the sweet cherry. Flowers, which form abundantly on one-year growth is white, diameter of 1.75-2.5 cm. Wild type fruits are red or nearly black, circular with sour pulp. The commonly grown cultivars of sour cherry are more hardy than sweet cherry and thus may be planted to extend the growing area to higher latitudes/altitudes (Chandler, 1942).

Cultivars

Salient characteristics of some important sweet cherry cultivars are given below.

Early Rivers

This is the finest early cherry cultivar which ripens in the beginning of May. Fruits are large, black, oval with a small stone. The pink flesh is deliciously flavoured. The bearing is usually moderate but regular. It is resistant to cracking.

Governor Wood

Large-sized fruits are heart-shaped, light yellow with red cheek. The quality of white flesh is fair and soft. It ripens just after Early Rivers and needs to be picked green. Being a regular heavy cropper, it is a profitable cultivar to grow. It is however, very susceptible to brown rot.

Bing

The fruit is of the highest quality but cracks readily in rain. The tree is upright and spreading and bears moderate to heavy crop of large fruits. The blossom buds are susceptible to low temperatures

Early Purple Black Heart

This is the earliest cultivar ripening during the third week of April. It is a good yielder, fruits being juicy and sweet. The keeping quality of fruit is, however, poor. Being an early cultivar, it fetches good price in the market.

Schrecken

This old German cultivar is the earliest sweet black cherry which ripens about the beginning of May. Fruit is of large size and sweet flesh is of good quality and fine flavour. The fruit is susceptible to cracking in rain and preharvest drop may also be severe in some years.

Emperor Francis

The fruit is large and dark red in colour having good keeping quality. It flowers early but otherwise a late cultivar. The tree is a regular and good bearer.

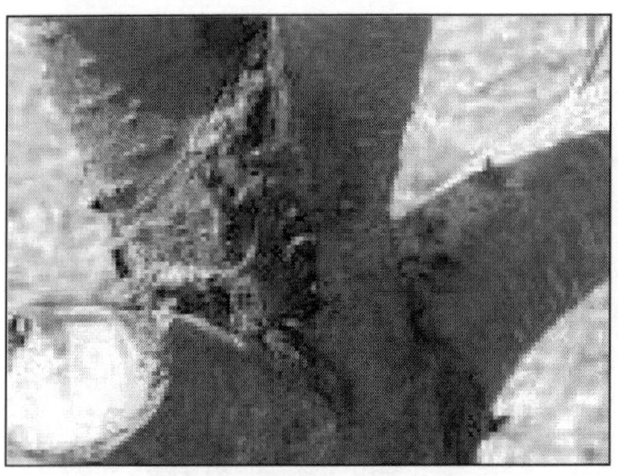

Bacterial Canker

Bing

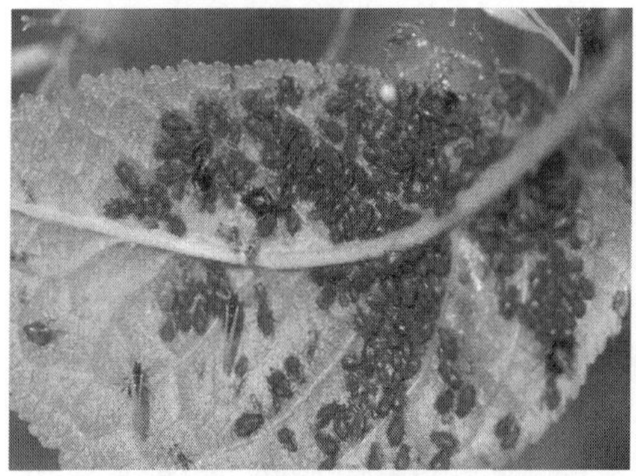

Black Cherry Aphid

Cherry Orchard

Compact Lambert

Compact Stella

Early Rivers

Montmorency

Grace Star

Sam

Verticilium Wilt

Royal Ann (Napoleon)

The fruit is large, oval, fairly attractive, and relatively firm. The flesh is sweet, juicy but somewhat bitter. It flowers late and fruit is ready to harvest by the end of June.

It may, however, be mentioned that the above sweet cherry cuitivars are old. They are vigorous, take long time to come into fruiting, susceptible to cracking and have average fruit quality. New cultivars such as Sunburst, Lapins, Compact Lambert. Compact Stella, Van, Starking Hardy Giant, Stella, Summit, Hedelfingen, Black Tartarian, Bigarreau Vista, Bigarreau Bing, Mona, Jubilee, Corum, Sue, Ulster and Salmo, which have given very good performance have been introduced in different cherry growing areas (Stancevic, 1982; Lapins, 1974). Stancevic (1982) has reported that the self-fertile sweet cherry cultivar Compact Stella made 33 per cent less growth than the Standard Stella and Compact Lambert was 40-50 per cent smaller than Standard Lambert. Both these semi-dwarf cultivars came into bearing early, cropped heavily and produced large fruits weighing 7.7 to 8 g, dark red with good quality suitable for fresh consumption and processing. Some cultivars, such as Sunburst, Lapins and Stella are self-fertile (Lane, 1979; Sansavini and Lane, 1983). Many new cultivars such as Windsor, Lapins, Sam, Sue, Victor, *etc.*, are resistant to crackng (Lane and Schmid, 1984; Lapins *et al.*, 1977). These two characters *viz.* self-fertility and resistance to cracking will result in increased cherry production.

Pomological Description of some Leading Cultivars

Compact Lambert

It is a radiation induced mutant of Lambert introduced at Summerland In 1964. The dwarf tree is about 80-90 per cent of the size of a standard cherry tree on a vigorous rootstock. The flowering is late but it bears heavily. Fruit size is good and its colour is black. The chance of fruit cracking is less as the fruit remains covered with foliage.

Compact Stella

This cultivar is also a radiation induced mutant of Stella evolved at Summerland in 1973. The tree is semi-dwarf and its final size is half of the parent cultivar. It comes into bearing early, carries heavy crops and is self-compatible. The fruit is black, large to medium heart-shaped to oval; flesh is medium firm and good to fair in texture and flavour. It is suitable for canning as well as dessert use. Fruit cracking due to rain is moderate. It is an universal pollenizer (Lapins, 1975).

Jubilee

It is a seedling of Lambert x Napa Long Stem Bing, introduced in 1964. It is an early cherry for local markets. Flesh is light red and slightly soft and flavour is flat when fruit set is heavy. It is resistant to rain cracking.

Salmo

A seedling from Lambert x Van introduced at Summerland in 1970. It is one of the hardiest cultivar which bears moderate to heavy crops. The fruit is large, fully black relatively firm and fair to good in flavour. The fruit is subject to moderate cracking.

Sam

It is a rather slow to start to bearing, but mature trees carry heavy crops. Fruit is medium in size, fully black, medium firm, coarse texture, fair flavour. It is suitable for canning. The fruit has high resistance to rain cracking. The crack resistance of the fruit and the cold resistance of the tree are its good features.

Sue

A seedling of Bing x Schmidt developed at Summerland in 1954. It is a heavy bearer and consequently fruit is uneven in size and maturity. Fruit is soft with a fair flavour. Fruit cracking is very low. It is good for canning and also for brining. It is promising as a very crack resistant and heavy bearing white sweet cherry.

Van

An open-pollinated seedling of Empress Eugenie introduced at Summerland in 1944. Tree vigorous and hardier than most cultivars, bears while quite young, and is verv productive. The fruit is large but becomes small on overloaded trees. The fruit is fairly attractive with its bright skin. The flesh is nearly black, firm fine and has good flavor. Fruit cracking is moderate.

Kristin

Kristin (Emperor Francis x Gil Peck) is a mid-season cultivar which produces large, black, high quality fruits, which are also resistant to cracking. It is a heavy bearer and winter-hardy. Although it produces good pollen, it is self-infertile (Way *et al.*, 1982).

Viscount

The fruits of Viscount (V 35024 x V 35029) are medium to large-sized, kidney-shaped, glossy, dark rcd, firm and attractive. Average fruit weight is 8.5g and the stone is free, constituting 4 per cent of the whole fruit weight. The flesh is of good quality, red and firm; juice dark red. It is highly resistant to rain induced cracking. It is also resistant to diseases caused by *Monilinia fructicola* and *Pseudomonas syringae* (Tehrani, 1984).

Sovereign™

Developed at Agriculture and Agri-Food Canada, Summerland, British Columia, A hybrid of Sweetheart X O.P. Dark red coloured fruit, medium large, red flesh, free stone, sweet-acid balanced flavour, self-fertile, however susceptible to rain induced splitting.

Black Star

A hybrid of Lapins X Burlat, developed by the University of Bologna, Italy. Fruit dark-red skinned, pink-flashed, mid-season cultivar. Very sweet, clingstone, self-fertile, very productive and almost resistant to rain-induced splitting.

Blaze Star

Another hybrid (Lapins X Durone) developed by University of Bologna, Italy, introduced in 1998. It is a mid-season, dark-red skinned, pink fleshed cherry. Sweet flesh, semi-clingstone, self-fertile, very productive and good resistance to rain-induced splitting.

Grace Star

A hybrid of Burlat X O.P. introduced in 2001 by University of Bologna, Italy. Like Blaze Star, it is also a mid-season, dark-red skinned, pink-fleshed cherry. Fruits are large, cordate, symmetrical, semi-clingstone, self-fertile, very productive and medium resistance to rain-induced splitting.

Lal a Star

A hybrid of Compact Lambert X Lapins, developed by University of Bologna, Italy and introduced in 1998. Medium- to late season, dark-red skinned, red fleshed cherry. Fruit medium large, firm flesh, pleasant flavour, clingstone, self-fertile and very productive.

Nugent™

Developed from Cornell University, USA, introduced in 2001. A hybrid of Germersdorfer X O.P. Late-season, yellow-skinned, yellow-fleshed, freestone, resistant to rain-induced splitting, very productive, relatively tolerant to cherry leaf spot and bacterial cankar.

Anderson™

It was also developed from Cornell University, USA. A hybrid of Wederscher Markt X O.P. It is a late-season cultivar, yellow-skinned, yellow-fleshed, freestone, self-infertile, very productive and relatively tolerant to bacterial canker.

Sentennial

Very late-season, dark-red, self-fertile cherry developed from Summerland, British Columbia. A hybrid of Sweetheart X O.P. Fruits are moderately large, round, firm and clingstone. Self-fertile and very productive.

Heizhenzhu

It is a sport of the local cherry variety Wupiyingtao growing in South China (Huang-Jie and Jiang-YanHai, 2000). It is resistant to fruit cracking and fruit drop. Fruits having black-red glossy colour, weight 6-10 g, flesh light-yellow, juicy, TSS 17-21 per cent and ascorbic acid content of 05 mg/100g.

Skeena

It was derived from the cross 2C-60-07 and 2C-38-32 (Kappel *et al.*, 2000). Fruits are large, symmetrical and Kidney-shaped, have a wine-red to mahogany skin colour, red to dark-red flesh and weigh on average 11.1g. Skeena has good resistance to rain-induced cracking.

Bargioni 137

Derived from the cross 1SF123 X Caccianese made at Verona, Italy (Bargioni and Bassi, 1999). It is an early-maturing, self-incompatible, producing medium to large white-fleshed fruits with a yellow and red skin colour. It is very well suited to processing and to mechanical harvesting. Resistant to fruit cracking following rain. Suitable pollinators include Napoleon and Durone della Maraca.

Justyna

This cultivar was developed from Czech Republic, as a cross between Kordia and Starking Hardy Giant (Blazkova and Hlusickova, 2007a). Fruits are big (9-10g), skin brownish-red, flesh red, suitable for both as fresh fruit and for processing. Moderately resistant to cracking but susceptible to spring frost.

Helga

Another hybrid (Early Rivers X Moreau) sweet cherry cultivar developed from Czech Republic (Blazkova and Hlusickova, 2007b). Fruits are big (7.5-8.5g), self-sterile, skin brownish-red, flesh red. It is also resistant to cracking but susceptible to late spring frost.

Qinzao

Chinese cultivar bred from O.P. seedlings of cv. Dajin. Matures at the end of June in Hebei province, China (LiBin *et al.*, 2004).

Qinxia

Seedling selection from Governor Wood (LiBin *et al.*, 2004). Fruits mature in early July, fruit weight 6.8-8.1g, red skin, soft and juicy flesh with TSS 15.3 per cent.

Quinhong

It was also bred from O.P. seedlings of cv. Dajin. Fruits mature in early July, fruits have a yellow skin with a pinkish cheek, very sweet, TSS 17.9 per cent (Li Bin *et al.*, 2004).

Sweet cherry cultivars of Japan includes:

Kouka Nishiki

Fruit weight 5-6g, skin colour reddish-yellow, good quality.

Satou Nishiki

This is the main cultivar of Japan. Fruit weight 6-7 g, skin bright yellowish-red, flesh yellowish-white, excellent quality.

Hokko

Fruit weight 6-7g, bright red skin, good quality.

Nanyou

Late cultivar, fruit weight 7-9g, skin yellowish red, flesh yellow, excellent quality.

Beni Shuhou

Late cultivar, fruit weight 8-9g, skin colour bright yellowish red, fresh yellowish-white, excellent quality.

Sour Cherry

The cultivars of sour cherry has been classified into two distinct groups on the basis of the colour of the fruit juice. Sour cherries having light red flesh and juice are generally known as Amarelles though the British often designate them as Kentish cherries and the French as Cerisier Commun. The fruits are pale red more or less flattened at the ends and less acidic than the dark coloured cherries. Montmorency and Early Richmond are the well-known cultivars of this group. Sour cherries of very dark colured fruits with reddish juice are generally known as Morello, but the French call them Griottes. The fruits are more spherical or cordate in shape than those of the Amarelles. Typical cultivars of this group are English Morello, Ostheim and Olivet.

Montmorency

It is an old cultivar which originated in the Montmorency valley in France. This is the only sour cherry cultivar which enjoys commercial importance. It responds to good care and feeding. The trees are hardy self-fertile and productive.

English Morello

The fruit is medium in size, dark red with dark-coloured juice. The tree is small, hardy, productive, but the fruit ripens unevenly.

Early Richmond

The trees are slightly hardier than that of Montmorency. Fruit ripens early, but the quality of fruit is just fair.

Meteor

This cultivar is a seedling of Montmorency X Vladimir originated in 1952. The plant is relatively vigorous dense, very, productive and above all self-fertile. The fruit is medium in size, roundish oval, bright and light red; the juice is colourless. The flavour is fair to good (Lapins *et al.*, 1977).

North Star

It originated as a seedling of English Morello X Serbian Pie No 1. The tree is productive though very small with thin and drooping branches. The fruit is medium in size, round and dark red with coloured juice. The flavour and texture are good. It matures about 10 days earlier than Meteor. It is being tested as a hardy and dwarfing intersrock for sweet cherries. It is an excellent, hardy, self-fertile cultivar for home garden (Lapins *et al.*, 1977).

Piramis

This dark-red, early ripening, sour cherry cultivar was developed from Hungary (Apostol *et al.*, 2005). Fruits are large, kidney shaped or a bit flattened round. It has a pleasant sour cherry aroma, with sweet acidic taste. It is highly suitable for fresh consumption, but also shows good quality when canned. It is tolerant to cherry leaf spot and flower blight disease.

Radunic *et al.* (2014) evaluated four local (Gomilicka, Rana Stonska. and Stonska Tugarka) and five introduced sweet cherry cultivars (Burlat, Garnet, Isabella, Summit and Van) at Croatia. They observed large differences in quality attributes between cultivars. Fruit of Isabella and Tugarka had mean weights > 7.5 g, while those of Burlat were less than 5 g. Fruit soluble solids content varied from 13.1 per cent (Tugarka) to 19.2 per cent (Van). The Garnet fruits had the highest L values (dark-red skin colour). Chroma and hue angle were higher for fruits of Summit, Isabella and Garnet and lower for other cultivars.

Climate

Temperature

The sweet and sour cherries differ in their climatic requirements as the latter can tolerate more cold than the former. At a temperature of -25° to -33°C, sweet cherry cultivars in the nursery suffered 26.6 per cent to 88.0 per cent bud damage whereas sour cherry cultivars showed only 17.2 per cent to 29.9 per cent bud damage (Petre, 1987). Sweet cherries are also more frost susceptible in the orchard than sour cherries. The cold-tolerant sweet cherry cultivars are Van, Stella and Uriase de Bistrita and sour cherry cultivars are Nana and Piti de Lasi (Petre, 1987). Chilling requirements for cherry cultivars generally range between 750 and 1,400 h (Seif and Gruppe, 1985). In general, sour cherries tend to have some what higher chilling requirement than sweet cherries (Thompson, 1996). Sour cherry requires 15°C mean temperature during the growing period. Sour cherries do not thrive well in warm climate while sweet cherry is averse to both severe cold and warm climate.

The cherry blossom is badly damaged by spring frosts. It is, therefore, necessary to select a frost-free site. Fruit set is adversely affected by wind on hill tops. Hill slope is a preferable site where the cold air may drain down. The valleys with an outlet for the heavy cold air to move out are also suitable sites for cherry growing. Selection of the proper aspect is also important in hills. Southern and southwestern aspects are usually warmer than northern or eastern aspects. Thus, on lower elevations, the trees grown on the southern aspect may suffer from drought in May and June which are usually the months with little precipitation. At higher elevations, however, the southern aspect proves advantageous.

Changzhu *et al.* (2011a) working with sweet cherry cultivars of China stated that fruit bud differentiation of sweet cherry occurs between 20-27°C. Lakatos *et al.* (2014a) observed that the start of the bloom in sour cherry had a close relation with the difference of average day and night temperature during the 30-day period prior to bloom. As the difference between day and night temperatures increased, the length of the bloom period decreased. Bloom was shorter when maximum temperature averaged between 13.5-14.5°C, 30 days prior to bloom. If the average maximum temperature was below 13°C or exceed 15°C during this period, the bloom period will be longer than 10 days. They also reported that the degree of self-fertilization influenced directly by temperative. Increase of maximum temperature reduced self-fertilization rate and a 1°C increase of maximum temperature reduced self-fertilization rate by 0.6 per cent. The most favourable minimum temperature was noted as 7.5-8.5°C. If the minimum temperatures were under 4°C or above 12°C, self-fertilization rate was reduced to a quarter of the value at 8°C. (Lakatos *et al.*, 2014 b) Mahmood *et al.* (2000) developed chill unit models for the sweet cherry cvs. Stella, Sunburst and Summit. By using the model, they predicted that chilling temperature of 3.2°C for 1131 hours was optimum for cv. Stella, 3.2°C for 1081 hours for cv. Summit and for cv. Sunburst, the optimum was estimated as 1214 hours at 3.7°C.

Late frosts in spring can damage sweet cherry blossoms and may affect yield. Frost protection candles are burnt to increase the air temperature.

Rainfall

Though a fairly even distribution of rainfall throughout the year is desirable, there are two periods when high rainfall may cause disaster, *viz.* at the time of flowering and towards the ripening period. Rain at flowering time results in heavy blossom wilt (caused by *Pseudomonus syringae*) as its spores gain entrance through the stigma of the opened flowers. Though some measure for control can be achieved by spraying, but the attack is always more severe under moist conditions. Rain at the time of fruit ripening causes cracking of the fruit to which most of the cultivars are susceptible to a greater or lesser degree. Preferably, it should be dry at the time of fruit ripening. High transpiration rates cause substantial size reduction in cherry fruits (Marshall, 1954). The high transpiration rate from the extensive leaf area restricts the availability of water supply to the fruits which become under-sized and shrivelled. Inadequate sunlight adversely affects the photosynthetic efficiency of leaves and thus fruit set is decreased (Marshall, 1954).

Soil

Cherries give better performance on well-drained, warm, deep, free-working, gravelly or sandy loam soils (Partridge and Veatch, 1936). Soil should be able to hold sufficient water for 2-3 weeks during the period of high water requirement. It should not be waterlogged as cherries do not tolerate wet feet (Veatch and Partridge, 1934; Tukey, 1940). Stoev *et al.* (1984) also observed premature withering of young trees of the cultivar Hedelfinger on *Prunus mahaleb* in patches having excessive and continuous moisture in the rhizosphere which contained clay in the bottom. The soil should have a high level of fertility. Some soil-related issues can be managed by choice of rootstock. For example, Mahaleb is used where drought tolerance is required and Mazzard is often used where poor drainage is known to occur.

Rootstocks

The main rootstocks for sweet and sour cherries all over the world are seedlings or clonal selections of *P. avium* and *P. mahaleb*. Wild selections of *P. avium* known as 'Mazzard', wild cherry and 'Gean' were used by the Greeks and Romans during 400 to 330 BC (Hedrick, 1915). *Prunus mahaleb* also known as St. Lucie or 'perfumed cherry' provided the French horticulturists the first alternative rootstock in 1768

(Perry, 1987). Mahaleb gave better performance with most of the sweet cherries on calcareous droughty soils of France. In England, Mahaleb dwarfed the sweet cherries but could not adapt to the soils. Subsequent work of Howe (1927) convincingly proved the superiority of Mazzard rootstock for sweet and sour cherries. However, Bryant (1940) and Coe (1945) demonstrated that the plants of sweet and sour cherries were more vigorous, productive, drought resistant and cold-hardy on Mahaleb rootstock in arid and gravelly soils of Colorado and Utah. As such, Mazzard has given better performance on heavier wet soils and Mahaleb on arid, gravelly soils.

There are reports that not only the plants of various sweet and sour cherry cultivars are less productive on FI2/1 but also the size of the fruit is reduced (Westwood *et al.*, 1976; Zahn, 1977). The growth and development of sweet cherry cultivars in relation to rootstock vary significantly. Timoshenko (1977) reported that bud take of sweet cherry cultivars on sweet cherries was 94.8 per cent, on Mahaleb 91.4 per cent and on sour cherries 72.6 per cent. The success of transplants was 54 per cent on Mahaleb, 46 per cent on sweet cherry and 18 per cent on sour cherry. Colt, a semi-dwarfing rootstock (*P. avium* x *P. pseudocerasus*), is being used in different countries. According to Pennell *et al.* (1983), Colt, *P. mugus* and *P. canescens* clones were precocious in cropping and high yielding.

Characters of Cherry Rootstocks

Colt (*P. avium* F299/2 x *P. pseudocerasus*)

It was developed in 1970s from East Malling Research Station, UK. Semi-dwarfing, resistance to cherry stem pitting, *Phytophthora* root rot and bacterial canker but is susceptible to crown gall. It is sensitive to cold winter temperature.

Mazzard (*P. avium*)

Due to same species of sweet cherry it is compatible with all sweet cherry cultivars. It is winter hardy and vigorous rootstock. Mazzard does well in a wide range of soils from sandy-loam to clay-loam.

Mahaleb (*P. mahaleb*)

More precocious and slightly less vigorous than Mazzard, drought tolerant but extremely sensitive to water-logged soils. Some sweet cherry cultivars (Chelan, Tieton) shows delayed incompatibility on Mahaleb.

F 12/1 (*P. avium*)

A clonal selection of 'Mazard', largely used in many cherry growing countries. Resistant to bacterial cankar and more vigorous than Mazzard. It is well suited in gravelly, calcareous and droughty soil.

Gisela Series (Developed from Giessen, Germany)

Gisela 5 (*P. cerasus* x *P. canescens*)

Dwarfing rootstock, reduce vigour by about 50 per cent or more than Mazzard seedlings. Suitable to use for high-density plantings of 1000-2000 trees/ha. However, it is not suitable for heavy soils and needs good drainage.

Gisela 6 (*P. cerasus* x *P. canscens*)

Relatively vigorous rootstock, suitable for planting of 750-1250 trees/ha, highly productive, slightly advanced flowering and ripening compared with Mazzard. It does well on a wide range of soils from light to heavy with good drainage.

Gisela 12 (*P. cerasus* x *P. canscens*)

It is adapted to a wide range of soils, both precocious and productive, producing early heavy crops. Good fruit size and quality are possible with proper pruning.

Maxima 14 (*P. mahaleb* x *P. avium*)

Open pollinated selection of 'Mahaleb', originated in Oregon, USA. It is precocious, semi-dwarfing and have resistance to iron induced chlorosis. It does well on wet soils.

Krymsk 5 (*P. fruticosa* x *P. lannesiana*)

Originated in Black Sea region of Russia. Semi dwarfing rootstock have adaptability in a wide range of soils and winter hardy. Trees are well anchored and do not need support.

Krymsk 6 (*P. cerasus* x (*P. cerasus* x *P. maackii*)

Dwarfing rootstock than Krymusk 5. It shows adaptability both in cold and hot climates.

Weiroot Series (Developed from Weihenstephan, Germany)

Weiroot 10

Hardy, adaptable to various soil and climates, tolerates heavy soils, semi-vigorous.

Weiroot 13

Good adaptability, hardy, tolerates heavy soils, vigorous.

Weiroot 158

Prefers fertile soils, not tolerant to drought, tolerant to water logging and lime, cold-hardy, dwarfing.

Adara

Developed at Spain, resistant to water logging, tolerant to high pH(8-9), lime soils, vigorous.

M x M Clones

Six promising clones, *viz.* M x M 2, M x M 14, M x M 39, M x M 46, M x M 60 and M x M 97 were investigated as rootstocks for sweet and sour cherry. Clone M x M 14 is the only dwarfing rootstock while others are semi-dwarfing to vigourous (Westood *et al.*, 1976). Generally, scions are more precocious and productive on M x M clones than on Mazzard or F12/1 (Stebbins and Cameron, 1984; Stebbins *et al.*, 1978).

OCR Clones

Two bybrid clones of *P. avium* x *P. mahaleb, viz.*, OCR-2 and OCR-3 have been under rootstock trial in Oregon (Westwood *et al.*, 1976). Montmorency on OCR-2 was found more yield efficient and smaller to the extent of 86 per cent than on F12/1 and continued to have more commercial potential. Scions of sweet cherry on OCR-2 were vigorous but precocious (Stebbins *et al.*, 1978).

Oppenheim

It is a hybrid seedling of *P. fruticosa* x *P. cerasus* which was selected by Plock (1973) at the Oppenheim Station. At this station, 12-year-old trees of Hedelfinger on Oppenheim rootstock were two-thirds the size of Fl2/1, well-anchored, productive and resistant to *Phytophthora* root rot (Cummins, 1984). It, however, suckered excessively (Hein, 1982),

Belgium Clones

Three clones, *viz.*, GM9 (*P. incisa* x *P. serrula*) also known as Grand Manil, GM 61/1 (*P. dawyckensis*) and GM 79, (*P. canescens*) have been identified at Gembloux among the exotic cherry species and interspecific hybrids. These clones have a wide range of compatibility with sweet cherry scion cultivars and reduce the size of plant by 30, 60 and 70 per cent on GM 9, GM 61/1 and GM 79, respectively, compared with F12/1.

Elta

(*P. lannesiana* x O.P) developed at Russia and introduced in 2004. It is a productive rootstock for sweet or sour cherries, medium vigour for sweet and sour cherries, easy to propagated by softwood cuttings.

Table 3.4: Attributes of Various Commercially available Cherry Rootstocks*

Variety	*Tree Size (Per cent of full size)*	*Precocity*	*Compatibility*	*Anchorage*
Colt	100	NO	Good	Good
Gisela 5	50-60	Yes	Good	Fair to good
Gisela 6	85-90	Yes	Good	Fair
Gisela 12	80-100	Yes	Good	Good
Krymsk 5	85-90	Yes	Not available	Good
Krymsk 6	65-70	Yes	Not available	Good
Mahaleb	90	Slight	Fair to good	Good
Maxma 14	100	Yes	Good	Good
Mazzard	100	NO	Good	Good

* Adopted from Lynn and Kaiser (2010).

The cherry rootstocks have been developed mainly by selection and screening of *P. mahaleb*, *P. avium* and *P. cerasus* populations. While this method is still in practice, greater degree of improvement is expected through interspecific hybrids.

The main objectives of rootstock improvement for both sweet and sour cherries are: (i) tree size reduction, (ii) increased scion precocity and cropping, (iii) easy clonal propagation, (iv) cold-hardiness, (v) compatibility (vi) adaptation to a wide range of soil, (vii) uniformity in performance, and (viii) disease and pest resistance.

It is important to know that while *P. avium* does not readily propagate vegetatively, *P. cerasus* does, and this is an important character to be kept in view for the development of high density orchards (Flore and Sams, 1979; Hartmann and Brooks, 1958). Future improvements in rootstocks from *P. avium* can be mainly for disease resistance, precocity and productivity.

Some clones have been selected from wild sour cherry population, but compatibility, lack of anchorage and excessive suckering are common problems. In Italy, clonal selections under the name CAB 6P and CAB 11 E have reduced the plant size by 20-30 per cent than FI2/1 rootstock. Similarly, in West Germany, selections W10, W11 and W13 have reduced scion vigour by 20 to 30 per cent (Schimmelpfeng and Liebster, 1979). The yield on GM series of rootstocks compared with F12/1 and Montmorency is given in Table 3.5.

Table 3.5: Effects of Rootstock on Montmorency Tree Size and Fruit Production after Seven Years at Gembloux, Belgium*

Rootstock	*Crown Vol. (m^3)*	*Yield (kg)*		*Number of Fruits per kg*	*Crown Vol. (Per cent)*
		Cumulative 1978-1980	*Yield Efficiency (kg/m^3)*		
GM9	9.2	6.4	0.15	323	32
GM8	10.5	15.4	0.38	265	36
Montimorency	23.6	26.8	0.30	252	81
GM15	23.0	18.9	0.18	270	79
GM101	24.2	25.8	0.25	306	83
F12/1	29.2	20.4	0.20	263	100

*Genetic background: GM9: *P. incise* x *P. serrula;* GM8: *P. pandora* x *P subhirtella;* GM 15: *P. campanulata* x *P incise*; GM 101: *P. cerasus.*

Bassi (2014) compare different dwarfing rootstock for cherry. The result showed advantages and disadvantages of dwarfing rootstocks. The worst results in terms of all most all parameter (including mortality, trunk cross-sectional area, production efficiency and fruit quality) were recorded with rootstock PHL-C, followed by Weiroot 158 due to considerable vigour, which showed the development and reduced fruit productivity and efficiency. Piku 1 showed optimal productivity and efficiency with both cultivars Kordia and Queen and had vegetative development intermediate between Giesla 5 and Giesla 6. Gisela 6 was most productive with Kordia. Gisela 3 proved too weak and unproductive. Domozetov *et al.* (2014) studied the effect of two clonal rootstocks (Gisela 5 and Maxma 14) on growth and yield of ten sweet cherry cultivars. They reported that the trunks of all cultivars on Maxma 14 have more diameter than on Gisela 5. Yield of cv. Hedelfinger was the highest followed by cvs. Katalin, Sunburst, Merchant and Naprum. The yield of cv. Kordia was the lowest. The cv. Summit had the highest fruit weight, followed by cvs. Sunburst and Namare. Gisella 5 rootstocks proved highly productive for cv. Bing in Oregon, USA and over the six years it produced about the five times cumulative yield than Mazzard (Facteau and Cahn, 2000).

Interstocks

To reduce the size of sweet cherry trees more dramatically, North Star sour cherry was suggested as interstem by Brase and Way (1950). Since then, North Star, Montmorency, Kansas Sweet and Redrich interstems have been used with sweet cherry scions and Mahaleb and Mazzard rootstocks to reduce the size of plants and make them more precocious (Larsen, 1970). In Central and Eastern Europe, interstocks are used to achieve dwarfing with the use of vigorous stocks suited to certain soil conditions, rigid climates or because they are tolerant to or resistant to disease. However, small fruit size has been the most common criterion of these interstem trees. Results from Poland shows that 'North Star' (*P. cerasus*) as an interstock with *P. avium* and several cherry cultivars yielded differently. It failed with cv. Burlat but performed well with cvs. Vega, Kordia, Buttners Red and Merton Premier. The cv. Stella when grafted on P1 (a seedling of *P. mahaleb*) rootstock and Gisela 5 as interstock showed higher yield than only on P-1 rootstock (Lichev *et al.*, 2014).

Propagation

The cherries are propagated either by budding or grafting on the seedlings or vegetatively propagated rootstocks.

Seed Propagation

Stratification is accomplished by placing the seed of cherries between the layers of moist sand, vermiculite, or in a mixture of sand and peat moss and then storing in a cool place at 2° to 4°C temperature for a period of 80 to 120 days for Mahaleb and 120 to 150 days for Mazzard seeds. (Hartmann and Beutel, 1979; Teskey and Shoemaker, 1972). However, before placing the seed for stratification, it should be soaked in water for about 24 hours. Seeds are kept in the stratification medium until they are planted in the nursery or the embryonic root pushes out the seed covering. During stratification period, it must be ensured that the stratification medium is moist and not excessively wet. The container used for stratification purpose should not be air-tight, as aeration is necesary during stratification. After stratification the seeds should be planted immediately, otherwise in case of high temperature the seeds may again enter into dormancy and fail to germinate. Studies conducted have revealed that induction of secondary dormancy during the stratification peirod enhances the seed germination by causing two consecutive peaks of protein synthesis (Suszka, 1976; Michalska, 1982; Michalski, 1982). Percentage germination of seeds of sweet cherry cultivar Red Mercier stratified for 30-60 days at 5°C was significantly increased by dipping the stratified seeds in 100 ppm GA_3. Similarly, good germination of sour cherry seedling rootstocks was obtained by soaking the seeds for 24 hours in 500 ppm to 700 ppm GA_3 before stratification (Osrrejko, 1969).

Vegetative Propagation

In cherries vegetative propation is mainly done by stooling, airlayering, cuttings, budding, grafting and micropropagation.

Stooling

In this method, the mother plant is cut back at the ground level before sprouting in spring. New shoots develop on the stub in the spring. When the shoots are 25 to 30 cm long, their bases are covered with a mound of soil or sawdust building the mound to a height of 20 to 30 cm as the shoots grow during the spring. Roots arise from the bases of the shoots and grow into the mound. By applying IBA at 7500 ppm on ringed portion of the F12/1 rootstock shoots during summer, 100 per cent rooting was obtained (Pandey *et al.*, 1983). The mound is to be kept moist throughout the summer. The shoots normally become well-rooted by the winter and are then lined out in the nursery rows to grow them for a year to produce strong plants for grafting or budding. Suckers of Mazzard (*P. avium*) and Mahaleb (*P. mahaleb* rootstocks are removed druing winter. If the suckers are well rooted and larger than 0.6 cm in diameter, then they can be grafted or budded during the next spring.

Cutting

Hardwood cuttings: Hardwood stem cuttings are made between late fall to early spring depending on the planting time. The length of cuttings is usually 12 to 35 cm,or more with basal cut about 1 cm below node and the top cut 1 cm above node. In areas where winter is mild, good results have been obtained by taking the cuttings in the late fall and treating them with lBA (100 ppm) for 24 hours. Treated cuttings are then stored at 21°C until roots appear and then they are planted in the nursery (Hartmann and Beutel, 1979). Jafari and Bouzari (2010) recorded highest rooting (46.6 per cent) of Gisela 6 cherry rootstock by using basal cuttings collected in January and treated with IBA at 3000 ppm.

Semi-hardwood cuttings: Semi-hardwood cuttings are leafy and are usually taken during mid or late summer. The rooting in such cuttings increased markedly if planted in mist chambers or in closed glass-covered frames having high humidity. Semi-hardwood cuttings of Gisela 6 rootstock cuttings showed 70 per cent rooting when terminal cuttings were collected in July and treated with IBA at 1000 ppm before planting of cuttings. (Jafari and Bouzari, 2010)

Softwood cuttings: Softwood cuttings consist of leafy stem cuttings made from the young, succculent spring growth of plants. Although propagation from softwood cuttings is not a common practice in cherries, good rooting has been obtained with the help of plant growth regulators in the mist propagating beds (Hartmann and Beutel, 1979).

Budding and Grafting

Budding is usually done from the last week of April (after the sap flow started) to the last week of May (before the onset of rainy season). T-budding is generally preferred as it has given better results. T or inverted T-method of budding has not showed any difference (Tukey and Brase, 1945). However, Poniedzialek (1984) reported that modified chip budding gave better results than T-budding.

For grafting, the scionwood is collected during the winter when the buds are dormant. Scionwood is stored by packing in sphagnum moss or wood shavings and then wrapping in moist paper or cloth. The moist pack is stored at temperature of 2-4°C. Cleft grafting is more frequently done towards the end of the dormant season or even after the stock has begun to sprout. However, if the grafting has been done early in the season, the grafted plants are stored as recommended for the scionwood and they can be planted in the nursery towards the end of the dormant season. Budding success and growth of sweet cherry cvs. Heidegger, Summit, Techlovan, Lapinis, Sylvia and Regina grafted on clonal rootstocks F12/1, P-HLA, P-HLB and Gisela 5 were studied in Poland by Sitarek and Grzyb (2007). They did not find any effect of rootstock on budding success, but tree growth in the nursery was found affected by rootstocks. Trees of all cultivars tested on P-HLA, P-HLB and Gisela 5 rootstocks were smaller and their trunks were thinner than on F 12/1 rootstock (used as control). The cv. Heidegger on P-HLA, P-HLB and Gisela 5 showed obvious symptoms of incompatibility. However, microscopic examination of the graft union did not show any visual dysfunction between scion and rootstock.

Micropropagation

Mass *in vitro* propagation of cultivars as well as rootstocks in both sweet and sour cherries has successfully been standardized. Micropropagation was, however, more suitable for sour cherries than sweet cherries (Khaak and Nusst, 1989; Paul and Feucht, 1985). Multiplication rates of the cultivars and rootstocks are similar in sour cherry but different procedures for micropropagation of cherries have been described by many researchers (Baraldi *et al.*, 1988; Cerovic and Ruzic, 1987; Khaak and Nusst, 1989). Invigoration procedure has been used by Ranjit and Kester (1988) to enhance rooting in difficult-to-root '46-1 Mazzard' rootstock. The tissue-cultured trees of sour cherry are, however less productive and have the low production efficiency (Rosati and Goggioli, 1989). Protocol variations are needed for the new dwarfing rootstocks, specially if they are inter-specific hybrids (Fidancy *et al.*, 2008). A new method of propagating Gisela 5, which is used in many countries, uses an *in vitro* growth medium with a multiplication rate of about 4 shoots per explants (Nacheva and Gercheva, 2009). HaiRong *et al.* (2012) standardized the protocol for tetraploid rootstock Y1. They observed that Woody Plant Media (WPM) was the best culture medium for the adventitious shoot regeneration. The optimal hormone combinations were 6-BA at 2.0 mg/l + IBA at 1.0 mg/l. Dark incubation for 14 days improve the regeneration rates. Adventitious shoots regeneration rate was 90 per cent when *in vitro* leaves were used as explants. Sedlak and Paprstein (2011) standardized rapid *in vitro* shoot multiplication and rooting of sweet cherry cv. Rivan. They obtained highest multiplication rate by using MS medium containing 2.0 mg/l BAP while the best results for rooting was noted with IAA at 1.0 mg/l. The dwarfing cherry rootstock PHL-A is a hybrid of *P. avium* and *P. cerasus*. Mahdavian *et al.* (2011) developed the *in vitro* propagation protocol for PHL-A. Shoot tips and axillary buds, derived from greenhouse grown one-year-old trees, were used as explants and established *in vitro*. In proliferation stage, multiplication rate of about 5.37 was achieved over a 4 week period using MS medium with 0.5mg/l BA. Rooting of shoots, with upto 100 per cent, were also obtained within four weeks on DKW medium.

Growth of Nursery Plants

High branches are induced in nursery cherry trees by retaining rootstock shoots early in the maiden year which suppress the development of low scion laterals. It was found that in trees carrying additional rootstock shoots the scion-shoots elongate slightly faster than in normally trimmed trees. Faster shoot elongation was accompanied by a 10-20 per cent reduction in the thickness of the stem near the shoot apex. When the extra shoots were either removed or detipped, scion shoot elongation declined rapidly and, after a delay of about 2 weeks stem thickness increased. (Harrison-Murray and Mitra, 1986).

Planting

The pits for the plants are usually of one metre dimension, but care should be taken to see that there is no hard pan or big stone either below or on its sides which may restrict the development of the roots. The pit is filled with a mixture of top soil and farmyard manure in I: 1 ratio. The pit may be filled at least up to 15 cm above the ground level, so that after watering it may settle down to the ground level. Before planting is done, the plants should be examined for their freshness, good root system and well-developed buds. Cut off all the broken or injured roots and treat the cut portion with a fungicide. The plant should be placed in the pit up to the same depth as it was in the nursery. The soil in the pit should be pressed around the plant followed by watering, and staking is required to keep the plant straight.

In areas where winter is not too severe, planting may be done in early winter, so that the root system is well established before the opening of the buds in spring. However, in areas where winter is very severe, planting may be delayed to late winter. In both the cases it is advisable to plant freshly dug plants, as the use of stored plants has not given good results.

Planting Distance

The planting distance in case of cherries is influenced by the fertility of soil, budding height and the rootstock used. In case of F12/1 which is a very vigorous rootstock and Mazzard seedlings, it will be safe to plant at a distance of 12 m. In case of Mahaleb seedlings, planting distance should be 9 m, while in the case of dwarfing rootstocks like Colt it may be 5 m. Vasilev (1982) recommends 6.5-7.0 x 5.0-6.0 m spacing for sweet cherries with spherical or elliptical crowns and 4.5-5.0 x 2.5-3.0 m spacing under hedge system. The spacing in sour cherries should be lesser by about one metre in each direction than sweet cherries.

The previously standard system of sweet and sour cherry culture with vigorous cultivars planted at wide spacings has become unprofitable due to difficulties in pruning, harvesting and spraying for crop protection. A very intensive cherry orchard with trees trained on a vertical axis, spaced 4 m between rows and 0.5-1.5 m along the row (up to 5,000 trees/ha) has been tested in France. All new shoots are shortened to 2-3 leaves in summer to promote basal flower bud formation until the 6th year of the trial. The results seem promising (Sansavini, 1982). In Romania, the sour cherry cultivars Schattenmorelle, Nana, Meteor, Crisana 11/6, Mocanesti 401, Mocanesti 6/7, Pitic, Timpurii de Cluj, Pandy 38, Cygany Meggy and Gorsemkric grafted on wild sour cherry and Mahaleb were planted in 1975. The sweet cherry cultivars were Van, Bing, Sam, Gill Pek, Chinook, Stella, Cerna and Compact Lambert grafted on the same rootstocks and planted in 1979. The density of the sour cherry plantings ranged from 600 to 2,500 trees/ha, spaced at 4 m x 1-4 m, and that of the sweet cherry plantings from 400 to 666 trees/ha with spacings of 5 m x 3-5 m. (Parnia *et al.*, 1986). In high and very high density orchards of sour cherry it was found that tree height should be limited to 2.5 - 3.0 m and the hedgerow width to 1.5 m. In the seventh year the very high density orchards trained as vertical cordons out yielded trained palmettes of the same cultivars (Table 3.6).

Table 3.6: The Effect of Spacing and Cultivar on Sour Cherry Yield in the 7th Year after Planting (1983)*

Canopy Shape	*Trees/ha*	*Mean Yield t/ha*	*Cultivar*	*Mean Yield t/ha*
Palmette with tied branches	625	10.62	Crisana 11/6	14.4
	715	9.15	Mocanesti 6/7	13.4
	833	10.32	Pitic	12.7
	952	11.32		
Mean		10.35	Mean	13.5
LSD 1 per cent = 0.96			LSD 1 per cent = 1.00	
Vertical cordon	1250	15.70	Crisana 11/6	14.4
	1428	15.00	Mocanesti 6/7	12.8
			Pitic	7.4
Mean		15.35	Mean	11.53
LSD 1 per cent = 2.93			LSD 1 per cent = 0.66	

* Parnia *et al.* (1986).

It is essential to understand the characters of the rootstock for maintaining a high density orchard in a particular environment and site conditions. In Hungary, Hrotko *et al.* (2009) classified rootstocks Cema, SL64 and Bogdany as vigorous; MaxMa 97, Pi-Ku1 and Tabe 1 as moderate vigorous and Edabriz, Gisela 5 and Prob as dwarfing stocks. They reported that cv. Axel, planted at 4m x 2m produced highest yield on Bogdany and on Cema, contray to Gisela 5, which produced only 50 per cent of Bogdany and on Cema. According to average yield/year production of the measured four years, SL64, Bogdany, Cema and Pi-Ku1 were found as the most effective rootstock. They observed same yield effects on Gisela 5 at higher densities (4 x1.5 m). The largest fruits were noted by using Bogdany rootstocks. They finally recommended SL 64, Bogdany, Cema and Pi-Ku 1 for high density plantation under Hungarian condition. In Europe, the new trends are to use new dwarfing rootstocks and increasing plant densities from 400-600 to 1000-1500 per hectare. These high density planting feature relatively short trees, early and uniform bearing, relatively uniform ripening, reduced overhead and a market edge (Sansavini and Lugi, 2014).

Training and Pruning

Pruning at Planting

Plants taken from the nursery are headed back at about 90 to 100 cm. Losses following planting are usually lessened if the branches are not cut back severely. If the top is to be thinned to strike balance between the root system and top, then remove whole branches rather than cutting them back severely. Cutting back of branches removes the largest and best nourished buds, which develop the leaves to give the young trees a good start. The wide-angled scaffold branches should be 20 to 25 cm apart and located spirally around the trunk. The selected scaffold branches be given a minimum of heading back, to have a good balance between them and the leader while forming a well-shaped tree. The cherries sprout very early in the spring requiring an early water supply. The new root system may form too slowly to provide sufficient water to the top and may account for the death of trees after they have leafed out. A moderate cutting back of the branches reduces the water requirement of the top in two ways: (i) it removes some of the transpiring surface; and (ii) it removes the terminal buds, which are first to burst, leaving buds which are slower in developing, thus retarding top growth.

Pruning after One Year of Planting

The central leader is retained and all other upright branches removed by cutting close to the trunk, only 3-5 wide-angled framework branches, 20 to 25 cm apart, and spirally around the tree are retained. This is important to avoid winter killing with the collection of snow in the narrow crotches. The lowest scaffold branch may be kept 40 to 60 cm above ground.

Pruning during Pre-Bearing Period

The formation of strong framework is necessary because the trees are susceptible to crotch splitting. Scaffolds should be headed back only when they grow very long and willowy and when it is desirable to have secondary shoots. When the tree grows larger, the leader is cut back to an outward growing branch.

Pruning of Bearing Trees

Sometimes pruning is done in bearing trees with a view to removing weak and unnecessary branches, keeping the centre of the tree open and generating new wood of suitable average length. In sweet cherry, fruit is borne laterally on spurs which arise on one-year old shoots. The spurs grow less than 2.5 cm in a year with a terminal leaf bud. Since the average productive life of the spurs is 10-12 years, less renewal wood is needed as compared with sour cherry and most of the other deciduous fruits.

In sour cherry annual shoot growth should average 10-20 cm long in fully grown bearing trees for maintenance of yield. Fruit buds of the sour cherry are produced laterally on one-year-old terminal growth while fruiting spurs are formed on 2 to 3 years old vigorous shoots and 4 to 5 years old less vigorous shoots. A single crop of fruit weakens the spurs and they are short lived. As such, most of the crop is borne on the periphery of the tree. The pruning should, therefore, comprise thinning out of the long branches (Teskey and Shoemaker, 1972).

Pruning should be done late in the season, after growth has started as the danger of infection by silver leaf disease caused by *Stereum purpureum* is then minimized (Anon., 1949). In fact some experienced growers leave this pruning until as late as May. Muller and Gehler (1985) also preferred April pruning to February for reducing tree vigour and better wound healing. In case of upright growing cultivars, such as Black Tartarian removal of upright growth and cutting to outward growing laterals cause spread of shoots. Similarly cultivars such as Early Rivers which produce long whippy leaders may need some tipping for 5 or more years. Webster and Shepherd (1984) also found that summer tipping of shoots in Van and Merton Glory consistently reduced mean shoot length and tree size. The number of floral buds on one-year-old shoots also doubled after tipping. Mika *et al.* (1983) reported that the yield is greatly affected on half-standard sour cherry trees trained for mechanical harvesting. Therefore, they suggested that the tree to be trained as half standards should be left free to develop a good branch system. Brunner (1985) obtained increased cumulative yield in sweet cherry trees with a combination of crown pruning in autumn (August) plus root pruning in spring.

In order to do away with the necessity of dwarfing rootstocks for high density sweet cherry planting, Ende and Jerie (1985) suggested the Tatura trellis system. In trials with cultivars Bing and Van on *P. mahaleb* rootstock at a planting density of 2222 - 3175/ha, the yields obtained after three years of planting were 31-39 and 9-10 t/ha, respectively. While a number of training systems are suitable for increasing sweet cherry planting density, the V and vertical axis systems are perhaps best suited for high density planting for sweet cherry orchards (Musacchi and Lugli, 2014). New ideas regarding tree shape include short trees grown using dwarfing rootstocks like Gisela series. Musacchi and Lugli (2014) reported from two ongoing cherry training system trial at Italy. The first trial (in year 3) compares the vertical axis, spindle and V training systems, the dwarfing rootstocks Giesela 5 and Gisela 6 in planting densities of 1905 and 5,714 trees/ha, and 11 cultivars. The second trial involves trees trained to vertical axis with cvs Kordia and Ferrovia at ultra-high densities of 5000 (4.0 x 0.5m) and 6,666 (3.0 x 0.5m) trees/ha, in two orchards, both at 6 year of age. Cumulative yield of cv. Ferrovia in the second trial at year 6 was recorded 46.6 t/ha. Cherry quality was also very high; 88.8 per cent of the fruit at the density of 5,000 trees/ha and 89.8 per cent at 6,666 trees/ha were over 28 mm in diameter.

Nutrition

Since fruit development and vegetative growth occur simultaneously in cherries, the plants are hard pressed with heavy demand for mineral nutrients. Unless application of nutrients is adequate and timely, the plant growth and fruit quality will be adversely affected. The amount of fertilizer to be applied is influenced by the age or size of the tree, soil type, cultural practices and anticipated fruit yield.

Role of Nutrients

Nitrogen

For each year of tree age 22.5g N should be applied. Thus a ten-year-old plant will require 225g N or 489g urea (Teskey and Shoemaker, 1972). Cooper (1962) suggested utmost care in avoiding high dose of N as it delays cropping and increases disease incidence. Some workers have suggested application of 150 g N for every 2.5 cm girth of the plant (Azad, 1981). The fertilizer should be applied in early spring at about 2-3 weeks before bloom, in a 60 cm wide under the canopy of the plant. As most of the roots (80-87 per cent) are at depths of 11-40 cm, placement of fertilizer and tillage should be shallow. Quast (1983) reported that the application of 90 kg N per hectare increased the N level in the leaves to 2.6 per cent of dry weight and soil N adequately. He obtained an increased yield of 24 per cent over control (unfertilized). The N application did not adversely affect the flower formation but increased fruit set. Georgiev (1984) reported that application of N to fully grown trees of Hedelfingen on *P. mahaleb* rootstock gave the highest yield. He further reported that split application in March and November (2/3 in March and 1/3 in November) produced better growth and yield. Foliar application of Urea had positive influence on tree growth and yield. It also improved fruit size (Drahosova *et al.*, 2011). In sweet cherries, accumulate nitrogen stores from the previous fall support early spring growth. To complement the N supply in the soil, growers sometimes used foliar urea application after harvest. To study the canopy distribution of N applied as urea foliar spray, Ayala *et al.* (2014) conducted an experiment at Chile using Bing/ Gisela 6 sweet cherry. 15^{N} urea was used. It was reported that labeled urea was taken up by the leaves and translocated to various organs. The highest N levels were observed in the fruiting sections. Spraying of urea between January and April showed that the stored N was recovered in the flowers, immature fruits and young leaves during the following spring. Most of the N translocation was basipetal. Urea sprays after fruit harvest constitutes an effective alternative to complement soil N supply in sweet cherry trees with Gisela 6 rootstock (Thielemann *et al.*, 2014). Leaf samples collected from 79 cherry orchards in Utah gave the mean value of N as 2.3 per cent and the plants showed no symptom of N deficiency (Christensen and Walker, 1964). According to Shear and Faust (1980), the safe range of N content in the leaves is 1.8 to 3.3 per cent.

Phosphorus

Phosphorus in the form of superphosphate, ammonium phosphate, or rock phosphate spread on the soil into the grass cover is fixed in the top 5 to 7.5 cm of soil and does not become available at the depth of the main root zone. Application of 9 kg superphosphate per tree per year for 3 years increased the P content of leaves more through the use of superphosphate on sod than on mulch although all phosphate treatments increased leaf P (Wander, 1943). Increasing the P content of the fruit had no effect on maturity or quality (Stanberry and Clore, 1950). Agarwal (1970) recommended use of 300 g P for a ten-year-old plant in Himachal Pradesh, India.

Potassium

Sweet cherry trees respond to K application when the K content of the leaves is below 0.75 per cent. Application of 900 g to 1125 g of K per tree increased the K content in the leaves from 300-400 per cent (Boynton, 1944). At least 0.80 per cent leaf K content is necessary to avoid potassium deficiency. Fruit size was improved when the leaf K content reached 1.23 per cent (Kwong, 1965) According to Shear and Faust (1980), the normal range of K in cherry leaves varies from I to 3 per cent. Potassium treatment @ 650g for a ten-year-old plant has been recommended in Himachal Pradesh (Agarwal, 1970). Regular soil-drenching of cherry trees with soluble potassium silicate (soluble 1 per cent potassium silicate (28 per cent Si) applied to the drip area, every 3 weeks from flowering has the potential to increase fruit firmness and stem pull force without affecting fruit size or sweetness negatively (Kaiser *et al.*, 2014)

Zinc

Zinc deficiency symptoms in cherries are characterized by the production of small, narrow, chlorotic leaves which are usually distorted or wavy (Azad. 1981). Foliar spray of zinc sulphate at 2.5 kg in 100 litres of water in early spring at 2-3 years' interval may correct its deficiency (Teskey and Shoemaker, 1972). Zinc deficiency was not observed when its leaf Zn content rose to 19 ppm (Christenseon and Walker, 1964), safe range being 15-70 ppm (Shear and Faust, 1980).

Boron

Boron deficiency may result in hard, shrivelled blotchy fruit. Deficiency symptoms appear when the leaf B content is less than 15 ppm, the optimum range is 20 to 60 ppm (Shear and Faust, 1980). Application of 112 to 225 g of borax in the soil under a large tree has been suggested to correct the deficiency (Teskey and Shoemaker, 1972). Nagy *et al.* (2014) working with different tart cherry cultivars, growing in sandy soil with low B content in Eastern Hungary reported that foliar born spray had inconsistent effect on fruit composition. It increased leaf B content in some cultivars but had not effect to others. Beppu *et al.* (2007) working with sweet cherry cv. Satohnishiki in warm areas of Japan reported that spraying of Boron at 500 ppm in early September prolonged ovule longevity and accelerated the elongation of pollen tubes in pistils in the following spring. Thus, the rate of fruit set in boron-treated trees was higher than that in the control trees.

Manganese

When leaf samples contain 17-18 ppm or less Mn, its deficiency appears as interveinal cholorosis. However, as the leaf Mn content increased to 59 ppm no deficiency symptom was observed (Christensen and Walker, 1964). According to Shear and Faust (1980), 20 to 300 ppm is the safe range of Mn content in the leaves.

Tissue Analysis

The best guide for nutrient requirement is the leaf analysis which indicates the status of different nutrients in the plant. The critical levels for various elements have already been established (Table 3.7). Roversi and Monteforte (2006) estimated the mineral uptake of six sweet cherry cultivars (B. Moreau, Ferrovia, Stella, Colafemmina, Droganova and Montagnola) grafted on Mazzard rootstock. They reported mineral uptake (Kg/ha/year) of 39-65 for N, 6-11 for P, 16-47 for K, 26-55 for Ca and 5-9 for Mg.

Table 3.7: Nutrient Content of Cherry Leaves

Nutrient	*Normal Leaves*	*Deficient Leaves*
N (per cent)	2.30[1]	NE
P (per cent)	0.14[1]	NE
K (per cent)	2.06-2.13[2]	0.39-1.82[2]
Ca (per cent)	0.7-3.0[4]	Ne [4]
Mg (per cent)	1.45[1]	NE[4]
Na (per cent)	0.05[1]	
Mn (ppm)	20-300[4]	<20[4]
Fe (pprn)	20-250[4]	NE
B (ppm)	75[1]	< 15[4]
Cu (ppm)	5[4]	Ne [4]
Zn (ppm)	13-44[3]	6-15[3]

1 = Compton (1959), 2 = Boynton (1944), 3 = Woodbridge (1955), 4 = Shear and Faust (1980).

NE = Not estimated.

Quast (1983) stated that fruit drop was much higher when leaf N content was 3.4 per cent compared with 3 per cent N. Van Petersen (1977) suggested leaf contents of 2.8 per cent N, 0.2 per cent P and 1.4 per cent K for optimal growth and yield. Fregoni *et al.* (1973) found very little differences in nutrient elements between leaves from current year shoots and spur leaves (Table 3.8). By relating the amount of nutrients applied in the previous year to the content shown by the leaf analysis, it is possible to determine the amount required for the current year. Soil analysis, on the other hand, indicates the nutrient status of the soil. Estimates of nitrogen, phosphorus and potash requirements for every 2.5 cm girth of the trunk and on the basis of plant age have also been made. In a healthy plant, annual shoot growth on the periphery of the tree should be about 30 to 40 cm.

Table 3.8: Nutrient Contents in Shoot Leaves and Spur Leaves of Cherries*

Nutrient Content (Per cent of dry matter)	*Shoot Leaves*	*Spur Leaves*
N	2.44	2.60
P	0.21	0.21
K	1.53	1.52
Ca	0.97	1.12
Mg	0.33	0.38

* Fregoni *et al.* (1973)

Irrigation

In cherries new growth takes place when fruits are developing during spring and early summer. As such water requirement in this period is heavy. However, one or two showers during this period may obviate the necessity of irrigation. From mid-June rains start and continue up to mid-September. Thus, only one irrigation after the rainy season will be enough. The resistance to drought is most on sour cherry rootstocks, intermediate on sweet cherry and the least on Mahaleb (Lishchuk and Eremeev, 1974).

Cherries are very susceptible to wet feet. So overwatering is not desirable. During the rainy season extra care should be taken to see that the drainage is proper and there is no waterlogging in the orchard. Irrigation should be given only when soil samples taken from a depth of 60 to 90 cm show dry soil. On most soils there is no need to irrigate if the soil sample can be squeezed into a ball that holds its shape when jarked slightiy If the soil crumbles after squeezing then it needs water otherwise not (Teskey and Shoemaker, 1972). Sweet cherry trees produced more than 20 per cent higher yield when irrigated at 80-85 per cent than at 60-65 per cent soil moisture level. In a sprinkler irrigated bearing orchard, 3-6 irrigations using 250-390 m^2/ha provided maximum yield and also improved fruit firmness (Alani, 1976; Yastreb, 1981). Blasse and Bringezu (1984) reported considerable increase in the yield of sweet cherry when irrigated before and during fruit ripening. The young plants also made increased growth. In drip irrigated orchards, highest yield and returns were obtained when soil moisture was not allowed to deplete lower than 70 per cent capacity. Irrigation increases the tree frost resistance of both cherries (Yastreb, 1988). A combined double-lateral drip irrigation and in-row wheat straw mutch production system is viable for sweet cherry in areas with water shortages and soil erosion (Long *et al.*, 2014).

Crop Load Management

With the commercial acceptance of precocious, size controlling rootstocks and productive self-fertile cultivars, cherry production took a giant step forward. These new innovations, however, are not without difficulties. Both the precocious rootstocks and many of the self-fertile varieties tend to cause heavey fruit set leading to a high fruit-to-leaf area ratio (F:LA) that limits the amount of carbohydrate movement into the fruit, detrimentally affecting fruit size (Whiting and Ophardt, 2005). Most North American Cherry growers manage crop load exclusively through pruning. However, successful crop load management through pruning alone becomes more difficult when less vigorous rootstock (eg. Gisela 5) are used, or when highly productive self-fertile cultivar (eg, Sweetheart) is cobined. Improved fruit quality and greater percentage of large fruits were observed by adjusting crop loads in Sweetheart/Mazzard sweet cherry. However higher crop value was found associated with large volumes of medium-sized fruits (Einhorn *et al.*, 2011). Stern *et al.* (2014) reported that spraying of Uniconazole (E-1-(4-chlorophenyl)-4, 4-dimetayl-2-(1,2,4-triazol-1-yl-1-penten-3-ol)) at 100-125mg/l in the form of 0.2-0.25 per cent Magic™ at 30 per cent full bloom, reduced the fruit set of 'Bing' cherry fruit. As a result, the time spent in hand-thinning was reduced by about 50 per cent and the crop load by about 40 per cent, and there was a considerable shift to larger fruit size (>26mm), provided the crop load was not too low. Carvalho *et al.* (2012) working with sweet cherries cvs. Brooks and Lapins on *P. mahaleb* at Protagal observed that crop load slightly influenced mean fruit weight, the soluble solids content and colour attributes. These parameters of fruit decreased only beyond 10,000 fruits per tree.

Weed Management

Cherry trees, particularly young trees do not compete with weeds well. Chemical weed control has been the predominant method of weed control used in orchards, however there are a number of other options that can be used either separately or in conjunction with chemical (herbicides) strategies. Mulching by spreading of large amounts of organic matter (such as straw, hay, wood chips or compost) along the tree line is becoming increasingly favoured in orchards. The benefits of these additions include regulation of soil temperature and moisture, weed suppression and the addition of organic matter in the soil (Mooney *et al.*, 2009). Mulching with geotextile suppressed weeds increase tree canopy volume and improve fruit quality (Blazkova *et al.*, 2009)

Rankova (2006) studied the effect of soil herbicides (napropamid, pendimethalin, terbacil and metolachlore) on the vegetative habits of Mahaleb cherry seedling rootstocks. The results obtained showed that napropamid-Devrinol 4F at 400-500 ml/a; pendimethalin-Stomp 33 EC at 400 ml/da and metachlore-Dual Gold 960 EC at 150 ml/da could be applied in the production of Mahaleb cherry seedling rootstocks. In organically maintained cherry orchards of Turkey, frequently observed weed species recorded were *Sorghum halepense, Cuscuta campestris, Lactuca serriola, Trifolium sp., Polygonum convolvulus, Cynanchum acutum, Cirisum arvense* and *Veronica hederifolia*. It is suggested to practice deep tillage and their roots removed from soil for perennial weeds while annual weeds should be controlled by hand weeding and tillage (Basaran and Ozdem, 2013).

Flowering and Pollination

Flowering

Sweet cherries grafted on vigorous rootstocks like F12/1 takes about 8 years to come in commercial bearing. However, on dwarfing rootstocks like Colt bearing commences in the third year. Flowers of sweet cherry are usually borne on spurs of 2-year and older wood, while very few are borne near the base of 1-year shoots. In sour cherry, flowers borne on 1-year wood and on spurs and there are 2-4 flowers per bud, Most of the productive fruiting spurs are produced on 2- to-3-years old wood. Few less productive spurs are also persist on 4- to-5-year old wood.

Chemical control of flowering in sour cherry increases cropping efficiency. As trees mature and encounter stresses, tree vigour decreases and often results in excessive flower bud formation on new shoots. These flower buds are pure flower buds that lack a vegetative meristem, and therefore the supporting nodes become blind after fruiting, resulting in reduced spur formation and loss of fruiting wood (blind wood synodrome). Gibberellic acid (GA_3) can markedly reverse the blind wood syndrome by reducing flower formation and promoting spur formation on new shoots. N-substituted phthacimides, particularly AC-377 also showed equivalent activity to GA_3 and have potential to ameliorate the blind wood syndrome and increase the cropping efficiency in sour cherry (Retamales *et al.*, 2014). Once the chilling requirement is satisfied, bud break occurs after a specific amount of heat has accumulated. Insufficient chilling caused delayed and erratic blooming, reduced fruit set and abscission of some buds. Hot weather, if persist during the early stages of floral initiation may cause abnormal flower development. For Mediterranean climate, characterized by mild winter and hot, dry summer of Muricia, Spain; Garcia *et al.* (2014) classified sweet cherry cultivars in groups of diverse sensitivity-to-chilling. Concerning the early and middle flowering periods cvs. Mister Early, Primulat, Sweetheart, Somerset and Sylvia had low sensitivity to chilling changes. For late flowering cvs. Early Lory, Grace Star, chelan, Garnet, Satin, Samba and Santina were found less sensitive to prior chilling changes. The spur elongates in almost a straight line for several successive years because the terminal bud is a leaf bud. GA_3 application reduces flowering of sweet cherries the year following treatment (Facteau, 1986).

Pollination

The flowers of sweet cherries are bisexual and hypogynous. The pollen grains are heavy and sticky and as such are not carried to any extent by wind (Baker, 1975). Honey-bees usually carry the pollen grains from the anthers of one cultivar to the stigma of another cultivar. However, pollination problem in case of sweet cherries is not as simple as in case of apple, pear, plum, almond, *etc.* It is not the question of planting 2 to 3 cultivars together for normal fruit set, because cultivars of most of the sweet cherries are not only totally incompatible with their own pollen, but also with the pollen of some other cultivars (Anon., 1949). Care should be taken to see that not only their flowering period overlaps, but they also do not have the same sterilty alleles. For example, if Merton Bigarreau, Schrecken, Van, Venus, Windsor which have the same sterility alleles are planted together, they will not set fruit. A cultivar with at least one different sterility allele, such as, Bing, Lambert, Victor, *etc.*, is required to be planted (Lapins *et al.*, 1977).

One of the most import events in sweet cherry cultivation is good pollination and fertilization for abundant fruit set. The successful completion of a sequence of reproductive events depends on supplies of viable, compatible pollen; an effective transfer of pollen when stigmas are respective; proper growth of pollen tubes and entering the ovule during the period when embryo sacs have matured and ovules are viable; and successful double fertilization followed by growth and development of the embryo and endosperm (Thompson, 1996). After fertilization, other phycilogical factors play a contributing role (Choi *et al.*, 2000).

As with other self-incompatible fruit tree species of the Rosaceae, sweet cherry exhibits the homomorphic and monofactorial gamelophytic self-incompatibility system, which controlled by the S-locus with multiple alleles (Crane and Lawrence, 1929; Tehrani *et al.*, 1992). With this S allele that carried by a pollen grain is the same as one of the two specific S allels carried by the somatic tissue of the receptor pistil. Thus, sweet cherry requires accurate pollination group knowledge to assure fruit set in commercial orchards. For effective cross-pollination, atleast 2-3 cross-compatible cultivars which has synchronous bloom times should be planted in an orchard and bees provided as pollinators (Choi *et al.*, 2000). The effective pollination period represents the period of ovule longevity minus the period required for pollen tubes to grow down the style and effective fertilization. In sweet cherry, this effective pollination period may be as short as 1-2 days or as long as 1 week depending on cultivars, environmental factors and the season (Guerrero-Prieto *et al.*, 1989, Toyama, 1980).

Crane and Brown (1937) classified cherries in 13 groups on the basis of their sterility alleles. The universal donors were placed in group XIII. Since then many new cultivars have been bred, and Lapins *et al.* (1977) classified new cultivars in XIII groups and gave O group to universal donors. A combined list of important sweet cherry cultivars classified into different groups on the basis of sterility alleles as prepared by Crane and Brown (1937), Teskey and Shoemaker (1972) and Lapins *et al.* (1977) is given in Table 3.9.

Table 3.9: Incompatibility groups for inter pollination.

Group I S_1 S_2 : Bauman's May, Bedford Prolific, Black Downton, Black Eagle, Black Tartarian, Early Rivers, Knights Early Black, Roundell Heart, Circassian.
Group II S_1 S_3 : Belle Agathe, Bigarreau Schrecken, Black Elton, Black Heart, Clauster Black, Victoria Black, Waterloo, Frogmore, Gil Peck, Lamida, Merton Bigarreau, Sodus, Van, venus, NY3 1599, NY 1725
Group III S_3 S_4 : Bigarreau Napoleon, Emperor Francis, Ohio Beauty, Bing, Lambert, Royal Ann, Star, Vernon, Schneiders Spate Knorpelkirsche
Group IV S_2 $S_{3]}$: Kentish Bigarreau, Ludwig's Bigarreau, White Bigarreau, Amber, Sue, Victor, Merton Heart, Velvet
Group V S_3 S_5 : Bohemian Black, Late Black, Turkey Heart
Group VI S_3 S_6 : Elton Heart, Governor Wood, Early Amber
Group VII S_4 S_5 : G.d. hedelfingen, Monstreuse Mezel, Black Republican, Bradbourne Black, Vic.
Group VIII S_2 S_5 : Noir de Schmidt, Peggy Rivers, Schmidt, Giant, Ursula
Group IX S_1 S_4 : Red Turk, Ursula Rivers, Black Giant, Chinook, NY 1495, Rainier.
Group X S genes unknown: Bigarreau Jaboulay, Black Tartarian
Group XI S genes unknown: None
Group XII S genes unknown: Noble
Group XIII S_2 S_4 : Florence Heart, Gaucher Black, Noir de Guben, Nutberry Black, Vic.
Group O universal donors : Stella, Vista, Vega, Seneca

There is confusion about the correct sterility allele constitution with respect to Windsor, Gold and Hudson cultivars.

Results of controlled pollination tests, PCR based S allele typing analysis and examining bloom times on sweet cherry at the New York State Agricultural Experiment Station, Geneva, USA, were compiled (Choi *et al.*, 2000). At least 20 pollen incompatibility groups were detected from currently used and newly developed cultivars (Table 3.10) and bloom times were classified into 5 groups (early, early-mid, mid, late-mid and late). Based on long-term field observations and laboratory results, a comprehensive sweet cherry pollination chart was developed (Figure 3.1). Pollination, fertilization, autogamy degree and fruit set of six sour cherry cultivars (Schatlenmorelle IR-2, Koral, Debreceni Botermo, Ujfehertoi Furtos, Karneol and Vowi) in growing condition of Poland was studied by Szpadzik *et al.* (2014). They reported that flowers of cvs. Schatternmorelle IR 2, Koral and Vowi had a higher amount of germinated pollen grains per stigma as well as the most fertilized pistils than other cultivars. Most of the germinated pollen grains per stigma and fertilized ovules were noted at the middle or the end of the bloom period, depending on season. The Hungarian cultivars (Debreceni Botermo and Ujfehertoi Furtos and Karneol) set only a few percentage of fruits by autogamy and a bit more by allogamy, and resulted in insufficient yields.

Table 3.10: S-genotypes and Pollen-incompatibility Groups of Sweet Cherry Cultivars/Selections as Tested by Controlled Pollination and PCR Based S-allele Genotyping Analyses

S-genotypes	*Cultivars/Selections*
S1S2	Early Rivers, Summit
S1S3	Regina* Van, Venus, Windsor
S1S4	Hudson, NY 1725*, Rainier, Republican, Sylvia*, Viscount
S1S5*	RN4R7T160**
S1S6*	Noble, NY 518, NY 8182
S2S3	Knights Bigarreau*, Vega*, Velvet, Victor, Viva, Vogue
S2S4*	Royalton, Sam, Schmidt, Vic
S2S6*	RN004R1T102** RN004R1T117**
S3S4	Bing, Buttners Spate Rote Knorpelkirsche*, Emperor Francis, Kristin, Napleon, Somerset, Turkey Heart B*, Ulster*
S3S5*	Burlat, Chelan, Moreau
S3S6	Gold, Governor Wood, Hartland
S4S5	RN004R7T163**
S4S6	RN004R2T175**
S5S6*	Early Lyons
S2Sz*	Cryalls Seedling, Guigne d'Annonay
S3Sy*	Schneiders
S4Sy*	NY 9801
S1Sx*	Noir de Guben, Seneca, Valera
S3Sx*	Hedelfingen, Nadino
S4Sx*	NY 1625

*: S-genotypes have been newly assigned or changed from previously classified as in Tehrani and Brown (1992) or Knight (1969); **: Seedling identification number. RN004 (Orchard identification) R7 (Row) T 156-176 (Tree).

Adopted from Choi *et al.* (2000).

Figure 3.1: Sweet Cherry Pollination and Bloom Time Chart in NYSAES*

Bloom Time	Early		Early Mid					Mid											Late Mid							Late	
	Somerset	*Viscount*	*Kristin*	*Viva*	*Chelan*	*Burlat*	*Republican*	*Royalton*	*Summit*	*Rainier*	*Napoleon*	*Valera*	*Ulster*	*Bing*	*Emperor Francis*	*Hartland*	*Schmidt*	*Van*	*Hedelfingen*	*Regina*	*Lambert*	*Windsor*	*Vogue*	*Sam*	*Gold*	*Hudson*
Somerset	X		X								X		X	X	X				▲	▲	X	▲	▲	▲	▲	▲
Viscount		X					X			X									▲	▲	▲	▲	▲	▲	▲	X
Kristin	X		X								X		X	X	X				▲	▲	X	▲	▲	▲	▲	▲
Viva				X															▲	▲	▲	▲	X	▲	▲	▲
Chelan					X	X													▲	▲	▲	▲	▲	▲	▲	▲
Burlat					X	X													▲	▲	▲	▲	▲	▲	▲	▲
Republican		X					X			X									▲	▲	▲	▲	▲	▲	▲	X
Royalton								X									X		▲	▲	▲	▲	▲	X	▲	▲
Summit									X										▲	▲	▲	▲	▲	▲	▲	▲
Rainier		X					X			X									▲	▲	▲	▲	▲	▲	▲	X
Napoleon	X		X								X		X	X	X						X					
Valera												X														
Ulster	X		X								X		X	X	X						X					
Bing	X		X								X		X	X	X						X					
Emperor Francis	X		X								X		X	X	X						X					
Hartland																X									X	
Schmidt							X										X							X		
Van																		X		X		X				

Bloom Time	*Early*		*Early Mid*					*Mid*											*Late Mid*							*Late*	
	Somerset	*Viscount*	*Kristin*	*Viva*	*Chelan*	*Burlat*	*Republican*	*Royalton*	*Summit*	*Rainier*	*Napoleon*	*Valera*	*Ulster*	*Bing*	*Emperor Francis*	*Hartland*	*Schmidt*	*Van*	*Hedelfingen*	*Regina*	*Lambert*	*Windsor*	*Vogue*	*Sam*	*Gold*	*Hudson*
Hedelfingen	▲	▲	▲	▲	▲	▲	▲	▲	▲	▲	▲								X							
Regina	▲	▲	▲	▲	▲	▲	▲	▲	▲	▲	▲							X		X		X				
Lambert	X	▲	X	▲	▲	▲	▲	▲	▲	▲	X		X	X	X						X					
Windsor	▲	▲	▲	▲	▲	▲	▲	▲	▲	▲	▲							X		X		X				
Vogue	▲	▲	▲	X	▲	▲	▲	▲	▲	▲	▲												X			
Sam	▲	▲	▲	▲	▲	▲	▲	X	▲	▲	▲						X							X		
Gold	▲	▲	▲	▲	▲	▲	▲	▲	▲	▲	▲					X									X	
Hudson	▲	X	▲	▲	▲	▲	X	▲	▲	X	▲															X

☐ All Compatible; X Incompatible; ▲: Compatible but different Bloom Sequence.

* Adopted from Choi *et al.* (2000).

Placement of Pollenizers

As discussed earlier, cultivars of one group can not fertilize each other. Only the cultivars of other group are able to fertilize, provided their flowering time overlaps and they also do not have the same sterility alleles. As such growers should carefully select cultivars from different groups to fertilize each other. For maximum fertilization, keeping in view the importance of each cultivar, and the availability of trees, planting is to be done in such a manner that plants of one cultivar face the other. In the planting arrangement shown in Figure 3.2(a), cultivar A has 60 per cent plants as it is commercially more important than cultivars B and C which are of the same importance. In Figure 3.2(b), it has been shown that in case two cultivars are of equal commercial importance and one cultivar is of less importance, then their percentages in the orchard can be kept as 40, 40 and 20 respectively. In Figure 3.2(c), it has been shown that if all the three cultivars are of equal commercial importance or if they are partially compatible cultivars of groups I, II and III, they can be planted to have 34 per cent, 34 per cent and 32 per cent plants of each cultivar.

3.2 (a) Planting plan for fully compatible cultivars of groups VI, VIII and IX where A is an imponant commercial cuiuvar

A		A	A	A	A	A	A	A	A	A	Percentage of plants
B		C	B	C	B	C	B	C	B	C	A=60
A		A	A	A	A	A	A	A	A	A	B=20
A		A	A	A	A	A	A	A	A	A	C=20
C		B	C	B	C	B	C	B	C	B	

3.2 (b) Planting plan for fully compatible cultivars of groups VI. VIII and IX where C cultivar is of less commercial Importance

A		B	A	B	A	B	A	B	A	B	Percentage of plants
B		C	B	C	B	C	B	C	B	C	A=40
A		B	A	B	A	B	A	B	A	B	B=40
C		A	C	A	C	A	C	A	C	A	C=20
A		B	A	B	A	B	A	B	A	B	

3.2 (c) Planting plan for partially compatible cultivars of groups I, II and III

	A	B	C	A	B	C	A	B	C	A	Percentage of plants
	B	C	A	B	C	A	B	C	A	B	A=34
	C	A	B	C	A	B	C	A	B	C	B=34
	A	B	C	A	B	C	A	B	C	A	C=2
	B	C	A	B	C	A	B	C	A	B	

Figure 3.2: Different Planting Plans for Effctive Pollination.

In case of sour cherries, yield is increased to an average of 67.8 per cent by planting self-fertile cultivars compared with 13.4 per cent in self-incompatible cultivars (Kramer and Scharicht, 1985).

Fruit Set

Open-pollination gives adequate fruit set for a commercial crop in cherries which ranges from 7 per cent to 50 per cent depending on the cultivar, tree management and climate (Guerrero-Prieto *et al.*, 1985; Roversi *et al.*, 1984; Roversi and Ughini, 1985). Frequent and high rainfall and low temperture during flowering often interfere pollination which can be overcome by increasing the number of bee-hives in cherry orchards. The number of hives recommended varies from three to five strong hives per hectare, depending upon local weather conditions during the pollination period. Bees exhibit very little activity below 12°C but, as temperature, rises to 20°C, there is a rapid increase in flight and forging (Thompson, 1996)

Working with nine sour cherry cultivars, Davarynejad *et al.* (2014a) recorded an average 11 per cent fruit set in open pollination compared with 9 per cent in artificial self-pollination and 2 per cent in natural self-pollination. The cv. Debreceni botermo was found as the best pollen, with 90 per cent overlap of flowering time. The highest fruit set (32.6 per cent) was recorded in cv. Oblacsinszka. Seasonal changes for fruit set and drop revealed four abscission peaks. The first fruit abortion wave appeared during the second week after pollination. Therafter, the second and third periods occurred during the third and fourth weeks after pollination. The fourth abscission peak was noted during the fourth weeks after pollination (Davarynejad *et al.*, 2014 b). Growth regulator application under unfavourable environmental conditions may compensate for poor pollination and increase the yield. The plant growth regulators not only increase the number of flowers, but also the percentage of fruit set (Ryugo and Sansavini, 1972a; Webster and Goldwin, 1984). Mixtures of plant growth regulator sprays containing GA_3 200 ppm+DPU (diphenyl urea) 300 ppm+NOXA (naphthoxy acetic acid) 50 ppm or NAA (naphthaleneacetic acid) 10 ppm, or 2, 4, 5-TP (2, 4, 5-trichlorophenoxypropionic acid) 10 ppm when applied in the spring increased fruit numbers in different sweet cherry cultivars (Goldwin and Webster, 1983). Increase in fruit weight and firmness of Bing and Lambert sweet cherries has also been reported by a spray of 20 ppm GA_3 (Facteau, 1986). On the contrary, Modlibowoska and Wickenden (1982a) reported that in sweet cherry cultivar Merton Glory, application of GA_3 alone was inadequate to induce significant increase in fruit yield. However, treatment with mixtures ot GA_3 (50 ppm) + 2, 4, 5-T (10 ppm) or GA_3 (50 ppm)+2, 4, 5-TP (10 ppm) with or without DPU (20 ppm) at cot split stage was more effective than at either petal fall or 2 weeks after cot split. Webster *et al.* (1983) stated that high-volume sprays of GA_3+2, 4, 5-TP, in the period between 50 per cent petal fall and cot split increased the yield of several sweet cherry cultivars (Table 3.11). Application of paclobutrazol in March at the rate of 0.8 g in an area of 2 x 2 metre weed-free soil under each tree of sweet cherry cultivar Early Rivers reduced the tree size by 28 per cent, the plants yielded more efficiently and fruit size was also larger than on control trees (Webster and Quinlan, 1986).

Table 3.11: Effect of Sprays of GA_3 (50 ppm) + 2, 4, 5-TP (10 ppm) at the Cot Split Stage on Yields of Hedge-Trained Van and Merton Glory Planted in 1971-72*

		Yield (kg/tree)				
		1975	*1976*	*1977*	*1978*	*Total*
Van						
	Sprayed	1.3	7.9	4.9	17.7	31.8
	Unsprayed	0.5	2.0	0.6	1.9	5.0
Merton Glory						
	Sprayed	I.{J	3.5	1.3	6.7	12.5
	Unsprayed	1.0	1.I	0.6	0.8	3.5

* Webster *et al.* (1983)

Fruit Growth and Parthenocarpy

Fruit Growth

The growth of cherry fruits follows a double sigmoidal curve, *i.e.*: there are three stages of fruit development. Epidermal fruit cells increase in number and size rapidly in the stage I, growth slows down in stage II, and thereafter increases tremendously in the stage III (Tukey and Young, 1939). During stages I, II and III the share of respiration is 32.7 per cent 70.9 per cent, and 19.9 per cent, respectively (Kappes and Flore, 1986). According to them, the need for higher respiration during stage II might be due to lignification and lipid synthesis during pit hardening and embryo development. Roper and Loescher (1987) established a positive correlation between fruit quality and leaf area per fruit. Increased leaf fruit ratio in both Lambert and Bing cultivars of sweet cherry has been reported to cause increased fruit firmness and weight (Facteau *et al.*, 1985).

Kondo *et al.* (2000a) working with Japanese sweet cherry cv. Satohnishiki noted an increase in IAA content of seeds at the early stage of fruit development [(24 days after full bloom (DAFB)] and before fruit maturation (33 DAFB) while its level in pulp rose at 22 DAF and at the beginning of fruit maturation (43 DAFB). Generally, an increase in IAA contents in seeds and pulp correspondent to changes in fruit diameter. The content of gibberellins (GA_s) increased gradually before fruit maturation (38 DAFB). Concentrations of trans-jasmonic acid (JA) and trans-methyl jasmonate (MeJA) in the pulp were found higher during early growth stage, then decreased towards harvest. In seeds, JA and MeJA concentrations gradually increased until harvest and the concentrations of both JA and MeJA were found higher in seeds than in pulp (Kondo *et al.*, 2000b).

The aroma component of sweet cherry fruit during development was studied (Xuet *et al.*, 2007). Aldehydes, alcohols and esters were found as the major constitutents. The contents of C6 aldehydes increased quickly to 84.16 per cent at the colour stage, then decreased as ripening proceeded and the content decreased to 59.20 and 55.58 per cent at the commercial and ripe stages, respectively. The aromatic aldehydes (benzaldehyde) increased as ripening proceeded and the maximum was found at ripe stage. Alcohols of sweet cherry fruit included C6 alcohols, aromatic alcohols and alcohols. The content of (E)-2- hexen-1-ol increased as ripening proceeded. Ester components included ethyl acctate, butanoic acid (butyric acid), hexanoic acid and ethyl ester, which increased as fruit ripened, Hexanol, (E)-2-hexenal, benzaldehyde, (E)-2-hexen-1-ol, ethyl acetate, hexanoic acid and ethyl ester were the characteristic aroma components of sweet cherry fruits. These aroma components started to form drastically at the colour stage and the main aroma was formed at the commercial stage, then turned bad at the ripe stage because of the presence of alcohol. The optimum hervest time of sweet cherry was therefore suggested during the commercial stage.

Many reports have shown that the application of alar, at 3000 ppm, after 2-3 weeks of flowering not only enhanced fruit ripening in different cultivars of sweet cherry but also enhanced their colour development (Modlibowska and Wickenden, 1982) Gil and Momberg, 1975: Ryugo and Sansavini, 1972b; Gil. 1974; Schurnachar and Fankhauser, 1968).Application of 2, 4-dichlorophenoxy propionde acid (2, 4-DP) or GA_3 to fruits at 15 and 23 DAFB delayed frui maturation (softening and anthocyanin accumulation) considerably (Kondo *et al.*, 2000a).

Parthenocarphy

Fruit growth in Montmorency sour cherry is highly dependent on seed development. Of the known plant hormones, only the gibberellins have significant activity in inducing parthenocarpy in sour cherry. Bukovac *et al.* (1985) demonstrated that a N substituted panthalimide induced parthenocarpic fruit development in sour cherry and that this biological activity can be enhanced markedly with NAA (Tables 3.12 and 3.13).

Table 3.12: Promotion of Parthenocarpic Growth of Montmorency Sour Cherry Fruit by Selected Growth Substances*

Sl. No.	*Treatment*		*Fruits Persisting (per cent)*		*Characteristics at Maturity*^{z}			
		Flowers Treated (No.)	*21 Days after Treat-ment*	*At Matu-rity*	*Fruit cheek diam (mm)*	*Fruit wt(g)*	*Endo-carp wt (mg)*	*Soluble Solids (Per cent)*
1.	Control, emasculated	16	0	0	...	...	....	...
2.	Control open-pollinated^z	...	...^{z}	...^{y}	21 a^x	4.3 a	182 a	13.1 a
3.	AC 94377, 1.3 x 10^{-2}M	14	100	0	18 cd	2.7 b	...^{w}	12.8 a
4.	NAA, 5.4 x 10^{-4}M	12	0	0	...	...	...	...
5.	GA_3, 5.8 x 10^{-3}M	36	42	38	...	...	...	...
6.	Treatment 3 + 4	26	73	21	19 bc	3.3 b	120 b	12.3 a
7.	Treatment 3 + 5	38	95		16 d	2.7 b	105 b	13.2 a

z: Five open-pollinated fruits were collected from each of 3 replications for comparison of fruit characteristics at maturity; X: Means within a column separated by Duncan's multiple range test, P = 0.05; y: Not determined; z: Missing value.

* Bukovac *et al.* (1985).

Table 3.13: Induction of Parthenocarpy in Sour Cherry by AC 94377 Alone and in Combination with NAA and 6-BA and Fruit Characteristics at Maturity*

Treatment[z]	*Flowers Treated (no)*[y]	*Fruit Set (Per cent)*	*Fruit Cheek Diam (mm)*	*Fruit Wt (g)*	*Endocarp Wt (mg)*
Control (open-pollinated)	599	28 bx	19 a	3.5 a	260 a
AC 94377	589	7 a	16 b	2.3 b	120 b
AC 94377+NAA	624	23 b	17 b	2.5 b	110 b
AC 94377+NAA+6-BA	547	37 b	17 b	2.6 b	110 b

z: Three applications at weekly interval of AC 94377. NAA and 6-BA at 3.3×10^{-3}, 1.1×10^{-4}M and 2.2×10^{-4}M,respectively; y: Five replications of about 120 flowers each, Fruit measurements at maturity based on 12 fruits/replication; x: Means within a column followed by similar letters are not significantly different at P=0.05. Tukey's ω test.

* Bukovac *et al.* (1985).

Maturity

The yield and the quality of the cherries are appreciably affected by the stage of maturity at which fruits are harvested. Early picking results in flat fruits and less yield as cherries usually make rapid development during the last few days before full maturity is reached. Allowing the cherries to become overripe also results in loss of weight and volume. Hence, the stage of maturity for harvesting is of great importance in maintaining consumer acceptance for both fresh and processed cherries (Marshall, 1954). However, it is not always possible to pick cherries at optimum maturity when the fruits are to be despatched to a distant market. It is, therefore, desirable that indices for the determination of degree of maturity should be employed as guides for harvesting of fruits at a time when the grower may get maximum price and the consumer a quality fruit.

Indices of Maturity

Fruit Firmness

Acceptable fruit firmness range at harvest measured with the firmTech 2 Fruit Firmness Tester was estimated as 2.56-4.71N for sweet cherry cultivars (Hampson *et al.*, 2014).

Colour

The colour development is the best guide, though it varies with the cultivar and the purpose for which the fruit will be used. For example, colour of Lambert for fresh fruit shipment should be somewhat maroon, optimum for fresh market should be dark colour and for canning fully mature (Teskey and Shoemaker, 1972).

Total Soluble Solids

In Oregon (USA) from first to last picking the juice of cv. Napoleon increased in TSS (total soluble solids) from 12.1 to 24.4 per cent and of Lambert from 11.3 to 22.2 per cent. The increase in TSS content was at a uniform rate, until the fruit had passed its prime (Hartman, 1925). In British Columbia (Canada), maximum desirable TSS by refractometer for different cultivars is for cv. Lambert-17 per cent, Deacon-18 per cent, Bing-18 per cent, Van-I8 per cent (Teskey and Shoemaker, 1972). Investigations have further shown that a deficiency of soil moisture and nitrogen decreased the level of soluble solids in the fruits (Kenworthy and Mitchell, 1952).

Acid Content

A gradual reduction of the acid content takes place during the ripening of sweet cherries. The cultivar Napoleon showed a reduction from 0.64 to 0.35 per cent and Lambert from 0.41 to 0.24 per cent (Teskey and Shoemaker, 1972).

Increase in Weight

Cherries gain in weight from the time they begin to mature until the fruit was fully ripe. In case of cvs. Napoleon and Lambert the increase in weight was 36.2 per cent. The cherries continue to gain weight even after attaining full maturity (Hartman, 1925).

Desert Quality

Fruits of early pickings of sweet cherries does not taste well. Late picked fruits may be either very sweet or insipid. So taste of the fruit may guide the time of picking. Serradilla *et al.* (2014) working with cv. Pico Colorado suggested harvesting at later stage (termed as stage 3) when the fruit showed the highest nutritional and functional quality.

Early picked cherries, in general are small, more or less flat in taste, soft in texture and subsequently display shrivelling. Those picked at a right stage have better size, colour, and appearance, firmer, more aromatic, tastier, good sugar/acid blend and with a pronounced cherry flavour. Those picked overripe show some discolouration when canned.

Fruit Cracking

Cracking of sweet cherry fruit due to rain near harvest is a major source of crop loss in the cherry industry. The disorder is characterized by a splitting of the outside layer of the cherry skin called cuticle. The splitting most commonly appears around the stem bowl, where water can accumulate, but is also seen on other areas of the cherry cuticle. The results are the same, fruit which no longer be sold for the fresh market.

In some areas with highly susceptible to cracking cultivars, such as Bing, losses from cracking may exceed 50 per cent of the crop. In individual orchards, losses of 85 per cent or more have been reported (Teskey and Shoemaker, 1972). Sweet cherry cultivars differ in their susceptibility to cracking because of differences in the rate of water absorption and capacity for expansion of the peripheral tissues to accommodate the increased fruit volume that results when water is absorbed. Demrsoy and Demrsoy (2004) studied the epidermal characteristics of fruit skin of eight sweet cherry cultivars (Turkoglu, Izmit, Honey heart, Arap, Otabatmaz, Bing, Bella di Pisfoia and 0900 Ziraat) in relation to fruit cracking. The cultivars 0900 Ziraat and Bella di Pisfoia having the thickest cuticle had the lowest cracking index. There was a negative correlation between the cuticle thickness and fruit cracking. There were no correlations relationship between fruit cracking and thickness of epidermis and subepidermis and number of subepidermal cell layers.

The basic mechanism causing this disorder, although not fully understood, seems to be the rapid increase in water absorption by the fruit. This increment on the water content of the fruit may be the result of the direct water absorption through the vascular system. Measham *et al.* (2009, 2010) suggested that the type of cracking (*i.e.*, stem, apical and side) may not be an expression of the severity of a single process, but may rather be the manifestation of the different water uptake pathways during the onset of the problem. Recent results from University of Tasmani, Australia have broadened the understanding of this fruit cracking phenomenon through a number of novel findings (Measham *et al.*, 2014). It was discovered that two water uptake pathways result in cherry fruit cracking and that each pathway results in particular crack types. Results from this study show that development of apical and stem end cracks are induced by skin surface wetting while deep cracks on the side of the fruit are induced by water moving via the vascular system. This knowledge provide the basis for developing more effective management strategies. They further identified that the extent and type of fruit cracking are impacted by cultivar, season and crop load. Nine cultivars were monitored over four years and it was established that each cultivar had a particular tendency for crack type. Crop load was found negatively correlated with fruit cracking. In addition, the amount or distribution of rainfall in the weeks prior to harvest did not significantly affect cracking incidence, yet season was highly influential, suggesting more than a direct rainfall effect. They also noted strong indications of diurnal rhythms and ambient environmental conditions influencing development of cracking.

Both the degree of permeability and plasticity of the fruit skin is subject to modification. Any treatment that decreases the rate of water absorption, or increases the capacity of the fruit tissues to stretch without rupturing reduces the amount of cracking. It has been reported that spray of $CaCl_2$ (calcium chloride) at 300g to 350g per 100 litre water applied at weekly interval before harvest checked the fruit cracking in Bing and Lambert (Larsen *et al.*, 1983: Tukey, 1984). Hermann and Feucht (1984) however, observed no relatronship between the morphology of fruit skin and its sensitivty to cracking. They further reported that spraying of $CaCl_2$ increased the calcium level in the fruits but did not affect cracking. Spraying of GA_3, (2000 ppm) on Hedelfingen and Ekero cultivars - 3weeks before harvesting reduced the amount if fruit cracking caused by heavy rainfall following drought (Nyhlen, 1976). Tabuenca (1985) reported that the spray of NAA (1 ppm) at 25-30 days before harvest reduced cracking in cultivars Bing, Marmotte, Mollar de Caceres and Napoleon by 25-55 per cent but had no effect on fruit cracking in cultivars Daiber and Van. Sprays of antitranspirant C 5-6432 at 1.5 per cent reduced fruit cracking in sweet cherry cultivar Durone dell Anella (Intrieri, 1974). Significant reduction in fruit cracking was reported by spraying with a lipophilic hydrophobic compound which formed a film over the fruit (Torres *et al.*, 2009). The use of physical and mechanical barriers to prevent rainwater to touch the fruit, such as polyethylene covers significantly reduce fruit cracking (Lugli *et al.*, 2001). Ma (2007) reported significant reduction of cracking by spraying of 0.7 per cent sodium silicate applied weekly (three times) from colour turning stage until 1 week before harvest. Silicon is believed to increase the elasticity of the cell wall and reinforcing structures, forming a protective layer that minimize the penetration of water and fungal pathogen. In the Pacific Northwest United States, due to high labour costs, when fruit cracking exceeds 25 per cent at harvest, fruits are not picked. Oregon State University Horticulture and Pharmacy Faculty have collaborated in producing and patenting a novel, elastic, organic biofilm, "SureSeal" which significantly reduced sweet cherry fruit cracking by up to 250 per cent (Kaiser *et al.*, 2014). Formulations of SureSeal are hydrophobic and consist of a copolymer of complex carbohydrates, phospholipids and calcium. Two sprays of 1 per cent SureSeal applied at straw colour and again ten days later, reduced fruit cracking considerably.

Cultivars Resistant to Cracking

The most effective way to avoid fruit cracking is to plant highly resistant cultivars. As desired resistant cultivars may not be available, selection may be made from the resistant, fairly resistant and moderately resistant cultivars listed below:

Resistant	*Fairly Resistant*	*Moderately Resistant*
Attika	Black Giant	Compact Lambert
Early Rivers	Black Republican	Compact Stella
Hedelfingen	Deacon	Corum
Lamida	Gil Peck	Merton Bigarreau
Lapins	Hudson	Mona
NY 1725	Jubilee	Rainier
OSC 6	Noble	Royal Ann
Regina	NY 1495	S-4 c-15-25
Sam	Kristin	Salmo
Sue	Vic	Summi
Victor	Ulster	Van
Windsor	Schmidt	Venus
Lapins	Star	
Viscount	Stella	

Harvesting

After pruning, harvesting of the fruits is the most expensive operation associated with the production of cherries. Harvesting of sweet cherry is a hand operation, for which labour is usually required, so as to begin picking before the fruit is fully mature, and complete it before the fruit is overmature. Sweet cherry is picked with tile stems (pedicels intact) unless they are to be delivered for processing. Pulling the cherries from the supporting stems causes a slight wound from which the exuded juice sticks on the surface of the fruits and adversely affects their appearance. Moreover, fruit decay starts from the wound (Marshall, 1954). Pitting damage is one of the most important causes of quality loss during the postharvest life of cherries. Candan *et al.* (2014) stated that pitting sensitivity is cultivar dependent and affected by temperature and drop height. They suggested that to reduce pitting the maximum drop height during harvest should be 15cm, where fruit temperature is usually above 20°C and 5 cm in the packing lines where the fruit temperature is between 10 and 2°C.

Nanii and Kotaman (1974) used ethrel to facilitate the mechanical harvesting of sweet cherries. They found that the effectiveness of ethrel (450 ppm) varied with the cultivar. Kollar and Scortichini (1986) found better removal of fruits at harvest with a Schaumann mechanical shaker than under control when sweet cherry trees were sparayed with ethrel (400 ppm) 2 weeks before harvesting. Ethrel produces more concentrated colour and even ripening of fruits. Pommier *et al.* (1987) recommend ethrel treatment 4 weeks before maturity instead of 2 weeks. Post blossoming spray of gibberellic acid (GA_3) or aminoethoxy vinylglycine (AVG) delayed harvesting time of cheery cvs. Stella and Colney (Webster *et al.*, 2006). Spraying of AVG at 50mg/l at 25 days of commercial harvet delayed harvesting of cherry cv. 0900-Zirrat grafted on *P. avium* (Cetinbas *et al.*, 2012)

Sour cherry is used for processing. It is, therefore, picked by pulling or stripping the fruits from the stems leaving the latter on the trees. This method of picking is more rapid than picking with stems. Moreover, it saves the labour cost of removing the stems in the processing units. Many processing units do not accept fruits with stems because of the time and expense involved in removing the stems Ethrel is used by some progressive growers to collect fruits by shaking the trees. Kiss (1982) reported that mechanical harvesting of sour cherries is cost effective only with traditionally cultivated (8 x 5 m spacing) trees whereas manual harvesting gives higher return in the intensive systems, *i.e.* bush (5 x 4 m), hedge (2.5 x 3 m) and spindle (5 x 2 m).

Yield

The yield of sweet cherry is extremely variable from year to year and poor yield is of frequent occurrence. The two main causes of these poor yields are inadequate fruit set and/or excessive fruitlet abscission. Although the processes affecting fruit set are not fully understood, it is thought that inadequate pollen transfer, poor quality pollen and/or ovules, slow pollen-tube growth and delayed fertilization may all seriously reduce fruit set. Cold, wet or windy weather at or around blossom time can, by reducing pollen viability, pollen transfer by insects, or speed of pollen-tube growth reduce fruit set. Low temperatures and conditions unfavourable for tree growth may also promote the abscission of fertilized fruitlets (Webster *et al.*, 1983).

Sweet cherry trees grow larger than sour cherry trees and bear heavily. A good yield is 6.3 to 9 tonnes or more per hectare on standard trees (Teskey and Shoemaker, 1972). However, in high density trials in France, 5-year-old plants of Bigarreau Hatif Burlat on SL-64 rootstock planted at 1 x 4 metre (2500 plants/ha) yielded 30 tonnes per hectare. In another trial in France, 5-year-old plants of Bigarreau Nepoleon planted at 50 cm x 4 m (5000 plants/ha) yielded 80 tonnes per hectare (Azad, 1981). In an investigation on 226 sweet cherry cultivars for 11 years,

Revin (1986) recorded the mean yield of high yielding cultivars as 57 to 92 kg/tree (or 13 to 21 t/ha). Chang *et al.* (1987) found that fruit number, fruit weight, number of lateral buds and spurs and fruit set were the most important yield components of sour cherry.

Storage

Cold Storage

The storage life of sweet cherry fruits are 18 days at 0°C and 85 per cent RH (Hriber *et al.*, 1986). Agulheiro-Santos *et al.* (2014) successfully stored 'Sweetheart' cherries for 21 days at 10°C and 95 per cent RH without significant loss in quality. The fruits should be cooled down immediately after harvest to reduce surface shriveling decay and keep greener stem with firm fruit (Manganaris *et al.*, 2007). Conte *et al.* (2009) observed that during cold storage, TSS decreases or remain constant while TA decreases, resulting an increase of TSS/acid ratio.

Modified Atmospheric Condition

Sweet cherry cvs. Napoleon, Karabour and Stella can be stored at 5 per cent O_2 and 5 per cent CO_2 (Eris *et al.*, 1993) while for cvs. Helden figen and Gemedroff, the recommended atmosphere is 3 per cent O_2 and 5 per cent CO_2 (Ionescu *et al.*, 1978). The cv. Sweetheart requires 5 per cent Co_2 and 10 per cent O_2 or 4 per cent CO_2 and 6 per cent O_2 (Meherluk *et al.*, 1997).

Modified Atmosphere Packaging

Storage of cherry fruits in modified atmosphere packaging (MAP) has been shown to be effective in extending the storage life. When handling and packing fresh sour cherry fruit, great care must be taken to assure adequate sanitation. Fruit have a short shelf life, as postharvest deterioration occurs very quickly if adequate precautions are not taken to prevent it. Reduction in the amount of colour, acid and firmness, incidence of decay and stem browning are potential benefits of MAP. Davarynejad *et al.* (2014), in their study with sour cherry cvs. Erdi Jubilem and Erdi Botermo, stored the harvested fruits at 0°C in MAP (15 per cent O_2, 10 per cent Co_2 and 75 per cent N_2) for 6 weeks. They observed that weight loss was reduced significantly under MAP compared with ambient air packaging. Fruits packaged with MAP films maintained higher flesh firmness than did control. During the postharvest period, both titratable acidity and pH increased slightly. Skin colour was lighter (higher 'L') and less blue (higher 'b') under MAP than usual air packing. Preharvest Ethepn application, however improved fruit colour (became darker) after 6 weeks of storage under MAP (Aryanpooya *et al.*, 2014). Zoffoli and Rodriguez (2014) selected 'Sweetheart' fruit from a packaging line and packed in three commercial modified atmosphere bags, including passive and active systems. The CO_2 and O_2 concentrations inside of active and passive MA bags were similar and varied between 6 to 9 KPa and 10-13 KPa after two days of packaging, respectively. It was reported that all MAP systems, reduced the incidence of decay, however the incidence of decay was not controlled after 3 days at 20°C. Remon *et al.* (2000) suggested to store 'Burlat' cherries following the procedure-harvest at red colour stage, package in air using low-density polyethylene (LDPE) bags and store at 2°C. Under these conditions, red coloured 'Burlat' cherries can be kept in an excellent commercial state for 3 weeks.

Effect of 1-MCP

Treatment with 1-MCP at 1.5 µl/l of cherry fruits at commercial maturity and storing at 0 ± 1°C and 90 ± 5 per cent RH can extend the storage life for 60 days. 1-MCP treatment reduced endogenous ethylene production, malondialdehyde content and polyphenol oxidase activity (QingRui *et al.*, 2011). Cold storage (0°C) of 1-MCP treated fruit maintain fruit quality and firmness, reduce respiration rate, prevent TSS and soluble protein from degradation. They also significantly affected the fruit cell wall metabolism, inhibiting the increase of water-soluble pectin and ionic-soluble pectin and the degradation of covalent soluble pectin and hemicelluloses (JianMei *et al.*, 2011). Pasalic *et al.* (2013) treated the 'Burlat' sweet cherry with 1-MCP at 0.05 to 0.5 ppm for 20 hours at 5°C. After treatment, the fruits were stored in the cold room (3°C) in normal atmosphere for 15 days, followed by 5 days storage at room temperature. They observed lowest fruit weight loss by treatment with 1-MCP at 0.5 ppm while the average fruit firmness was the lowest in the fruits treated with 0.05 ppm 1-MCP. The TSS content of fruit was highest with 1-MCP at 0.1 ppm.

Calcium Treatment

Postharvest dipping of fruits in 40g/l solution of $CaCl_2$ (containing 0.25 g/l Citowelt non-ionic surfactant) for 2 hours at ambient temperature (20°C) maintain fruit firmness after 30 days in storage (Traversi and Massignan, 1987). Wang *et al.* (2014) working with sweet cherry cvs. Sweetheart and Lapins reported that $CaCl_2$ mixed in hydro cooling water (0.2 to 2 per cent at 0°C for 5 min) increased fruit tissue Ca content, reduced respiration rate, ascorbic acid degradation, membrane lipid peroxidation and enhanced total phenolic content and total antioxidant capacity. This resulted increase in fruit firmness and pitting resistance and decrease in titratable acidity loss and decay of both cultivars. Pedicel browning was inhibited by $CaCl_2$ at 0.2 and 0.5 per cent, but increased by higher rates at 1.0 and 2.0 per cent.

Effect of Coating

Freshly harvested sweet cherry fruits can be stored for 15 days at room temperature without any change in quality by treatment with Semperfresh at 3.0 per cent. It was found effective to reduce weight loss, improved firmness, ascorbic acid content, retain acceptable flavour, taste and colour during storage (Ali *et al.*, 2014).

Pests

European Cherry Fruit Fly (*Rhagoletis cerasi* L.)

The is the most important pest of sweet and tart cherries. Once the skin of fruits becomes soft enough to penetrate, adult female insert egg with their ovipositor, and larvae develop inside the fruits. The result is 'wormy' fruit which is not marketable. It is difficult to determine whether a fruit is infested until the larva exists through a hole that it chews. Insectiside treatments should be timed to prevent adult mating and/or female fruit flies laying eggs in developing fruit. To monitor the adults of fruitfly pheromone traps and yellow sticky traps are used (Ozolina and Apenite, 2012). Bioinsecticides controlled cherry fruit fly up to acceptable infestation levels when the pest pressure is not very high. Organophosphate insectide gave the highest efficacy in reducing fuit infestation (Caruso *et al.*, 2012). Hohn *et al.* (2012) evaluated the efficacy of insectides thiamethoxam (Actara) and actamprid (Gazelle SG) to compare with dimethoat in commercial orchard trials on 20 Swiss fruit farms. Dimethoat and thiamethoxam were each applied once three weeks before harvest. Actamiprid was applied twice an interval of 14 days and with a waiting period of again two weeks. All three insecticides resulted in a sufficient efficacy and therby damage was below the limit required in practice. Insect exclusion netting (1.3 mm net) was found equally effective in protecting the crop from damage if the nets were kept closed from colour change of the fruit (from yellow to pink) until harvest. Ground covers and mulches around the base of trees can prevent larvae from burrowing in to the soil to complete development into the pupal stage. There are some natural enemies that will attack fruit fly life stage, such as parasite wasps that lay eggs on larvae within fruit, but control has not been shown to be significant.

Flat Headed Peach Tree Borer (*Sphecnoptera lafertei*)

During the sixties this borer appeared as a serious pest on cherry and peach trees in Himachal Pradesh, India. Conventional trearment of individual borer holes on the main trunk with fumes of petrol or PDCB (*para*-dichlorobenzene) was not effective and the pest continued to multiply unchecked. Subsequently, swabbing treatment of the stem and main trunk to a height of about 2 metre from the ground with high concentration of one of the peristent insecticides like aldrin, dieldrin, parathion or diazinon was found effective against freshly hatched larva. The first treatment is given in February-March with 0.5 per cent concentration of insecticide followed by 0.25 per cent insecticidal concentration of swabbs during May-June, August and October (Anon., 1976).

Black Cherry Aphid (*Myzus cerasi*)

This sucking insect or plant louse occurs in large numbers on sweet cherry and causes serious injury unless controlled. The aphids attack the leaves as soon as the buds burst and they continue to feed as long as the growth remains succulent. Infested sweet cherry leaves curl and cluster, then they turn yellow and die. The aphids also feed on the fruit and secret a sticky honeydew that makes the fruit unfit for sale unless it is washed (Davidson, 1973)

Winged adults migrate the plants in the mustard family, their summer hosts. Controlling weeds, especially mustard types, helps reduce insect survival. Natural predators like, ladybird beetles and parasite wasps are also useful. Dormant oil sprays can be used during delayed dormant period to control overwintering aphid eggs. Neem oil products and potassium laureate (insecticidal soaps) applied directly to the aphids can reduce populations. Spraying of rogor, metasystox or parathion 1 ml per litre of water before or at bud-swell stage, has been suggested to control the pest. In case of severe attack, the spray is repeated after petal fall.

San Jose Scale (*Aspidiotus perniciosus*)

San Jose scale infests many fruit trees except sour cherry. Insects suck the sap from the bark, leaves, and fruit. When they are abundant, they may almost cover the entire bark, and by sucking the sap weaken and eventually kill the tree. Infested fruit has a spotted or mottled appearance, usually with a reddish area around each scale (Davidson, 1973). Spraying plants with mineral oil at 3-6 ml/100 l water during dormancy (Singh, 1976) is suggested.

Diseases

Bacterial Canker (*Pseudomonas syringae pv. syringae*)

This disease is most serious in sweet cherries, but it also affects all other stone fruits. It causes bare branches or leggy condition by extensive killing of leaves and flower buds and spurs on 2-year or older wood. Cankers, dark sunken areas from 12 mm in size around dead buds or spurs to a very large areas on main branches exude gum. The bacteria destroy the conducting tissues just below the bark. Dark brown streaks of dead tissue may extend up and down the branches for a considerable distance beyond a visible canker. Leaf and flower spurs or entire branches wilt and die. In the spring, leaf infection develops as small purplish spot surrounded by a ring of light green tissue. These spots become irregular in shape, turn brown and drop out leaving the leaves very tattered (Davidson, 1973). The cultivars Rainer and Regina appear more resistant than Sweetheart and Bing (Spotts *et al.*, 2010). Growers should make heading, pruning and scoring cuts only during dry weather. Pruning in summer can help to keep infection low. Field observations indicate that stressed trees show higher level infection than those with a moderate amount of vigour.

Only limited success in controlling bacterial canker has been achieved by use of copper sprays from early leaf fall until early spring, and sprays of streptomycin have been even less effective (English *et al.*, 1979). Agarwal (1970) recommended the application of Mashobra paste during dormancy on cleaned, localized wounds to control the disease. If possible, resistant cultivars such as Lambert, Poznanska and Bladorozowe should be grown (Lyskanewska, 1976).

Verticillium Wilt (*Verticillium albo-atrum* and *Verticillium dahliae*)

This disease causes wilting and defoliation mainly in trees of less than 10 years old. It is most common in sweet cherries, apricot and peaches. Infected trees remain stunted and unproductive for several years. The disease is sometimes fatal in cherries. The first indication of infection is the wilting of foliage on the whole tree or more commonly, from a part of the tree. Later, the leaves turn pale green or yellow and drop. Leaf fall may be sudden and is often the first symptom noticed. The trunk, branches, and twigs show no external evidence of the disease. However, a cross section of an infected branch will show a ring or partial ring or dark brown wood.

The fungus lives in the soil, where it may persist for 15 years or more. Infection takes place through the feeding rootlets. It goes into the trunk and then into the branches where it may girdle the stem or clogg one side of the branch (Davidson. 1973).

It is always desirable to select well-drained soil for planting cherry orchards. This will help in minimizing many soil-borne diseases. It is also advisable not to plant potatoes, tomatoes, egg plants, peppers, strawberries and raspberries in a young orchard, as they favour increase of fungus in the soil. A thick layer of grass or legume cover crop is recommended in areas where this disease is a problem. Before planting, the pits may be fumigated with 3 per cent formalin.

Crown and Root Gall (*Agrobacterium tumefaciens*)

In this disease, the crown and roots are usualy affected, though aerial galls are also not rare. The disease is recognized by the wart-like tumors or swellings at the crown or on the roots. The galls are at first light in colour and soft in texture but later turn dark and become hard. They may vary from the size of a pea to that of a soft ball and are usually irregular in shape. The bacterium of this disease lives in the soil and attacks crown and roots that have been injured (Davidson, 1973).

It is advisable to purchase plants from a disease-free nursery. The plants may be dipped in a 0.4 per cent copper sulphate solution for one hour before planting. In case of infected plants apply Chaubattia paste after superficial heating of the localized lesions with blow lamp. Use of antagonistic species of bacterium named *Agrobacterium radiobacter* strain K1026 when applied to a wound inhibits the development of crown gall caused by *A. tumefaciens*.

Cherry Leafspot (*Cercospora circumscissa*)

The disease causes early defoliation of sweet and sour cherry trees. Round, light brown necrotic spots with brownish red edges appear on the leaves which enlarge and later drop out leaving shot holes. The pathogen overwinters as stromata on debris of diseased leaves. In the spring, these stromata develop characteristics conidiophores and conidia. Four applications of benomyl or thiabendazole from mid-April at monthly interval have been found effective in controlling the disease (Sztejnberg, 1986).

Brown Rot (*Monilinia fucticola, M. laxa*)

Brown rot is one of the most important disease of sweet and Duke cherries in temperate fruit-growing regions. Small round spots develop all over the affected fruits causing severe losses particularly during hot and humid weather conditions at harvest. Cox *et al.* (2011) reported that planting suffering from fruit rot (*M. laxa*) had fruit lesions that began as pale brown, soft lesions with indiscriminant margins that covered 15-85 per cent of the fruit serface area. The disease is controlled by adopting sanitation practices in combination with protective fungicide programmes (Zehr, 1982). At the end of petal fall, a wettable sulfur fungicide can be applied to the tree to provide partial protection of developing tissue.

Viral Diseases

These diseases are systemic troubles, the infective agent being present in the sap of trees. Generally, the X disease affects only a part of the tree. Healthy and affected fruits can develop side by side, even in the same cluster. Sweet cherries on *P. mahaleb* rootstock wilt and decline rapidly after infection (Davidson. 1973). Angela, a sweet cherry cultivar is resistant to this disease (Wadley. 1975).

Mottle leaf causes yellowish mottling of the leaves and serious stunting of Bing cherries. Most of the other cultivars are highly tolerant to this viral diseases.

Breeding and Improvement

Cherry is a cross-pollinated crop. The phenomenon of cross pollination has resulted in a considerable diversity which has been utilized for its improvement. Most of the breeding work in cherries has been done in England, France. USA, Canada. USSR. Germany, Sweden, Italy. Romania and Australia. Through regular hybridization, starting from the year 1900 and up to 1971, about 84 cultivars have been released allover the world (Fogle, 1975). Similarly, some important rootstocks, such as, F12/1, Colt, SL-64 and SL-405 have been bred and are being used. A number of rootstocks of MM (Mazzard x Mahaleb), GM (Grand Manil) and OCR series have been released. In India, *P. cornuta* gave better performance as a rootstock with sweet cherry cultivar Emperor Francis, compared with *P. cerasoides* (Singh *et al.*, 1971).

Qualitative improvement and a significant acceleration of breeding sweet cherry is possible using marker assisted selection. The genetic research of agronomically important traits in sweet cherry are being carried out by different imstitutes of the world. Molecular studies are still ongoing, but already known genes that determine self-compatibility, self-incompatibility, fruit size, fruit quality *etc.* have been documented.

Objectives for Breeding of Cultivars

Resistance to Rain Cracking

All over the world rainfall during the period of cherry ripening results in severe fruit cracking. In India, also where sweet cherry is grown in some pockets of northwestern hils, pre-monsoon rains cause heavy fruit loss through cracking. It has been observed that cultivars with soft flesh are resistant to cracking. As such soft flesh cultivars can be utilized in breeding crack-resistant cultivars, Though the majority of crack-resistant cultivars, such as, Early Rivers, Sam, Victor and Windsor are chance seedlings, some crack-resistant cultivurs such as, Sue and Lapins have been produced through hybridization (Sansavini and Lane, 1983).

Fruit Quality

Diameter of sweet cherry fruit is an important character in the choice of the consumers. The size of sweet cherry fruit is considered to be a quantitative trait and as a result, it is difficult to analyze using classical genetic methods. The colour of exocarp and the fruit varies in the sweet cherry. Inheritance of colour in sweet cherry has been studied. The factor of flesh colour (F) and the main factor of skin colour (A/a) are key determinants of fruit colour when red colour shows incomplete dominance over yellow. The third factor of skin colour (B/b), which may shows blush, is epistatically masked by a dominant allele *A*. Wunsch *et al.* (2014) discovered the differential expression of *PavMYBIO* in white and red cultivars of sweet cherries affecting the anthocyanin levels.

Cold Hardiness

Due to fluctuations in temperature at the time of flowering, fruit set is adversely affected. It is therefore, desirable to breed cold-hardy cultivars. The rain-shadow districts of Kinnaur, and Lahul and Spiti in Himachal Pradesh, India are ideal for cherry cultivation. Since, 'Bing' has been reported to impart hardiness to its progenies, it can be utilized in breeding cold-hardy cultivars (Tehrani and Kerr, 1968). Russian and Canadian scientists are using *P. fruticosa* for improving winter hardiness of sour cherry (Enikejew, 1968; Bors, 2005).

Self-compatibility

Most of the sweet cherry cultivars are self-incompatible. As such one or two cross-compatible cultivars are normally planted with the main cultivar to get good fruit set. However, in case of bad weather at the time of flowering, bee activity is very much restricted and they fail to do pollination. Less or lack of pollination resulted in poor crop. It is, therefore, desirable to breed self-compatible cultivars which will not only help in increasing fruit produciton, but will also do away with the necessity of pollenizers. Cultivars such as Sunburst and Lapins which are self-fruitful have been produced from the cross between Van x Stella (Sansavini and Lane, 1983). This gives a clue to the possibility of producing more self-fruitful cultivars by crossing different desirable parents. Following the description of the first six cherry S-alleles (S_1-S_6) (Mathews and Dow, 1969), biochemical and molecular analyses indicated the presence of another 10 *S*-alleles (Boskovic and Tobutt, 2001). Further 12 alleles (S_{17}-S_{22} and S_{27}-S_{32}) have been identified in Belgium and UK wild cherries (DeCuyper *et al.*, 2005; Vaughan *et al.*, 2008). While hundreds of sweet cherry cultivars and wild accessions have *S*-genotyped from Eastern Europe to North America, only limited data are available on accessions native to the centre of origin of this species (Ipek *et al.*, 2011).

Since sour cherries are mainly grown for processing the greatest emphasis is laid on producing productive cultivars. Extension of harvest period and adaptability to mechanical harvesting are the other major objectives. Desirable plant characteristcs are dwarf stature, heavy spur development and bearing only on spurs (Iezzoni, 1984).

Characterization of Cultivars

Simple sequence Repeats (SSRs) were used to evaluate sweet cherry cultivars of Iran. A total of 120 alleles were detected by 11 SSR primar pairs, with an average of 10.91 putative alleles per primer combination which ranged from 4 to 18 for maximum and minimum alleles that were detected with primars. The size of alleles ranged from 105 to 291 bp. Most of the cultivars were grouped according to their pedigree. The results indicated that sweet cherry has a high level of genetic variation in Iran and therefore Iran might be regarded as one of the major locations in the world where a wide variety of this crop is distributed (Abdossi *et al.*, 2011)

Objectives for Breeding of Rootstocks

The main rootstocks in use all over the world are either seedlings or selections of Mazzard and Mahaleb. However, in Himachal Pradesh and hills of Uttar Pradesh, India some nurseries are producing sweet cherry plants on 'Paja' (*P. cerasoides*) which is incompatible with sweet cherry. Due to the demographic pressure, the best land all over the world is being utilized for cereals and other high income crops, while only marginal soils are used for cherries (Perry, 1987). Keeping on the agroclimatic conditions in view, the breeding of rootstock is being done with the following objectives.

Adaptability to Drought and Wet Soils

Most of the area which can be brought under sweet cherry in the northwestern Himalayas is rainfed and of marginal fertility. As such it is desirable to introduce Mahaleb based rootstocks, such as, SL-64 and SL-405 which are drought resistant and highly productive (Azad, 1981). Introduction of new hybrid rootstocks of M x M (*P. avium* x *P. mahaleb*) series which are semi-dwarfing to dwarfing may also prove very useful under rainfed conditions of the northwestern Himalayas (Westwood *et al.*, 1976). Colt, which is a dwarfing rootstock may be used where irrigation could be given. At the same time, it will be useful to study the rootstock performance of 'Jaman' (*P. cornuta*) which has already been reported as a compatible rootstock for sweet cherry cultivar Emperor Francis (Singh *et al.*, 1971).

In the wet and humid conditions of the northeastern Himalayas, Mazzard selection F12/1 is liekly to perform well as rootstock. Some clones of *P. avium* x *P. cerasus* have also been found to be tolerant to wet soils (Parnia *et al.*, 1982). According to Webster (1980), *P. cerasus* can also be tried as a rootstock for wet and heavy soils. *P. nepaulensis* and *P. jenkinsii* which are wild species in the northeastern Himalayas, may also be tried as rootstocks for sweet cherry (Randhawa, 1987).

Dwarfing

Interspecific hybridization has proved very useful in producing dwarfing and precocious rootstocks. Colt (*P. avium* x *P. pseudocerasus*). M x M clones (*P. avium* x *mahaleb*), Oppenheim (*P. fruticosa* x *P. cerasus*) and G.M. clones (*P. incisa* x *P. serrulata*) are some of the examples of dwarfing and precocious rootstocks which may be introduced for trial in other countries producing cherries. Walther and Sauer (1985) produced dwarf *P. avium* mutants using cobalt-60 however, most of the mutant were found unstable chimeras. James *et al.* (1987) produced dwarf mutants of Colt rootstock using colchicines. The hexaploid (6n = 48) Colt showed 10 per cent less vigorous and showed no difference in yield efficiency of cv. Lapins (Hrotko *et al.*, 2006).

Resistance to Diseases and Pests

There is an urgent need to look for resistance to fungal, bacterial and viral diseases in other species of *Prunus*, and incorporate the same in the commercial cherry rootstocks. Substantial tree losses have occurred in California on Mahaleb rootstocks from *Phytophthora* root-rot pathogen where soils have poor internal drainage (Mircetich *et al.*, 1976). Search for resistance against root-rot pathogen in *P. avium* and *P. cerasus* needs to be intensified. Similarly, incidence of *Prunus* stem pitting was repeatedly found in sweet cherry trees propagated on Mahaleb and Stockton Marella, while no symptom was found in trees on Mazzard in the same orchard (Mircetich *et al.*, 1977). Since, Mazzard selection F12/1 and *P. cerasus* have shown tolerance or resistance to bacterial canker (Cameron, 1971; Larsen and Tukey, 1982), crown and root-rot due to *Phytophthora* spp. (Mircetich *et al.*, 1976) and to root-knot nematodes (Day and Tufts, 1944), these may be utilized in breeding programme to further improve the new dwarfing and semi-dwarfing rootstocks. Most sour cherry cultivars are susceptible to leaf spot. The tetraploid interspecific hybrids 'Almaz' and 'Paljus', though have resistance to leaf spot but show low fertility. The tetraploid wild species *P. maackii* and the diploid *P. canescens* could be used as resistance donors for sour and sweet cherry breeding (Schuster, 2004).

Important Sweet Cherry Cultivars Released from different Countries after 1990

Country	*Cultivar*	*Parentage/Origin*
Australia	Sir Douglas	Stella x Vega
	Sir Hans	Stella x Vega
	Sir Tom	Black Douglas x Stella
	Dame Nancy	Stella open pollinated
	Dame Roma	Black Douglas x Stella
	Sir Don	Black Douglas x Stella
Canada	Cristalina	Star x Van
	Samba	2S- 84-10 x Stella
	Sandra Rose	2C-61-18 x Sunburst
	Santina	Stella x Summit
	Satin	Lapins x 2N-39- 05
	Skeena	2N-60-07 x 2N-38-32
	Sonata	Lapins x 2N-39-05
	Sovereign	Sweetheart Open Pollinated
	Staccato™	Sweetheart Open Pollinated
	Stardust ™	2N- 63-20 x Stella
	Sweetheart	Van x Newstar
	Summer Jewel	2C-61-18 x 2D-28-30
	Symphony	Lapins x Bing
	Tehranivee	Van x Stella
	Vandalay	Van x Stella
China	Qinzao	Dajin Open Pollinated
	Qinxia	Governor Wood Open Pllinated
	Qinhong	Dajin Open Pollinated
France	Bellise	Starking Hardy Giant x Burlat
	Earlise® Rivdel®	Starking Hardy Giant x Burlat
	Ferbolus	Hedelfinger x Reverchon
	Fercer	Starking Hardy Giant x O.P.
	Ferdiva	Fercer x O.P.

Country	Cultivar	Parentage/Origin
	Ferdouce	Rainer x Fercer
	Fermina	Vittoria x Clone INRA
	Ferobi	Burlat x Fercer
	Fertard	Sunburst Open Pollinated
	Folfer	Fercer x O.P.
Germany	Alma	Rube x Allers Spate
	Bianca	Rube x Allers Spate Knorpel
	Erika	Rube x Steckmanns Bunte
	Johanna	Schneiders Spate Knorpel x Rube
	Nadino	Spansche Knorpel Open Pollinated
	Namare	Grobe Schwarze Knorpel Open Pollinated
	Namati	Bopparder Kracher Open Pollinated
	Namosa	Farnstadter Schwarze Open Pollinated
	Naprumi	Hedelfinger x St. Charmes
	Oktavia	Schneiders Spate Knorpel x Rube
	Regina	Schneiders Spate Knorpel x Rube
	Valeska	Rube x Stechmanns Bunte
	Viola	Schneidero Spate Knorpel x Rube
Hungary	Alex	Van x John Innes 2420
	Aida	Moldvaifekete x OP of Germersdorfi
	Anita	Trusenszkaja x OP of Germersdorfi
	Kavics	Germersdorfi Orias x Budakalaszi
	Petrus	Burlat x Stella
	Sandor	Burlat x Stella
	Vera	Ljana x Van
Italy	Black Star	Lapins x Burlat
	Blaze Star	Lapins x Durone Compattodi Vignola
	Early Star[R] Panaro 2	Burlat x Stella
	Grace Star	Burlat Open Pollinated
	Lala Star	Compact Lambert x Lapins
	Sweet Early® Panaro 1	Burlat x Sunburst
	Bargioni 137	1SF 123 x Caccianese
The Czech Republic	Marta	Kordia x Early Rivers
	Tamara	Krupnoplodnaja x Van
	Techlovan	Van x Kordia
	Vanda	Van x Kordia
	Vilma	Kordia x Vic
	Justyna	Kordia x Starking Hardy Giant
The USA	Anderson™	Wederscher Markt Open Pllinated
	Andy G's Son	Sports of Early Burlat

Country	Cultivar	Parentage/Origin
	Black Gold™	Starks Gold x Stella
	Black York™	Giant x Emperor Francis
	Blushing Gold™	Yellow Glass x Experor Francis
	Chelan	Stella x Beaulieu
	Columbia™	Stella x Beaulieu
	Earlisweet	Stella Open Pollinated
	Early Garnet™	Garnet x Ruby
	Early King	Sport of King
	Early Robin™	Whole tree mutation of Rainier
	Giant Ruby™	Large Red x Ruby
	Glacier™	Stella x Burlat
	Glenare	Tulare Open pollinated
	Glenred	Tulare x Brooks
	Glenrock	Tulare Open Pllinated
	Late Garnet™	Large Red x Garnet
	Liberty Bell™	(Rainier x Bing) x Stella
	Minnie Royal	Seedling 6HB480 Open Pollinated
	Nugent™	Germorsdorfer Open Pollinated
	Olympus™	Lamber x Van
	Red Crystal	Chance seedling
	Redlac	Budsport of Rainier
	Royal Dawn	Seedling 32G500 Open Pollinated
	Royal Kay	Seedling 13HA431 Open Pollinated
	Royal Rainier	Stella Open Pollinated
	Scarlet	Chance seedling
	Sequoia™	Unnamed seedling Open Pollinated
	Simcoe	Stella x (Hollander or Starking Hardy Giant)
	Sunset Bing™	Branch Mutation of Bing
	Tieton	Stella x Early Burlat
	White Gold™	Emperor Francies x Stella

Important Sour Cherry Cultivars Released from different Countries after 1990

Country	Cultivar	Parentage/Origin
Germay	Achat	Koroser x (Fanal x Relleriis 16)
	Coralin	Kelleriis 16x (Koroser x Schatten morelle)
	Gerema	Kelleriis 14 x Open Pollinated
	Habunt	Valleska x Sunburst
	Hamid	Kordia x Regina

Country	*Cultivar*	*Parentage/Origin*
	Jachim	Koroser x Safir
	Jade	Koroser x Rohrigs Weichsel
	Karneol	Koroser x Schatlenmorelle
	Korund	Koroser x Schattenmorelle
	Maliot	Grobe Lange Lotkirsche x Rote Mai
	Morima	Koroser x Reinhardts Ostheimer
	Narana	Knauffs x Souvenir de Charmes
	Rubellit	Koroser x Schattenmorelle
	Spinell	Koroser x (Fanal x Kelleriis 16)
	Topas	Fanal x Kelleriis 16
Hungary	Debrecenibotermo	Landrace selection
	Eva	Local selection
	Kantorjanosi	Local selection
	Maliga emleke	Pandy x Eugenia
	Petri	Local clone selection Ujfehertoifutos
	Piramis	Local selection
Russia	Ideal	*P. chamaecerasus* x *P. pennsylvanica*
	Jubilfenaja	Ostheimer Weichsel Open Pollincated
	Komsomolskaja	Ideal x Tschernij Orel (sweet cherry)
	Krassa Swera	Vladimirskaja ranaja x Winklers Weibe (*P. avium*)
	Nesjabkaja	Ideal x Krassa sewera
	Poljevka	Ideal Open Pollinated
	Polshir	Ideal x Plodorodnaja
	Studentskaja	Schukovskaja x Schirportep tschernaj
	Tamaris	Schirpotrep x Tschernaj Open Pollinated
	Turgenjevka	Schukovskaja Open Pollinated
The USA	Danube™	Pandy x Nagy Angol
	Suda	Schattenmorelle Open Pollinated
	Surefire	Borchert Black Sour x (Pichmorency x Schattenmorelle)

Genetic Transformation

Of more than 30 cherry species, genetic transformation has been reported only for some genotypes, including sour cherry (*P. cerasus*), chokecherry (*P. virginiana*) and the cherry rootstocks 'Rose' (*P. subhirtella* autumno), 'Gisela 6' (*P. cerasus* x *P. canescens*), 'Colt' (*P. avium* x *P. pseudocerasus*), 'Inmil' (*P. incisa* x *P. serrula*) and 'Damil' (*P. dawyckensis*). The majority of cherry species, including sweet cherry are still considered recalcitrant to transformation (Song, 2014). Gene delivery has not been found to be a barrier for cherry transformation. Two major methodologies, *Agrobacterium* and biolistic mediated transformation, have been attempted to deliver foreign genes into cherry cells. Compared to the biolistic method, ease of handling, and allowing for production of transgenic plants with single copy insertions, have made it favoured option for cherry transformation. Stable transformation of cherry species (and confirmed by molecular analyses) is currently limited to only a few genotypes. Da Camara Machado *et al.* (1995) obtained the first transgenic cherry rootstock 'Rosa' and stable transformation was confirmed by northern blot analysis. Based on GUS assay and Southern blot, stable transformations have also been reported for chokecherry, 'Montmorency' sour cherry and 'Gisela 6' cherry rootstock (Song and Sink, 2006, 2007; Dai *et al.*, 2007). Both *A. tumefaciens* and *A. rhizogenes* have enabled efficient gene delivery into cherry species (Da Camara Machado *et al.*, 1995, Dolgou, 1999, Song and Sink, 2006, 2007; Dai *et al.*, 2007). Most recently an inter fering RNAs (*RNAi*) vector targeted to prunus necrotic ringspot virus (PN RSV) has been transformed into the hybrid cherry rootstocks. This *RNAi* strategy, utilizing transformed rootstocks to achieve virus resistance in the scion, minimizes concerns about 'transgenic flow' and 'foreign protein production' in commercial scion cultivars, and it is therefor a potential approach for engineering virus resistance for fruit bearing cherry genotypes without direct transformation of the fruiting genotype itself (Song, 2014). This is an excellent example of what future, success in cherry genetic engineering will require availability of desirable target genes and reliable transformation system.

References

Agarwal, G.S. (1970). Fmg. Guide. Dept. Agric., Himachal Pradesh, Shimla, India.

Agulheiro-Santos, A.C., Palma, V.I., Rato, A.E., Machado, G., Lozano, M. and Gonzalez-Gomez, D. (2014). Acta Hort., 1020: 101-10

Alani, K. (1976). Erwerbsobstbau, 18: 188-90.

Ali, A., Masud, T., Ali, S., Abbasi, K.S. and Hussain S. (2014). J. Biodiv. Environ. Sci., 4: 252-63

Anonymous, (1949). Plums and Cherries. Ministry of Agri. and Fish. His Majestys Stationary Office. London. Bull.119 pp. 46-69.

Anonymous, (1969). The Wealth of India. VIII. Council of Scientific and Industrial Research. New Delhi, pp. 262-64

Anonymous, (1976). Res. Rept. on Pome, Stone and Nut Fruits. Project Coordinator,ICAR. Luckuow,India. Tech. Doc. No. 9. pp. 442-44.

Apostol, J., Klinesek, P. and Szani, Z (2005). Kertgazdasag Hort., 37: 49-53.

Aryanpooya, Z., Davaryanejad, G.H. and Lakatos, L. (2014). Acta Hort., 1020; 121-26.

Ayala, M., Banados, P., Thielemann, M. and Toro, Ra. (2014). Ciencia e investigacion Agraria, 41: 71-80.

Azad, K.C. (1981). Recent Techniques in Optimization of Fruit Production in European Countries Study Tour Rept. Dept Hort. Himachal Pradesh. Shimla, pp. 33-44.

Bailey, L. H. (1914). The Nursery Book. Macmillan. New York. pp. 206-208.

Bailey, L.H. (1924). Manual of Cultivated Plums. Macmillan. New York.

Baker, P.H. (1975). Pollination and Fruit Set in Tree Fruits. Publications Office. Dept of Agric. Parliament Building. Victoria.

Balazkova, J. and Hlusickova, I. (2007a). Vedeeke Prace Ovocnarske, 20: 145-46

Balazkova, J. and Hlusickova, I. (2007b). Vedelke Prace Ovocnarske, 20: 149-50

Ballistreri, G., Continella, A., Gentile, A., Arnenta, M., Fabroni, S. and Rapisarda, P. (2013). Fd. Chem., 140;630-38

Baraldi, R., Manne, C., Rosati, G., Rosati, P. and Golinelli, A. (1988). Acta Hort., 227: 405-407.

Bargioni, G. and Bassi, G. (1999). Informatore Agrario, 55: 68-69

Basaran, M.S. and Ozdem, A. (2013). Proc 4th Int. Sci. Symp. 'Agrosymp-2013' Bosnia and Herzegovina, 3-6, October, 2013. pp. 211-16

Basi, G. (2014). Informatore Agrario, 70: 41-44

Ben, J. (1988). Zeszyry Naukowe Akadcrnii Rolniczcjirn. Hugona Kullataja VII Krakowie. Habiliracyjna. No.124. p 72.

Benson, L. (1957). Plant Classification. Raytheon Education Co. Boston, USA.

Beppu, K., Fujimoto, K. and Kataoka, I. (2007). Tech, Bull. F. Agric. Kagawa Univ., 59: 55-58

Blasse, W. and Bringczu, A. (1984). Gartenbau, 31: 177-79.

Blazkova, J., Hlusickova, I. and Drahosova, H. (2009). Vedecke Prace Ovocnarske, 21: 95-108

Boynton, D. (1914). Proc. Amer. Soc. Hort. Sci., 44: 11-33.

Brase, K. D. and Way, R.D. (1950). Rootstocks and Methods, Used for Dwarf Fruit Trees N.Y. Agric. Expt. Sta. Bull. No. 783.

Bryant, L.R. (1940). Proc. Amer. Soc. Hort. Sci., 37: 322-23.

Bukovac, M. J., Rctamales, J. B., Xu Shao-Ying and Matsui, H. (1985). HortSci., 20: 1083-85

Cameron, H. R. (1971). Plant Dis. Rep., 55: 421-23.

Caruso, S., Ladurner, E., Benuzzi, M., Tamagnini, E., Granchietti, A. and Sacchetti, P. (2012). 10BC/WPRS Bull., 74: 177-82

Carvalho, M.L., Dias, C., Barrisca, C., Branca, M. and Melo-Abreu, J.P. de (2012). Acta Hort., 932: 187-93

Cerovic, R. and Ruzic, D. (1987). Pl. Cell Tissue and Org. Cult., 9: 151-57

Chadha, T.R. (1957). Indian J. Hort., 14: 119-22.

Chandlcr, W.H. (1942). Deciduous Fruits. Lea and Febigcr, Philadelphia, USA.

Chang, L.S., Lezzoni, A.G. and Flore, J.A. (1987). J. Amer. Soc. Hort. Sci., 212: 247-51.

Changzhu, Z., Jianfu, J., Huiqin, Z., Ming, X., Shuping, Li., Miaomiao, L., Hong, G. and Jinbao, F. (2011). J. Fr. Sci., 28; 1005-11

Childers, N.C. (1983). Modern Fruit Science. Hort. Publications,Gainesville. Florida, USA.

Christensen, J. V. (1975). Acta Agric. Scandanavica, 25: 11-13.

Christensen, M.D. and Walker. D. R. (1964). Proc. Amer. Soc. Hort. Sci., 85: 112-17.

Coe, F.M. (1945) Cherry Rootstocks. Utah Agric. Expt. Sta. Bull. No.319.

Compton, O.C. (1959). Proc. 17th. Int. Hort. Cong. Md., I. Abstr. No.29.

Conte, A., Scrocco, C., Lecce, L., Mastronatteo, M. and Delnobile, M.A. (2009). Innovative Fd. Sci. and Emerging Technol., 10; 564-71.

Cooper, W.E.S. (1962). The A.B.C. of Fruit Growing The English Universities Press Ltd. London.

Cox, K.D., Villani, S.M., Raes, J.J., Freier, J., Faubert, H., Cooley, D. and Clements, J. (2011). Pl. Disease, 95: 1584-85

Crane, M.B. and Brown, A.G. (1937). J. Pomol. Hort. Sci., 15: 96-116.

Crane, M.B. and Lawrence, W.J.C. (1929). J. Pomol. Hort. Soc., 7: 276-301.

Cummins, J N. (1979). Fruit Var. Hort. Dig., 27: 74-84.

Cummins, J.N. (l184) Compact Fruit Tree, 17: 57-63.

Da Camara Machado, A., Puschmann, M., Pithringer, H., Kremen, R., Katinger, H. and Laimer da Camaro Machado, M. (1995). Pl Cell Rept., 14: 335-40.

Dai, W., Magnusson, V. and Johnson, C. (2007). HortSci., 41: 140-42.

Davarynejad, G.H., Szabo, Z., Persely, S., Szabo, T. and Nyeki, J. (2014a). Acta Hort., 1020: 181-84

Davarynejad, G.H., Nyeki, J., Szabo, T., Lakatos, L. and Szabo, Z. (2014b). Acta Hort, 1020: 185-89.

Davarynejad, G.H., Aryanpooya, Z. and Persely, S.Z (2014c). Acta Hort., 1020: 143-47.

Davidson, T.R. (1973). Diseases, Insects and Mites of Stone Fruits. Agric. Canada Publication, 915.

Day, L.H. and Tufts, W.P. (1944). Calif. Agr. Exp. Sta. Cir., 0359.

Demrsoy, L. and Demrsoy, H. (2004). Pakistan J. Bot., 36: 725-31.

Dolgov, S.V. (1999). Transgenic Trees (ed. Y.P.S. Bajaj), Springer, New York pp 28-39.

Domozetov, D.Y., Christov, N.D. and Sotirov, D.K. (2014). Acta Hort., 1020: 435-38.

Drahosova, H., Blazkova, J. and Hlusickova, I. (2011). Vedecke Prace Ovocnarske, 22: 45-60.

Einhorn, T.C., Laraway, D. and Turner, J. (2011). HortTech., 21: 546-53.

Ende, B. Van Den. and Jerie, P. (1985). Compact Fruit Tree, 18: 95-101.

English, H., Dubin, H.J. and Schick, F.J. (1979). Plant Dis. Rep., 998-1002.

Enikejew, H.K. (1968). Proc. Symp. on Cherries and Cherry growing. Bonn

Eris, A., Turk, R., Ozer, M.H. and Celik, E. (1993) Proc. Int. CA. Res. Conf. pp. 534-42

Facteau, T.J (1986). Good Fr. Grower, 37: 27-28.

Facteau, T.J. and Cahn, H. (2000) Good Fr. Grower, 51: 48-50.

Facteau, T.J., Rowe. K.E. and Chesturur, N.E. (1985). Scientia Hort., 27: 247-62.

Fidancy, A., Burak, M., Erenoglu, B. and A. Kcay, M.E. (2008). Acta Hort., 795: 409-12

Floor, J. (1957). Euphytica, 6: 49-53.

Floore, J. A. and Sams, C. (1979). Compact Fruit Tree, 12: 85-86.

Fogle, H.W. (1975). In: Advances in Fruit Breeding. (eds. Jules Janick and James N. Moore) Purdue University Press. West Lafayette, Indiana, pp. 348-66.

Fregoni, M., Visai, C. and Zioni, E. (1973). Bull. Istituto di Coltivazioni Arboree del Univerpita Cattolica S Cuore. Piacenzas, Italy, pp 1-36.

Garcia, F., Frutos, D., Lopez, G., Carrillo, A. and Cos, J. (2014). Acta Hort., 1020: 191-96.

Georgiev, S. (1984). Gradirnarsk i Lozarska Nauka, 21: 10-16

Gerhardt, F. (1942). Proc. Amer. Soc. Hort. Sci., 40: 119-23.

Gil, S. G. (1974). Cicncia e Investigation Agraria, 1: 187-98.

Gil., S.G., and Momberg, L.A. (1975). Ciencia e Investigacion Agraria, 2: 77-83.

Goldwin, G.K. and Webster, A.D. (1983). J. Hort. Sci., 58: 505-506.

Guerrero-Prieto, V.M., Vasilakakis, M.D. and Lombard, P.B. (1985). HortSci., 20: 913-14.

Gurrero-Prieto, V.M., Vasilakakis, M.D. and Lombard, P.B. (1989). HortSci., 29: 913-14.

HaiRong, W., XiaoJuan, Z., Jiawei, W. and Qingzhong, L. (2012). J. Fruit Sci., 29: 49-52

Hampson, C.R., Stanich, K., Mckenzie, D.L., Herbert, L. Lu, Ran, Li, J. and Cliff, M.A. (2014). Postharvest Biol. Technol, 91: 104-11.

Harrison-Murray, R. and Mitra, S.K. (1986). Rept. East Malling Res. Stn. for 1985. p.98.

Hartman, H. (1925). Proc. Amer. Soc. Hort. Sci., 21: 79-86.

Hartmann, H.T. and Beutel, I.A. (1979). Propagation of Ternperte Zone Fruit Plants. Division of Agric. Sci., Univ. of Calif. Leaflet. No. 21103.

Hartmann, H.T. and Brooks, R.M. (1958). Proc. Amer. Soc. Hort. Sci., 71: 127-34.

Hedrick, U.P. (1915). The Cherries of New York. Albany. N.Y.

Hein, K. (1982). Proc. 21st. Int Hort. Cong. Hamburg, Abst., No.1116.

Hermann, M. and Feucht, W. (1984). Erwerbsobstbau, 26: 1-11.

Hilling, K.W. and Tezzoru, A.F. (1988). J. Amer. Soc. Hcrt. Sci., 113: 28-93

Hohn, H., Walder, R., Muhlenz, I., Samitez, J. and Linder, C. (2012). Rev. Suisse de Vitic. Arbori. Et Hort., 44: 152-58.

Hooker, J. D. (1879). The Flora of British India. L. Reeve and Co. Ltd., Kent. UK.

Howe, P. (1927). N.Y. State Agric, Exp. Sta. Bull., No. 544.

Hriber, J., Pitake. D., Levicnik, M. and Plcstenjak, A. (1986). Jugoslovensko Vocarstvo, 20: 643-46.

Hrotko, K., Magyar, L. and Gyeviki, M. (2009). Acta Hort., 825: 245-50.

Huang-Jie and Jiang-YuanHai, (2000). South China Fruits, 29: 51

Iezzoni, A.F. (1984). Fruit Var. J., 38: 121-25.

Intrieri, C. (1974). Informatore Agrario., 30: 16497-500.

Ionescu, L., Milin, K., Batovici, R., Panait, E. and Maraineanu, L. (1978). Lucrari Stiintific, 9: 43-51

Jafari, M. and Bouzari, N. (2010). Seed and Pl. Prod. J., 26: 343-57.

Jianmei, W., XiuDong, Qi, Junfeng, G. and XianQuiu, Z. (2011). J. Fd. Agric. Environ., 9: 118-22.

Kabu, K.L. and Dalal, M.A. (1976). Res. Rept. on Pome, Stone and Nut Fruits. Project Coordinator ICAR Lucknow. Tech. Doc. No.9. pp. 78-87.

Kaiser, C., Christensen, J.M., Whiting, M,D., Azarenko, A.N. and Long, L.E. (2014a). Acta Hort., 1020: 339-46.

Kaiser, C., Fallahi, E., Meland, M., Long, L.E. and Christensen, J. M. (2014b). Acta Hort., 1020: 477-88.

Kappel, F., Lane, W.D., MacDonald, R.A. and Schmid, H. (2000). HortSci., 35: 306-307.

Kappes, E.M. and Flore, J.A. (1980). Acta Hort., 114: 123-27.

Kelebek, H. and Selli, S. (2011). Int. J. Fd. Sci. Technol., 46: 2530-37.

Kenworthy, A.L. and Mitchell, A. E. (1952). Proc. Amer. Soc. Hort. Sci., 60: 91-90.

Khaak, E.R. and Nusst, Yu.O. (1989). Sadovodstvo i Vinogradarstvo, 1: 27.

Kiss, L.Z. (1987). Kerteszeti Egyetem Kozlernenyei, 46: 261-68.

Kolesnikova, A.F. (1975). Serektsiya i nekotrie niologicheskie osobennosti vishni V stredrei polose RSFSR.

Kolesnikova, V. (1966). Fruit Biology. Mir Publishers, Moscow.

Kollar, G. and Scortichirn, M. (1986). Rivista della Ortoflorofrutticoltura Italiana, 70: 85-95.

Kromer, S. and Schricht, W. (1985). Gartenbau, 32: 21.

Kwong. S.S. (1965). J. Amer. Soc. Hort. Sci., 86: 115-19.

Lakatos, L., Dussi, M.C., SZabo, Z., Soltesz, M., Nyeki, J., Szabo, T. and Davarynejad, G.H. (2014a). Acta Hort., 1020; 293-302

Lakatos, L., Dussi, M.C., Skabo, T., Szabo, Z., Soltesz, M. and Nyeki, J. (2014b). Acta Hort., 1020: 277-85.

Lane, W.D. (1984). J. Hort. Sci., 4: 87-89.

Lane, W.D. and Schmid, H. (1984). Canad. J. PI. Sci., 64: 211-14.

Lapins, K.O. (1974). British Columbia Orchardist, 14: 4-5.

Lapins, K.O. (1975). Fruit Var.J., 29: 20.

Lapins, K.O., Denby, L.G. and Lane, W.D. (1977). Fruit Tree Cultivars in British Columbia. Agric. Canada Publication. No. 1609.

Larsen, F.E. (1770). Fruit Var. Hort. Dig., 24: 40- 44.

Larsen, F.E. and Tukey. R.B (1982). Rootstock for Sweet Cherries. Washington State Univ. Coop. Extn. Bull No.11: 50.

Larsen, F.E., Fritts, R. Jr., Patten, K. and Patterson, M.E. (1983). Good Fruit Grower, 34: 26-28.

LiBin, Z., Tongkun, W., GuiSen, L., XueDong, W. and Yulin, C. (2004). China Fruits, 4: 13-14

Lishchuk, A.I. and Ercrneev, G.N. (1974). Byulleten Gosudrstvennogo Nikitskogo Boranicheskogo Sada, 1: 41-13.

Long, L.E., Yin, X., Huang, X.L. and Jaja, N. (2014). Acta Hort., 1020: 331-38.

Lugli, S., Sansavini, S. and Monari, W. (2001). Frutticoltura, 3: 24-31

Lynn, E.L. and Kaiser, C. (2010). Pacific Northwest Pb., PNW619

Lyskancwska, M.K.(1976). Plant Dis. Rep., 60: 465-69.

Ma, J.F. (2007). Italus Hortus, 14: 1-13

Mahdavian, M., Bouzari, N. and Abdollahi, H. (2011). Biharean Biol, 5: 86-90

Mahmood, K., Carew, J.G., Hadley, P. and Battey, N.H (2000). J. Hort. Sci. Biotech., 75: 602-606.

Mangnnaris, G.A., Ilias, I.F., Vasilakasis, M. and Mignani, (2007). Int. J. Refrigeration, 30: 1386-92.

Marshall, R E. (1954). Cherries and Cherry Products. Inter Science Publishers, Inc., NY. USA.

Measham, P.F., Bound, A., Gracie, J. and Wilson, S.J. (2009). Crop and Pasture Sci., 60: 1002-1008.

Measham, P.F., Gracie, A.J., Gracie, A.J., Wilson, S.J., and Bound, A. (2010). Adv. Hort. Sci., 24: 243-48.

Measham, P.F., Gracie, A.J., Wilson, S.J. and Bound, S.A. (2014). Acta Hort., 1020: 217-22.

Mecheriuk, M., Mc kenzie, B., Girad, B., Molys, A.L., and Weintraub, Stocking, R (1997). J. Fd. Quality, 20: 189-98.

Michalska, S.(1982a). Arboretum Kornickie, 27: 311-32.

Michalski, Z. (1982b). Arboretum Kornickie., 27: 333-45.

Mika, A., Czynczyk, A., Cieslak, H. and Jackiewicz. A. (1983). Prace Doswiadczalne Z Zakresll Sodownictwa, 24: 113-22.

Mircetich, S.M., Schreader, W.R., Molle, W.J. and Micke, W.E. (1976). Calif. Agric., 30: 10-11.

Mircetich, S.M., Mollar, W.J. and Nyland, O. (1977). Plant Dis. Rep., 61: 931-15.

Modlibowska, I. and Wickendcn, M.F. (1982a). J. Hort. Sci., 57: 413-22.

Modlibowska, I. and Wickenden, M.F. (1982b). J. Hort. Sci., 57: 423-29.

Mooney, A.M., Hetherington, S.D. and Bright, J.D. (2009). NSW DPI Management Guide,19th Edition, Australia

Muller, B. and Gehier, G. (1985). Gartenbau, 2: 148-49.

Musacchi, S. and Lugli, S. (2014). Acta Hort., 1020: 489-46.

Nacheva, L. and Gercheva, P. (2009). Acta Hort., 825: 261-68.

Nagy, P.T., Persely, S., Szabo, Z. and Nyeki, J. (2014). Acta Hort., 1020: 347-51

Nanii, V.S. and Kotarnan, E.M. (1974). Sadovodstvo, Vinogradarstvo I Vinodelie Moldavii, 8: 11-13.

Nyhlen, A. (1976). SYR Information, 18: 18-20.

Olden, E.J. and Nyborn, N. (1968). Hereditas, 59: 327-45.

Ostrejko, S. (1969). Sadovodstvo, 2: 24-25.

Ozolina, P.L. and Apenite, I. (2012). 10BC/WPRS-Bull, 74: 183-87.

Pandey, D., Tripathi, S.P. and Sinha, M.M. (1983). Haryana J. Hort. Sci., 12: 66-67.

Parnia, C., Dumitrescu, G., Rosca, C. and Prahoveanu, G. (1986). Acta Hort., 160: 167-76.

Parnia, P., Mladin, G.H., Dutu, I. and Stanciu, N. (1982). Proc. 21st. Int. Hort. Cong. Hamburg, Abstr., 1050.

Partridge, N.L. and Veatch, J.C. (1936). Michigan Agric. Expt. Sta. Circ. Bull. No.155.

Pasalic, B., Zabic, M. and Bosancic, B. (2013). Proc. 4th Int. Sci. Symp. Agrosym 2013, Jahorina, Bosia and Herzegovina, 3-6 October 2013. Pp. 108-13

Pathak, R.K. and Awasthi, D.N. (1976). Res. Rep. on Pome, Stone and Nut Fruits. Project Coordinator lCAR, Lucknow, Tech. Doc. NO.9. pp. 248-51.

Paul, L. and Feucht, W.(1985). Gartenbau., 50: 268-73.

Pennell, D., Dodd, P.B., Wester, A.D. and Matthews, P. (1983). J. Hort. Sci., 58: 51-61.

Perry, R.L. (1987). In: Rootstocks for Fruit Crops. (eds. Roy, C. Rom and Robert F. Carlson). Wiley Interscience Publication, John Wiiey and Sons, N.Y., pp. 217-64.

Petre, L. (1987). Cercetari Agronomice in Moldova, 20: 107-12.

Piock, H. (1973). Mitteilungen, 23.137-40.

Pommier, P., Bienfait, D. and Lycaon, B.F. (1987). Arboriculture Fruitiere, 34: 49-54.

Poniedzialek, W. (1984). Zeszyty Naukowe Akademii Rolniczej w Krakowie, Ogrodnictwo, 12: 104.

QingRui, Y., Lei, W., FuQing, Li., JunLian, Ma and Zide, Z. (2011). Frontiers of Agric. China, 5 : 631-36.

Quast, P. (1983). Mitteilungen des obstbauversuch-ringes des Alten Landes, 35: 224-25.

Randhawa, S.S. (1974). Evergreen Orchards in Nepal, 1M, Nepal, Kathmandu, pp. 27-40.

Randhawa, S.S. (1987). Wild Germplasm of Pome and Stone Fruits. Directorate II, New Delhi 110012.

Ranjit, M. and Kester, D.E. (1988). J. Amer. Soc. Hort. Sci., 113: 150-54.

Rankova, Z. (2006). Bulgarian J. Agric. Sci., 12: 429-433

Redunic, M., Spika, M.J., Strikic, F., Ugarkovic, J. and Cmelik, Z. (2014). Acta Hort., 1020: 385-88.

Remon, S., Ferrer, A., Marquina, P., Burgos, J. and Oria, R. (2000). J. Sci. Fd. Agric., 80: 1545-52.

Retamales, J.B., Bulatovic-Danilovich, M. and Bukovac, M.J. (2014). Acta Hort., 1020: 223-28.

Revin, A.A. (1986). Shornik Nauchnykh Trudov po Prikaladnoi Botanike, Genetike i Selektsii, 104: 18-24.

Roper, T.R. and Loescher, W.H.(1987). HortSci., 22: 1273-76.

Rosati, P. and Goggioli, D. (1989). Scientia Hort., 39: 201-209.

Roversi, A. and Monteforte, A. (2006). Acta Hort, 721: 123-27.

Roversi, A. and Ughini, V. (1985). Rivista di Rivista di Frutticoltura edi Ortofloricoltura, 47: 35-43.

Roversi, A., Ughini, V. and Tavelia, M.(1984). Rivista di Frutticoltura edi Ortofloricoltura, 46: 49-54.

Ryugo, K. and Sansavini, S. (1972a). J. Hort. Sci., 47: 173-78.

Ryugo, K. and Sansavini, S. (1972b). Italia Agricola, 109: 121-30.

Sansavini, S. (1982). Poc. 21st Int. Hort. Congress. Hamburg, 1: 182-91.

Sansavini, S. and Lane, W.D. (1983). Rivista di Frutticoltura edi Ortofloricoltura, 45: 55-57.

Sansavini, S. and Lugi, S. (2014). Acta Hort., 1020: 411-37

Sausavini, S. and Grazia, A. (1984). Gartenbau., 33: 177-78.

Schimmelpfeng, H. and Liebster, G. (1979). Gartenbau., 44: 55-59.

Schumacher, R. and Fankhauser, F. (1968). Schweiz. Landw. Forsch, 7: 128-47.

Sedlak, J. and Paprestein, F. (2011). Vedecke Prace Ovocnarske, 22: 151-57.

Seif, S. and Greeppe, W. (1985). Acta Hort., 149: 71-75.

Sekre, L. (1988). Gartneryrket, 78: 292-93.

Shear, C.B. and Faust, M. (1980). Hort. Reviews, 2: 142-63.

Singh, M. (1976). Himachal Hort., 12-12: 9-21.

Singh, R.N., Randhawa, S.S. and Gupta, P.N. (1971). Indian J. Hort., 28: 196-98.

Sitarek, M. and Grzyb, Z.S. (2007). Acta Hort., 732: 345-49.

Sive, A. and Resnizky, D. (1987). Alon Hanotea, 42: 269-77.

Song, G.Q. (2014). Acta Hort., 1020: 89-97

Song, G.Q. and Sink, K.C. (2006). Plant Cell Rep., 25: 117-23.

Song, G.Q. and Sink, K.C. (2007). Acta Hort., 738: 683-89.

Spotts, R.A., Wallis, K.M., Serdani, M. and Azarenko, A.M. (2010). Pl. Disease, 94: 345-50.

Stanberry, C.O. and Clore, W.J. (1950). Proc. Amer. Soc. Hort. Sci., 56: 40-45.

Stancevic, A. (1982). Jugoslovensko Vocarstvo, 16: 113-16.

Stebbins, R.L. and Cameron, H.R. (1984). Fruit Var. J., 38: 21-23.

Stebbins, R.L., Tjienes, J.R. and Cameron, H.R. (1978). Fruit Var. J., 32: 31-37.

Stern, R.A., Raz, A. and Goldway, M. (2014). Acta Hort, 1020: 513-19.

Stoev, T., Vasilev, V.N. and Antanasov, A.T. (1984). Gradinarska i Lozarska Nauka, 21: 13-21.

Suszka, B. (1976). Arboretum Kornickie, 21: 257-70.

Szpadzik, E., Jadczuk-Tobjasz, E. and Lotocka, B. (2014). Acta Hort., 1020: 257-63.

Sztejnberg, A. (1986). Plant Disease, 70: 349-51.

Tabuenca, M.C. (1985). Anales de la Estacion Experimental de Aula Dei, 17: 209-13.

Tarasenko, M.P. (1958). Agrobiologia, 5: 127-29.

Tehrani, G. (1984). HortSci., 19: 451.

Tehrani, G. and Brown, S.K. (1992). Pl. Breed. Rev., 9: 367-88.

Tehrani, G. and Kerr, E.A. (1968). Proc. on Cherries and Cherry growing, Bonn.

Teskey, B.J.E. and Shoemaker, J.S. (1972). Tree Fruit Production. The AVI Publishing Company, Inc. Westport, Connecticut, pp. 238-93.

Thielemann, M., Toro, R. and Ayala, M. (2014). Acta Hort., 1020: 353-61.

Thompson, M. (1996). In: Cherries: Crop Physiology, Production and Uses (eds. A.D. Webster and N.E. Looney). CABI Pub; UK. pp. 223-241

Timosheuko, S.E. (1977). Trudy Stauropol' skogo N II. S-Kh. 27: 39-49.

Torres, C.A., Yuri, J.A., Venegas, A. and Lepe, V. (2009). Acta Hort., 825: 116-23.

Toyama, T. (1980). Fruit Var. J., 34: 2-4

Travcrsi, D. and Massignan, L. (1987). Notiziario di Ortoflorofrutticoltura, 13: 107-13.

Tukcy, H.B. and Young, J.G. (1939). Proc. Amer. Soc. Hort. Sci., 36: 269-70.

Tukey, H.B. (1984). Good Fruit Grower. 35: 46-49.

Tukey, H.B. and Brase, K. (1945). Proc. Amer. Soc. Hort. Sci., 27: 88-92

Tukey, H.D. (1940). New York State Agric. Expt. Sta. Circ. No.188

Van Petersen. A (1977). Tidsskr. Planteavl. 81: 346-50.

Vangdal, E., Slimstad, R. and Sekse, L. (2006). Veg. Crop. Res. Bull., 65: 169-75.

Vasilev, V.N. (1982). Gradinarska i Lozarska Nauka., 19: 8-15.

Veatch, J.O. and Partridge, N.L. (1934). Michigan Agric. Expt. Sra. Spec. Bull., N0. 57

Vidrih, R., Hirbar, J. and Sekse, L. (2014). Acta Hort., 1020: 165-72.

Wadley, B N. (1975). Fruit Var. J., 29: 70.

Walther, E. and Jauer, A. (1985). Acta Hort., 169: 97-104

Wander, I. W. (I941). Proc. Amer. Soc. Hort. Sci., 42: 1 -6.

Wang, T., Xie, X. B. and Long, L.E. (2014). Fd. Chem., 160: 22-30.

Way, R.D., Livermore, K.G., Lamb, R.C. and Ystaas, J. (1982). HortSci., 17: 986-87.

Webster, A.D. (1980). Acta Hort., 114: 108.

Webster, A.D. (1984). Compact Fruit Trees. 17: 103-17.

Webster, A.D. and Goldwin, G.K. (1984). Acta Hort., 149: 217-23.

Webster, A.D. and Quinlan, J.D. (1986). Acta Hort., 179: 577-78.

Webster, A.D. and Snepherd, U.M. (1984). J. Hort. Sci., 59: 175-82.

Webster, A.D., Modlibowska, I. and Wickenden, M.F. (1983). Acta Hort., 17: 285-92.

Westwood, M.N. (1978). Temperate Zone Pomology. Freeman. San Francisco.

Westwood, M.N. and Wann, F.B. (1954). In: Nutrition of Fruit Crops. (ed, N.F. Childers). Horticultural Publications. Rutgers University. New Brunswick, pp. 151-73.

Westwood, M.N., Poberts, A.N. and Bjornstad, H.O. (1976). J. Amer. Soc. Hort. Sci., 101: 268-69.

Whiting, M.D. and Ophardt, D. (2005). HortSci., 40: 1271-75.

Wills, R.B.H., Sriven, F.M. and Greenfield, H. (1983). J. Sci. Fd. Agric., 34: 1383-89.

Woodbndge, C.G. (1955). Canad. J. Agric. Sci., 34: 345-51.

Wunsch, A. (2014). Proc. 7th Int. Rosacea Genomic Conf. Seattle, p. 95

Xu, Z.I., YuanMao, J., FuTian, P., NaiBo, He., YanJu, Li. and Dengchao, Z. (2007).

Scientia Agric. Sinica, 40: 1222-28.

Yastreb, G.V. (1985). Sadovodstvo i Vinogradarstvo, 6: 11-15.

Yastreb, G.V. (I981). Michunna, 33: 55-59.

Zahn, F.G. (1977). Mittcilungen des Obstbauversuchsriugcs des Aircn Landes. 32: 97-106.

Zehr, E.I. (1982). Pl. Disease, 66: 1101-105

Zoffoli, J.P. and Rudriguez, J. (2014). Acta Hort., 1020: 115-19.

Chapter 4

Peaches

The Peach (*Prunus persica* (L.) Batsch.) along with its smooth-skin mutant, the nectarine, is a temperate juicy fruit of excellent appearance and quality. It comes in the market early in the season, particularly from low chilling cultivars grown in warmer regions. The grower can benefit from the relatively higher market prices at this juncture due to the scarcity of other fruits. The first good crop of peach is obtained within 4-5 years of planting, which is sooner than majority of the other temperate fruits. Peach culture is, therefore, expanding on a faster pace in many countries.

Composition and Uses

Fruit composition of two representative cultivars, each of peach and nectarine, is being given in Table 4.1 (Wills *et al.*, 1983). Sugars are found in abundance in the ripe fruits but starch is absent. The total sugar percentage in ten cultivars varies from 6.58 to 9.50 (Rathore, 1984). Sucrose is the major sugar while fructose and glucose are present in appreciable quantities almost in equal proportions. Nectarines are richer in organic acids than peaches, the dominant acid being malic followed by citric and quinic, respectively (Wills *et al.*, 1983). In comparison with other fruits, peaches are rich sources of protein as almost all-essential amino acids are present (Gopalan *et al.*, 1987).

Peaches are also a good source of vitamins and minerals being rich in carotene, thiamine, riboflavin and niacin among the vitamins and potassium, sodium, calcium, magnesuim, iron and zinc of the minerals present (Wills *et al.*, 1983). The pectic substances are about 0.86 per cent while tannin is only in traces (Anon., 1969). Several lactones are responsible for characteristic peach flavour. Spencer *et al.* (1978) reported that volatile compounds responsible for the clingstone peach odour are predominantly esters, monoterpenes, and γ-and α-lactones and

Table 4.1: Proximate Composition of Peach Fruit (g/100^{-1} edible portion)*

Cultivar	*Water*	*Protein*	*Fat*	*Sugars*			*Starch*	*Dietary Fibre*	*Organic Acids*	*Ash*	*Energy*** (kJ)*
				Fructose	*Glucose*	*Sucrose*					
Peach											
Elberta	88.5	0.8	0.1	1.0	1.3	4.2	0	2.3	0.9	0.4	135
Redhaven	90.9	1.0	0.1	0.8	0.7	3.1	0	1.8	0.6	0.4	103
Nectarine											
Goldmine	87.1	0.8	0.1	1.3	1.6	6.7	0	2.5	1.2	0.3	188
Nectared	86.2	1.7	0.1	1.3	1.5	5.1	0	2.2	1.3	0.3	176

Cultivar	*Organic Acids (mg 100g^{-1})*			*Titratable Acidity (meq^{H+} 100g^{-1})*	*Soluble Solids (^{o}Brix)*
	Malic	*Citric*	*Quinic*		
Peach					
Elberta	430	240	200	8.9	12.0
Redhaven	310	120	180	8.8	8.5
Nectarine					
Goldmine	810	230	120	12.9	13.3
Nectared	590	470	230	16.6	12.3

Cultivar	*Vitamins (mg 100g^{-1} edible portion)*					*Minerals (mg 100g^{-1} edible portion)*					
	Vitamin C	*Carotene (μg)*	*Thiamine*	*Riboflavin*	*Niacin*	*K*	*Na*	*Ca*	*Mg*	*Fe*	*Zn*
Peach											
Elberta	7	1500	0.01	0.03	1.0	190	2	6	6	0.1	0.1
Redhaven	6	915	0.01	0.05	1.1	180	2	7	7	0.2	0.1
Nectarine											
Goldmine	13	97	0.02	0.04	1.1	240	1	11	6	0.2	0.1
Nectared	11	1405	0.02	0.03	1.4	230	0	8	8	0.1	0.1

** Energy (kJ) = (g protein 17) + (g fat 37) + (g sugar as monosaccharide 16) + g organic acid 10)

* Wills *et al.* (1983).

differences among peach cultivars were due more to the relative concentration of esters and monoterpenes than to the γ-and α-lactones. In a comparison of low and high quality genotypes, Robertson and Meredith (1988) found that low quality fruit contained about seven times more total phenolics than high quality fruit. Glucose and sorbitol were at higher levels in low quality fruit, whereas fructose was more in high quality genotypes.

Peaches are grown both for fresh market and processing. Being highly perishable, fruits are difficult to transport and store. Processing is largely restricted to canning but on a limited scale, pickling, freezing and other preservation methods are also used. Poor quality fruits are used in wine or brandy making. The consumption of peach fruit juice is also increasing rapidly in the form of nectars, fruit drinks and breakfast drinks. Soft ripe fruits are especially suitable for juice preparation (Woodroof *et al.,* 1975). The quality of canned peaches is releated to the optimum maturity as half ripe fruits give better quality of canned product. The cultivar Golden Bush is best suited for canning followed by Foster and Crawford's Early but Prince of Wales and Hales Early are unfit for this purpose (Garg *et al.,* 1975). The browning in canned peaches can be controlled by dipping of fruit slices in 3 per cent chloride solution prior to canning (Garg *et al.,* 1978). Peach nectar is prepared by blending the puree with sugar syrup and acidifying with citric acid (Tressler and Joslyn, 1961). Freestone peaches of large size, high sugar content, rich golden yellow flesh and pleasing flavour can be converted into dried peaches by sun or artificial drying (Anon., 1969).

Peach kernels contain fat 39.9 per cent, protein 31.2 per cent, crude fibre 14.8 per cent, sugars 2.2 per cent, starch 3.6 per cent, and mineral matter 2.7 per cent (Anon., 1969). Peach kernel oil is utilized in food, cosmetics and pharmaceutical preparations. The cake contains 7.9 per cent nitrogen and can be used as a fertilizer or cattle feed after debittering. Peach flowers and leaves are purgative and anthelmintic.

Origin and Distribution

All evidences prove the origin of peach in China, where its culture dates back at least to 3000 years (Li, 1984; De Candolle, 1967; Hedrick, 1917). Three wild species are still commonly found there. *Prunus davidiana* is an ornamental tree growing wild in northern China and is used as rootstock. The other two species *P. mira* and *P. ferganensis* are indigenous to Tibetan plateau and Sinkiang province, respectively (Childers, 1975). The variability of peach is very high in the North West of China comprises of Lanzhou, Gansu Province to the West including the Tarim basin bordered by the Tien Shan, Pamir and Karkoram mountains. *P. kansuensis and P. davidiana* are native on the East side of this area and *P. jerganensis* is native on the West side. It is likely that nectarine originated in this area (Faust and Timon, 2011). According to Zeven and Zhukovsky (1975), the primary centres of peach diversity are montane areas of Tibet and southwest China and secondary diversity centres are Iran, Central Asia, Caucasus, Crimea, Moldavia, Italy, Spain and California. Besides *Prunus persica,* some other species, *viz., P. jacquimontii, P. undulata, P. salicina,* occur commonly in the interior Himalayas (Hooker, 1946). Another species, *P. behimi,* considered a natural hybrid of almond and peach, finds its use as rootstock for almond, peach and plum (Randhawa, 1987).

Although it is not exactly known that how and when the peach reached Persia. Li (1984) postulates its spread through traders during the Zhou Dynasty (about 1100-221 BC). De Candolle (1967) has expressed the possibility of carrying over the stones of peaches from China to Kashmir, Bokhara and Persia. Its spread from Persia to Mediterranean basin and other regions is authenticated by its name, which was given by early Greek and Roman writers. The peach was planted in Greece by 332 BC and was mentioned by Virgil in 70 to 19 BC (Li, 1984). Its further spread was brought about by the Romans throughout their realm and to northern Africa (Hesse, 1975). Spanish conquerors introduced peach to the American continent. Cultivars from a breeding programme at Gainsville, Florida, started in 1952 (Sharpe and Sherman, 1975) extented peach production in tropical highlands of northern Thailand, India, Peru, Brazil and Israel.

Peach is commercially grown around the world between 25° and 45° latitudes above and below the equator (Childers, 1975). It is commercially cultivated in countries like USA, Italy, France, Japan, Argentina, Australia, Mexico, Korea, USSR, West Germany, Portugal, New Zealand, Spain, Greece, South Africa, Turkey, Canada, Yugoslavia, Chile, India, Austria, *etc.*

Production

Peach and Nectarine production in the top ten producing countries are presented in Table 4.2. In India, peach cultivation is confined to the mid-hill zones of the Himalayas extending from Jammu and Kashmir to Khasi hills at an altitude of 1500 to 2000 m ASL. Low chilling peach cultivars are also grown in sizeable area in the submountainous regions and western part of the coutry, *i.e.*, Punjab, Haryana. Delhi and Uttarakhand.

Table 4.2: Production of Peach and Nectarine in Top Ten Producing Countries

Sl.No.	*Name of the Country*	*Production (tones)*
1.	China, mainland	1,20,00,000
2.	Italy	13,31,621
3.	USA	10,58,830
4.	Greece	7,60,200
5.	Spain	7,47,200
6.	Turkey	5,75,730
7.	Iran	5,00,000
8.	Chile	3,25,000
9.	Argentina	2,90,000
10.	Egypt	2,85,194

FAOSTAT (2014).

Taxonomy

Peach belongs to the family Rosaceae, subfamily Prunoideae with 8 basic and 16 somatic chromosome number (2n=16). The large genus *Prunus* includes peach, nectarine, almond, plum, prune, apricot, cherry and many species, used only as rootstocks or ornamentals. The genus has further been grouped into three subgenera, *i.e. Amygdalus, Prunophora* and *Cerasus,* in which flower buds do not contain leaves but only flowers and they are always lateral and never terminal. Since the drupe consists entirely of a developed ovary, both pollination and fertilization must take place to provide necessary stimulus for fruit development (Westwood, 1978).

Peach and almond belong to subgenus *Amygdalus* which is characterized by sessile or short stalked flowers, opening before the leaves, fruit tomentose (glabrous in nectarine); stone pitted or smooth, leaves conduplicate in bud; buds 3 in each axil, the lateral ones being flower buds.

Prunus persica is the common botanical name in use, the other names, *viz., Amygdalus persica, Persica vulgaris* were also in use during early years (Hedrick, 1917). *P. persica* is a small to medium-sized (3-8 m height), upright spreading, open topped, deciduous tree with a rather deep root system. The trunk bark, which generally splits, is dark brown rather black rough; branches medium, stocky, spreading and have large oval lenticels: young shoots are smooth and pinkish in colour. The leaves are simple, relatively large (6-12 cm long), oblong-lanceolate, acuminate, serrulate, glabrous above, pubescent beneath along midrib when young, petiole glandular; stipules fimbriate. Flowers appear in the first week of March and continue up to the third week. Vegetative and flower buds- are borne in twos or threes, sometimes singly in the axil of the leaves. Flowers are numerous, nearly sessile, white or pink, appearing before the leaves. Fruits are variable in size, shape, colour of skin and flesh, sulcate, glabrous, usually with fuzzy skin; flesh juicy, sweet or mildly subacid; stone large, free or clinging, deep pitted corrugated, sutured and hard (Westwood, 1978). Most of the cultivars have bitter kernel, while few have sweet kernels. The other important species related to peach are as follows:

Species	*Origin*	*Tree*	*Characters*
P. davidiana	China (north-east)	Tall	Tolerance to drought, used in breeding to improve disease resistance
P. ferganensis	China (west)	Medium	Wide variability in fruit, resistance to powdery mildew
P. kansuensis	China (north east)	Dwarf and bushy	Poor fruit quality, resistant to frost
P. mira	China (far west)	Tall	Variable fruit colour, shape and shize, used as seedling rootstock.

Nectarines are a group of peaches in which the fruit surfaces do not have fuzzy skin. Nectarines are usually smaller in size and have greater aroma. These have been developed from peach as bud sports in which the single gene responsible for fuzziness is lacking. Because of smooth skin, nectarines are more susceptible to russeting, mechanical bruishing and thrips damage.

August Prince

Autumn Crunch

Babcock

Bacterial spot

Brown Rot

Elberta

Peach Flowering

Peach Leaf Curl

Peach Leaf Curl

Peach Orchard

Sweet Scarlet

Spring Crest

Vinegold

According to their morphology and geographical distribution, the peach cultivars have been divided into four ecological groups, *viz.*, Chinese, Central Chinese, Western Chinese, and Yellow-fleshed from Europe. But Li (1984) has recognized only three groups, *i.e.*, Northern, Southern and European or Persian group.

Cultivars

Choice of suitable cultivars for any region is governed by factors, such as, (i) type of market to be served, (ii) distance to market, and (iii) adaptability to the local soil and climatic conditions, (Childers, 1975). For table purposes, the cultivars should be yellow-fleshed, freestone, regular producer and relatively free from fuzz. The cultivars which ripen before the onset of rains are ideally suited for commercial plantations. For canning purposes, the fruit should have yellow flesh, freestone, small non-splitting pit, good symmetrical size and should mature evenly. However, in recent years, clingstone cultivars are preferred for canning purposes. For dehydration, white-fleshed sweet cultivars having freestone kernels are preferred.

The important peach cultivars being grown in the world for various purposes are:

Table Cultivars

Elberta, J. H. Hale, Babcock, Cardinal, Dixigem, Redtop, Redhaven, Rio Oso Gem, Candor, Red Globe, Baifeng.

Canning Cultivars

Cortex, Carolyn, Dixon 1, Halford, Fortuna, Crawford's Early, Golden Bush, Stuart, Vivian, Veteran.

Low chilling cultivars are preferred in Australia, India and Florida (USA). These are Early Amber, Early Grande, Flordabelle, Flordagold, Flordaking, Flordasun, Flordaqueen, Rio Grande, Sharbati and May Gold (George *et al.*, 1986; Sherman *et al.*, 1986; Singh *et al.*, 1984).

The nectarine is a non-pubescent (fuzzless) peach having the identical growing and fruiting characteristics as the peach. Its important cultivars under cultivation are Annqueen, Cherokee, Late Le Grand, Nectared 4, Sun Grand, Sunlite, Sunred, Sunrise, and Sunripe (George *et al.*, 1986; Sherman *et al.*, 1986).

Brief Descriptions of some Important Cultivars

Alexander

An excellent early season cultivar, ripening in the last week of May to first week of June and is a medium bearer. Fruit medium to big in size, round with unequal sides, cavity deep, skin smooth, beet-root-purple with some patches of pod-green colour, flesh soft, greenish white, juicy, very sweet, aromatic and freestone; keeping quality-medium to poor.

Babcock

Babcock (Strawberry Peento) was introduced in 1933 at California (Brooks and Olmo, 1972). The fruit is small to medium, round to ovoid spherical having ventral suture prominent. The skin is light pink with a deep red blush and flesh is white, very sweet with freestone. It ripens 2 weeks before Elberta. Tree is medium-large, vigorous and very productive.

Baifeng

Its origin is in Japan but has become an important mid-season cultivar in China (Li, 1984). White-fleshed, soft, melting, clingstone fruit is of excellent quality. The tree is productive and well adapted to cold conditions. Heavy thinning is required for ensuring good fruit size.

Candor

It originated from a cross (Redhaven x Early Red Fre) in Carolina (Brooks and Olmo, 1972). Fruit is medium, round in shape with pointed blossom end and deep stem end cavity (Rathore, 1984). The skin is bright red on rich yellow ground and deep red near the blossom end. Flesh is yellow, texture fine, firm, excellent flavour and freestone. Tree is small, very heavy bearer, regular, self-fruitful and has medium chilling requirement.

Dixigem

It resulted from a cross (Admiral Dewey x St. John) in 1944. The fruit is medium, round with a bright yellow skin having light blush (Brooks and Olmo, 1972). Flesh is light yellow, freestone and good in quality and texture. It matures about a month before Elberta. Tree is vigorous, productive and needs about 950 hours chilling.

Early Amber

It is a low chilling cultivar (350 hrs) bred in Florida. Fruit ripens early, medium size, skin yellow with dark red blush (Rathore, 1984). Flesh is yellow, clingstone and of very good quality. Tree is medium, vigorous, productive and blooms early in the season.

Elberta

This is an important peach cultivar. Fruits are mainly used for canning purposes. Fruit is large in size, form round, regular and symmetrical, skin smooth, pale yellow with a red splash. Flesh is firm, yellow, juicy, very sweet in taste, excellent quality, freestone.

Feichengtao

It is a famous cultivar of northern China, being given as a tribute to the emperors (Li, 1984). Fruits weigh an average of 340 g, the largest reaching 680 g. White flesh is firm, good in texture and clingstone. Tree is upright, vigorous, having single bud at most nodes. The floral buds are cold susceptible.

Flordaqueen

It was evolved by R. H. Sharpe in Florida as a medium chilling cultivar requiring 550 hours of chilling (Brooks and Olmo, 1972). Fruit is yellow with light red blush, round and yellow flesh is firm, clingstone and fair in quality. Tree is heavy bearer.

Flordasun

A low chilling cultivar developed in Florida. It is an early maturing cultivar which ripens in the last week of April. Fruits medium-sized, freestone, round shape with red blush on the surface, pulp yellow, acidic sweet taste with 11.5 per cent soluble solids.

J.H. Hale

This is an old peach cultivar in commercial production, probably of Elberta parentage originated in Connecticut. Fruit is large, yellow with firm flesh of excellent quality. Freestone peach is good for both fresh market and canning. It is a self-unfruitful cultivar.

Maygold

It is a cross of Sunhigh and Southland, selected in Georgia. Fruit is medium-large, ovate, skin yellow, more than half red. Flesh is yellow, firm but melting, good flavour and clingstone (Rathore, 1984). Tree is vigorous. productive, precocious, self-fruitful and has medium chilling requirement (650 hours). It is an early promising cultivar, fruit ripening in the second week of June.

Nectared 4

The parentage of this nectarine involves Candoka peach and Flaming Gold and New Jersey N 14 nectarines. Fruit is of good size. quality and colour. This mid-season cultivar is very productive and promising.

Redglobe

It originated in Maryland from (Admiral Dewey x St. John) x Fireglow corss (Brooks and Olmo, 1972). Medium-sized fruit is round, attractive due to a bright red blush and suitable for freezing, canning and long shipment. Yellow flesh is fine-textured, firm, melting with a small pit. Tree is vigorous, good bearer and has chilling requirement similar to Elberta.

Redhaven

A cross of Halehaven and Kalhaven. Redhaven is an important and wildely planted cultivar. Its outstanding characteristics are firm flesh, which enables easy picking and handling, excellent red colour, good size and heavy and dependable cropping. It ripens about 30 days before Elberta, having chilling requirement of 950 hours.

Rio Oso Gem

It originated in California, parentage unknown. Fruit is large, round to elongated, with an attractive skin colour. Flesh is yellow, fine texture, firm, good quality and freestone. It resembles J.H. Hale and ripens a week before Elberta requiring 850 chilling hours.

Sharbati

It is a chance seedling selected at Saharanpur, India (Singh, 1960). It is a late cultivar and fruits ripen in the second week of June. Fruits are of medium size, clingstone and round oblong in shape. Pulp is white, soft, juicy and aromatic with a TSS of 13.5 per cent. It is a heavy yielder and most popular cultivar of western Uttar Pradesh, India.

Spring Crest

Fruits are small to medium and round, with skin showing a red blush on yellow, moderately firm, melting, and semifree stone when fully ripe. Quality is good. It is an early variety.

Sunred

It is a low chilling nectarine bred by R. H. Sharpe in Florida. Fruits are small in size, round, bright red skin and of excellent dessert quality. Flesh is yellow, firm and semi-freestone when fully ripe. Ripening occurs in the first or second week of May. Tree is large, heavy bearer, requiring 300 chilling hours.

Sunrise

Fruit of this nectarine is globose and medium thick skin is tough having mahogany-red colour (Brooks and Olmo, 1972). Flesh is yellow, firm, crisp and mildly subacidic. It is semi-freestone and ships well. The cultivar is medium in vigour, upright and bears a good annual crop.

Veteran

A cross between Vaughan and Stark's Early Elberta, originated in Canada. Fruit is medium to large, round-oblate and is harvested 10 days before Elberta. Its main qualities are very good fruit bud and tree hardiness, heavy production and suitability for both fresh and canning purposes. Flesh is, however, soft and skin lacks red colour. The chilling requirement is also high being 1150 hours (Brooks and Olmo, 1972).

In India, most of the cultivars introduced from peach growing countries have been acclimatized and recommended for commercial orcharding. Only a few cultivars are the resultant of seedling selection or planned hybridization. The popular cultivars in the hills are Alexander, Alton, Elberta, July Elberta, Halberta Giant, J. H. Hale, Prince of Wales, Golden Bush, C.O. Smith and World's Earliest (Kabu and Dalal, 1976; Mishra and Srivastava, 1972; Singh, 1963). The cultivation of low chilling cultivars is, however, gaining popularity currently in the subrnountainous regions and the plains of northern India. The cultivars like Sharbati, Safeda, Saharanpur 6, Safeda Early Cream have been found promising under Pantnagar conditions (Singh and Sirohi, 1976). Nauriyal and Misra (1982) observed better performance of Flordasun and Sunred in comparison to local cultivars in the valley areas of central Himalayas. Singh *et al.* (1984) found that the cultivars Flordasun and Shan-e-Punjab were the earliest to ripen giving the heaviest yield, while Sharbati was the latest to ripen but had the highest total soluble solids content

New Peach Cultivars

USA

Augustprince: (Sunprince x BY87P943), yellow-fleshed, late-season, freestone peach, fruit bright red with attractive yellow ground colour, moderately resistant to bacterial spot.

Candyprinces: (Unnamed peach x Tropic Beauty), yellow-fleshed, early, clingstone peach, fruit large, dark red, flesh firm, crisp and melting.

Galactica: (NCN-4 x Hangchow), white-fleshed, saucer-shaped, freestone peach. Fruits bright red blush, excellent texture, flavor and aroma, susceptible to bacterial spot.

Gulfcrest: (FL 92-8C x Spring Baby), yellow-fleshed, mid-chill, clingstone peach, fruit large, very deep red mottled over a dark pink background, flesh firm, crisp and melting.

Snow Duches: (Snow Princess x Crimson Lady), white-fleshed, firm, freestone peach. Fruit large, very deep red smoothly blending to a moderate pink back ground, flesh firm, tough, melting.

TruGold: (Redhaven x Veefreeze), yellow-fleshed, freestone peach. Fruit medium large, red skin with dark red stripes produces seedlings with identical fruit and tree type.

Vista Snow: (174LE309 x 2LD470), white-fleshed, firm, early season, clingstone type, Fruit red over yellow, globose, medium; flesh firm, flavor mild.

Gulf Prince: (Aztecgold x Oro A), moderate chill peach (400 chill units) large attractive fruit with 50 per cent red skin over a deep yellow ground, fruit sweet, firm, non-melting, clingstone flesh, resistant to bacterial spot.

Australia

Sweet Scarlet: (39EB706 x Sweet Gem), regular and productive, bearer of large, firm, yellow-fleshed, clingstone fruits with an attractive red skin, good handling and storage qualities.

France

Flatpreety: Yellow-fleshed, semi-clingstone flat peach, released in 2010. Fruit medium, flat, rounded, bright red over colour on yellow-orange ground colour, flesh white with slight red pigmentation near the stone; flesh juicy, semi-sweet, adopted to areas with 350-1200 h chilling requirement.

Flat Queen: White-fleshed, melting, semi-freestone flat peach, fruit medium, uniform, flat, rounded, thick skin tenacious to the flesh, skin orange red homogenous blush over yellow orange ground colour, flesh white, juicy, semi-sweet, spring frost resistant.

Crispregal: Yellow fleshed, clingstone peach, fruit very large, red blush of surface over yellow-orange ground colour, flesh yellow, firm, juicy, semi-sweet.

Flatprincess: White-fleshed, semi-clingstone, saucer-shaped peach. Fruit large (150-200g), uniform orange-red overcolour of yellow-orange ground colour, flesh white, juicy, semi-sweet, aromatic.

Julienice: Yellow-fleshed, clingstone peach, very large fruit (260-290g), rounded, red over colour on orange ground colour, flesh yellow with a slight star-shaped red pigmentation around the stone, firm, juicy, long shelf life.

Italy

Lucrezia: Yellow fleshed, melting, freestone, very early ripening peach. Fruit small (65-70g), round, red blush over yellow ground colour, flesh yellow, medium firm, melting, free stone, good flavor, highly productive.

Rosella: (Rubiette x Spring Lady), white-fleshed, melting, clingstone, very early ripening peach. Fruit small (75-80g), red blush on white ground colour, flesh white, firm, melting, highly productive.

Nadia: (Suncrest x IF7030087), yellow-fleshed, melting, freestone peach. Fruit very large (200g), round, red blush on yellow ground colour, flesh yellow, firm, melting, good flavor, self fertile, resistant to brown rot.

Bolero: (Cresthaven x Flamecrest), yellow-fleshed, melting, freestone peach resistant to bacterial spot, fruit very large (180-200g), round, uniform, red blush on yellow ground colour, flesh firm melting, good flavor.

Bordo: (Rich Lady x May Crest), yellow-fleshed, melting, clingstone, early ripening peach. Fruits round, large (150-170g), red blush on yellow ground colour, good flavor, highly productive.

Pink Ring: (Flavtrop x Snow Queen) x Stark Saturn, white-fleshed, mid-season, semi-clingstone, sub-acid, saucer-shaped peach. Fruit medium, flat, red over white ground colour, flesh medium-firm, melting, sub-acid.

Sweet Ring: (Spring Lady x Stark Saturn), yellow-fleshed, early-season, semi-clingstone, sub-acid, saucer-shaped peach. Fruit medium-small, red over colour yellow ground colour, flesh yellow, medium-firm, juicy, melting, sub-acid.

Canada

Vinegold: (NJC95 x Veecling), yellow-fleshed, non-melting, clingstone processing peach. Fruit large, round, red blush over dark yellow ground colour, flesh dark-yellow, firm, non-melting, good quality. Moderately resistant to bacterial spot, brown rot and Cytospore canker.

Virgil: (Veecling x NJC95), yellow-fleshed, clingstone, processing peach. Fruit medium large, uniform dark-red over colour on golden yellow ground colour, flesh golden yellow, productive, moderate resistant to Cytospora canker, bacterial spot and brown rot.

Vulcan: (Veecling x NJC 95), yellow-fleshed, non-melting, clingstone, processing peach. Fruit medium, dark red blush over a yellow orange ground coulour, Yellow-orange flesh, firm, good processing quality, resistant to bacterial spot, brown rot and Cytospara canker.

Brazil

Turmalina: Fruits round, firm flesh, smooth skin, suitable for canning. The fruit has good flavor, high productivity and good adaptation to region with 300-400h accumulated chilling (Nakasu and Raseira, 1999)

Jubileu: It is a canning, clingstone cultivar, with dark flesh and large fruits. It has a low chilling requirement (250-350 hours), released in 1998 (Raseira and Nakasu, 1998)

Marfim: Table peach variety, sweet taste, white, juicy, firm flesh, low chill cultivar (350 hours) released in 1998 (Raseira *et al.*, 1998).

Mexico

Arkangel: (A298 x 23D), yellow-fleshed, clingstone peach. Fruit medium, bright-red over colour, golden-yellow ground colour, flesh deep-orange, firm, non-melting, resistant to powdery mildew.

Armoore: (H2-22 x Vinegold), yellow-orange fleshed, sub-acid, clingstone peach, Fruit medium-large, bright-red over colour of golden-yellow ground colour, firm, non-melting, good flavor, resistant to oxidative browning.

South Africa

Autumn Crunch: yellow-fleshed, late season, canning clingstone peach, Fruit large, round, bright yellow, firm, non-melting, chilling requirement 600-800 h.

Cascade: Yellow-fleshed, non melting, medium-late-season, canning clingstone peach. Fruit large, round, orange-yellow colour, flesh orange-yellow, non-melting, medium-sweet, chilling requirement 400-600 h.

Cederberg: Yellow-fleshed, melting, midseason, freestone peach, suitable for drying, fruit large, medium red marbled overcolour with yellow ground colour, melting, sweet, low acid.

Uruguay

Moscato Blanco: White, firm, late-season, mid-chill, melting, freestone peach. Fruit medium-large, round-hearted, red blush, flesh white, very good flavor.

Moscato Delicia: Yellow, very firm, late, mid-chill, freestone melting peach. Fruit large, red blush, flesh yellow with abundant red fibers, good flavor.

Moscato Largo: Yellow, firm, mild-chilling requirement, freestone peach. Fruit medium to large, red over colour, flesh firm, good flavor and aroma.

China

Jiuan: Seedling selection of 'Okubo'. Average fruit weight 237g, round in shape, flesh white with intermediate red pigment, hard melting and freestone, TSS 12.5 per cent, productive (Xiao *et al.*, 2011)

Ruipan 20: (Fantasia x Ruipan 4), late ripening flat peach (*P. persica* var *platycarpa* Bailey). Fruit is flat, average weight 254.7g, white fleshed, hard melting, sweet (TSS 13.1 per cent), and freestone (JianBo *et al.*, 2011)

Zaohongtao: An early maturing peach cultivar bred by artificial pollination (JiaLin *et al.*, 2011). Self fruitful, high and stable yield, rose-red colour, average fruit weight 136.6g, flesh white, melting, TSS 10.5 per cent.

Qiuyue: (Dabaifeng x Baixianglu), late maturing, clingstone, fruits are round in shape, average weight 208g, skin white covered with red blush, flesh white, melting, very sweet (TSS 13.5-14.5 per cent), good flavor (ZhengWen *et al.*, 2011).

21 Shiji: An inbred line derived from F1 of the cross Changli Dongtao x Shananxi Huanggantao (LiBin, 2000). It has large fruit, attractive appearance, productive, clingstone, creamy-white ground colour with a red blush, flesh pure white, non-melting, TSS 13-16 per cent, good canning quality.

Xumi: Xumi was obtained as a chance seedling (Zhongkui *et al.*, 1999). Fruits are large (average weight 197g), creamy-white skin covered with red blush, flesh creamy-white, semi-melting, very sweet (TSS upto 20.7 per cent), highly productive.

Qiulu: Seedling selection of 'Huanglu'. The fruit is orange, non-melting, sweet-sour and has good quality. It is an early bearing high yielding, canning cultivar. The canned product has orange colour, fine luster and delicate fragrance (Haichun *et al.*, 2010).

Qilu Juhong: (Hanxiangmi x Taowang 99 and Dongxuemi). Late maturing cultivar, fruit shape is nearly round, fruit weight 375-500g, skin colour bright-red with light-yellow background, flesh fine, sweet flavor and good quality (TSS 14.4 per cent), The cultivar is highly productive (Xuesen *et al.*, 2010).

Table 4.3: Important Peach and Nectarine Cultivars of Japan

Cultivar	*Fruit Weight (g)*	*Flesh Colour*	*Stone Adhesion*	*Quality*
Peach				
Hakuho	250	White	Cling	Excellent
Akatsuki	250	White	Cling	Excellent
Kawanakajima Hakuto	260	White	Cling	Excellent
Hikawa Hakuto	230	White	Cling	Good
Asama Hakuto	290	White	Cling	Excellent
Shimizu Hakuto	290	White	Cling	Excellent
Ogonto	300	Yellow	Cling	Excellent
Okubo	270	White	Free	Good
Nectarine				
Flavortop	250	Yellow	Free	Good
Fantasia	270	Yellow	Free	Good
Shuho	260	Yellow	Cling	Excellent

Japan

Most of the Japanese peach varieties have melting flesh and more than 80 per cent of peach production is for use as table fruit. The popular cultivars are Hakuho, Akatsuki, Kawanakajima Hakuto, Hikawa Hakuho, Asama Hakuto and Shimizu Hakuto. Most peach cultivars are white-fleshed and is a freestone cultivar. The main nectarine cultivars are Shuho, Fantasia and Flavortop; Canning peach cultivars like Kanto-2, Flavor Gold and Myojo were developed. The characters of some important peach and nectarine cultivars of Japan are presented in Table 4.3.

The National Horticultural Research Institute, Korea developed six cultivars between 1983 and 1998. The important characters of the cultivar are presented in Table 4.4.

Table 4.4: Characteristics of Peach Cultivars Developed at the National Horticulture Research Institute, Korea

Cultivar	*Flesh Color*	*Skin Color*	*Weight (g)*	*Soluble Solid (per cent)*	*Pollen*	*Hair*
Baekmijasaeng	White	White	180	11.0	Plenty	Yes
Cheoanhong	Yellow	Red	250	12.5	Plenty	No
Jinmi	White	Red	270	13.0	Plenty	Yes
Yumyeong	White	Red	300	12.0	Plenty	Yes
Baekhyang	White					Yes
Changhowonhwangdo	Yellow					Yes

Climate

Temperature

Peach is originally a temperate-zone plant, meaning thereby that the cold temperature has profound effect on its growth and productivity. Its commercial production areas mainly lie between latitudes of 30° and 40° N and S, though it is now grown almost all over the world extending from 10° to 49° N and 18° to 45° S latitudes (Edwards, 1987; Layne, 1987). The limiting factors in peach cultivation are the minimum winter temperatures, chilling hours, spring frosts, hailstorms, high humidity and desiccating winds during summer. However, the most important limiting factor in temperate region is the lack of flower bud hardiness either to low dormant temperatures or to frost or freeze conditions during spring (Hesse, 1975).

In the early fall, twigs and branches contain large amount of reserve carbohydrates which may help to maintain hardiness and provide energy for production of flowers and leaf tissues in the spring (Dowler and King, 1966). In early fall, the primordial flower is hardy, the bud scales and stem tissues around it are more tender, and serious injury may occur to trunk and crotch. In the early stages prior to bloom, the receptacle may become more tender than the ovary. When the style is extending rapidly it is most tender and is first to be killed. At the small green fruit stage, the ovule may be more tender than the ovary tissue resulting in seedless fruit which may or may not drop (Proebsting, 1975). Freeze or cold damage to peach trees occurs in several forms in south eastern United States. A sudden drop from -6.6°C for 2 or 3 days, often accompanied by snow following a warm period (15.5-24°C) from November to January can result in severe injury to the trunk, crotch and limb of trees; similar injury can also occur in February and March. This happens during sudden subfreezing weather followed by the usual warm period during which the growth processes (sap rising) have started (Clayton, 1975). At a temperature of 0.4°C, considerable damage to the linear shoot growth and flower buds were recorded in cv. Alexander (Pandey, 1977).

Since peach flower is bud hardy even between -23° to -26°C, absolute hardiness of the whole plant becomes less important (Hesse, 1975). Nectarines, in general, are more susceptible to low tempertures than peaches (Nitrantsky, 1987). A rapid method of assessing winter hardiness in peach genotypes has been developed by Yokovleva (1985) involving the measurement of electrical impedance in the primary cortex of one-year-old shoots. In most of the cultivated *Prunus* spp., at very low temperatures (-20°C), a freez-avoidance mechanism has been observed in which cold sensitive tissues, flowers primordia and xylem ray parenchyma get supercooled (Layne, 1984; Quamme *et al.*, 1982). Supercooled tissue water in the absence of ice nucleators does not freeze remaining in supercooled state in fully acclimated shoots and buds. But if the temperatures drop below - 25°C, this water freezes suddenly causing the death of tissues and plant.

The application of potassium gibberellate (Edgerton, 1966) or gibberellins between 80 and 100 ppm in late August and September increases the bud hardiness (Proebsting and Mills, 1978). Corgan and Widmoyer (1971) also found that the application of gibberellic acid (200 ppm) increased the hardiness in peach near bloom. Stembridge and Gambrell (1969) recorded cold hardiness of peach flowers during the following bloom period as a result of fruit thinning with the help of 3-CPA. The application of potassium gibberellate before leaf fall retards floral development in the following spring and provides cold protection although there is little increase in bud hardiness in the winter (Stembridge and Lorue, 1969). Marini (1986) stated that compared with dormant pruning, summer topping slightly reduced cold hardiness of flower buds. Fall application of ethephon increased the peach flower bud resistance to low temperature stress by delaying the bloom for one week (Durner and Gianfagna, 1988).

Most peach cultivars need specific chilling hours at 7.2°C for proper foliation and bloom in the spring. Conversely, they fail to foliate and fruit satisfactorily on growing in warmer regions. Scalabrelli and Couvillon (1985) investigated the effect of temperature and bud type on rest completion in Redhaven peach. They found that 7.2°C was more effective in releasing the buds from dormancy than 3° or 2°C. The flower buds had a longer chilling requirement than lateral vegetative buds, whereas the terminal vegetative buds the least. Prolonged chilling (2040 h) resulted in a decrease in the time to vegetative bud opening regardless of chilling temperature. But high temperature prevailing during December-January causes bloom abnormalities resulting in failure to set fruits (Weinberger, 1954). Weinberger (1956) stated that the high temperature during winter not only prolonged the rest period of peach buds but also counteracted the influence of chilling. Couvillon and Erez (1985) observed complete chilling negation in Redhaven cuttings exposed to 8 h diurnal cycles of 20° or 21°C but not when the cycles were of 2 or 4 hours. They found that no chilling negation occurred after 12 days' exposure to 23°C if three-fourth of the chilling requirement had accumulated before the exposure. Erez and Couvillon (1983) successfully counteracted the effect of high temperature by evaporative cooling of buds during dormancy with overhead sprinkling. The lack of required shared variable response which, however depends on cultivars. Delayed flowering (by 6-17 days), extended flowering periods (4-10 days), weak vegetative growth and high heterogeneity of fruit size were observed in north and centre of Mediterranean Tunisia (Ghrab *et al.*, 2014).

Chilling requirement is the quantity of accumulated temperature (hours) below 7.2°C which is necessary to condition buds for normal growth after the rest. Inadequate chilling results in delayed and sporadic foliation, deformed and non-viable flower parts and flower bud abscission. There may be crop failure and weakening of plant if the delayed defoliation is very serious (Bowen, 1971). Chilling requirements of most of the peach cultivars have been determined (Brooks and Olmo, 1972; Bowen, 1971; Weinberger, 1955). The relationship between cultivars chilling requirement (CR) and planting altitudes at Taiwan was studied by Shyikuan *et al.*, 2000). They stated that low chill peach genetypes (requiring 150 chilling units) are well adapted to low altitude growing conditions (85 masl), whereas high-chill peach genotypes (requiring 850h) perform well at high altitudes (1950 masl). For every 2.7m in altitude, 1 chilling unit is accumulated. Chilling requirement could partially be substituted by 100 ppm GA spray on Elberta (Bowen, 1971).

After the completion of chilling requirement, the time of bloom is determined by post-rest heat accumulation which depends upon the base temperature of a cultivar. Werner *et al.* (1988) compared the bloom date of Redhaven and Western sand cherry clone CG3-24. They suggested that the basis for difference in time of bloom was due to difference in the base temperature for heat accumulation and was not related to chilling requirement. Redhaven had a lower base temperature for heat accumulation than CG3-24. Souza *et al.* (2011) evaluated base temperature, thermal sum at different phonological stages, phonological phase duration, yield and seasonality of one nectarine and fourteen peach cultivars at Brazil. They considered phenoligical phases were pruning-sprouting; sprouting-flowring, from swollen bud to open flower, flowering-fruiting, from petal fall to medium-sized fruit, and ripening. Minimum basal temperatures (Tb) were found as: a)pruning-sprouting, 8°C, irrespective of the cultivars, b) sprouting flowering, 10°C, except for cv. Cascata 968, which required 8°C Tb, c) flowering-fruiting, 12°C, except for cv. Oro Azteca, which required 14°C Tb, (d) ripening, 14°C, except for cvs. Sunblaze, Diamante, Mejorado and Precochinho with 12°C Tb. For most cultivars, the maximum basal temperature were 30°, 34°, 34° and 28°C for phases pruning-sprouting, sprouting to flowering, flowering-fruiting and ripening, respectively. The cultivars differed for their total cycle and for the accumulated thermal sums which varied from 245 days and 1881.4 degree days, respectively for cv. Oro Azteca to 144 days and 1455.7 degree-days respectively for cv. Precohcinho.

In contrast to cool winter, warm to hot summer is favourable for the development of quality peaches as the fruits of many cultivars growing in cool summer conditions become astringent (Hesse, 1975).

Other Climatic Factors

Rainfall or irrigation is desirable during summer season for good peach crop but high humidity is conducive to the attack of pests and diseases. A dry climate with low humidity for 2 to 3 weeks before harvest is particularly ideal for nectarines. The ripening nectarine fruits, due to lack of pubescence, are readily affected by brown rot fungi and insects.

Amount of sunshine exposure is another important consideration in peach growing. Sunlight governs the rate of food manufactured by the leaves which in turn influences the size and quality. The fruit colouration enhances by good exposure to the sun. Areas situated in deep valleys or where cloudy weather prevails for longer durations are not fit for peach culture.

Areas experiencing frequent hail storms should not be selected for peach cultivation. Hails can be destructive to the fruit, impairing the quality or almost destroying it completely if the stones are large. It shatters the leaves and is also destructive to the bark; bruising and splitting it and stunting the general growth of the plant. The cv. Summerset was reported as a susceptible peach cultivar to late spring frost at flowering (71.5 per cent flowers frozen); while cv. Glohaven was found considerably more resistant (25.5 per cent damaged flowers) in Jugoslavia (Coric, 1997)

Strong winds which blow more or less continuously over a site are definitely undesirable. Dry winds during blossoming period adversely affect the fruit set and lowers the productivity. There is also a tendency in young trees to grow one sided (lop sided) and lean with prevailing wind.

Soil

In areas, subject to freeze damage, even the best peach soil would not make profitable planting. When frost protection cannot be obtained from large bodies of water, the site should be on rolling land and above the natural frost pocket (Hoffman, 1975).

In India, generally in mid-hills (800-1500 masl), northern aspect of slope is preferred for peach plantation. The land which is not too steep for efficient orchard operations is the most desirable site. Deep valleys are not the ideal sites because cold air settles in these areas and frost and freeze injuries are very common. On the steep land peach should be planted in the mid portion because soil at ridges are drier and less fertile while those of bottom are prone to low temperature injuries.

Type

Peach can be grown in varying soil conditions. However, for commercial plantation deep (one metre) sandy loam soil rich in organic matter is ideal. Hoffman (1975) emphasizes that physical properties of soil profile are more important than its fertility. Many of the small rootlets which constitute the absorbing surface of the root system are normally lost during dormancy. This absorbing surface must be replaced in early spring to supply nutrients and water for the rapidly expanding leaf surface, bloom development and for shoot growth. During this time not only water is needed in an available form in the soil, but there must be oxygen-containing air in the soil so that roots can respire and thus release energy that is used by them when they grow and take in water and nutrients.

Oxygen is essential for production of new rootlets. In fact, oxygen is much more important for the production of new rootlets than for maintenance of life in existing roots. Soils are likely to be poorly aerated if they have less than 10-20 per cent of their volume as pore space. A good peach soil may hold a reservoir of available soil moisture within root zone equivalent to 15 cm of rainfall (Hoffman, 1975). Such a reserve of moisture would carry trees through a prolonged dry spell without damage from drought. Loamy soils have more capacity for holding available moisture than light-textured soils.

The pH of the soil should vary from 5.8 to 6.8. Acid or saline/sodic soils should not be considered for peach cultivation. Besides the top soil, subsoil is also very important criterion for its suitability. If subsoil is hard, impervious, trees may grow satisfactorily for few years, but subsequently they may become weak and die.

Drainage

Amongst temperate fruits, peach plants are highly susceptible to waterlogged situations and prefer perfect drainage. Peach trees do not tolerate wet soils particularly during growing season. Submergence of rootsystem for even few days during the growing season when temperatures are high eventually results in death of roots and ultimately the plants. For the same reason high water table and a hard impervious soil layer need to be avoided as they cause yellowing of foliage, premature leaf fall and cessation of growth, and death of the tree in severe cases (Teskey and Shoemaker, 1972). Waterlogging injury is associated with the metabolism of the cyanogenic glucoside, prunasin, in *Prunus* spp. In absence of oxygen, prunasin gets hydrolized to hydrogen cyanide, which is autotoxic. The tolerance of a *Prunus* sp. depends on its ability to hydrolize prunasin (Salesses and Juste, 1971).

Rootstocks

Selection of proper rootstocks for peaches is an important consideration in commercial orcharding. Rom (1975) described the characteristics of an ideal rootstock for peaches, which may vary from region to region. These are, (i) soil adaptability, (ii) climate adaptability, (iii) pest resistance, (iv) disease resistance, (v) stock-scion interactions including vigour control, and (vi) nursery handling features. From practical point of view, there is no one rootstock which can solve all problems or be ideal for all situations. Thus for each production area, locality or site, a rootstock must have some characteristics to assure productive trees in that particular situation.

World situation on peach rootstocks has been reviewed by Layne (1987) describing the characteristics of presently used rootstocks (Tables 4.5 and 4.6). Seeds for raising rootstocks are obtained from three sources, *i.e.*, from wild types, from commercial cultivars and from special rootstock selections (Layne, 1987). Rootstocks raised from wild peach seeds lack uniformity and are genetically variable, whereas from the seeds of known cultivars they are more uniform and their performance is predictable. Peach x almond hybrids. (GF 556, OF 677) are important peach rootstocks in Europe and Mediterranean basins. They are clonally propagated and are useful on alkaline soils due to their resistance to chlorosis. On heavy, wet, or waterlogged soils, plums and plum hybrid rootstocks are commonly used (Layne, 1987). Several other *Prunus* spp. such as *P. davidiana. P. amygdalus, P. armeniaca, P. insititia, P. salicina, P. tomentosa, and P. besseyi* are also used as rootstocks for peach in specific situations (Rom, 1983). *Prunus besseyi* and *P. tomentosa* are the dwarfing rootstocks (Rom, 1983; Laync, 1987), but these are poorly compatible with peach. Rootstocks developed in Romania were selected to meet the demand of an expanding industry (Parnia *et al.*, 1988). The rootstock T-16 (*P. persica*) was selected from a local peach population in Dobrogea on hilly land with high Ca content. This is a vigorous stock, resistant to diseases and tolerant of high soil Ca. The rootstock De Bale (*P. persica*) is resistant to drought, vigorous and adapted to wide range of soils. De Bale is resistant to spring frost, vigorous and self-fertile while Oradea 1 (*P. persica*) is well adapted to a wide range of soils. In Japan, native peaches distributed in Nagano, such as 'Nagano Yaseito' and 'Ohatsu Momo' are the most popular rootstocks.

Greenpac: Originated in Spain in 2009 [(Felinem (almond x peach) x Cadaman (peach x davidiana)]; good compatibility with wide range of peach and nectarine cultivars, resistant to lime induced chlorosis and *Meloidogyne incognita.*

Castore: Originated in Italy in 2006 [GF557 (peach x almond hybrid) x OP]; adapted to fertile, permeable soils; prone to water logging; tolerant to lime induced chlorosis.

Felinem: Almond rootstock, compatible with peach, adapted to calcareous soil, resistant to broad range of nematodes.

Garnem: Almond rootstock, compatible with peach; characters similar to Felinem.

Monegro: Almond rootstock, compatible with peach; characters similar to Felinem.

Rootstock Adaptations

Temperature

Besides the scion cultivar, cold hardiness of rootstocks is equally important in cold regions so that the root systems withstand the soil temperature below -8°C (Layne, 1987). He determined the cold hardiness of root systems of peach seedling rootstocks and found Siberian C, Chui Lum Tao, Baily and Harrow Blood to be hardy, whereas Gold Drop and Elberta as tender. Several other rootstocks were also found cold-hardy.

Table 4.5: Characteristics of some Seed Propagated Prunus Rootstocks of Peach and Nectarine*

Rootstocks	*Nursery Characteristics*					*Orchard Characteristics*					*Disease Resistance*			*Nematode Resistance*	
	Germination	*Uniformity*	*Seedling vigor*	*Bud take*	*Resistant to suckering*	*Climatic adaptation*	*Chlorosis*	*Anchorage*	*Suckering*	*Compatibility*	*Vigor induction*	*Crown gall*	*Collar rot*	*Root knot*	*Root lesion*
Rootstock of *P. persica*															
Bailey	4	4	3	4	2	(3)	?	3	4	4	3	?	?	(1)	
ChuiLumTao	3	4	3	3	3	?	?	3	4	(3)	3	?	?	1	2
Elberta	2	3	3	3	3	2	?	3	4	4	3	1	?	1	1
GF 305	4	4	4	3	3	?	2	3	4	4	4	2	?	1	2
Halford	3	3	3	4	3	3	?	3	4	4	3	1	?	1	1
Harrow Blood	3	2	2	3	3	(3)	?	3	4	4	3	1	?	1	1
Lovell	3	3	3	3	3	3	?	3	4	4	3	1	1	2	1
Nemared	3	3	4	(3)	4	(2)	?	3	4	4	4	?	?	4	2
Rubira	4	4	4	(3)	3	(2)	?	3	4	(4)	4	3	?	1	2
Rutgers Red Leaf	3	4	4	3	3	2	?	2	4	2	3	1	?	1	3
Siberian C	4	3	3	3	2	2	2	2	4	3	2	1	?	1	(2)
TzimPee Tao	4	4	3	3	3	?	?	3	4	(3)	3	?	?	?	(3)
P. persica* x *P. davidiana															
Nemaguard	3	3	4	3	3	2	2	3	4	3	4	3	4	3	4

Rootstocks	*Nursery Characteristics*					*Orchard Characteristics*					*Disease Resistance*			*Nematode Resistance*	
	Germination	*Uniformity*	*Seedling vigor*	*Bud take*	*Resistant to suckering*	*Climatic adaptation*	*Chlorosis*	*Anchorage*	*Suckering*	*Compatibility*	*Vigor induction*	*Crown gall*	*Collar rot*	*Root knot*	*Root lesion*
P. instititia* x *P.domestica															
St. Julien Hyb.															
No. 1	3	3	3	3	(2)	?	3	3	2	2	3	4	(1)	?	?
No. 2	3	3	3	3	(2)	?	3	3	2	2	2	3	1	?	?
P. americana	(3)	(2)	(2)	(2)	1	?	?	1	1	1	(1)	?	?	?	?
P. besseyi	3	(2)	(2)	(2)	1	?	?	(2)	1	1	1	?	?	4	?
P. tomentosa	(3)	(2)	(2)	(2)	1	?	?	(2)	1	1	1	?	?	(2)	?

Rated on a scale where 1 poor, 2 fair, 3 good, 4 excellent, ( ) tentative rating, ? unknown or unreported.

* Modified after Layne (1987).

Table 4.6: Characteristics of some Clonally Propagated Prunus Rootstocks for Peach and Nectarine*

Rootstocks	*Nursery Characteristics*						*Resistance of*		*Effect of Scion*			*Others*
	Ease of Propagation		*Resistance to*		*Length of Period for Shield Budding*	*Ease of Trans-planting*	*Water*	*Iron-induced Chlorosis*	*Suck-ering*	*Vigor*	*Fruit Quality*	
	Hard-wood Cuttings	*Soft Wood Cuttings*	*Peach Rust*	*Crown Gall*								
P. persica* x *P. amygdalus												
GF677	1	2	5	2-3	5	5	2	5	5	5	3-4	
GF	1	2	5	3-4	5	5	1	5	5	5	3-4	Resistant to root-knot
P.insitita												
St.Julien GF655.2	4	3	2	4	1	3	3	3	3	2	5	Concentrated fruit maturity
P.domestica												
Brompton	2-3	NR	3	NR	2	3	4	3	3	3-4	3-4	Susceptible to collar rot
GF	2	2	3	5	2	4	3	3	4	4	3-4	Susceptible to GLSV
P. domestica* x *P. spinosa												
Damas GF 1869	3	4	4	4	4	1	4	4		3	4	Incompatible with many nectarines and some peaches

Rating were on a scale of 1 – poor to 5 – good, NR – not rated.

*Bernhard *et al.* (1979).

Soil Moisture

Peach seedling rootstocks show best performance on well-drained loam soils. For fine textured soils and waterlogging conditions, plum species and plum hybrid rootstocks are more suitable as they are resistant to waterlogging conditions. These include Damas OF 1869, St Julien d'Orleans, St Julien Hybrid No.1, St. Julien Hybrid No.2 and St. Julien GF 655.2 (Layne, 1987).

Soil Reaction

Peach seedling rootstocks perform quite well under slightly acid soil. GF 1869, OF 677 and St. Julien Hybrids 1 and 2 adapt well to alkaline soils tolerating iron-induced chlorosis.

Soil Fertility

Rich soils enhance excessive tree vigour enhancing susceptibility to cold and perennial canker. Vigour promoting rootstocks like Nemaguard and OF 677 should be avoided on very fertile soils, instead vigour reducing rootstocks such as Lovell and St. Julien GF 655.2 should be preferred (Layne, 1987).

Nematodes are more harmful to peach than other pests and diseases. As a result of breeding and selection, Nemaguard, Shalil, S-37, Okinawa, Higama and Nemared rootstocks resistant to root-knot nematodes have been evolved (Rom, 1983).

Rootstock Relationship to Yield and Quality

Rootstocks exert a considerable influence on the scion behaviour right from nursery to orchard but the latter is more important. A number of rootstock-influences like tree growth, nutrient uptake, time of bloom, yield, cropping efficiency, fruit size, adaptability and disease response have been enumerated by Layne (1987). When Redhaven peach was grafted on ten different rootstocks, the mean yields on OF 677, OF 43, peach seedlings PS A5 and GF 305 were 96.4, 90.4, 85.5, 82.0 and 79.1 kg/tree, respectively (Marangoni *et al.*, 1985). Nitrantsky (1987) observed the variation in yield of Redhaven cultivar on 18 rootstocks from 60.12 per cent to 166.30 per cent. In a comparison of four rootstocks, *viz.*, GF 305, Rubira, Montclar and Higama for the cultivars Red Top and Fontasia, OF 305 was found to be the most vigorous and highest yielding rootstock; Rubira was a less vigorous rootstock but gave high yield and quality fruits (Edin and Garcin, 1986).

Mizutani *et al.* (1985) compared the effect of vigorous (*P. persica* wild form) and dwarfing rootstocks (*P. japonica* and *P. tomentosa*) on the growth characteristics of Ohkubo trees. They found that tree height, trunk diameter and annual shoot growth were smaller in the trees on dwarfing rootstocks than in those on *P. persica,* but faster fruit maturation and higher soluble solids were obtained in the trees on dwarfing rootstocks. Rootstock *P. persica* gave the highest yield and larger fruits. Guerriero *et al.* (1988) identified seven peach rootstocks of low or medium vigour suitable for high density planting systems. The results obtained by Guerriero *et al.* (1988) over the first 3 years with 24 rootstocks of various genetic origin allowed the identification of 7 superior rootstocks of diverse vigour levels. Two rootstocks selected (I.S. 5/22 and Mr. S. 2/8) proved particularly interesting, being low or medium-low in vigour, they are suitable for use in high density planting, particularly in very fertile soils. The medium-vigour stocks, Manson and Rancho Resistant performed well in this trial. Manson and P x B 299 can be used in semi-arid conditions where little water is available for irrigation, Mr. S2/8 is suited to waterlogged condition and I.S.5/22 to calcareous soils. Of the rootstocks studied those of medium to high vigour proved well-adapted to the planting distances of the trials (4.5 x 2.0 m; 1111 trees/ha). Highest yields were obtained from trees on Manson and Rancho Resistant and OF 677 (Table 4.7).

Table 4.7: Rootstock Effects on Cumulative Yield and Individual Fruit Weight of Springcrest Peaches in Third, Fourth, and Fifth Seasons in the Orchard*

Rootstock	*Cumulative Yield (kg/tree)*	*Productivity Index (10 x kg cm²)*	*Fruit Weight (g)*
Manson	31.8	5.0	106.4
Rancho Resistant	30.7	5.0	107.4
Amandier (GF 677)	30.5	4.0	107.9
PB 299	27.9	4.6	108.4
I.S. 5/29	27.4	4.0	105.1
P.S.A6	27.4	4.1	110.1
P.S.A3	27.3	4.0	115.1
P.S.A5	26.8	4.2	117.6
GF 557	26.0	3.6	106.5
P.S.B2	25.9	3.6	107.2
Damas (GF 1869)	25.2	4.4	101.0
I.S. 5/31	24.1	4.1	101.8
Mr. S.218	23.3	4.0	109.5
I.S. 5/23	22.2	4.0	104.2
I.S. 5/22	21.3	6.6	105.2
I.S. 5/21	21.2	3.7	94.4
I.S. 5/19	20.8	3.9	104.3
GF 43	20.1	3.1	102.9
I.S. 5/8	19.1	4.5	112.0
I.S. 5/18	17.4	3.4	102.3
I.S. 5/26	16.3	3.8	100.7
I.S. 5/24	15.1	4.1	96.0
I.S. 5/20	14.7	2.9	100.6
I.S. 5/14	13.9	2.6	96.0

*Guerriero *et al.* (1988).

Werner and Young (1987) after studying the causes of low foliar calcium of the trees on Siberian-C rootstocks, suggested that reduced Ca levels in the scion were due to reduced uptake of Ca by Siberian C roots and not to blockage of Ca transport at the graft union.

Propagation

Peaches are commercially propagated by budding and grafting on seedling peach, plum and apricot rootstocks and rarely on clonal interspecific peach hybrids (Layne, 1987).

Seed Propagation

Peach seedlings are the principal peach rootstocks all over the world and the seed for this purpose is obtained either from wild types or commercial cultivars or from species rootstock selections (Rom, 1983). Mature fruits from seedling trees are collected during August-September and the seeds are extracted. Carlson and Badizadigam (1967) suggested to consider the following factors to get the optimum germination of peach seeds: (i) harvesting of fruit at optimum maturity, (ii) removal of fleshy pulp from the pit as soon as possible, (iii) drying of pits slowly, (iv) not to allow fruits with pits or pits alone to be submerged in water, (v) not to allow pits to overheat while going through the cleaning and drying processes, and (vi) the seeds after drying should be stored in dry cool place until sowing time.

Peach seeds do not germinate readily and require low temperature for stratification which is usually done under moist conditions at about 5°C for 70 to 90 days. Stratification period varies with the genotypes, and prolonging of the period induces secondary dormancy (Guerriero and Scalabrelli, 1985). They stratified the seeds of various rootstock lines at 5°C for 30-103 days and maximum germination in most of the lines occurred after 60 to 74 days' stratification. All seedlings grew normally when Okinawa seeds were stratified for 30 days between layers of damp cotton at 5-10°C. Nemaguard seeds need 75 days' stratification for the best germination (Nasir and Mohamed, 1987). Lovell seeds probably require the longest after-ripening treatment of 120 days (Teskey and Shoemaker, 1972).

Removal of seed testa and GA treatment reduced stratification period enhancing the germination. Sharbati seeds without testa gave higher germination than with testa (Chopra *et al.*, 1987). Chang and Werner (1984) proposed a relationship between seed germination and respiration during stratification. They found increased respiration rate during stratification with the removal of seed testa and likewise low chilling lines exhibited a higher respiration rate than those from higher chilling lines. Baz (1986) obtained the highest germination (96 per cent) by storing Mit Ghamer peach seeds at 5°C for 45 days but almost similar germination (95 per cent) occurred when the seeds were soaked in GA_3 at 100 ppm for 24 hours and cold stored only for 15 days. Donohoe and Walker (1957) reported that gibberellin was effective in terminating the rest period of Elberta seeds. The non-chilled Nemaguard seeds treated with GA_3 germinated 80 per cent and after two weeks chilling 100 per cent (Mehanna *et al.*, 1985).

On the basis of their investigation, Gianfagna and Rachmiel (1986) opined that GA_3 application before stratification increases the germination only if the seeds have already received the cold treatment for a minimum period. They further stated that the major changes in gibberellin-like activity are less associated with the germination process and more with the growth and development of seedlings. During stratification at low temperature, basic type gibberellins decreased and acid types increased. It was confirmed that breaking of the seed dormancy depends upon increase in the concentration of acidic type gibberellins (Sharma and Singh 1978, 1980 and 1981). On the contrary, Mehanna and Martin (1985) reported non-involvement of inhibitors and suggested that the seed-coats provided a physical barrier to germination. The embryo from a non-chilled seed was not dormant and germinated rapidly after removal of the seed-coat. Baz and Salama (1984) also recorded rapid germination after the testa removal without any additional treatment, whereas in seeds with intact testa dormancy was broken in 75 days at 7°C compared with 210 days at room temepature. They chromatographically partitioned and bioassayed the testa extract and found that the germination of cold-stored and warm-stored seeds had a negative relationship with endogenous growth inhibitors and total phenols which decreased faster under cold storage.

The pits are generally sown in the raised beds during October-November eliminating storage after ripening and planting the sprouting embryos in the spring. Pits are placed 5 cm deep and 10 cm apart in rows. Rudikova (1988) sowed the fruit stones at a spacing of 70 x 40, 25 or 16 cm and obtained large number of standard transplants with increasing density though the percentage of standard l-year-old-transplants decreased. Raising of seedlings directly in the nursery is preferred to germinating the pits first in the shallow beds and then transplanting to the nursery (Teskey and Shoemaker, 1972). Irrigation and removal of weeds are necessary for getting the good stand of seedlings in the nursery.

Vegetative Propagation

Budding and Grafting

Peach is generally propagated by shield or ring budding and occasionally by splice or cleft grafting (Pathak and Pandey, 1976). The techniques of its propagation have been described by a number of workers (Nauriyal, 1967; Thapar, 1960). The stocks trained into single stem should be budded when the sap flow is at the peak. Budding during September was reported to be superior than in May-June in the western hills of India (Srivastava, 1966). Ring budding during April-May and T or shield budding from June to September gave good success (Thapar, 1960). However, under subtropical conditions maximum percentage of success was observed in January (Sharma and Singh, 1979). The success in grafting of subtropical peaches can be obtained up to 90 per cent with the use of polythene strip as tying material (Sandhu *et al.*, 1983). Kanwar and Kahlon (1983) got 98 per cent success by grafting in early November.

Cutting

Hardwood Cuttings

Propagation through cuttings is the easiest and cheapest of all the methods of multiplication. In recent years, it has gained importance in the context of high density planting and clonal propagation of rootstocks. The raising of scion cultivars on their own roots even eliminates the need of rootstocks. Couvillon (1985) reviewed the propagation and performance of peach trees raised from cuttings for high density planting. According to him, insertion of hardwood cuttings in the field after treating with IBA is the least expensive method of rooting peach cuttings. Plants obtained by rooting of cuttings are more tolerant to drought and absorb or translocate more Ca than scions on Lovell, Halford, Nemaguard or Siberian C rootstocks. Tree growth, cold tolerance, fruit production and bacterial canker resistance do not differ between trees on their own roots and those on Lovell rootstocks. .

Peach has been raised successfully by hardwood cuttings after dipping them for 10 seconds in IBA 1000 ppm and rooting for 6 weeks with bottom heat at 23°C (Issell and Chalmers, 1979; Robitailie and Yu, 1979). The rooting was found better in mist when the basal hardwood tissues were dipped for 10 seconds in an IBA solution of 100 to 500 ppm before insertion into mixture of equal parts of sphagnum peat moss and perlite over underlying gravel (Robitailie and Yu, 1980). Hardwood cuttings were also rooted easily with 5000 or 10000 ppm IBA (Couvillon *et al.*, 1986). The treatment of IBA 1000 ppm and basal wounding further increased the rooting in hardwood cuttings (Scalabrelli *et al.*, 1984; Seyit and Couvillon, 1983; Sen and Couvillon 1983; Slack, 1978). Yu (1985) reported that peach could be propagated throughout the year by dormant hardwood cuttings, except for the softwood cuttings which required IBA treatment of high concentration (5000 to 10000 ppm) by quick dip method for four seconds and propagated in mist chamber. Scalabrelli and Couvillon (1986) described the method of rooting hardwood cuttings of Redhaven peach in the field and reported that rooting and establishment of the cuttings inserted in November were better than December or January.

Semi-Hardwood Cutting

Peach can also be multiplied by semi-hardwood cuttings taken from shoots which had been etiolated for 35 days after bud break (Nam and Kim, 1985). The rooting in semi-hardwood cuttings of peaches was found to be improved with IBA at 1500-3500 ppm and by basal wounding (Couvillon and Erez, 1980) or with the help of misting (Erez, 1975). The success in rooting of semi-hardwood cuttings is not dependent on cultivar but requires a well-drained rooting media (Couvillon *et al.*, 1986). Babaer *et al.* (1984) noted that softwood cuttings of peach treated with IBA at 20° to 30°C improved rooting and their subsequent development but either rise (40°C) or fall (15°C) in the temperature reduced the rooting capacity. The treatment of subapical cuttings with 50 ppm IBA for 24 hours and storage in moist sand for 30 days for callusing were found beneficial in improving the rootability (Jawanda *et al.*, 1979). However, Chauhan and Pundir (1972) obtained better response of combined treatment of 5000 ppm each of IBA + NAA by quick dip method in improving rootability of subterminal cuttings. While 50 ppm IBA alone showed promising response in cv. Sharbati (Chauhan and Maheshwari, 1970). Heighway (1984) successfully rooted the softwood cuttings of dessert peaches and nectarines by employing IBA 1000 ppm and perlite, vermiculite, polystyrene rooting medium and intermittent mist. Apart from IBA treatment, weekly sprays with a mixture of BA + NAA also increased the percentage of rooted cuttings (Gur *et al.*, 1986).

Considerably few attempts have been made for multiplication of rootstocks by cuttings. The cuttings taken from one-year-old shoots of temperate peach rootstocks were propagated successfully with the help of IBA 2000 and 2500 ppm and bottom heat treatment (Pandey and Upadhyay, 1981). Loreti *et al.* (1985) obtained the best rooting (95 per cent) in hardwood cuttings with IBA at 1000 ppm applied to PS B2 in October and to GF 677 in December; bottom heat generally gave best results.

Stooling

Stool layering of seedling plants and Behmi, a wild strain, has been attempted successfully by treating the shoots with 2500 ppm IBA (Mishra *et al.*, 1984; Pathak *et al.*, 1976).

Micropropagation

Both scion cultivars and rootstocks propagation *in vitro* have been standardized. Jona and Vigliocco (1985) obtained rosette formation within 10 days in bud explant of the cultivars Springtime and Dixired when placed on Zuccherelli medium with 3.5 µM BA or Lepoivre and Quorin medium with 5 µM BA. The formed plantlets could be either transferred to a rooting medium or to a proliferating medium to continue the propagation cycle. Dixired responded to treatments better than Springtime. Best rooting in 8 cultivars was, however, observed on half-strength MS medium supplemented with NAA rather than IAA or IBA. Cytogenetic analysis of micropropagated plants revealed that all were diploids.

Propagation of the peach x almond hybrid GF 677 in MS medium was best with low auxin concentration (Ruzic *et al.*, 1984). Nemaguard rootstock gave better survival on solid MS than in liquid MS medium and overall growth of explants was better at media pH 5.8 than at pH 5.2 (Reeves *et al.*, 1983). Morini and Concetti (1985) investigated the possibility of propagating PS B2 rootstock by *in vitro* culture and obtained the best result by culturing the shoots on MS medium for the elongating phase and on WPM (Woody Plant Medium) for rooting. Barbosa *et al.* (1985) tried embryo culture *in vitro* for breeding early peach cultivars and got higher percentage of seed germination *in vitro* than either in stratified or non-stratified seeds.

Planting

Preparation of Land

The site of orchard should be properly cleared of the bushes and weeds. It is advisable to plough or dig the whole plantation area and the arrangement of the drainage be made in advance of planting. In hilly areas, if slopes have a gradient higher than 45°, contour planting system should be adopted, otherwise terrace should be made by raising the retaining walls. In this case, the slope of the terrace should be kept inwards in order to facilitate the soil conservation. The rectangular system of planting should be adopted. Pits of approximately 1 x 1 x 1 m size at a distance of 4-6 m are dug in the month of September-October. The pit is well-filled by adding 30-40 kg well-rotten FYM. At planting time 40-50 g aldrin is added in the individual pits to avoid any damage from insects. As far as possible, new sites should be selected for establishing orchards in order to avoid the peach replant problem.

Replanting

The peach replant problem refers to all the causes of poor growth encountered in replacing old orchard with new one on the same site (Clayton, 1975). Old orchard sites are often used due to non-availability of new ideal sites. Trees in the second or third planting on old peach land either fail to grow well or may decline and die within few years. Number of factors may be responsible for the problem. These may be toxin (s) released from decomposition of peach roots, freeze injury, bacterial canker, root rots, nematodes, crown gall and rootlet decay.

As far as root toxicity is concerned, amygdalin in the cells of peach root bark is decomposed by certain fungi and bacteria. These toxic decomposition products, HCN and benzaldehyde, can kill peach rootlets or roots, adversely affect growth or even kill the young peach replants. These chemicals are not formed and released in all soils or in all areas. Peach root toxicity does not account for difficulty in establishing peach trees in most old peach orchard soils where all the roots have completely decayed.

Planting Distance

With the advent of high density planting and meadow orcharding, planting distance has assumed greater significance making it very diffiicult to give a general recommendation. To obtain early and higher returns, it is desirable to plant temporary trees along with the permanent trees, the former being removed in phases. Hartmann and Hill (1983) investigated the response of peach trees to various distances, *viz.*, 6 m between rows and 6, 4.5, 3, 2.25 and 1.5 m spacing within rows. Cumulative yield increased as spacing decreased, but closer spaced trees required both dormant and summer pruning. The net profit was highest at 1.5 m spacing. In hedgerow planting (2x1 m), alternate row of temporary plants should be removed by the 5th year so as to turn it into 4x1m spacing (Mitchell and Chalmers, 1983a). Bargioni *et al.* (1984), however, recommended 4.5 x 1.5 m spacing in view of the higher cost of establishment, higher labour requirement for pruning and the difficulties in maintaining the canopy in trees planted at further close spacings.

In meadow orchards, self-rooted scions are planted at 1 x 1 m or 1.2 x 0.9 m, accommodating 10000 or 9260 trees/ha (Bellini *et al.*, 1985; Coston *et al.*, 1985a).

Generally, one-year-old plants are procured from nurseries for planting which should neither be too tall nor too small. One-year-old plants are usually branched while June-budded plants are whips or have only few branches. Difference in plant vigour gets eliminated within few years in the orchards. Bare-rooted plants are supplied by the nurseries during the dormant season. Roots of these plants should not be allowed to become dry. If the planting is delayed due to some reasons, the roots should be buried in a well-drained moist soil.

High yield and low unit cost of production have been recognized as important components of profitable fruit growing (Hutton *et al.*, 1987). As a result, high density planting and meadow orcharding are gaining significance day by day. The important components of these intensive cultivation systems are increasing planting densities, dwarfing rootstocks, compact scion cultivars, planting of self-rooted cultivars, efficient tree structuring by training and pruning, use of growth retardants and efficient orchard management through trickle irrigation, weedicides, nutrition, integrated control of pests and diseases and mechanization. High tree densitites coupled with proper training and pruning, however, form the basis of these intensive growing systems giving exceedingly higher and earlier yields than conventional orchards (Bassi *et al.*, 1985; Sansavini *et al.*, 1985; Ende and Chalmers, 1982). Under high density planting, peach tree management requires more attention and care because fruiting shoots must be renewed each year, unlike the long living fruiting spurs of apples and pears. An even light distribution over the tree is essential to obtain uniformly ripened fruits with intense colour, good size and quality. In Po valley of Italy, high density planting of peaches at 1000-2000 trees/ha yielded a total of 200 tons/ha during the first 6 years of production after planting. This is equivalent to 10-year production record with medium-density planting (Sansavini, 1982). The extra cost involved in these sytems is recovered by their higher productivity (Hatzetheodorou and Hatzehariese, 1983). Gambrell (1975) reported that trees planted as close as 1.8 m apart in rows, produced well, lived for 12 years before being pulled out. Close planting had no detrimental effect on tree vigour and longevity. He suggested that trees should be trained starting from the first year by removing all lower laterals. The second and third years are very critical and the desired laterals should be selected. He recommended that after 3rd year 'mold and hold' pruning system should be practised. Hayden and Emerson (1975) concluded their results based on average yield of three cultivars, Redhaven, Glohaven and Madison in high density orcharding, that the two scaffold vase, the central leader fans and the Belgian fence had all given substantial increase in yield as compared with an 8-year-old standard peach planting of the same cultivars. Based on the cumulative early yield, new training systems, particularly for high density orcharding in peaches have been advocated.

Planting Time and Setting

Peach planting is commonly done in late winter or early spring. In moderate climate and unirrigated areas, late fall and early winter plantings give the roots ample time to establish before top growth begins. Where severe freezes occur, planting is best delayed until early spring. Since bare-rooted plants are planted, a hole of approximately the same size is made in the centre of the pit to ensure natural spread of the roots. The saplings should be planted to the same depth as it was in the nursery. Soil arround the plant should thoroughly be pressed followed by preparation of a ring around it. Immediately after planting it should be irrigated and 10-15 cm thick mulch of hay, oak leaves or FYM spread around it. Besides solid planting, peaches are also used as fillers in widely spaced fruits such as apple, pear and citrus. A peach is planted in the centre of a square of permanent plants.

Growth Regulation in High Density

For completely mechanized production, the system of growing plants under ultra density (about 10 000 trees/ha) is known as meadow orcharding. Mostly self-rooted cultivars are planted which are subjected to annual postharvest pollarding (Bellini *et al.*, 1985). All the investigations have shown higher yields by this system so much so that the yield of first year harvest equals to the yield of mature trees at a conventional spacing (Horton, 1985; Dallorto *et al.*, 1984; Young and Crocker, 1982). The size, shape and appearance of the fruits also show an improvement (Bellini *et al.*, 1985; Young and Crocker, 1982). Erez (1986) suggested a combination of paclobutrazol and pruning treatments to reduce pruning and increase yield in intensive peach meadow orchards. Autumn or winter application of paclobutrazol at 25 to 50 mg/tree in soil was found the most effective.

Martin *et al.* (1987) have enumerated many effects of soil application of paclobutrazol. These are reduction of canopy growth, increase in the number of vegetative and floral buds in the lower portion of the tree, elimination of the need for summer pruning, increase in hand thinning time in 2 out of 3 years, increase in fruit yield and size, advance in fruit maturity, decrease in leaf K and increase in leaf Ca. It also increased the cumulative yield of Flavorcrest peach from 30.1 in the controls to 41.8 t/ha for three years' experimental period. Tonutti *et al.* (1986) reported increased yield but delayed ripening and early leaf drop with paclobutrazol treatments. The effectiveness of paclobutrazol, however, depends both on application method, time and concentration (Li *et al.*, 1988). Soil treatment was more effective than foliar application (Li *et al.*, 1988; Blanco, 1986). According to Erez (1986) and Li *et al.* (1988), the effect of paclobutrazol was more persistent at higher concentrations (1000 ppm or 50 mg/tree) compared with lower concentrations (500 ppm or 5 mg/tree). Chlormequat (CCC) application generally increased shoot length and preharvest fruit drop in peaches (Mansour *et al.*, 1981). It further delayed fruit growth and ripening but improved fruit quality, especially at 2000-3000 ppm.

Mefluidide, [N-{2,4-dimethyl-5-(trifluoromethyl) sulfonyl amino phenyl} acetamide] a growth regulator which decreases GA-biosynthesis, inhibits apical meristematic growth accompanied by increased axillary branching (Arnold *et al.*, 1981). In peach cv. June Gold, application of 0.24-1.92 g/l of mefluidide resulted in shorter and narrower trees with smaller trunk and increased axillary branching as the concentration increased (Table 4.8). Two annual spring applications of 0.48-0.96 g/l mefluidide could result in adequate tree growth suppression without excessive phytotoxicity (Arnold *et al.*, 1983).

Table 4.8: Effect of Mefluidide on Lateral Branching of 'June Gold' Peach[z*]

Parameter	*No. Months After Application*	*Untreated Check*	*Mefluidide g/liter*			
			0.24	*0.48*	*0.96*	*1.92*
No. laterals within distal 30 cm	1	3.2(0.6)[Y]	3.7(0.5)	5.4(0.6)	4.8(0.6)	6.8(0.6)
	4	3.2(0.6)	3.8(0.0)	5.5(0.6)	4.8(0.6)	6.8(0.6)
No. laterals beyond intial terminal	1	2.2(0.8)	3.9(0.7)	6.3(0.8)	2.3(0.8)	4.5(0.8)
	4	3.9(1.2)	8.0(1.l)	7.8(1.2)	7.0(1.2)	11.2(1.3)

z: Only the 0.24 and 0.48 g/1 concentration were retreated; y: Figures in parentheses represent one standard error of the mean.

* Arnold *et al.* (1983).

The effect of gibberellic acid application at various stages of growth in early and late peach cultivars has also been investigated. Powell and Howell (1985) studied the effects of girdling or scoring the scaffold branches of young bearing peach trees (7 cultivars). Girdling consistently advanced harvest (3-10 days), increased fruit size (10-25 per cent) and yield (10-55 lb/tree). Scoring gave similar results in Bicentennial, June Gold and Redhaven but only fair results in other cultivars.

Pruning and Training

Training

Pruning of young plants is termed as training which is done to shape or build a strong framework of the trees in order to support maximum crop when they reach bearing stage. The aim is to utilize the available space and sunlight to the maximum extent for the production of quality fruits. Among the conventional training systems, *i.e.*, central leader, modified leader and open centre, the last one is usually adopted to train peach trees. In Indian situations, where plenty of sunlight is available, peach trees should be trained in the form of modified central leader system with 4-5 scafflod branches (Thapar, 1960). On the contrary, when sunlight exposure is a limiting factor, vase or open centre system of training is advocated. Training begins right from the planting.

Modified Leader System

Generally, one-year-old whip without a single branch is planted in the dormant season. Immediately after transplanting the tip of the plant is headed back 60 to 75 cm above the ground level. In the following summer most of the buds on main branch will sprout. Buds sporuting up to 30 cm from the ground should be pinched off soon after their appearance in order to develop clean stem (Armstrong, 1975; Christ, 1975). Three or four well-spaced buds projecting in opposite directions are retained. Care should be taken to develop wider angled crotches by projecting the developing buds towards the outer periphery. If summer pruning has been done as mentioned above, little pruning is required in the winter season. The selected branches are headed back to a bud projecting to the outer direction. In the absence of summer pruning, branches up to 30 cm from ground are thinned out. Three to four well-spaced branches with wide crotches are retained and the rest is thinned out. As a principle, the weaker branches are pruned heavier and the vigorous lighter

In the 2nd summer, the unwanted buds again should be pinched off. In winter all the weak, diseased, dead and interfering shoots are removed. Similarly, laterals which will grow towards the centre of the tree or straight to the sky should also be completely thinned out. The selected branches are then headed back to a bud projecting in the outer direction. In the subsequent years light pruning is required to attain the desired shape. All the selected branches should be spaced 15 to 20 cm apart vertically along the trunk. The central leader is also terminated in a bud which develops in the form of a side branch. From the third year onwards, some thinning out is practised to keep the tree fairly open and vigorous for admitting more light.

Table 4.9: Relationship among Canopy Volume, Fruit and Leaf Number in Loadel Cling Peach at 6th and 8th Years*

Training System	*Fruits (No/tree)*	*Fruit Weight (g)*	*Fruit Size (cm)*	*Canopy Volume*		*Units (per cent)*	*Fruit Density No. Fruit*		*Specific Leaf Density*	
				Tot. (TVU)	*Fruiting (PVU)*		*PVU*	*TVU*	*No. Leaves/ Fruit (PVU)*	*No. Leaves/ Fruit (TVU)*
6th year										
Early palmette	490	85.1	17.8	69	40	58.0	12.3	7.1	13.2	15.1
Delayed vase	516	94.2	17.9	87	50	57.5	10.1	5.7	14.7	19.9
Free central leader	575	104.3	18.8	88	56	63.6	10.1	6.3	13.3	17.2
6th year										
Early palmette	961	93.4	18.9	87	65	74.7	14.8	11.0	17.3	21.2
Delayed vase	913	101.7	19.7	139	89	64.0	10.3	6.6	19.1	23.5
Free central leader	735	119.9	20.5	107	65	60.7	11.3	6;9	20.9	25.5

PVU: Productive volume units; TVU: Total volume units; Canopy volume unit = 1/4 m^3 (90 x 30 x 90 cm).

*Sansavini *et al.* (1984).

Open Centre System

The training system has become very popular being very yield-efficient (Hutton *et al.*, 1987; Young and Jese, 1985; Ende and Chalmers, 1982). Trees are spaced 6 m in between rows and 1, 1.35 or 2 m in a row, resulting into a planting density of 1666, 1234 or 853 trees/ha. Two limbs of each tree are trained across the interrow space at an angle pinched off. In the following dormant season, if the summer pruning has been carried out, tipping of the selected branches with a bud projecting towards outer periphery is required. If the summer pruning has not been done, three or four branches arising in opposite directions with wide-angled crotches are selected and headed back. The unwanted branches are thinned out. In the second year, five to seven secondary limbs on the main scaffolds are selected and headed back. On secondary scaffold, diseased, weak and dry shoots are thinned out. Two or three strong secondary laterals are usually sufficient at this stage. The main stem is kept clear up to 45 cm above the ground. In the thinning operations, all the side branches which grow towards the ground, centre of the tree or straight to the sky should be removed.

V-shaped Tatura Trellis

Two limbs of each tree are trained across the interrow space at an angle of 60 from the horizontal, forming a V-shaped canopy (Hutton *et al.*, 1987). The canopy is supported by a permanent trellis constructed of high tensile galvanized steel fence posts. Major secondary scaffolds develop along each primary limb forming fruiting canopy.

High Density Vase

In this system trees are planted 3 m apart in the row (725 trees/ha) and trained similar to standard orchard trees except that the scaffolds are tipped in the winter to stiffen them and the trees are summer sheared to limit tree size when they had filled the space in the rows.

Modified Leader Fan

The trees are planted 3 m in the row and sheared once or twice in the summer with the shearing knife to prevent scaffold branches from developing into the row middles. Central leader with lateral scaffolds extending in the row only produces a flat tree about 1.3 m thick and 3 m wide.

Two Scaffold Vase

Trees planted at 3 m in the row are trained with an open centre and a single scaffold is allowed to develop in each direction in the row. This also produces a flat tree about 1.2 m thick and 3 m wide, but with different scaffold structure.

Pillar

Trees are planted 2.7 m, 1.8 m and 0.9 m apart in the row, and trained to a central leader with no scaffold branch. The tree shape approaches a cylinder. Small branches, tipped during the summer, fruit in the second year and are stubbed back to or nearly to the central leader after fruiting.

Slender Spindle Bush

The Slender Spindle Bush type system is commonly used for compact sized trees, especially grafted on dwarfing rootstock. However, it is difficult to apply this system to trees grafted on vigorous rootstock by winter pruning. Such practices only cause the trees to repeat imbalanced vegetative-oriented growth every year (Mizutani, 2011).

Belgian Fence

Trees are raised on 1.2 m spacing at a 45° angle in the planting hole with the top inclined in the row. Then vertical upright shoots are allowed to develop from this inclined trunk, which are summer tipped and sheared to form a flat wall of fruiting wood four feet thick. A variation of the Belgian fence is to dig holes 2.4 m apart and plant two trees per hole each inclined at 45° in opposite directions in the row.

The most widely used peach training system in Italy (Sansavini, 1983) differ little in their theoretical principles but do differ in orchard design, free spacing and, consequently, in the method of pruning and development of the tree frame. In fact, during the early years of developing tree shape, winter pruning is reduced when compared with older training systems. The peach cv. Loadel on Damas 1869 spaced at 5 x 3 m and trained to the early palmette, delayed vase and free-central training systems that were pruned uniformly showed higher yields for the delayed vase and free-shape training systems than for the palmette and hedgerow systems (Sansavini *et al.*, 1984). The free-shape system had significantly greater yield than the palmette system and these high yields were evident in the free-shape training system at the beginning of fruiting. The fruit and leaf distribution within the canopy shows that the high yield produced by free-shaped trees at the end of the 6th year were due to a greater number of fruit per tree (575 as opposed to 490 with the palmette).

Some other training systems which have performed well are classical goblet with severe pruning, dense goblet with moderate pruning, Kecskemeti hedge or Italian palmette, spindle, free standing palmette and double row opposed planar A-shaped canopy (Hutton *et al.*, 1987; Carelli *et al.*, 1986; Kappel, 1985; Taylor, 1985; Bencat and Matuskovic, 1982). The cultivars which showed the best response to ultradense planting are Ouromel 2, Ouromel 3, Joia 1, Joia 2 and Flordaprince (Dall'orto *et al.*, 1984). High yielding cultivars for meadow orcharding are Bicentennial, Camdam, Redhaven and nectarine Sunred (Bellini *et al.*, 1985; Coston *et al.*, 1985a). Early ripening cultivars should, however, be preferred to for meadow planting because delayed topping in late cultivars results in the retardation of shoot growth which ultimately reduce the production of flowers in the following years (Matta *et al.*, 1987). In Europe, the peach training systems have undergone an evolution in their form, with both research advances and with the passage of time. In the 1950s, the open vase dominated, while in the 1980s and in the 1990s the Central Leader and the Y were implemented (Grappadelli and Sansavini, 1998). The central Leader produces branches starting at 50cm from the soil, and the branches have wide angles with the Central leader; moreover the plant should not be allowed to grow beyond 3m in height (Errani and Barroni, 2000). According to Minguzzi (1993), the Central Leader shape of tree develops when vigorous shoot tips are allowed to develop and branches are pruned and trained in a conical shape with all the vegetation developing below the initial 50 cm being eliminated. The Y training system is adapted to high tree densities and early high yields, but is expensive to maintain because of increased labour costs and is difficult to maintain the production of fruiting wood as the trees age and grow larger. This system is, however, ideal for family run farms where there is no problem with supply of hand labour (Grappadelli and Sansavini, 1998). To obtain the Y it is necessary to prune newly planted trees 17 cm from the soil surface and let two vigorous shoots develop opposite to each other. In the second year, each branches (scaffold) are trained to an angle 30 to 35° from vertical (Minguzzi, 1993). Rufato *et al.* (2004) reported similar trunk diameter; fruit yield and canopy size but better quality of fruits in Y system.

Mahhou and Elghazzali (2000) compared different training systems at Morocco. They obtained higher yield from trees trained to goblet and palmette systems than those obtained from Tatura, Y and central leader systems.

Pruning

Pruning of Young Trees

Within 2-3 years of planting in the orchard, peach trees form a good framework consisting of main scaffold and secondary branches. At this stage only light to moderate pruning is required to keep the centre open and the branches properly spaced. Water sprouts and badly placed branches are also to be removed. Secondary branches closer to the trunk need to be removed and secondary scaffolds be headed back to outside laterals. Thereafter, the dormant pruning comprises thinning out of bearing wood to limit the crop and a light heading back to maintain the desired height of the plant. Padlog *et al.* (1983) compared the effect of moderate (long) and severe (short) pruning on the initial yield of peaches. Long pruning not only gave higher yields for 7 harvests than short pruning, but also increased precocity as in the first harvest yields were 12.95 and 2.64 kg/tree in the former and the latter, respectively.

Peach growers in the Southeastern United States have used the open Centre on vase-shaped pruning technique to train trees for more than 100 years. Due to increased production costs and managed competition, have gradually shifted to new management techniques including new cultivars and tree training methods. The growers have adopted training systems such as the perpendicular V (Kearney V or KAC-V) and quad-V (Dejong *et al.,* 1994), which require high tree numbers per hectare. Compact columnar genotypes are suitable for high-density orchard systems to take advantage of the unique branching pattern of these phenotypes to maximize both orchard production and labour efficiency.

Aslikyan (1983) pruned young peach trees trained as goblets by (i) the standard method, shortening l-year-old shoots by half to two-thirds of their length (control), (ii) shortening by a quarter to a third with thinning out 10-20 per cent of shoots growing inwards and removing competing shoots, (iii) shortening by a third to a half and thinning out, and (iv) shortening by two-thirds to four-fifths and thinning out 25-30 per cent of shoots. He found the best balanced tree growth and highest yield by light to moderate pruning (treatment ii and iii) and reduced yield by severe pruning (treatment iv) in comparison to control.

Pruning of Mature Trees

Bearing peach trees require heavier and regular pruning than other temperate fruits which in principle is done to maintain a balance between growth and fruiting. Peach tree bears fruit laterally on the previous season's growth and once a growth has fruited it will never bear again in its life. This is the main cause of tip bearing, particularly in the orchards where scientific pruning is not practised. The main objective of pruning of a bearing peach tree is to reduce the unproductive parts having one or two buds which in turn will form new fruiting branches in the following season (Srivastava, 1965). If a bearing peach tree of 10-12 years of age is making 25-30 cm of shoot growth on the majority of the older branches, it is sufficient for maintaining optimum productivity. Pruning consists of both thinning out and heading back of the shoots.

Pruning intensity plays an important role in modifying the growth and productivity of peach cultivars. Moderate pruning results in greater yield (Pandey, 1977). Light pruning up to 12 buds results in higher fruit setting (Singh, 1983). Results of various experiments have shown that during early bearing the selected shoots should be lightly pruned by removing 20-30 per cent linear growth after corrective pruning of the tree. After 7 to 8 years of bearing, the natural vigour of the tree is diminished. The corrective pruning should be followed by removal of linear extension growth to the extent of 70-80 per cent. The remaining 50 per cent of selected shoots should be lightly pruned as mentioned above for early bearing. Pandey (1977) suggested that moderate and light pruning should be combined in the alternate branches to achieve both attributes in a single bearing peach tree. After thinning out of superfluous branches (corrective pruning) the next step is to reduce the fruit buds in order to produce well-distributed crop. The fruiting shoots should be lightly pruned and alternate ones severely headed back to get sufficient yield and following year's growth, respectively. Moderate pruning will encourage optimum production while severe pruning will reduce the crop load and encourage linear shoot growth for production in the following year. In the subsequent years, the pruning severities in these two branches are changed and the tree can be maintained in growing and fruiting condition. Kumar *et al.* (2010) examined different intensity of pruning on three peach cultivars. They observed that fruit yield decreased with the increase in severity of pruning in Flordaking and Saharanpur Prabhat peaches, whereas, medium pruning treatment gave highest yield in Flordasun peach. Fruit weight, size, total soluble solids, sugar and acid content of fruit were found significantly increased by pruning in Flordasun peach. In cv. Flordaking medium pruning and in cv. Saharanpur Prabhat heavy pruning caused significant improvement in physics-chemical characters of fruit.

In order to avoid tip bearing in the later stages, some of the old shoots should be thinned out and few headed back which have already fruited in the previous season. This will guard off the usual rejuvenation practice, involving heading back of old, tip bearing or barren trees.

Time of Pruning

Peaches are pruned every year in dormant season but, of late, pruning in summer has also gained some importance (Mika, 1986). Marini and Rossi (1985) made a comparative study on pruning of Sunqueen trees by giving three treatments, *viz.*, dormant pruning in March, summer pruning in late July and summer topping in late July. They found lower yield, fruit size and crop values by summer topping than dormant or summer pruning. The crop values were $ 16 749, $ 16 389 and $ 14 854/ha for dormant pruned, summer pruned and summer topped trees, respectively. Pruning time is also governed by the size of orchard, available labour, and climate. In small orchards, pruning should be done just prior to starting of growth in early spring, but in large orchards it has to commence earlier in order to accomplish the pruning of all trees by due time. In canker prone areas, pruning needs to be delayed to late winter or early spring (Teskey and Shoemaker, 1972).

Dormant Pruning

Peach should be pruned after the onset of dormancy during winter months. It is generally practised during December-January. However, in cooler areas it should be delayed up to middle of February in order to avoid low temperature injury to the twigs and flowering buds (Pandey, 1977). Pruning operations should be completed within a specified period because flowering may be delayd by early or late pruning (Sandhu and Singh, 1984).

Summer Pruning

Summer pruning consists of pinching or shearing or heading, all accomplishing a gross removal of current season's growing tissues. It includes removal of only a, part of the growth that occurred in the current season. In general, about half of this growth is removed during summer pruning.

The pruning is done when approximately 20-25 cm of growth has occurred and it may be necessary to repeat the pruning a month later. Summer pruning is helpful in restricting the tree vigour, particularly for high density plantation. It reduces current-season shoot length, diameter, and increases lateral shoot growth (Rom and Ferree, 1984). But it does not suppress the shoot growth more than dormant pruning (Marini, 1984). Mizutani (2011) however, working with early maturing peach varieties growing on vigorous rootstocks and trained as slender spindle bush reported that summer shoot heading back and thinning favours reducing tree vigour and light penetration near the tree, and heading back keeps alive shoots and buds near the trunk which otherwise weaken or die back due to apical dominance and/or shading. In modern peach orchards, pruning in late summer is suggested, which is very useful to obtain a better light distribution in the canopy and a more efficient carbon allocation to fruiting shoots, preventing and reducing the need for winter pruning (Neri and Massetani, 2011). Marini (1985) reported that summer topping of mature open-center trees in June and/or July caused more than doubled light levels in the interior portion of the tree for the remainder season, but light levels at the tree periphery were not enhanced. This can be mechanized through the use of currently available mechanical devices. Summer pruning increases the incidence of diseases such as peach canker and it requires high standard of general management for successful peach orcharding.

Nutrition

Nutritional requirements of peaches are comparatively higher than other fruits because it bears on previous year's growth making the production of annual growth imperative for fruit production. Recoupment of the wood removed every year in pruning is also necessary. Fertilizers are, therefore, applied annually except following a crop failure (Teskey and Shoemaker, 1972). A large number of factors, *i.e.*, the type of soil, its fertility, climate, cultivar, planting density, shoot growth, and irrigation facilities determine the amount of fertilizers to be applied. If an young peach tree makes approximately 40-45 cm and a bearing tree 20-25 cm annual shoot growth with dark green foliage, the fertilizer application is not required (Childers, 1975).

Role of Nutrients

Nitrogen

The deficiency symptoms of N first appear in leaves which become small, deformed and defoliate prematurely (Weinberger and Cullinan, 1936). The leaf also become yellow showing purplish red spots (Davidson and Blake, 1936). Under low N condition the fruits become small, brightly coloured and ripen earlier. Conversely, abundant N availability enhances vegetative growth, *i.e.*, greater fresh and dry leaf weights, darker green colour and thicker trunks (Ritter, 1957). It increases the yield, number of pickings, but delays maturity. Excessive vegetative growth late in the season is prone to winter injury (Johnston, 1952). Arora *et al.* (1999) reported that flowering intensity, fruit set and fruit weight were directly associated with leaf N content. Mrvic (1998) studied the effect of timing of N fertilizer on peach cv. Suncrest. The trees were fertilized with NPK (77:56:168) in autumn, and 81 kg N/ha was applied as calcium ammonium nitrate (CAN) in March, and June or autumn. The most balanced soil N content was achieved by repeated application of CAN (March and June). The yield and efficiency of N taken up by the tree showed distinct improvement by repeated application of CAN. In peach cv. Andross, application of 15.5 N units (Kg 0.1/ha) supported by 15-15-15 and 26-0-0 slow release fertilizes caused adequate N in the soil for the whole vegetative period for tree growth and development and the attainment of high productivity (Koutinas *et al.*, 2010).

Nitrogen, when applied at 125 or 250 kg/ha annually to Redhaven peaches, accelerated tree growth and yield but delayed fruit ripening and slightly reduced fruit quality (Janie, 1979). Delay in ripening and lack of colouration by high rates of N application have also been reported (Ashley, 1931; Lou, 1931; Nightingale *et al.*, 1930). Increased N application augments the percentage of water soluble and total N and reduces-total sugar contents of fruit. Application of N increases the yield but adversely affects the ascorbic acid content of fruit (Wittwer and Hibbard, 1947). Fruit acidity and total soluble solid contents are enhanced by high nitrogen rates (Schneider *et al.*, 1958).

In a study of rootstock seedlings in nutrient solution, the growth was found to be maximum when $N0_3$: NH_4 ratio was 50:50 or 25:75 but poorest when NO_3 was the only source. In a comparison of 1 per cent urea sprays and ammonium sulphate application at 225-900g/tree, EI-Banna *et al.* (1981) found that urea had little effect on yield, whereas ammonium sulphate (900g) increased the yield from 28.1 kg in the control to 34 kg/tree in the first year and from 26.6 to 30.9 kg/tree in the second year of experiment.

Phosphorus

Phosphorus-deficient trees make restricted growth, produce leathery leaves with abnormally red petioles and short internodes with leaf arrangement in whorls. The fruits are flat in shape, small in size, coloured and poor in quality. P-deficient trees showed less growth and flower buds than trees receiving NPK fertilizers (Veerhoof, 1948). On the basis of orchard surveys, Lilleland and Brown (1942) concluded that leaves from better orchards had lower P content while poor tree growth was often associated with high leaf P.

The range of leaf P recorded in different orchards was from 0.14 to 0.27 per cent. P treatment results in larger, glossier and darker green leaves than control (Lilleland, 1935). Hirai *et al.* (1961) reported that P application induced greater fruit bud differentiation and development but use of excessive N retarded it. Sato *et al.* (1953) suggested foliar spraying of 1.0 per cent mono-ammonium phosphate to correct P deficiency. P application should, however, be only made when low soil content is indicated by its analysis. Low temperatures reduce root growth and rate of absorption and thus create the condition of P deficiency (Teskey and Shoemaker, 1972).

Potassium

Short, thin shoot growth, intense red colouring of shoots, small light green leaves, wrinkling of midrib, inward and upward rolling of leaves and low fruit set have been described as symptoms of K deficiency (Boynton, 1944). Severe marginal chlorosis of leaves due to lack of K has also been reported (Bollard, 1953). Elberta shoots, when deficient in K under sand culture experiments, were spindly and with longer internodes, and leaf tips showed chlorosis similar to Fe deficiency (Weinberger and Cullinan, 1936). Less than 1 per cent K in leaves is an indication of its deficiency (Lilleland and Brown, 1941; Cullinan and Waugh, 1939). K deficiency usually occurs in trees grown on light soils with pH 6.0 to 7.0 or on soils high in Na content (Lilleland, 1946). There is a positive relationship between K of leaf and soil but the yield or shoot growth is not related to either of leaf K or soil K (Griffin, 1959). Mulching results in higher leaf K content than unmulched trees and mulched trees retain leaves longer in fall producing more flowers (Baker, 1941). Chandler (1933) and Brinkley *et al.* (1958) did not find any effect of K fertilizers on growth and yield of peaches. It also did not affect firmness, acidity or keeping quality of Elberta fruits (Weinberger, 1929, 1934). Griffin (1959), however, reported that K fertilization increased fruit bud formation, fruit set, fruit size and hastend maturity but decreased the soluble solid content of the juice. Humbert (1968) recorded an increase in size, weight and yield of fruits by K fertilization.

Calcium

On the basis of nutrient culture experiments, Ca deficiency symptoms have been reported in peaches. In the absence of Ca, the trees were dwarf with short internodes and the leaves exhibited cholortic patches which gradually truned brown and extended (Cibes, 1950). Under severe Ca deficiency, marginal chlorosis was followed by necrosis in leaves, and shoot tips dried in some seedlings (Edwards and Horton, 1979). Roots became brown and fruits were smaller, greener and firmer and had less soluble solids and red blush than at higher Ca levels (Abdulla and Childers, 1973). When Ca levels were increased in treatments, the leaf Ca content rose but fruit Ca declined. Moderate Ca application enhances leaf N and retards leaf K than the higher Ca rates (Gammon and Shoemaker, 1963). Crisosto *et al.* (2000) reported that foliar calcium sprays applied every 14 days starting 2 weeks after full bloom and continuing until 1 week before harvest did not reduce surface discolouration or increase fruit quality of mid-or late-season peach and nectarine cultivars growing in Tulare country, California. However, Ca content in the skin of the fruit was found to increase in most of the cultivars.

Magnesium

Absence of Mg in sand culture causes interveinal yellowing or leaves with veins remaining dark green. Interveinal necrosis occurs later and leaves abscise (Vcinberger and Cullinan, 1936). Cibes (1950) reported dark green water-soaked area, along the margin of older leaves which soon turned greyish white and abscissed due to lack of Mg. Under field conditions, marginal chlorosis and interveinal spotting of leaves caused by Mg deficiency were corrected by magnesium sulphate application (McClung, 1953).

Sulphur

Weinberger and Cullinan (1936) reported the sulphar deficiency symptoms as small leaves with red tints developed near terminal tips which gradually scorched and withered. Many fine roots developed which were brown in appearance. Since fertilizers, irrigation water, pesticides and fungicides contain enough S to meet the requirement of peaches, separate application of this element is not needed.

Zinc

Zinc deficiency in peach orchards has been reported from all over the world (Mitra, 1988). Zn deficiency causes crinkling, waving and chlorosis of terminal leaves. Rosette of small leaves develops at terminals and the leaves may be without stalks, small and rigid (Teskey and Shoemaker, 1972). Woodbridge (1954) reported the ranges of leaf Zn content, *i.e.*, severe chlorosis 5.3 to 8.0 ppm, moderate chlorosis 10.7 to 10.8 ppm, slight chlorosis 7.9 to 13.4 ppm and healthy leaves 9.9 to 30.4 ppm.

Foliar Zn sprays in summer are more effective in removing Zn deficiency than either soil application or dormant sprays (McClung, 1954). Zinc sulphate lime also controls rust besides correcting the Zn deficiency. Singh *et al.* (1986) obtained a reduction in leaf chlorosis and increase in leaf Zn content by spraying $ZnSo_4$ (0.4 per cent), Zn-EDTA (0.1 per cent) and Zn-EDPTA (0.1 per cent), on Sharbati peach trees.

Iron

Lime-induced iron deficiency causes interveinal chlorosis while veins remain green. In severe conditions, the entire leaf turn whitish yellow before drying. Die back of twigs, small fruit size, poor skin colour, low yield and poor fruit quality are other consequences of Fe deficiency (Shen and Hsiang, 1949; Guyot, 1948). The chlorosis appears first on terminal leaves (McGeorge, 1949). The iron-deficient trees are more susceptible to low temperature (Thorne and Wan, 1950). Fe deficiency has been corrected by foliar spraying, soil drench application, solution injection into trunks and dry salt injection into trunks (Bennet, 1931). Significant reduction in leaf chlorosis of Sharbati peach was obtained by spraying of $FeSo_4$ (0.4 per cent), Fe-EDTA (0.1 per cent) or Fe-DTPA (0.1 per cent). The treatments improved the fruit quality and enhanced the leaf Fe content (Singh *et al.,* 1986). Briones and Montes (1986) reported best results with 16.8 g Fe-EDDHA/cm trunk diameter, which increased foliar Fe concentration to 84 ppm. Tree injection of $FeSo_4$ (12-24 per cent) regreened trees, enhanced bloom, regenerated vegetative growth, increased fruit set and improved yield and fruit quality (Yoshikawa, 1988).

Boron

B deficiency symptoms have been described as small leaf size, shortened internodes, dying back of twigs, browning of the embryo, internal fruit necrosis, reduction in fruit set and yield and abnormal lenticel development (Kim *et al.,* 1969; McClung and Clayton, 1956). Sometimes B deficiency symptoms resemble cold injury. B content of leaves and fruit indicates its status in the plant clearly. Kamali and Childers (1970) established B content of the peach leaves as follows: severe deficiency 10 ppm; deficient 11-17 ppm; optimum 31-59 ppm and high 60-80 ppm.

Boron deficiency can be corrected by annual application of sodium tetraborate or boric acid at 13.5 kg per acre in the soil or foliar spray but soil application is preferred over foliar spray (Foster and Moisey, 1959). Application of B causes early ripening, greater fruit size, but reduces red colour development and delays blooming (Williams and Veerhoff, 1948). Kavvadias *et al.* (2012) noted maximum yield (135 Kg/tree) of 8 years old peach cv. Andross on GF677 by soil application of boron while foliar application of boron in combination with calcium at flowering, fruit set and fruit growth stages showed fruits to be less affected by cracking.

Toxicity of B is shown by necrotic lesions on stems above the leaf axils, withering and dying back of tips of current-season's growth, development of cankers along stem with gumming, distortion of fruit with sharp sunken areas and maldevelopment of kernels (Eaton, 1944; McLarty and Woodbridge, 1950).

Copper

Andersen (1932) observed cessation of terminal growth, rusetting and twig die back due to Cu deficiency. In the event of Cu deficiency, the leaves first becorne dark green and subsequently turn yellowish green between the veins scattered green spots giving a mottled appearance (Cibes, 1950). In severe cases, the entire leaf becomes chlorotic with small reddish brown or copper coloured dots on its surface. Soil application of copper sulphate proves effective in controlling copper chlorosis and regreening of leaves.

Manganese

Interveinal chlorosis occurring due to Mn deficiency should be corrected by an early summer spray of $MnSo_4$ (Teskey and Shoemaker, 1972).

Aluminium

Aluminium concentrations (0-100 ppm) in nutrient solution reduced proportionately the dry weight of stem, roots and leaves of cv. Lovell peach seedlings (Kirkpatrick *et al.*1975). The concentrations of Ca, Mg, Mn and P in plant tissues decreased with increasing Al concentration in the solution. Advanced stages of Al toxicity were characterized by collapse of the midrib terminal die back and defoliation of the seedlings. The roots died back and new roots developed as irregularly shaped cylinders (Edwards *et al.*, 1976).

Fertilizer Recommendations

Generally, N alone or in combination with K improved fruit yield, whereas P failed to give the same response. Havis and Gourley (1937) reported the response of only nitrogenous fertilizers on nine-year-old Elberta peaches. Dotti (1942) concluded that ammonium nitrate had an appreciable effect on tree growth but P and K without N did not give the same response. Liwerant (1959, 1960) obtained 102.4 per cent increase in yield as compared with control by application of 40 kg N, 27.5 kg P, and 50 kg K per acre for a period of 14 years. The yield declined by 36 per cent with PK as compared with NPK treatment. Degtyar (1981) reported the highest yield by the application of N:P:K at 225:90:225 kg/ha. On the basis of macroelement contents and their distribution in peach trees, Stassen (1987) suggested that peach tree requires about 140 kg N, 10 kg P, 80 kg K, 75 kg Ca and 18 kg Mg/ ha to produce a crop of 25 t/ha.

Application of 40-50 kg farmyard manure yields good effect (Srivastava, 1974). Singh (1969) recommended a fertilizer dose of 55 to 65 kg N, 55 to 65 kg P_2O_5 and 110 to 135 kg K_20 per hectare for bearing orchards. Moharam and El-Deen (2011) reported highest yield and better fruit equality of cv. Early Grand by application of 50 per cent NPK and 50 per cent requirement of NPK by olive solid waste under supplemental irrigation, compared with application of NPK as mineral fertilizers only. Singh and Srivastava (1978) reported a combination of 20:15:15 g of N:P:K per year age of the plant for optimum growth and production. They also suggested that the rates should be increased annually and after ten years a constant dose be given. In subtropical peach cv. Sharbati, Sharma (1976) found beneficial effect of 40:60:120 g N:P:K application per year age of the plant. Further, Yamdagni and Jindal (1979) recommended for application of 400 g N, 100 g P and 400 g K for a bearing peach. Khan *et al.* (2000) recorded maximum fruit weight (194.1), fruit set (69.6 per cent) and yield (37.9 kg/tree) in peach cv. 6-A at Peshwar, Pakistan by annual application of 1000g N and 1200g each of P and K.

Application of micronutrients has also shown beneficial responses in peach production. Sandhu *et al.* (1980) obtained higher yield by 0.5 per cent $ZnSo_4$ as foliar spray or incorporation of 1 kg/tree as soil application. Spray of 0.2 to 0.3 per cent of borax and copper sulphate has also been recommended to improve the productivity and correct the deficiency (Anon., 1975). Chopra *et al.* (1982) reported that spray of boric acid (0.2 per cent) at 15 day interval between pea and maturity stage of fruit growth and development increased the fruit Ca level from 0.19 to 0.28 per cent and shelf life of fruit by 6 days. The quality of fruit improved with the application of Fe and, Zn chelates. However, Zn proved more effective than Fe (Singh *et al.*, 1986). Folliar spraying of boric acid (0.1 per cent), zinc sulphate (0.5 per cent) and ferrus sulphate (0.5 per cent) twice (after petal fall and again after 15 days) was reported to improve fruit size, fruit weight and yield of peach cv. Sharbati (Yadav *et al.*, 2013). Ali *et al.* (2014) in their experiment with peach cv. Early Grand at Peshawar, Pakistan reported that when soil was deficient in P, Zn, Fe and B, application of a combination of Zn + Cu + Fe + Mn + B as foliar spray increases fruit yield, improved fruit size and quality.

Time and Method of Application

Peach is a deciduous plant and flowers during February-March. Hence the plant must have sufficient nutrition built up before that. Required amount of farmyard manure or compost must be applied in December-January (Srivastava, 1974). The three-fourth dose of nitrogen and full dose of phosphorous and potash should be given prior to dormancy break. Remaining one-fourth of nitrogen should be given immediately after fruit harvest.

Phosphatic and potassic fertilizer should be applied in trenches of 20 - 25 cm width and 10-15 cm deep made beneath the tree canopy at a distance of 1-2 m from the main trunk. The nitrogenous fertilizers may be spread on the ground under the tree canopy and incorporated in the soil by gentle raking and light irrigation.

Tissue Analysis and Nutrient Status

In peach, the leaf analysis has been proved the best criterion to assess the nutritional requirement of a plant (Magnani *et al.*, 1984). For this purpose leaf sample should be collected between early June and end of August from the mid portion of current-season non-bearing shoots (Khera *et al.*, 1981). Chapman (1964) recommended sampling of leaves including petioles from mid portion or near base of current season's terminal growth 12 to 14 weeks after full bloom. The most stable period is during stage III (final fruit swell) when there is the least change in the percentage of various elements (Batjer and Westwood, 1958). The percentage of N, P, K, Zn and Cu were highest in early June and declined for the season as a whole with a more or less pronounced tendency to level off in late July and August. Percentages of Ca, Mg, Fe and B increased for the season as a whole but remained relatively constant during the later part of the season (McClung and Lott, 1956). The ranges of nutrient levels in peach leaves determined by Westwood (1978) are given in Table 4.10.

Table 4.10: Nutrient Levels in Peach Leaves Indicating the Ranges from below Normal to Excess*

Nutrient Level	*Per cent Dry Weight*						*ppm Dry Weight*			
	N	*K*	*P*	*Ca*	*Mg*	*Mn*	*Fe*	*Cu*	*B*	*Zn*
Below normal	2.0	1.0	0.08	0.20	0.18	20	40	1	30	10
Normal	2.8	1.5	0.12	1.0	0.24	25	50	4	35	18
Above normal	3.8	3.0	0.30	2.5	1.0	200	400	SO	80	100
Excess	4.5	4.0	0.70	3.0	2.0	450	500	100	100	200

* Westwood (1978).

Status of one nutrient in leaves affects the status of other nutrients. Higher dose of nitrogen nutrition increases N but decreases P and K, whereas low N results in reduced N but high P and K contents in leaves. A positive correlation was reported between P application and leaf P and K contents but a negative correlation between applied P and N contents, whereas positive correlation is reported between applied K and leaf N, P and K contents (Sharma and Singh, 1982). On the basis of orchard surveys, the optimum values for leaf N, P, K, Ca and Mg for high yielding trees were estimated as 3.37, 0.20, 1.46, 1.16 and 0.35 per cent, respectively (Sato, 1955).

High levels of pruning, thinning and irrigation generally increase the N, P, K, Ca and Mg contents of peach (Kenworthy *et al.*, 1959; Morris *et al.*, 1962). Stojkovska *et al.* (1972) reported that irrigation increases the nutrient contents of leaves, shoots, and fruits in both fertilized and unfertilized soils. Archibald and Bradt (1964) found reduction in N and K contents of leaves from the orchard under sod cover.

Irrigation

Water, besides being a constituent of plant structure, is also a medium for biochemical reactions. Evidently water stress influences all the life processes of the plants. Greater yields and fruit size in peaches from irrigation resulted in economic benefits (Zini, 1984; Daniell, 1984). Irrigation further improves growth of trees and their survival, lowers levels of winter injury and perennial canker and increases annual cumulative marketable yield (Layne and Tan, 1984). Storchous (1986) also reports 2- to 3-fold increase in productivity and enhanced shoot growth and trunk girth due to irrigation. It improves productivity by increasing fruit set and often fruit size and decreasing fruit drop. Late cultivars benefit more than early ones (Zini, 1984). Modrate water stress (-0.8 MPa) for 16 days from 80 days after full bloom increased levels of sorbitol, sucrose, and total sugars as well as the activity of sorbitol oxide in fruit. Water stress also induced an increase in ABA in the fruit. It is thought that water stress-induced ABA accelerated sugar accumulation in peaches by activating sorbitol (Kobashi *et al.*, 2000). Irrigated fruits ripen 2-3 days earlier than non-irrigated fruits and can be harvested in one picking while two pickings are required for the latter (Feldstein and Childers, 1957). Evert (1988) reported that centre-pivot irrigation increased the temperature and reduced frost injury. The percentage of live fruits was almost double on the irrigated than the adjacent non-irrigated trees.

Irrigation is particularly necessary during dry and hot summer which depletes the soil moisture and increases the rate of evapotranspiration. This season usually coincides with the rapid vegetative and fruit growth in peaches culminating into a high demand of water. A temporary water stress may lead to early closure of stomata during the day resulting in the reduction of carbohydrate production and fruit growth (Westwood, 1978). It affects adversely shoot extension and trunk cross-sectional area increases and may cause shedding of developing fruits in extreme cases. Natali *et al.* (1985) subjected the container-grown trees of the cv. Redtop to two water deficit stress cycles, *i.e.* first stress cycle (at pit hardening stage) and second stress cycle (at third stage of fruit growth). A direct relationship between leaf water potential and fruit growth was noted and thus fruits on stressed plants were smaller than non-stressed ones, particularly in the second stress cycle. Adequate soil moisture to the fruit nearing maturity has also been suggested by Overholser *et al.* (1941). But on the basis of experimental evidence, Larson *et al.* (1988) concluded that postharvest irrigation had no significant effect on the subsequent fruit yield or fruit size excepting slight dealy in fruit maturity due to postharvest water stress.

Mitchell and Chalmers (1983a) suggested that irrigation techniques were important for the management of high density peach orchards because they were able to restrict root spread by irrigating through trickle or microjet outlets. They (1983b) also reported that one-third reduction in the normal requirement of water supply resulted in smaller high yielding trees having reduced need for summer pruning. Further optimum irrigation rate and half, quarter and one-eighth of the optimum rate had no differential effect on yield and fruit size but vegetative growth and tree size decreased progressively with reduced irrigation. On the basis of similar studies, Chalmers *et al.* (1984, 1985) proposed Regulated Deficit Irrigation (RDI) management system for dense peach orchard enumerating various advantages like control of tree size, low pruning cost, early bearing and high yield, increased flowering, reduction in both biennial bearing and poor fruit set, control of fruit quality to defined standards, with no additional cost using one-third irrigation water. However, soil conditions can affect the productive response of the trees to RDI. In peach, RDI maintains or enhances fruit production in shallow soils (Girona *et al.*, 2003) or in conditions of small soil volume (Boland *et al.*, 1994), but may decrease production in deep soils when the recovery of tree water status is delayed after full irrigation, particularly, if water stress extends into the final stage (*i.e.* stage III) of fruit development (Girona *et al.*, 1993; Goldhamer *et al.*, 2002). In shallow soils, a combined RDI technique, where deficit irrigation was applied during the 'resting' stage (*i.e.* stage II) of fruit development and at post-harvest was found most effective at increasing water savings without reducing yield in lightly pruned peach trees. In deep soils, single RDI regimes (stage II of fruit development) reduced irrigation by 13-24 per cent while combined (stage II of fruit development) reduced irrigation by 13-24 per cent. While combined (stage II and post harvest) regime reduced it by 23-35 per cent, without affecting yield (Girona *et al.*, 2005). Water deficit must be avoided during Stage III of fruit development to achieve large fruit size (Gonzalez-Dugo, 2011).

Flood, furrow, sprinkler and trickle (drip) methods of irrigation are more commonly employed depending on the availability of water, topography and type of the soil and financial resources. Drip system is preferred to sprinkling on clayey soil because of good lateral diffusion (Zini. 1984). Likewise. lonescu and Grurneza (1986) found drip system more efficient and economical giving better root spread than furrow system. Drip irrigation improved tree growth, development and also markedly increased the yield in peach (Semash and Storchous, 1987; Black *et al.*, 1977). Winter hardiness of peach cultivars also increased by drip irrigation (Storchous, 1984).

Depending on the time, frequency and amount of rainfall and moisture content of the soil peaches are usually irrigated in summer season at an interval of about a fortnight. Irrigation on more scientific basis can be applied at 70 per cent pan evaporation or at 60, 70 and 80 per cent field capacity. Excess irrigation may cause the loss of nutrients through leaching and may gradually result in the rise of water table. High water table may limit the development of roots and enhance the soil alkanity (Teskey and Shoemaker, 1972). Paltineanu *et al.* (2011) suggested that to save water, irrigation under water stress should be applied so that soil water content is allowed to decrease below 50 per cent of the total available water within a soil depth of about 1m. The rootstock should allow development of the rooting system down to the above depth to allow water uptake from a bigger soil volume.

Soil Management

Intercropping

A trial on intercropping in peach cv. Sharbati Safeda indicated that cultivation of cowpea showed better response. It induced higher fruit set and increased yield while okra and onion intercropping had adverse influences on plant vigour and production (Kaul and Mitra, 1976). In an experiment, intercropping with soybean and pineapple was found beneficial which resulted in increased plant vigour and leaf N content by 2 per cent (Ghosh and Chand, 1984). The merits of cover crops enumerated by Teskey and Shoemaker (1972) are: retarding soil erosion, runoff and leaching; maintaining the organic matter of the soil; increasing water infiltration; serving as an indicator of moisture and nutrient deficiencies; checking growth in the fall and thereby encouraging maturing of the tree tissues.

Sod Culture

Permanent sod covers should be avoided because they result in poorer tree growth, lower production and increased insect-injured fruit. In a high density orchard, growth rate of plants was retarded up to 12 per cent by permanent turf relative to temporary cover (Layne and Tan, 1988). Welker and Glenn (1988) compared the effect of four soil management systems, *viz.*, cultivated, herbicide treated, mown-turf and killed turf, on growth of young peach trees and changes in soil characteristics. They found that tree growth and fruit yield were greatset when trees were planted and managed in a killed turf system. Killed turf system prevented the depletion of soil organic matter, increased water infilteration rates, aggregated stability, macroporosity and microbial respiration rates compared with conventional systems. Mowing down of a sward like a lawn is definitely the poorest system of management (Welker, 1983). But permanent turf strips of *Festuca rubra* in the interrows combined with trickle irrigation in the tree rows proved the best treatment combination in the management of the orchards' floor (Layne and Tan, 1988).

Weed Control

Weeds in orchards compete with trees for water and nutrient. This competition becomes more harmful in the critical period of summer, when the demand for water is very high. Herbicides have been found very handy in controlling weeds. They have the residual effect and therefore control the weeds for a longer period. Daniell and Hardcastel (1969) reported that simazine at 4.5 kg/ha, paraquat at 570g/ha applied thrice and dichlobenil at 3.40 and 6.8 kg/ha were most effective and showed an excellent weed control. The annual weeds are controlled effectively by the application of simazine or terbacil, whereas for perennial weeds at 570 g to 1.5 kg/ha gives best result (Bowen and Dorman, 1974). The weed control in peaches was also recorded with the application of oxyfluorfen at 2.2 kg per hectare (Daniell, 1981; Dolby and Putnam, 1980) or in combination with glyphosate (West *et al.,* 1983). Sharma (1986) reported that atrazine and diuron at 5 and 6 kg/ha and oxyfluorfen at 0.2 and 0.3 kg/ha reduced the monocot and dicot weed population in peach orchards. Venig (2008) in his experiment with different herbicides on 4-year-old peach plantation stated that Roundup (glyphosate) at 0.5 l/ha, Galant Super (haloxytop-methyl) at 1.5 l/ha, Fozal (glyphasate) at 3 l/ha or Fusilade forte (fluazifop -P) at 1.5 l/ha were safe and showed no phtotoxic effect on the trees. Gautam and Chauhan (1984) found effective weed control in peach nursery by application of simazine, atrazine, diuron and terbacil each at 3-5 kg/ha. The weeds in nursery were also checked with oxyfluorfen at 0.5 kg/ha followed by diuron at 2.0 kg/ha (Sharma, 1986). In young nectarine orchard, terbuthylazine at 4 kg/ha gave the best weed control and tree growth but produced yellow edging on old leaves (Hartley, 1987).

Sod Culture Plus Strip Weed Control

This system has become very popular because it combines the advantages of both sod culture and chemical weed control. It is an improvement over permanent sod in which herbicides are applied along the tree rows. It eliminates the need for hand hoeing beneath the trees, reduces competetion between the trees and grasses for water and nutrients. Sod culture stripes facilitate the movement of sprayers and harvesting equipments and checks the soil erosion. In a 4-year trial, Dudzinski (1985) got the highest yield and best quality fruits under permanent sword in the interrows with herbicide strips along the rows of trees.

Flowering

Peach is a precocious tree commencing bearing in second or third year after planting. It bears on the shoots developed in the previous year from terminal and axillary vegetative buds. Generally three buds arise at a node the side buds being floral and central vegetative though all combinations of buds ranging from purely floral to vegetative may occur at nodes (Teskey and Shoemaker, 1972). Time of flower bud initiation of all low, medium and high chilling cultivars was found between 15 July and 8 August in 1984 and between 1 and 20 July in 1985 (Raseira and Moore. 1987). The earlier initiation in 1985 was considered due to drought stress. They concluded that, (i) chilling requirement is not strongly associated with time of flower bud initiation, (ii) the rate of morphological development of flower primordia is not associated with earliness of flower bud initiation, and (iii) there is no relationship between time of flower initiation and either time of bloom or time of fruit ripening of the cultivars. Maria *et al.* (1987) also opined that chilling requirement is not strongly associated with the time of flower bud initiation.

Flowering period in peaches is determined by the completion of chilling requirement and prevailing temperature during early spring. Low chilling cultivar Flordasun sometimes blooms in late January at Shimla, India, if the temperature has been abnormally higher, but normally it flowers slightly earlier than other medium and high chilling cultivars blooming in March. The early bloom can be damaged by low temperature frost in spring which is a limiting factor in growing low chilling cultivars in cold region. In subtropical conditions, several factors can cause an erratic production of peaches. The occurance of high temperature on the beginning of the bloom is one of them. Such temperatures can damage the stigma and pollen germination, causing a decrease on fecundation and fruit set (Zanandrea *et al.*, 2011). Autumn application of ethylene releasing chemical CGA-15281 at 2000 and 4000 ppm or ethephon at 100 and 250 ppm delayed flowering in cv. Redhaven by 3-9 days (Coston *et al.*, 1985a). Gianfagna *et al.* (1986) also successfully delayed flowering of Cresthaven peach by 3-7 days with October application of ethephon at 200 ppm.

Peach produces enough flowers every year but has a tendency to bloom lighter after a heavy crop. Excess flowering can be reduced by the application of GA, (Clanet *et al.*, 1979). Raff and Hallpike (1983) obtained 20 per cent reduction in flowering of cv. Redhaven by the application of 200 ppm GA_3 in early February. Treatment with paclobutrazol (PBZ) as soil drench in August increase flower numbers in branches in cv. Kardi. The treatment also increased final fruit set. Application of PBZ in March was found more effective in reducing shoot elongation (Samani, 2014).

Pollination and Fruit Set

Autogamy has been reported by Weinbaum and Erez (1983) in all the peach and nectarine cultivars studied. Sometimes the fruit set was even higher in bagged than open-pollinated flowers. Practically, all peach cultivars are self-fertile except J.H. Hale, June Elberta (Mikado), Halberta, Candoka, Chinese Cling and Giant (Childers, 1975). A self-sterile cultivar needs to be planted in double rows, alternating with two rows of self-fertile cultivars. Bees in the orchard at bloom provide a greater assurance for cross-pollination, particularly during adverse weather conditions. Adverse weather condition during pollination and fertilization period adversely affected fruit set and may cause 50-80 per cent of the fruit to remain unfertilized. If temperatures are below 12°-15°C during bloom and/or the 7-10 days following bloom, poor pollination and pollen tube growth will occur. This situation resulted low percentage of fertilized ovule and a large number of fruit abscission.

Sharma (1984) studied the pollen viability and fruit set of five peach cultivars and reported ranges of 36.37 to 92.18 per cent and 30.24 to 55.62 per cent, respectively. The fruit set in peach cv. varies between 36-85 per cent while it varies between 40-95 per cent in cv. Freestone (Filiti *et al.*, 1982, Visai *et al.*, 1985; Weinbaum and Erez, 1983). Lott and Simons (1966) described ten morphological stages of flowers from bloom to style abscission and found stage (X) characterized by the abscission of the style as suitable physiological reference point for cultural practices like fruit thinning. Spraying of GA_4+_7 at 5-10 ppm was reported to increase fruit set of peach cvs. Shan-i-Punjab, Flordasun, Flordared and nectarin cv. Sun Red (Singh *et al.*, 2000).

Thinning of Fruits

In peach, flowering and fruit set are generally more than required for commercial crop adversely influencing the fruit size and flower initiation for the subsequent crop. Fruit thinning is therefore, necessary to produce marketable sized fruits, to prevent limb breakage of overloaded branches, to promote early ripening, to sustain winter hardiness and to stimulate flower initiation for the next year's crop. It is, however more prudent to reduce the crop by adequate pruning in dormant season rather than resorting to heavy fruit thinning (Teskey and Shoemaker, 1972).

Degree of Thinning

Four criteria, *viz.* leaf to fruit ratio spacing between fruit, total number of fruits per tree and fruit size prediction are adopted to estimate the amount of fruit thinning in a peach orchard. Thinning escalates the leaf to fruit ratio, thereby making available more photosynthates and moisture supply per fruit. Availability of more leaf area increases the fruit size but not in the same proportion resulting in reduction of total yield (Westwood, 1971). Teskey and Shoemaker (1972) showed augmentation in size and sugar content of Elberta fruits from 10 to 75 leaves per fruit and indicated the possibility of further increases with more leaves. They stated that thinning also increases the differentiation of more fruit buds as shoots with 20 leaves per fruit have about 6 times more fruit buds than shoots with 10 leaves per fruit. Enhancement in leaf area, however, decreases the red skin colour of the fruit (Kinmann. 1941). Since fruit thinning reduces the yield, it is essential to maintain a proper balance between fruit size and yield. The number of leaves to be retained per fruit after thinning varies with the cultivar, tree vigour and soil condition. The appropriate number being 30-40 leaves per fruit (Teskey and Shoemaker, 1972).

In the past when hand thinning was common, fruits were thinned at 15 to 20 cm apart from each other. To obtain early ripening, better fruit size and higher yields in cvs. Redhaven, Cardinal and Dixired, thinning should be done soon after full bloom having fruits at least 15 cm and not more than 20 cm apart (Havis, 1962). In low chilling cultivars, the fruit should be thinned out so as to secure 10 to 15 cm spacing (Singh, 1969). Early small-fruited cultivars require lesser spacing than large-fruited, late maturing cultivars.

Since relatively small fruits remain smaller than other fruits even at harvest, thinning on the basis of fruit size and total number of fruits per tree has become more common rather than on a particular spacing between fruits. Westwood (1978) stated the method for predicting the fruit size at harvest. In view of this prediction, the total number of fruits to be retained for a proper fruit size and yield can be determined and all other small malformed and diseased fruits should be thinned out. The factors like soil fertility, irrigation, fertilizer application, pruning and fruit set are also to be taken under consideration while thinning.

Chemical Thinning

Chemical thinning has gradually replaced hand and mechanical thinning which are slow, costly and varying in effect. Large fruits are generally get removed by mechanical shakers. Most important factor in favour of chemical thinning is its low cost. Due to the difference in tree age and vigour, variable results of over or under thinning are often obtained in chemical thinning and it may also cause foliage injury. A large number of chemicals, *viz.*, DNOC, 3-CPA, NAA, GA_3, ethephon, carbaryl and thiourea, have been tried for thinning with varying degree of success, but the most widely used chemical is ethephon. DNOC was used as a full bloom thinner before 1960. Application of 300 ppm 3-CPA gave adequate thinning of peach fruits (Buchanan *et al.*, 1970). Thinning has also been accomplished by spraying 2 to 3 per cent thiourea or 40-100 ppm 2,4.5-T or 50 ppm 2,4-D (Erez, 1975). These treatments increased fruit size without impairing the quality (Savic and Stankovic. 1977). Beta-chloroethyl-methyl-bis-benzyloxy-saline (CGA 15281) applied at 500 or 1000 ppm at an ovule length of 8.9 mm caused about 50 per cent and 70 per cent leaf abscission. respectively, in peach Elberta, also both caused excessive fruit abscission (Table 4.11). The 200 ppm treatment at an ovule length of 15.2 mm also caused excessive fruit and leaf abscission (about 50 per cent). Fruit size increases in the ethephon treatment may have been the result of thinning (Byers, 1978). Blanco (1987) reported fruit abscission with the application of 2000 ppm paclobutrazol in cv. Sundanell-I but not in Redhaven. Heaviest thinning was obtained in Early Red Free and Red Cab cultivars with 150 ppm GA_3, in Nectared Red 4 with 50 ppm GA_3 and in Early Amber with 600 ppm carbaryl when applied 21 days after full bloom (Bay-karn. 1984). GA_3 spray (100-200 ppm) at red calyx and full bloom stages increased fruit drop and growth of the remaining fruits (Vitagliano *et al.*, 1985). Urea sprays significantly reduced fruit set increasing the final fruit weight (Zilokh *et al.*, 1980).

Schiaparelli *et al.* (1982) recommend 45-55 ppm ehtephon spray at 25 mm fruitlet width in processing peaches and at 20 mm fruitlet width in dessert peaches for effective thinning. Redhaven peaches gave 56.6-60.1 per cent fruit thinning at 100 ppm ethephon spray compared with 68.5 per cent in manually thinned control (Hatzehariese, 1981). Yields were 89.8, 97.0, 92.1 and 106.9 kg/tree in 10-11, 12-13, 14-15 mm fruitlet diameter stages and control, respectively. In South Australia, depending upon the thinning requirement, ethephon spray (40 to 100 ppm) was found very effective when applied 39-53 days after full bloom (Gathercole, 1982). In Flordasun, desired thinning has been accomplished by spraying 100 ppm ethephon at petal fall which increased the fruit size, fruit weight, pulp/stone ratio, and also improved the fruit quality (Chahil *et al.*, 1980). Sharma and Gautam

(1981) recommended spraying of 250 ppm ethephon on July Elberta. Ethephon at 100-200 mg/l can be used to induce adequate levels of fruit abscission of Redhaven peaches when sprayed at 45-50 days after full bloom (Taheri *et al.*, 2012). Sarkadi and Lordancescu (2012) suggested spraying of ethrel at 250-350 ppm at 25 days after fruit set for peach cvs. Spring Lady and Maja, and two nectarine cvs. Caldesi 2000 and Nectaross. Stan *et al.* (1985) obtained good thinning in cv. Cardinal with 240 ppm flordimex (46 per cent ethephon) applied at full bloom, and in cv Redhaven with 240 ppm flordimex or 30 ppm NAA applied at 8-8.5 mm ovule length. Baroni *et al.* (1986) compared the three treatments of thinning, *i.e.* (i) hand thinning, (ii) integrated chemical thinning with a suboptimal concentration (40 ppm of ethephon) and finished manually, and (iii) not thinned. Treatment (iii) produced many non-marketable fruits. The effect of treatments (i) and (ii) were similar in decreasing the total yield but increasing the yield of marketable fruits. The cultivars showed variation in their response to treatments. Fallahi *et al.* (1998) stated that spraying of hydrogen cyanamide (Dormex, 50 per cent a.i.) at full bloom at 1.610 mg/l significantly reduced fruitset in peach cv. Flavorcrest.

Table 4.11: Effect of CGA 15281 on Fruit Thinning and Fruit Size of cv. Elberta*

Treatment[z]	*Conc. (ppm)*	*Date Applied*	*Ovule Length (mm)*[y]	*Per cent Fruit Remaining*	*Fruit Diameter (cm)*	
					June 13	*August 2*
				June 13		
Control				35a[x]	3.5a	5.0a
Ethephon	50	May 10	8.90	30a	3.6a	5.5b
CGA 15281	500	May 16		20b	3.9b	6.4e
CGA 15281	1000	May 10		11c	3.9b	6.7c
CGA 15281	200	May 23	15.21	14bc	3.9b	6.3c

Z: Full bloom: April 16; Y: Ovule length of 50 randomly selected fruit from an adjacent row for each spray date; X: Mean separation in Columns by Duncan's multiple range test, 5 per cent level.

*Byers (1978).

Hand and Mechanical Thinning

Hand thinning during bloom is a common practice in peach orchards (Myers, 1986). Growers partially thin at bloom to obtain beneficial effects on fruit size and complete the process after the danger of frost is over. During the pink to full bloom stage, blossoms are easily removed by hand or by using various types of brushes. A mechanical bloom thinning technique using suspended flexible strands of rubber belting or rope moved through trees on a frame movement on a fork lift was tested on Early Loring peach in West Virginia. USA (Baugher *et al.*, 1988).Heaviest thinning occurred in the upper two-thirds of the tree canopy. Shoot orientation and flower position did not affect bloom removal. Maximum bloom removal occurred at full bloom when the thinning machine passed through a tree at least four times (Table 4.12). Schwartz *et al.* (1999) working with peach cv. Diamante at Brazil observed that

Table 4.12: Effect of Various Mechanical Bloom-Thinning Conditions on Flower Removal Fruit Set, Hand-Thinning Requirement and Fruit Size*

Treatment (No).	*Passes (No.)*	*Strand Type*	*Travel Speed (Km/hr^{-1})*	*Development Stage*	*Measurements Before Follow-up Thinning*		*Follow-up Hand Thinning Time[x] (min/tree)*	*Fruit Diameter at Harvest[x] (cm)*
					Flower (Fruit) removal [z,x] (per cent)	*No. Fruit/cm^2 limb CSA[y,x]*		
1	2	Belt and rope	1.6	Pink	25	23	25	5.65
2	4	Belt and rope	1.6	Pink	36	17	27	5.59
3	2	Belt and rope	1.6	Full bloom	35	14	22	5.54
4	4	Belt and rope	1.6	Full bloom	51	11	22	5.66
5	2	Belt	1.6	Full bloom	49	11	29	5.65
6	4	Belt	1.6	Full bloom	52	13	27	5.65
7.	4	Rope	1.6	Full bloom	54	9	24	5.65
8.	4	Oscillating belt and rope	1.6	Full bloom	45	11	20	5.70
9	8	Belt and rope	1.6	Full bloom	59	10	24	5.68
10	2	Belt and rope	1.6	30 days after bloom	18	15	25	5.55
11	4	Belt and rope	1.6	30 days after bloom	10	19	30	5.52
12	2	Belt and rope	1.6	Full bloom	26	14		
13	2	Belt and rope	3.2	Full bloom	38	11		
14		Belt and rope				18	28	5.57
		Hand thinned control						
		LSD 0.5			10	6	7	0.11

z: Determined day of thinning; y: Determined following physiological drop; x: Number of observations represented by mean value: columns 6 and 7, 12; column 8, 6; column 9.60.

*Baugher *et al.* (1988).

fruit size, weight and quality were better with hand thinning than with chemical thinning.

Thinning Time

Early maturing cultivars having shorter fruit development period should he thinned as early as possible before the first fruit drop, even blossom thinning is more effective than late thinning in enlarging the fruit size. Filiti and Costa (1986) recommended an early ethephon (50 ppm) treatment for an early cultivar Dixired.

Thinning time of midseason and late cultivars is less critica1 and can be delayed until the first drop occurs. Late fruit thinning is always cheaper than early thinning. Besides cultivar, the method of thinning-hand, mechanical or chemical, the thinning chemical and the stage of fruit growth are other decisive factors of thinning time. In Collins and Redhaven cultivars, the appropriate time of ethephon application is 35-45 days after full bloom or when fruits are 10-15 mm in diameter (Lekson, 1981). Svoboda and Bazani (1983) determined the suitable thinning time for peaches and recommended the fruit width (17 to 22 mm) as an easier measure than seed length for thinning with ethephon. The critical time for spraying ethephon is just before pit hardening and time of application is more important than the dose (David and Nigicser, 1986). Stembridge and Gambrell (1969) also recommended ethephon treatment at the end of fruit growth stage I or during the stage II for effective thinning. Thinning was greater when ethephon was sprayed from 12 mm seed length to pit hardening than at earlier developmental stages (Gambrell *et al.*, 1983).

Fruit Growth and Development

The development of peach fruit follows a typical double sigmoidal growth curve (Sandhu and Dhillon, 1983; Sharma, 1982; Westwood, 1978). Initial and final periods of rapid growth are separated by a middle period of comparatively slow growth. Durations of these three growth periods vary in different cultivars. Nevertheless, there is less variation in the duration of first and third periods than in the second period of reduced growth which is barely noticeable in early maturing cultivars (Hesse, 1975). The durations in three growth periods were 39, 28 and 42 days in cv. Safeda and 35, 14 and 35 days in cv. Flordasun, respectively (Kumar and Chitkara, 1983; Sandhu and Dhillon. 1981). Teskey and Shoemaker (1972) described the fruit development during the three growth periods as under:

I. Peach embryos develop very slowly few weeks after fertilization, whereas pericarp and seed grow very rapidly, the stone attaining almost full size. The duration of first growth period in cv. Elberta is of about 30-40 days.

II. Second period of retarded growth is due to slow growth of mesocarp (flesh) only but the endocarp and embryo develop rapidly by accumulating dry weight. Lignification of endocarp (pit hardening) is an important feature of this growth period comprising about 4 weeks in Elberta.

III. Third growth period is the most important from growers point of view because the maximum amount of flesh (over 60 per cent as dry weight) accumulates during this period particularly near maturity. Moisture and nutrient supplies must be maintained in such a way that these are available abundantly to the fruit in the final swell, characterized by increase in size, dry weight and sugar content of the flesh with a corresponding decrease of dry weight, carbohydrates and ash in the stone. The seeds of early maturing cultivars do not develop fully at the time of fruit maturity. This is a 6-week period in Elberta.

With the increase in fruit size, the primary cell wall grows in the expansion phase. Cosgrove (1999) observed that the growing wall behaves like a network of inextensible cellulose microfibrits laterally linked together via complex matrix of flexible polysaccharides (glycans and pectin) that may be bind to cellulose and to each other. Pectins that are loosely attached to the wall through ionic calcium bonds increase in amount as peach fruit reaches full size (Brummel *et al.,* 2004).

In peach, fruit growth rate varies with location within a tree (Genard, 1992). This variability depends on source and sink relationship. Fruit proximity to assimilate sources is one of the fectors. The fruit proximal to the leafy zone had greater growth than the more distal ones. Moing *et al.* (2000) studied the synthesis of organic acids during fruit development of two peach cvs. Fantasia and Jalousia, having fruits with high and low organic acid content, respectively. They observed the malate (malic acid) content was higher in Fantasia than in Jalousia at the end of the first rapid growth stage (50 days after bloom, DAB). Malate and citrate contents were higher in Fantasia than in Jalousia during the second rapid growth stage (from 100 DAB to maturity). The expression of phosphoenol pyruvate carboxylace (PEPC, EC 4.1.1.31) which is involved in organic acid synthesis showed that PEPC mRNA levels, and protein levels on a total soluble protein basis, peaked at 23 and 108 DAB in Fantasia. In Jalousia, they were low at 23 DAB and reached levels similar to Fantasia at 108 DAB. It is believed that PEPC may participate in the control of organic acid accumulation during fruit development in the normal-acid fruit of Fantasia. However, mechanisms other than organic acid synthesis might account for the difference in acidity between normal acid and non acid peach fruit.

Changes of fruit structure and sugar contents during the fruit growth and development in peach cv. Yumyeong (popular peach varieties in Korea) was studied by MyongDong *et al.* (2000). They observed two rows of vascular tissues at bloom, and the number and the position of internal vascular bundles were consistent during fruit growth. However, the number of the vascular tissue increased and distribution was found irregular in the flesh tissues. The tissues between the inner integument and the internal vascular bundles showed different development characteristics from other parenchyma cells which consisted of small and defuse cells containing tannins. Such observations suggested that the stones consisted of inner epidermis and cells in the internal vascular tissues. The outer epidermis consisted of single layer cells at flowering differentiated into 1-2 layers by horizontal cell division with in 14 days after full bloom. At 30 days after full bloom, the epidermis consisted of 5-6 layers by vertical cell division. The cell leyers constituting the outer epidermis were gradually decreased to 1-2 layers at maturity. The observations on the changes in the epidermis confirmed that some of the cells comprising the hypodermis of peach fruits were originated from the cells of the outer epidermis. Tylosis was observed from 35 days after full bloom and the size and number of tylosis were increased until full maturity. The sucrose content showed a sharp increase from 50 days to 120 days after full bloom then decrease slightly.

Metabolic profiling during peach fruit development and ripening reveals the metabolic networks that underpin each developmental stage. Principal component analysis of peach metabolic composition showed that the early developmental stages were characterized by a substantial decrease in protein abundance and high levels of bioactive polyphenols and amino acids, which are substrates for the phenylpropanoid and lignin pathways during stone hardening. Sucrose levels shared a large increase during development, reflecting translocation from the leaf. In contrast to early developmental stages, a decrease in amino acid levels is coupled to an induction of transcripts encoding amino acid and organic catabolic enzymes during ripening (Lombardo *et al.*, 2011).

Maturity

Stages of Maturity

When peach fruit is fully developed, just prior to start of ripening, it is said to be mature but ripeness indicates that it has attained all the physical, physiological and biochemical characteristics making it ready for use. Maturation occurs only on the plant, whereas ripening may take place before or after harvest. Between these two physiological conditions of the fruit, there are five stages of horticultural maturity which are very important from harvesting point of view (Teskey and Shoemaker, 1972). Depending upon the cultivar and prevailing temperature, the fruit passes from one maturity stage to another in 3-4 days.

HARD: When the fruit is almost of full size, very hard and green. If the fruits are picked at this stage, they will not attain proper eating quality on ripening, being bitter and astringent in taste.

FIRM: The ground colour begins to change and fruit yields very slightly to pressure. For a distant market, the fruit can be picked but the fruit quality will still remain inferior on ripening.

FIRM-RIPE: Noticeable change in ground colour occurs and the fruit can be pressed in a cupped hand. The fruit is fairly good in taste but has not attained the best eating quality. It can be stored for a long period.

RIPE (TREE-RIPE): This is a fully tree-ripe stage when fruit yields readily to pressure and is in prime eating condition of the cultivar. The fruit can be sold only in local market as it can neither be stored nor transported to distant markets.

SOFT (SOFT-RIPE): The fruit has lost its firmness and storage life. It will not withstand any transportation although has high dessert quality, especially in some juicy cultivars.

Maturity Indices

Picking of fruits at proper maturity is a very important operation determining the profitability of an orchard. Apart from the orchard factors, transport, market, and utilization are also to be taken into account before deciding the picking stage.

A large number of maturity indices such as days to maturity, calender date, fruit size, firmness, sense of touch, pit discolouration, freeness of pit, taste, ground colour, sugar, acidity, starch, sugar to acid ratio, and multiple linear regression analysis, have been assessed by various researchers working on different cultivars (Delwiche, 1987; Joshan and Chauhan, 1982; Smith, 1985; Kumar and Chitkara, 1983; Teskey and Shoemaker. 1972). The days required from flowering to maturity vary in different cultivars, locality, N supply and year to year. The cv. Flordasun takes 81 days, Alexander 86 days, midseason cv. Flordared 110 days and late season cv. Babcock 122 days and Elberta 127 days (Chander and Khajuria, 1981; Sharma. 1981; Dhuria *et al.*, 1978; Sud *et at.*, 1978; Pandey, 1977). A more accurate prediction of maturity can be made by calculating degree days on the basis of daily maximum temperatures during fruit growing period. The calender dates of maturity in previous years can also serve as an index (Kumar, 1982). Peach fruit size increases tremendously (up to 29 per cent) near the optimum maturity (Fisher and Britton, 1940). It enhances both the grade of fruits and yield (Upshall, 1943). An early picking before maturity reduces the yield as well as fruit quality. A grower is, however, interested in early harvest because he gets a higher price evading the glut in the market.

When approaching maturity, fruits lose firmness which can be measured by a pressure tester. Pressure readings, however, differ with the cultivar and position of the fruit tested. A reading of 8.6 to 6.4 kg has been suggested to be appropriate for indicating the maturity of Elberta fruits (Westwood, 1978). Firmness can also be determined by hand pressure. Pit discolouration and freeness of pit are good maturity indices only in specific cultivars. Pits get separated from flesh in freestone cultivars on maturity.

The amount of total soluble solids, non-reducing sugars, total sugars, acidity and starch have been found as good indices of maturity in cv. Safeda (Kumar and Chitkara, 1983). With the advancing fruit maturity, the sugar content enhances and acidity declines. Therefore, sugar to acid ratio can be tested for fruit maturity. The ripe fruits have high sugar to acid ratio but the interval period of estimation of sugar and acidity should not be more than 3-5 days (Comin, 1956). Fruit maturity can also be judged by tasting, especially for local market. Smith (1985) used multiple regression analysis for prediction of commercial harvest dates of various peach cultivars based on the prediction equations of degree day summation and date of full bloom. Ground colour variations in conjunction with flesh firmness are one of the best maturity indicators except in full red cultivars. The leaf-green ground colour of the peach fruit first changes to light-green, light-yellow and then develop yellow and orange-yellow in yellow-fleshed cultivars and straw-white to cream in white-fleshed types. The rate of colour change is proportional to the ripening rate (Teskey and Shoemaker, 1972). Delwiche and Baumgardner (1983) stated that ground colour is a better harvest indicator than flesh firmness. Delwiche (1987) found the grading performance satisfactory by using a peach ground colour maturity chart.

Rood (1957) indicated that the main indices for peach maturity are ground colour change from green to straw colour, flesh firmness and days from full bloom. Besides these, just prior to maturity, there is swelling of flattened side of peach (Childers, 1975). For most of the cultivars, the fruit size has been found to be an important factor for ascertaining the harvesting date. Late yellow peaches can be harvested in a relatively harder stage without deteriorating the fruit quality. In cv. Redhaven, the indices of maturity are firmness, ground colour, the soluble solids/ acid ratio, titratable acidity and flesh colour.

Harvesting

Peaches are one of the earliest fruits in the season to reach the market, hence there is a tendency among the growers to pick rather immature fruits and send the same for disposal. Green, hard, starchy and shrivelled fruits never fetch a good price in the market. In order to get premium price and reduce the losses during packaging and transportation, peaches should be harvested after the attainment of harvest maturity. Since peaches mature unevenly, it is essential to pick the fruits more than once. Harvester has the capacity to collect 60-80 tonnes of fruit per day from the trees trained on the Tatura trellis. Mechanical harvesting, however, results in considerable fruit losses. If the peach orchards are located closer to the consumption centres, tree-ripe fruits can be harvested twisting by hand and taking much care in packing and handling of the fruits. This system is particularly adopted for nectarines as these are used for table purpose. It is important to pick nectarines as mature as possible and the fruits should reach the consumer in good condition. As ripening proceeds on the tree, the fruit size and quality improve daily.

Mechanical Harvesting

In most of the developed countries, because of increasing labour cost and high density planting systems, equipments have been developed for mechanical harvesting of peaches, particularly for processing cultivars. In California, clingstone peaches are mechanically harvested with 'limb' and 'shock-wave' type trunk shakers. Ende *et al.* (1985) recommended the Trel-Pik harvester for harvesting peaches for processing. The harvester has the capacity to collect 60-80 tonnes of fruit per day from the trees trained on the Tatura trellis. Mechanical harvesting however, results in considerable fruit losses due to mechanical injury and picking of immature fruits due to uneven ripening (Bilanski and Menzies, 1984). For the success of mechanical harvest, the total mechanical system and orchard practices have to be properly planned and managed (Teskey and Shoemaker, 1972).

Postharvest Handling

Peaches being highly perishabe require the most careful handling to avoid cuts and bruises. The picking containers should be padded and lined with cushion materials. Every care should be taken in filling and emptying the picking containers. Immediately after harvest fruit should be handled quickly so that these are either stored at a cool place or marketed. The shelf life of peaches is prolonged by packing the fruits at stage II (35.50 per cent pink fruits) and treating with 1000 ppm $KMnO_4$ (Sandooja *et al.*, 1986). Preharvest application of calcium nitrate (1-1.5 per cent) increases storage life of peaches for 6-9 days and reduces the weight loss, respiration rate, and incidence of diseases during storage (Gupta *et al.*, 1984; Singh *et al.*, 1982; Gautam *et al.*, 1981). Application of GA_3 (800-1000 ppm) at full bloom also extends, storability and reduces the weight loss and rotting during storage (Dhillon *et al.*, 1982).

Peach fruits should be cooled immediately after harvest so as to retard ripening process and growth of rot pathogens. Precooling of peaches has been found effective in increasing the keeping quality. It can be done either by forced air in storage room or in special precooling chamber or tunnel or using vacuum ice-chilled water, sea water, brine or by employing cold air. Vacuum cooling and hydrocooling have been found suitable for peaches. Rapid cooling is beneficial because it reduces metabolic activity at lower temperature. Hydrocooled fruits develop more yellow ground colour than uncooled fruits (Schneider, 1956). In order to control brown rot and *Rhizopus* in the storage, botran (300 ppm) and tutone (0.25 per cent) should be added to the cooling water (Ridley and Sims, 1966). The rot caused by *Rhizopus* does not grow below 4.8°C (Albrigo, 1975). Sometimes. peach fruits also receive hot water and defuzzing treatments.

Postharvest Diseases

These diseases in peach fruit occur due either to contamination and infection that occur during the growing season or during harvesting, handling, packing and transportation. The following are the important diseases of peach fruit at the postharvest stage.

Brown Rot (*Monilinia* sp.)

The earliest symptoms of disease are small, water-soaked spots on the surface of fruits. Within 24 hours, the skin and the affected flesh becomes brown to black. In advanced stages, ashy grey tufts of the fungus with tan grey, powdery spore masses often in concentric rings, cover the affected areas. Within 3-4 days room temperature the entire fruit decays, but with the skin remaining intact. Later, these fruits shrivel and become mummies. Brown rot spreads rapidly from diseased to healthy fruits (Harvey *et al.*, 1972).

Refrigeration after harvest is the most commonly used method for controlling brown rot. Hydrocooling is a rapid method of removing field heat (Wells and Bennet, 1976). Combined hot water and fungicidal treatment have also proved effective to control the brown rot. Dipping the fruit in 50°C water for 3 minutes destroys spores on the fruit and prevents infection (Wells, 1970). Botran (DCNA, 250-500 ppm), benomyl (150-250 ppm) and bavistin (750 ppm) as dip treatments have been reported effective to control the disease (Wells and Gerodits, 1972; Wells, 1970; Eckert, 1969).

Whisker's Rot

It is the most important postharvest problem of peach in Himachal Pradesh, India. The first symptom of the disease is a circular tan area surrounding healthy skin. Within a few hours whole skin in the affected area becomes tan to brown. Later on, white growth of mould appears near the centre and rapidly covers the affected area and turns black. Finally, the mould growth turns grey and black spores are produced.

Precooling and refrigeration in storage and transit is a good control measure (Sommer, 1983; Matsumoto and Sommer, 1967). Hot water treatment of peaches at 50-54°C for 2-3 minutes has proved effective to control the disease (Wells, 1970; Smith, 1962).

Blue Mould Rot (*Penicillium expansum*)

This rot occurs on injured or overripe fruits of peach and nectarine. Small tan coloured spots occur anywhere on the fruit surface. The skin over the spots being moist readily breaks or slips from the fruit. As the spot enlarges, white tufted cushions of mould appear near the centre, later on producing masses of powdery blue green spores.

Rapid precooling and storage at 0°C are effective in reducing losses from the blue mould rot. Controlled atmosphere storage of peaches at 0°C with 1 per cent O_2+ 5 per cent CO_2 reduces blue mould rot slightly (Anderson *et al.*, 1969).

Grey Mould (*Botrytis cinerea*)

It rarely occurs on freshly harvested and marketed peaches, but infects in long storage or transit. In the early stages, grey mould rot is characterized by a light brown, firm spot with a distinct margin. In advanced stages, the spots become dark brown and may be leathery. Internally, the tissues are water-soaked.

Prompt cooling after harvesting greatly reduces the development of grey mould rot. Atmosphere with CO_2 level of 10 per cent reduces decay and extends storage life (Kader and Morris, 1977).

Black Mould Rot (*Aspergillus niger*)

Early symptom of the disease is the development of small, tan, slightly sunken spots at various places of the fruit. Later on white growth of mould appears near the centre of the spots and spreads over the affected area. Mould produces abundant spores, black in colour and sooty in appearance.

Losses from black mould can be reduced by careful handling to avoid injury to the fruit, avoiding long storage periods and using sanitary measures during storage and transportation.

Green Mould (*Alternaria alternata*)

This rot is found on injured or overripe fruits. The spots due to the infection are greenish, firm, slightly sunken and moist and usually affect only surface of the fruits. In advanced stages the affected areas are covered with olive-green spores and rot extends deeper in the flesh. Careful handling to reduce injuries during harvesting, packing and transportation reduces *Alternaria* rot.

Yield

Yield is a highly variable character which is influenced to a great extent by the productivity of a cultivar, age of plantation, planting density, climate, soil and management of the orchard. In general 7 to 10 tonnes/ha yield can be obtained from adult trees in a conventional orchard. Ende *et al.* (1987) reported 78 t/ha yield in the third year of Tatura trellis planting from cv. Loring at 5 x 1 m spacing.

Ripening

Peach harvesting coincides with the prevailing high seasonal temperatures which accelerate ripening and decay process during marketing. Ripening process declines to half with each reduction of 5.6°C from 21.1°C to 4.4°C, *i.e.*, if fruits ripen in 3 days at 21.1°C, it will ripen in 6 days at 15.5°C. The fruits after picking should, therefore, be immediately placed under cool conditions. Peaches ripen with good flavour and aroma at temperature above 15°C, with undesirable flavour at 10°C and break down internally instead of ripening at 4.4°C (Teskey and Shoemaker, 1972). Like other climactric fruits, the ripening process of peaches is regulated by System II ethylene for initiation and control of ripening (Lelievre *et al.*, 1988), Crisosto and Kader (2004) stated that ethylene production in peaches varies greatly from mature but unripe fruit (0.10µl/kg/hour) to ripe fruit (up to 160µl/kg/hour). However, Haji *et al.* (2003) indicated that stony- hard peaches like Manami, Odoroki and Yumyeong did not produce ethylene and their flesh did not soften (4.2 kg/cm^2) during ripening over an observation period of 8 days. Lu *et al.* (2008) also recorded fruit firmness of 3.8 kg/cm^2 on 9 day and ethylene production rate only ranged from 0.78 to 2.13 µl/g/h in cv.Yinggetao. The average respiration rate of peach fruit during ripening is classified as moderate when compared with other stone fruit (Wills *et al.*, 2007). The respiration rate (CO_2 production) at the climacteric peak has been reported to vary from 64 to 110 mg/kg/h at 20°C depending on the genotype (Crisosto and Kader, 2004).

Zhang-Bo *et al.* (2010) reported that during post harvest ripening of peach cv. Yulu at 20°C, the concentrations of n-hexanal, (E)-2-hexenal, (E)-2-hexenol, and (Z)-3-hexenal decreased, whereas the production of (Z)-3-hexenyl acctate, gamma-hexalactone, increased with fruit ripening. Lactones showed a clear pattern concomitant with the climacteric rise in ethylene production with gamma-decalactone being the principal volatile compound at late ripening stage.

Effect of Growth Substances on Maturiy and Ripening

The plant regulators alar (SADH) and ethephon have almost similar effect on peach fruit maturation and ripening but they differ in the time of application, the former is sprayed at pit hardening stage and the latter is applied just before harvest. Thus alar influences at fruit maturation stage and ehephon at ripening stage. The ultimate effect of both the chemicals are advanced and uniform ripening, less number of pickings, improvement of internal flesh and skin colour and fruit quality (Lavatti *et al.*, 1986; Gyuro and David, 1978; Morris *et al.*, 1978; Gambrell and Rhodes, 1967). Spraying of alar (500 ppm) at pit hardening in cv. Alexander resulted in increased fruit retention, fruit size and harvesting of marketable fruits (Sinha *et al.*, 1983). Preharvest application of ethephon at 100 to 700 ppm advanced the maturity in late cv. Sharbati by 3 to 12 days, while at lower concentration (100-400 ppm) it increased the total soluble solids and total sugars (Sandhu and Singh, 1984).

Storage

Handling and commercialization of peach fruit are limited by rapid softening and overall ripening, which results in short shelf life potential. For this reason, refrigeration is one of the main tools used to extent commercial life of produce. Harvested fruits should be precooled by hydrocooling or forced air cooling before storage. The precooling temperature varies between 1°C-10°C, depending on the the time required to place the fruit boxes in storage.

Cold Storage

The ideal peach storage temperature is -1°C to 0°C and 90-95 per cent relative humidity. Ulrich (1974) opined that precooled peaches should be stored at 0°C and 85-90 per cent relative humidity for obtaining commercial storage life of 28-36 days. Cold storage of peaches under inappropriate conditions leads to a range of chilling-induced disorders, which are manifested by textural disorders such as woolliness or mealiness and by abnormal flesh colourations such as browning or bleeding (Lurie and Crisosto, 2005). The severity of chilling injury (CI) is influenced by storage temperature and peaches stored at 5°C showed more CI than those kept at 0°C. (Crisosto *et al.*, 1999).

Coatings

Edible coatings, a new strategy used to extend shelf-life and to improve fruit quality of whole fruits and fresh cut produce. Chitosan is one of the most promising coating materials for extending shelf-life of fruits because of its good film-forming property and broad antimicrobial activity. Maintenance of fruit quality has been achieved by using chitosan in peach (Li and Yu, 2001). El Badawy (2012) recorded decrease in weight loss and decay percentage of Florida Prince peach in storage by treating the fruit with chitosan at 1 per cent. The effect was more pronounced when it was combined with $CaCl_2$ at 4 per cent.

Controlled Atmospheres

In controlled atmosphere storage containing 5 per cent CO_2 + 1-2 per cent O_2 at 0°C, peaches can be stored up to 42 days (Kader, 1983, Sharples, 1980) Kajiura (1975) reports benefits from 3 per cent CO_2 and 3 per cent O_2 in lowering O_2 and raising CO_2 in the storage atmosphere delayed or prevented the appearance of mealiness, internal reddening, and flesh browing (Murray *et al.*, 2007). Brecht *et al.* (1982) recommended that peach cvs. Loadel and Carolyn should be stored for up to 4 weeks under 2kPa O_2 + 5kPa CO_2 at 1.1°C while cvs. Andross, Klant and Halford should be stored for shorter storage period. Exposure to 10 kPa O_2 + 10 kPa CO_2 for 6 weeks has been reported to delay CI in nectarin cvs. Fantasia, Flavortop and Flkamekist (Lurie, 1992). ZhiHong *et al.* (2010) evaluated the dynamic controlled atmosphere (DCA) and controlled atmosphere (CA) during storage-life at 0-1°C for 70 days and subsequent shelf-life at 20°C for 3 days of peach cv. Okubo. They observed that both DCA and CA treatments significantly inhibited fruit rot and flesh browning, kept firmness, alleviated the harm to flesh tissue and chilling sensitivity, and improved quality after storage. However, better storability was noted under

DCA than CA. Crisosto *et al.* (1999) stated that fruit size, storage atmosphere and temperature all influenced storage life and chilling injury. Small fruits (~125 g) in air at 0°C had longer market life than large fruits (~275 g). At both 0°C and 3.3°C, large 'Elegant Lady' and 'O-Henry' fruits had a longer market life under CA (5 per cent CO_2 + 2 per cent O_2 or 17 per cent CO_2 + 6 per cent O_2) storage than under air storage. However, at 3.3°C, small 'Elegent Lady' fruits in CA showed flesh browning. CA (1.0 kPa O_2 + 3.0 kPa CO_2) at -0.5°C 'Eldorado' peach was found more firmer, internal browning incidence than air stored fruits. Ethylene removal during storage improved retention of fruit firmness (Cretta *et al.*, 2000). Ortiz *et al.* (2012) in their study with 'Rich Lady' peaches, picked the fruits at commercial maturity, stored at 2°C and 92 per cent RH under air or CA (3 kPa O_2 : 10 kPa CO_2) for 3 or 15 days, and subsequently kept in air 1 day at 7°C. They did not observed any significant differences in the flesh farmness between CA and air stored fruits. Firmness of these fruits was mainly dependent on the yield of covalently-bound pectin in the cell wall, rather than total amount of cell wall materials.

Intermittent Warming

Research has shown that intermittent warning (IW) is an efficient procedure to inipibit chilling injury (CI). Ben-Arie (1970) reported that when peaches and nectarines were stored at 0°C for 9 weeks and then warmed to room (18°C and 20°C) severe browning results, while peaches held in CA storage (1 per cent O_2 and 5 per cent CO_2) at 0°C maintain good storability when the temperature temperature is increased to 18.3°C for 2 days. Fernandez-Trujillo and Artes (1997) showed peaches stored at 2°C and placed at 20°C for 24h every 6 days had a longer shelf-life, better fruit colour and juiciness than peaches stored at 2°C with no warming periods. Wang *et al.* (2003) in their studies with peach cvs. Luhuajiu, Zhanghuashou and Qingzhou DongXuemi at China, exposed the fruits for 24h at 20°C following cold storage every 7, 14 and 21 days. They observed that IW for 24 h at 20°C every 7 days reduced chilling injury but the degree of fruit softening was high, thus shortening storage life. IW every 14 days was found as the most effective warming protocol because it prevented CI and maintained the fruits original texture, quality and flavor.

1-MCP

1-MCP showed potential to control ripening and increase of shelf life of peaches. Kluge and Jacomino (2002) working with peaches cv. Aurora -1 at Brazil observed that fruits treated with 1-MCP (100-900 nl/l for 12 hours at 25°C) had better firmness and less ground colour loss than non treated fruits, when stored at room temperature (25°C) for 6 days. Although, 1-MCP treatments have proved useful in delaying peach softening, in many cases this response was limited to the treatment and a few days afterwards, or upon removal from subsequent cold storage (Liguori *et al.*, 2004; Ziliotto *et al.*, 2008). Ortiz *et al.* (2011) reported that CA storage (3kPa O_2; 10KPa CO_2 at 0°C) and 1-MCP treatment (1µl MCP for 24 h at 1°C) were effective in delaying the softening of 'Tardibelle' peach during shelf life at 20°C to stimulate commercial shelf life following cold storage. It was observed that CA storage partially inhibited the activities of some pectolytic enzymes and increased the content of pectic polysaccharides, possibly related to delayed softening. 1-MCP treatment caused increased pectin, methylesterase activity, but other pectolytic enzymes were inhibited, resulting in an enrichment of chelate-soluble pectin in the cell walls.

Pests

A number of pests damage peaches but only the most common ones are mentioned here.

Peach Leaf Curl Aphid *(Brachycaudus helichrysi)*

It is the most serious pest of peach in all the peach growing countries of the world. The aphid sucks sap from the buds and sprouting foliage causing curling. yellowing and thickening of leaves. The activity of aphid synchronizes with the new growth during February-March and the affected leaves become distorted and crumpled. Floral buds also become weak resulting in poor setting and fruits fall off prematurely. In the absence of proper leaf functioning the entire tree becomes weak.

The alternate host plants are golden rod (*Erigeron canadensis*) and a wild ageratum (*Ageratum conyzoides*). The reproduction is sexual as well as parthenogenetic. In early November late viviparous females fly from golden rod to fruit trees and deposit nymphs on leaves and in bud axils. The nymphs develop into aptcrous, oviparous females in about a month and are fertilized by migrating males. The fertilized females lay minute eggs singly on buds. After producing 3 to 4 generations, the aphids return to the golden rod (Butani, 1979).

Prebloom application of insecticides preferably 7-10 days before flowcring or at pink bud stage with fine nozzle to give a thorough spray on plant surface particularly in the bud axils is necessary to control the aphid. For proper control, another post bloom spray is given 7-10 days after petal fall. Any insecticide like methyl demeton, dimethoate, parathion, fenetrothion or diazinon at 0.1 per cent will serve the purpose. The granular insecticides. *i.e.* temik, furadon and thimet are also effective in checking the aphid population. Gupta *et al.* (1986) obtained the best control with aldicrab 10G at 150g/tree applied to the soil below the tree canopy before bud burst.

Peach Green Aphid (*Myzus persicae*)

Green peach aphid is cosmopolitan in distribution and attacks a number of fruits. Aphids suck the sap of the leaves, flower buds and young fruits. The flower buds and young fruits shrivel and drop down. The aphid is responsible for the transmission of a number of viruses (Butani, 1979). Aphids overwinter in egg stages and hatching takes place in early spring. The new adults feed on swelling buds and reproduce parthenogenetically. After several generations on stone fruits, winged forms move to other host plants.

Insecticides are quite effective in controlling the pest population but they destroy parasites and predators also. Adoption of integrated practices is more desirable. Al-Antary and Khader (2011) recorded good control of aphid by spraying of Calypso followed by Karate and Evisect. Yajun *et al.* (2011) stated that peach aphid in Beijing area of China developed resistance to nearly all frequently used pesticides. They suggested that β-cypermethrin should not be used and the chlorphyrifos should be avoided, and abamectin, emamectin benzoate and imidacloprid could be used alternatively. Both water and hydroethanol extracts of neem leaves at 10 per cent were found effective to control peach aphid (Mondedji *et al.*, 2014).

Peach Scale Insect (*Eulecamuim coryli*)

It is widely distributed in all the temperate peach growing regions. Eggs are laid by females in the month of April, which hatch out and crawlers during May to July. The crawlers mostly settle on leaves and the migration of nymphs from leaves to twigs takes place during July to December. The females are formed during the month of February. Males which appear during March-April survive only for 20-48 hours while females die after oviposition and thus the pest completes one generation in a year (Gupta and Joshi, 1976). The crawlers and nymphs suck the sap and thus restrict the vigour and production adversely.

Spraying of dimethoate, phosphamidon, parathion, *etc.*, @ 0.03 per cent in June is quite effective. Application of lime-sulphur 1:20 with water during winter is also recommended.

Oriental Fruit Moth (*Grapholita molesta*)

It is a serious pest of peaches nectarines and quince. The larvae feed on succulent terminal growth during the initial spring and subsequent autumn vegetative growth flushes causing withered or dead shoots. The female adults oviposit directly on fruits that are at least halfway through their development. Newly hatched lavae are known for quickly tunneling beneath the surface and into the fruit. As the fruit matures, gum often entrance wounds

Application of *Bacillus thuringiensis* serovar Kurstaki Strain EC 2348 (formulated product Lepinox Plus®) was found effective in controlling oriental fruit moth in Italy (Ladurner *et al.*, 2012). Simultaneous use of toxic bait (Biofruit 3 per cent + Malathion 500 EC, 200 ml in 100 l) and mating disruption using sexual pheromone (Splat Grafo) in 1000 points/ha showed effective to minimize the pest population in Brazil (Harter *et al.*, 2010).

Fruit Flies (*Dacus* spp.)

Fruit flies are the most destructive pests of subtropical peaches and are also common in valley areas. The critical period of attack is usually of short duration starting from fruit maturity till ripening. During summer the maggots, which emerge from the eggs laid under the skin of fruits are seen in flesh. The attacked fruits develop brownish patches due to rotting, and drop from the plant.

Two sprays of 0.075 per cent fenthion emulsion are quite effective and spraying should be given in the month of June at 10 days interval (Gupta and Joshi, 1976). However, considering the problem of residual toxicity, attempt should be made to cultivate early cultivars like Candor and World's Earliest which escape from severe attack.

In April-May, on the appearance of adult flies, a bait consisting of 0.1 per cent malathion and 1 per cent sugar should be sprayed on foliage and other resting places of flies.

Tarnished Plant Bug (*Lygus lineolaris*) and Green Stink Bug (*Nezera viridula*)

Both these bugs feed on rapidly developing fruit and cause the fruit to be scarred and malformed, a condition known as 'catfacing'. Fruit gummosis and dark green water-soaked areas are also caused by stink bugs. Johnson *et al.* (1985) tested the efficacy of 7 insecticides against both pests and achieved the most effective control with permerthrin at 0.1 Ib a.i. and chlorpyrifos at 0.25 and 0.5 Ib a.i./100 gallon water, applied at 14-day interval beginning from pink flower bud stage. The quality of the fruit (colour, size and smoothness of surface) was not affected by any of the insecticides.

Peach Twig Borer (*Anarsia lineatella*)

The borer is a chocolate-brown worm which bores into actively growing twigs and fruits. The affected shoots wilt and dry up while fruits rot (Butani, 1979). Gum exudes near the entrance holes in fruits. Secondary infection of bacteria and fungi also occurs.

To control this pest, the affected shoots and fruits, should be removed and destroyed and parathion or diazinon (0.1 per cent) combined with the dormant oil spray for scale control to be sprayed. The same insecticides without oil should be sprayed again during petal fall and in late May (Butani, 1979). Pickel *et al.* (2006) suggested two bloom sprays of the entomopathogenic bacterium *Bacillun thuringiensis* which provides very reliable control of peach twig borer.

Stem Borers (*Sphenoptera lafertei* and *S. dad-khani*)

Adult beetles feed on foliage and lay eggs singly scattered all over the tree. On hatching, young grubs bore into the stem under the bark making minute galleries. Thereafter, grubs pupate in the small chambers made deep in the wood. As a result, bark splits, gum globules ooze out of the stem and leaves turn pale. The tree becomes stunted and may die in case of severe infestation.

The trunk should be sprayed three or four times with parathion or azinphosmethyl to kill the larvae. Seriously infested plant parts should be removed and destroyed.

Nematodes

Nematodes pose problems with peach rootstocks in many parts of the world by adversely affecting the establishment, growth, fruit production and tree longevity of peach orchards (Layne, 1987). Out of the nineteen species of nematodes found associated with peaches, three belong to root-knot group (*Meloidogyne*), seven to the root-lesion group (*Pratylenchus*), six to the ring nematode group (*Criconemoides*), and three to the dagger nematode group (*Xiphinema*). Beside directly influencing the plant growth, many nematodes act as vectors of viruses. The dagger nematode group transmits tomato ring spot and peach rosette viruses (Powell, 1984). Widespread and very harmful is the ring nematode (*Criconemoides xenoplax*). In USA, the presence of this nematode is closely related to the so-called 'Peach Tree Short-Life (PTSL)' Syndrome (D'Errico and Giacometti, 2012). Root lesion nematodes (*Pratylenchus* spp.) are a serious problem in cooler climate, whereas root-knot nematodes (*Meloidogyne* spp.) create the same situation in warmer regions.

Control programmes for nematodes in peach orchards involve the application of nematicides, soil fumigation and nematode-resistant rootstock. The use of resistant rootstocks is the most convenient and reliable method of combating the nematode problem. More research work has been done to evolve root-knot nematode-resistant rootstocks than any other nematode group. As a result, several rootstocks of each resistant to root-knot nematodes have been developed. Souza *et al.* (2014) reported that mume is highly resistant; Okinawa, Capdebascq, Aldrigui and Tsukuba are resistant and Flordaguard is moderately resistant rootstock for nematode *Melodidogyne enterolobii*. Guardian rootstock performed better in sites with high ring mematode population. Nemaguard, Shalil, S-37, Okinwa, Nemared (Rom, 1984), S-37 and Shalil rootstocks are resistant to *M. incognita,* while Nemaguard and Okinawa are resistant to *M. incognita* and *M. javanica* (Hesse, 1975). Research work is, however, in progress to develop resistant rootstocks against root-lesion and ring nematodes (Layne, 1987). McKenry and Kretsch (1987) studied the response of various soil treatments and irrigation on peach trees and got the best nematode control and improvement in tree growth and fruit production from methyl bromide treatment (450 kg a.i./ha). High irrigation regime also counteracted the detrimental effects of root-knot nematode.

Diseases

Peach Leaf Curl (*Taphrina deformans*)

Leaf curl is the most common disease, specially on seedling plantation. Higher temperatures during late half of February and March along with high humidity are favourable for infection. The disease appears soon after the bud burst and the affected leaves are deformed, thickened, puckered, rolled and reddened. These are brittle and waxy in texture. With advance of season, leaf colouration becomes less pronounced followed by disappearance of chlorophyll and appearance of powdery coating on the upper surface which turn brown and shed prematurely.

The incidence can be controlled by spraying lime sulphur 1:15 or 1:20 (before bud burst) and 1:40 (after bud burst). Spraying of benlate 0.075 per cent (Bose and Sindhan, 1973) or captan 0.2 per cent is also very effective. First spraying should be done before bud burst and another after full bloom. Bordeaux mixture, copper oxychloride and cupric-hydroxide are equally effective in controlling the disease (Tate and Mespel, 1985).

Bacterial Spot (*Xanthomonas arboricola pv. pruni*)

Bacterial spot attacks leaves, twigs and fruits. Small water-soaked areas develop on leaves which eventually die and drop. Similar spots are also observed on fruits.

Thus, it may defoliate the tree or severely degrade the fruit. The disease forms cankers on twigs which are the source of initial infection in the following spring. The inoculum is desseminated by rain and wind. The optimum temperature for growth and development of the disease is 29.4°C. There is no good control of the disease but vigorous trees are less susceptible. The most effective control is the use of dodine-captan mixture (Powell, 1975).

Brown Rot (*Monilinia fructicola*)

The fungus infects blossoms, twigs and fruits. Masses of ash-grey powdery spores appear on the surface of infected parts. Infected blossoms wither and die. The fungus extends causing wilting and cankering of young branches. Rotted fruits or mummies left on the tree or on the ground become a source of infection subsequently.

Blossom infections are largely controlled by sprays of fungicides such as sulphur, difolatan (150g/100 l water), captan (300g/100 l water) applied just before and during bloom. Fruit infections are also controlled by spraying the fungicides just before the harvest. Insects attacking fruit increase brown rot infection and they should be checked. All mummies should be removed and the soil be ploughed before flowering. Preventive measures are particularly necessary for nectarines in which the control of brown rot is more difficult.

Bacterial Gummosis (*Psuedomonas syringae*)

Circular to elongated soaked gumming lesions arise on bark or outer sapwood and fruits. In severe incidence, gummy pockets are commonly found on main branches and lateral shoots. Gummosis can be controlled by scrapping the gum pockets and painting with Mashobra paste consisting of non-sterile dihydrosteromycin sulphate as active ingredient (Agrawala, 1961). Copper oxychloride (300g/100 l water) should be sprayed after leaf fall.

Silver Leaf Disease (*Stereum purpureum*)

It is a serious disease in cooler climate. The affected foliage show peculiar hay-grey lustre approaching more to the colour of lead. On cutting the branches, the wood exhibits brownish discolouration. The affected trees should be uprooted and burnt. Iron sulphate @ 250 to 500g per tree should be applied in the roots followed by irrigation:

Powdery Mildew (*Sphaerotheca pannosa*)

It is a serious disease of peach attacking leaves, young shoots and twigs. The young leaves are coated with a thick layer of mycelium but only white patches appear on old leaves (Pathak, 1980). Terminally growing shoot and fruits are also covered by a white powdery layer. In case of severe infection fruits first turn pink then dark brown. The germination of conidia takes place between 21° and 27°C (Weinhold, 1961). The disease is controlled by spraying with sulphur compounds.

Viral Diseases

The occurrence of viral diseases varies greatly in different peach growing regions of the world. Peach yellows and little peach are the two important viruses which cause severe loss to peaches in North America. Yellow is so well-known that growers keep it under control by prompt removal of diseased trees. Red suture, resembling yellows and little peach in some respect is caused by strains of those viruses. Peach wart and calico both are serious and are controlled with the use of healthy bud wood. Peach yellow bud mosaic is caused by the same virus that causes tomato ring spot. It is transmitted by nematode and is, therefore, difficult to control. Maur dwarf, asteroid spot, golden net, line pattern, mottle, willow twig viral diseases are controlled with the use of disease-free budwood programme.

The peach virus diseases, ring spot and necrotic spot are seed borne. Stubby twig and stem pitting are new viruses. Rapid advances have been made in the control of peach viruses and virus-like diseases through improved bud-wood programme (Cochran, 1975).

Physiological Disorders

Sun Scald

This is the most destructive physiological disorder of peach. It causes severe damage to the exposed trunk and main scaffold branches. Shading of branches considerably reduces the incidence. Painting of exposed surface with lime paste and shading by wrapping straw or hay around the trunk and thicker branches is quite effective in mitigating the problem.

Split-Pit and Gumming

In some cultivars split-pit and gumming cause a great loss to the growers because the affected fruits become unmarketable. Splitting occurs at the joint of dorsal and ventral sides of the fruit. According to Teskey and Shoemaker (1972), the conditions favouring the large fruit development are also conducive to split pit which happens mostly at the time of pit hardening.

Sometimes, gum exudes from the fruit in large quantity making it unfit for consumption. It may fill the entire pit cavity surrounding the kernel. Seed abortion is frequently associated with gumming (Regland, 1934). When heavy rains occur after a long dry period, splitting and gumming have been found accentuated. The exact causes of these problems are yet to be determined.

Breeding and Improvement

The position of cultivars under commercial cultivation keeps on changing more rapidly in peaches than any other fruit as one time popular cultivars are replaced by newly evolved improved ones. Elberta and J.H. Hale are no longer being planted. Short juvenile phase of the plant and regular release of a large number of superior cultivars have made this possible. Brooks and Olmo (1972) compiled a register of the released cultivars of almost all fruits and nuts and it is interesting to note that peach (including nectarine) contains the maximum number of cultivars among them. The cultivar development in peach has reached an advanced stage even then there is scope for further improvement because of conceptual changes in fruit culture like high density planting, meadow orcharding, and mechanization. Availability of great genetic variability in peach gene pool, however, makes it possible to evolve cultivars to meet all the cultural requirements of the crop.

History of Improvement

From China, peach first spread to Mediterranean regions, Europe, America and then to all other temperate and sub-temperate parts of the world. A good account of its early distribution and development has been given by Hesse (1975) and Scorza *et al.* (1985). Improvement was based on the selection of open-pollinated seedlings, introduction of promising types from other countries till-the rediscovery of Mendel's law of inheritance (Cullinan, 1937). The genetic resources of the peach in the USA mainly consisted of indigenous seedlings, introduction from European countries and China (Chinese Cling). Important early cultivars were Early Crawford, Oldmixon Cling, Belle, Elberta, J.H. Hale and Hiley. In the twentieth century, improvement programmes on peaches have been taken up in many countries including USA, Canada, Italy, USSR, South Africa and Australia which have yielded tangible results (Hesse, 1975; Weaver, 1968; Olmo, 1965). The earliest formal breeding in Asia was initiated in Japan about 60 years ago followed by in China in 1970's. Breeding for development has also started in Thailand, Korea and India. 'Shanghai' Suimitsuto' (='Chinese Cling') has played a key role in the breeding of Japanese cultivars (Yamamoto *et.al.*, 2003). In the past five decades, the peach improvement programmes have followed- three directions, *i.e.*, breeding for superior high chilling cultivars, high quality low chilling cultivars, and pests and ecological tolerant rootstocks. As a result, a number of good cultivars and rootstocks have been released which have found worldwide acceptance. Interestingly most of the public peach breeding programme in Europe and USA has been reduced or closed and an increase of private sector breeding programme resulted release of several important varieties from USA, France and Spain in the last thirty years (Byrne, 2005).

Breeding Objectives of Scion Cultivars

The prime aim of a peach breeding programme is to evolve cultivars which give high yield of good quality fruits in an appropriate season. Tree structure and response to close planting, pruning, irrigation and fertilizers and hardiness to pests, diseases and ecological factors are the yield contributing factors. Effects of these factors on fruit production vary from locality to locality. If in any locality, the yield and quality of fruits are hampered by cold climate breeding for cold-hardiness will be included as a priority objective. In contrast for warm climate low chilling cultivars are to be developed. Here are some common objectives of peach breeding programmes, the priority being accorded on the basis of the intensity by which an yield contributing factor is operating in a locality for which the cultivars are to be bred.

(i) Development of cultivars giving high yield at low cost of production is of paramount importance in order to increase the profitability. Hansche and Beres (1979) have shown that triple income can be obtained by densely planted orchards due to higher yields than standard orchards.

(ii) High yields are usually a result of closer planting for which appropriate tree structure is required. Instead of standard, dwarf or compact trees or semi-dwarfs are to be evolved because they are more amenable to the demands of high density planting such as pruning and mechanization (Scorza, 1984).

(iii) Breeding of high quality fruit bearing cultivars is a major concern because the consumer standards are increasing rapidly. Fruit quality involves shape, size, skin colour, flesh colour, firmness, texture, freedom from loose fibre and non-browning of the flesh besides the nutritive value. Improvement of all these traits is needed to produce the fruits of high quality.

(iv) Fruits fetch a good price at the extremes of maturity season. Therefore, extension of maturity periods by breeding early and late cultivars is another objective of peach breeding. Earliness is a more wanted trait than lateness.

(v) Peach-hardiness is very important consideration in breeding which has to be incorporated in the cultivars for successful peach growing. The hardiness comprises cold-hardiness, resistance to diseases, pests and nematodes and resistance to wet conditions, drought, and mechanical injury.

(vi) Broadening of the genetic base of the cultivars should also be attempted in order to safeguard against epiphytotics and to augment the scope of genetic advancement.

(vii) The inheritance of various characters is required to be investigated in detail so as to manipulate the major genes for the improvement of peaches. Cooperation of molecular biologists will be useful in studying *Prunus* genetics, gene transfer systems and synthesis of gene combinations.

Breeding Systems

Outcrossing, inbreeding and backcrossing have been used in peach breeding on large scale whereas interspecific hybridization, mutation and protoplasm systems have been rarely adopted. Unrelated parents, such as. Early Crawford and Chinese Cling for improving fruit quality, and Red Ceylon, Jewel and Hawaiian for developing low chilling cultivars were crossed with the indigenous cultivars. Notable examples of inter specific hybridization is the release of peach x almond hybrids, GF 557 and GF 677, which are used widely in Europe as rootstocks for peach and almond. Interspecific hybridization has not been, however, successful for the development of scion cultivars (Hesse, 1975).

Unlike most other fruits, inbreeding has played a vital role in the peach breeding. Most breeding programmes after initial outcrossing restorted to selfing in order to maintain a high degree of desirable traits in the progeny which was otherwise not possible due to their segregation. Peach does not exhibit inbreeding depression markedly due to its high selfing nature (Hesse. 1975). Even after inbreeding up to F_7, Lesley (1975) found variability in some characters though there was considerable homozygosity for few characters at F_4, to F_6. Backcrossing has been found essential in improving the fruit quality factors in brachytic dwarfs and, likewise. it will be applied in future to incorporate pest or disease resistance into commercial cultivars (Hesse, 1975). Both bud sports and artificial mutation have not been as fruitful in peach as in other fruits (apples and grapes). Showy flowers and early ripening have been observed due to bud sports. Hough and Weaver (1959) discovered mutant forms of Fairhaven and Elberta on chronic radiation in gamma field. Hough *et al.* (1964) obtained linkage breaking and early ripening with X-irradiation.

Biotechnology needs to be utilized for increasing the pace of peach breeding. Ramming (1985) has standardized the technique of in-ovule embryo culture which can be used in breeding programmes for the improvement, of early maturing genotypes.

Inheritance of Specific Characters

Hesse (1975) has reviewed the inheritance of specific characters in peaches in detail providing a list of simply inherited characters but most of the important characters like fruit size, shape, red skin colour, juiciness, acidity, soluble solids, cold hardiness, chilling requirement and time of maturity are controlled by multiple genes. Furthermore, the characters are mostly phenotypically variable, environmentally influenced, interlinked and lack precision in quantitative measurements. Incomplete dominance and pleiotropism are also involved.

Dominance of small fruit size over large fruit size was reported earlier (Blake. 1934: Connors, 1923). Size being a product of height, width and depth, its data need logarithmic transformation before statistical analysis. Weinberger (1955) suggests the control of fruit size by multiple genes without complete dominance. Likewise, the dominance of low fruit acidity over the high acidity reported by Connors (1923) was not substantiated by subsequent studies as it was found to have a continuous range of values (Ryugo and Davis, 1958). On the basis of a study of hybrid populations, Yoshida (1970) concluded that sweet (low acid) peaches are the outcome of cumulative dominant genes.

Cold-hardiness consists of wood, flowerbud and bloom-hardiness, all of which are inherited independently, meaning thereby that a genotype may be wood-hardy but not flowerbud-hardy or *vice versa*. Flowerbud-hardiness is more important than either of wood or bloom-hardiness (Hesse, 1975). *Prunus davidiana* is wood-hardy but bud tender like Elberta and J.H. Hale. Redhaven and Veteran cultivars are known for bloom-hardiness. Transmission of poor fruit quality by bud-hardy parents (Chili and Greensboro) is another handicap in using them for breeding of cold-hardiness. It is also correlated with chilling requirement of the cultivar as high chilling clones are less susceptible to bud mortality due to cold (Mowry, 1963).

Investigations on heredity have revealed that characters like ripe date, full bloom date, amount of bloom, and fruit suture diameter have high heritability while crop load, fruit acidity, soluble solids and fruit firmness have low heritabilities (Hansche *et al.*, 1972). High heritability is an indication of a large genetic gain of a trait under selection per generation.

Correlation studies in peaches have given valuable information which needs to be utilized by the breeders. Sweetness is highly correlated with the fruit quality but soluble solids are negatively correlated with fruit size (Hesse. 1975).

Breeding for Rootstocks

Rather limited attempts have been made on breeding of peach rootstocks probably because peach seedling itself has been serving as a good rootstock. Rootstock selections made from the seedlings of late cultivars solely for this purpose in various countries are Siberian C and Harrow Blood in Canada; Bailey, Rutgers Red Leaf and Nemaguard in the United States; GF 305 and Rubira in France and southern Europe; by A 1,2,3 and 4 in Czechoslovakia and Ohatsumomo in Japan (Layne, 1987). Siberian C is both dwarfing and cold-hardy rootstock but is susceptible to waterlogging and nematodes.

More intensive systems of orchard management including closer planting, new training systems and drip irrigation require rootstocks adjusting to this purpose. Hence, the objectives of rootstock breeding in peaches should be to evolve rootstocks which are able to control tree vigour, produce heavy-crop of high quality fruits for many years, resistant to nematodes, pests and soil borne diseases and tolerant to cold, heat, drought and wet conditions. They should be easy to propagate and compatible with majority of peach cultivars. With the availability of a large number of *Prunus* species for hybridization with *P. persica* it may be possible to develop such rootstocks (Rom, 1983). Several size controlling clonal rootstocks of peach and almond hybrids ('Adarcias', 'Castore', 'Polluce', 'Sirio'); plum x peach hybrid ('Ishtara'), plum rootstocks ('Adesoto 101', 'Montizo', 'Penta' 'Tentra') and *P. salicina* x peach hybrid ('Controller 5') have been developed.

Beside, selection breeding was carried out to evolve clonal and nematode-resistant rootstocks. Bernhard (1965. 1949) hybridized peach with almond and selected two clones. GF 677 and GF 557 (GF stands for Grande Ferrade), which are clonally propagated and useful for alkaline, dry and wet soils. Riggoti NO.2 rootstock has been selected from *P. persica* x *P. davidiana* hybrids (Riggoti, 1942). It is propagated by seeds but produce the uniform seedlings which impart vigour to the scion cultivar. Plum hybrids, Damas GF 1869 (*P. domestica* x *P. spinosa*), natural hybrid), Marianna GF 8-1 (*P.domestica* x *P. munsonianai*) are better adapted to heavy soils (Layne, 1987). Peach plum hybrid (*P. belsiana* x Yunnan) Myran is tolerant to drought, poor soil, root-knot nematodes and armillaria and verticillium diseases (Rom, 1983). Peach x *P. davidiana* hybrids and 'Tsukuba' series of rootstock from Japan have been reported to show some tolerance to water logging.

With the discovery of root-knot nematode-resistant seedlings of Shalil and Yunnan, the work on nematode-resistant rootstock began (Hutchins, 1936). Nemaguard, apparently a hybrid of *P. persica* x *P. davidiana,* is a widely used root-knot nematode-resistant rootstock which is immune to *M. incognita* and field resistant to *M. javanica.* 'Garfe' X 'Nemared' that are resistant to root-knot nematodes and tolerant to calcareous soils have been released (Felipe, 2009). Other peach-almond and peach x *P. davidiana* hybrids resistant to root-knot nematodes are 'Barrier 1', Cadaman', 'Hansen 536' and 'Hansen 2168' (Edin and Garcin, 1994; Kester and Assay, 1986).

Sharpe *et al.* (1970) stated that the resistance to *M. incognita* is simply inherited as a dominant factor but resistance to *M.javanica* is based on duplicate, independent dominant factors. The association of low levels of endogenous cytokinin with resistance has been suggested by Kochba and Samish (1971, 1972). Resistance to peach leaf curl is determined by a polygenic system (Monet, 1989). The inheritance of powdery mildew resistance varies with its source. Pascal *et al.* (2010) described is as a single dominant gene from the peach 'Pamirskij 5'; two loci, one controlling high resistance, the other medium and low resistance from *P. ferganensis* (D'Bov, 1975).

Genome

The peach genome sequence has been recently made available to the scientific community. This represents the culmination of a long process that started less than two decades ago with the release of the first marker based linkage maps. The process has advanced rapidly with the studies of molecular diversity, detection of genome position of major genes and quantitive trait loci, development of large DNA sequence collections, transcriptome and proteome analyses, comparative genomic studies, construction of physical map, and development of databases where researchers can access information. The growth of genomic knowledge has been partly due to the simplicity of peach genome: short (230 Mbp) diploid, and distributed on eight pairs of chromosomes. It's usually short intergeneration period (2-4 years) and selfing mating behavior, plus a dynamic peach scientific community that has often collaborated in the development of the necessary tools, have also facilitated in constructing a robust sequence of its complete genome (Arus *et.al.*, 2012).

The GDR (http://www.rosaceae.org) is the central access point for the whole genome sequence of peach. Peach genomics information are also available in ESTreeDB (http://www.itb.cnr.it/estree/); Phtosome (JGI, http://www.phytozome.org.peach) and IGA (http://services.appliedgenomics.org/gbrowse/prunus_public/).

References

Abdulla, D.A. and Childers, N.F. (1973). J. Amer. Soc. Hort. Sci., 98: 517-22.

Agrawala, R.K. (1961). Himachal Hort., 2: 49-51.

Al-Antary, T.M. and Khader, B. (2011). Adv. Environ. Biol., 5: 3599-604

Albrigo, L.G. (1975). The Peach-Varieties. Culture. Pest Control. Storage, Marketing. Hort. Publ. New Brunswick. N.J., pp. 446-68.

Ali, A.,Perveen, S., Shah, S.N.M., ZengQiang, Z., Wahid, F., Shah, M., Bibi, S. and Majid, A. (2014). Amer. J. Pl. Sci., 5: 1258-64

Andersen, F.G. (1932). J. Pom and Hort. Sci., 10: 130-46.

Anderson, R.E., Parsons, C.S. and Smith, W.L. J. (1969). Aust. Inst. Agr. Sci. J., 22: 293-96.

Anonymous, (1969). The Wealth of India, VIII, Publications and Information Directorate, CSIR, New Delhi, pp. 274-79.

Anonymous, (1975). Fruit Industry of Himachal Pradesh in retrospect and prospect. Agro. Eco. Res. Centre, Himachal Pradesh Univ., Simla.

Archibald, J.A. and Bradt, O.A. (1964). A.R. Out. Hort. Exp. St. Prod. Lab., pp 20-26.

Armstrong, W.D. (1975). The Peach-Varieties, Culture, Pest Control, Storage, Marketing. Hort. Pub., New Brunswick, New Jersey, pp. 206-8.

Arnold, C.E., Aldrich, J.H. and Cornell, J.A. (1981). HortSci., 16: 170-71.

Arnold, C.E., Aldrich, J.H. and Martin, F.G. (1983). Acta Hort., 137: 145-52.

Arora, R.L., Tripathi, S. and Singh, R. (1999). Indian J. Hort., 56: 286-94

Arus, P., Verde., I., Sosinski, B., Thebentyayeva, T. and Abbott, A.G. (2012) Tree Genet. Genomes, DOI:10.1007/S11295-012-0493-8

Ashnley, T.E. (1931). Proc. Amer. Soc. Hort. Sci., 28: 28-33.

Aslikyan, V.T. (1983). Sadovodstvo Vinogradarstvo i Vinodelie Moldavii, 3: 20-21.

Babaer, V.I., Movchan, T.C., Orlov, P.N. and Shara Futdinov, K.H.V. (1984). UDO, i. Regu Rosta, V.Masco USSR pp. 75-81.

Baker, C.E. (1941). Proc. Amer. Soc. Hort. Sci., 39: 33-37.

Barbosa, W., Dall'orto, F.A. and Ojuma, M (1985). Bragantia, 44: 465-72.

Bargioni, G., Loreti, F. and Pisani, P.L. (1984). Rivista della Ortotlorofrutticoltura Italiana, 68: 423-32.

Baroni, G., Tonutti, P. and Ramina, A. (1986). Rivista della Ortotlorofrutticoltura Italiana, 70: 215-25.

Bassi, D., Brigheni, G. and Nardi, V. (1985). Acta Hort., 173: 339-47.

Batjer, L.P. and Westwood, M.N. (1958). Proc. Amer. Soc. Hort. Sci., 71: 116-26.

Baugher, T.A., Elliott, K.C., Blizzard, S.H., Walter, S.I. and Keiser, T.A. (1988). HortSci., 23: 981-83.

Bay-Karn, A. (1984). M.S. Thesis. Inst. Natural Sciences, Univ. of Cukurova, Adana, Turkey.

Baz, A.G.I.O. (1986). Annals of Agri. Sci., Moshtohr, 24: 183-94.

Baz, A.G.I.O. and Salama, M.A. (1984). Annals of Agri. Sci., Ain-Shams Univ., 29: 1003-14.

Bellini, E., Cirnato. A. and Mariotti, P. (1985). Acta Hort., 173: 329-38.

Ben-Arie, R.(1970). Phytochem., 19: 2553-55

Bencat, P. and Matuskovic, J. (1982). Vedecke Prace Vyskumneho Ustavu Ovocnych a Okrasnych Drevin V Bojniciach, 4: 11-20.

Bennet, J.P. (1931). Calif. Agr. Exp. St. Cir., 321: 12.

Bernhard, R. (1949). Rev. Hort. (Paris) 121: 97-101

Bernhard, R. (1965). Proc. Int. Peach Congr. (Verona) pp. 365-78.

Bilanski, W.K. and Menzies, D.R. (1984). Proc. Int. Symp. Fr. Nut and Veg. Harv. Mech., Israel, 1983, pp. 357-63.

Black, J.D.F., Mitchell, P.O. and New Green, P.N. (1977). Aust. J. Agri. and Animal Husb., 17: 342-45.

Blake, M.A. (1934). N.J. Agr. Exp. Cir, No. 303.

Blanco, A. (1986). Acta Hort., 179 : 573-74.

Blanco, A. (1987). J. Hort. Sci., 62: 147-55.

Boland, A.M., Mitchell, P.D., Goodwin, I., and Jerie, P.H. (1994). J. Amer. Soc. Hort. Sci., 119: 1157-62.

Bollard, E.G. (1953). N.Z.J. Sci., Tech., 35: 29-44.

Bose, S.K. and Sindhan. G.S. (1973). Progressive Hort., 4: 4-8.

Bowen, H.H. (1971). HortSci., 6: 11-15.

Bowen, H.H. and Dorman, T.E. (1974). Texas Agric. Expt. Stn. Circ., No. MP-1I49.

Boynton, D (1944). Proc. Amer. Soc. Hort. Sci., 44: 31-33.

Brecht, J.K., Kader, A.A., Heintz, C.M. and Noorna, R. C. (1982). J. Food Sci., 47: 432-36.

Brinkley, A.M., Rogers, E. and Green, F.M. (1958). Colo. Agr. Exp. Sta. Tech. Bull. No. 64.

Briones, G. and Montes, M.E. (1986). Proc. Trop. Region. Amer. Soc. Hort. Sci., 23: 28-30.

Brooks, R.M. and Olrno, H.P. (1972). Register of New Fruit and Nut Varieties. 2nd edition, Univ. of California Press, Berkley.

Brummell, D.A., Dalcin, V., Crisosto, C.H. and Labavitch, J.M. (2004). J. Exp. Bot., 55: 2029-39

Buchanan, W.D., Biggs, R.H., Blake, J.A. and Sherman, W.B. (1970). J. Amer. Soc. Hort. Sci., 95: 781-84.

Butani, D.K. (1979). Insects and Fruits. International Book Distributers. Rajpur Road, Dehra Dun, India.

Byers, R.E. (1978). J. Amer. Soc. Hort. Sci., 103: 232-36.

Byrne, D.H. (2005). *Hort. Technol.*, 15:494-500

Carlson, R.F. and Badizadigam, M. (1967). Mich. Agric. Exp. Sta. Quart. Bull., 49: 276-82.

Centibas, M.; Butar, S. and Koyuncu,F. (2012)Ege Universiti Ziraat Fakultesi Dergisi, 49: 103-106

Chahil, B. S., Grewal, S.S. and Dhau, A.S. (1980). Punjab Hort. J., 20: 70-73.

Chalmers, D. J., Mitchell, P.D. and Jerie, P.H. (1984). Acta Hort., 146: 143-49.

Chalmers, D. J., Mitchell, P.D. and Jerie, P.H. (1985). Acta Hort., 173: 283-88.

Chander, S. and Khajuria, H.N. (1981). Haryana J. Hort. Sci., 10: 174-76.

Chandler, R.F. (1933). Proc. Amer. Soc. Hort. Sci., 30: 07-09.

Chang, S. and Werner, D.J. (1984). J. Amer. Soc. Hort. Sci., 109: 42-45.

Chapman, H.D. (1964). Indian J. Hort., 21: 97-119.

Chauhan, K.S. and Maheshwari, L.D. (1970). Indian J. Hort., 27: 136-40.

Chauhan, K.S. and Pundir, J.S. (1972). Indian J. Agric. Sci., 42: 769-71.

Childers, N.F. (1975). Modern Fruit Science, Horticulture Publication, Rutgers Univ. Brunshrick, N.J.

Chopra, H.R., Jawanda, J.S. and Sandhu, A.S. (1987). Punjab Hort. J., 27: 42-47.

Chopra, S.K., Chadha, T.R. and Tripathi, S.N. (1982). Proc. 21st Int. Hort. Cong., Hamburg, 1: 124.

Christ, E.G. (1975). In: The Peach-Varieties, Culture, Pest control, Storage, Marketing. Hort. Pub. New Brunswick, N.J. pp. 228-30.

Cibes, H.R. (1950). M.Sc. thesis, Rutgers Univ., N.J., USA.

Clanet, H., Salesses, J.S. and Bernard, J.L. (1979). Arboriculture Fruitiere, 26: 27-36.

Clayton, C.L. (1975). In: The Peach-Varieties, Culture, Pest control, Storage, Marketing. Ho;t. Pub. New Brunswick, New Jersey.

Cochran, L.C. (1975). In: The Peach-Varieties, Culture, Pest control, Storage, Marketing. Hort. Pub. New Brunswick, New Jersey, pp. 363-65.

Comin, D. (1956). Ohio Agr. Exp. St. Spec. Cir., No. 93.

Connors, C.H. (1923). Proc. Amer. Soc. Hort. Sci., 19: 108-15.

Corelli, L. Brighenti, G., Palara, U., Ravaioli, F. and Sansavini, S. (1986). Rivistao di Frutticoltura e di Ortofloricoltura, 48: 55-60.

Corgan, J.N. and Widmoyer, F.B. (1971). Proc. Amer. Soc. Hort. Sci., 96: 54-57.

Coric, D. (1997). Jugoslovensko Vocarstvo, 31: 73-76

Cosgrove, D. (1999). Pl. Mol. Biol., 50: 391-17.

Coston, D.C., Krewer, G.W., Elkner, T.E., Williamson, J.G. and Sims, E.T. J. (1985b). J. Amer. Soc. Hort. Sci., 110: 874-77.

Coston, D.C., Sims, E.T. J. and Williamson, R.E. (1985a). Acta Hort., 173: 323-28.

Couvillon, G.A. (1985). Acta Hort., 73: 271-82.

Couvillon, G.A. and Erez, A. (1980). HortSci., 15: 41-43.

Couvillon, G.A. and Erez, A. (1985). J. Amer. Soc. Hort. Sci.,110: 579-81.

Couvillon, G.A., Pokerney, F.A. and Scalabrelli, G. (1986). Acta Hort., 160: 413-20.

Cretta, M., Antunes, P.L., Brackmann, A. and Nakasu, B.H. (2000). Ciencia Rural, 30: 73-79.

Crisosto, C.H. and Kader, A. A. (2004). Peach. In: Agriculture Handbook No. 66.

Crisosto, C.H., Day, K.R., Johnson, R.S. and Garner, D. (2000). J. Amer. Pom. Soc., 54: 118-22

Crisosto, C.H., Garner, D, Cid, L. and Day, K.R. (1999). Calif. Agric., 53: 33-36

Crisosto, C.H., Mitchell, F.G. and Ju, Z.G. (1999). HortSci, 34: 1116-18.

Cullinan, F.P. (1937). In: USDA Yearbook of Agriculture, 1937 pp. 665-702.

Cullinan, F.P. and Waugh, J.G. (1939). Proc. Amer. Soc. Hort. Sci., 37: 87-94.

D Errico, G. and Giacometti, R. (2012). IOBC/WPRS Bull, 74: 97-103

D'Bov, S. (1975). Genetica i Selektjcva, 8:267-71

Dall'orto, F.A.C. and others (1984). Agronomico, 36: 298-302.

Daniell, J.W. (1984). Proc. 34th Ann. Mtg. South Weed. Sci. Soc., Abstr. No.126.

Daniell, J.W. and Hardcastel, W.S. (1969). Proc. 22nd Ann. Mtg. South Weed Sci Soc., pp. 236.

David, M. and Nigicser, T. (1986). Acta Hort., 179: 681-86.

Davidson, O.W. and Blake, M.A. (1936). Proc. Amer. Soc. Hort. Sci., 34: 247-48.

DeCandolle, A. (1967). Origin of Cultivated Plants, Hafner publishing Co. New York.

Degtyar, L.A. (1981). Sadovodstvo Vinogradarstro i Vinodelie Moldavia, 11: 12-15.

DeJong, T.M., Day, K.R., Doyle, J. F. and Johnson, R.S. (1994). Hort. Tech., 4: 362-67.

Delwiche, M.J. (1987). HortSci., 22: 87-89.

Delwiche, M.J. and Baumgardner, R.A. (1983). J. Amer. Soc. Hort. Sci., 108: 1012-16.

Dhillon, B.S., Saini, H.S. and Singh, S. (1982). Indian Fd. Pack., 36: 31-33.

Dhuria, H.S., Sud, G., Kar, P.L. and Tripathi, S.N. (1978). Haryana J. Hort. Sci., 7: 31-34.

Dilley, D.R. (1983). Postharvest Physiology and Crop Preservation. Plerun Press, New York.

Dodd, M.C., Hartmann, P.E.O. and Kock, V.A.De (1986). Dec. Fr. Grower, 36: 517-20.

Dolby, J.M. and Putnam, A.R. (1980). Proc. North Western Seed Control. Conf., 35-92.

Donohoe, C.W. and Walker, D.R. (l957). Science, 128: 1178-79.

Dotti, F. (1942). Sta. Sperin. Ortofrutt. Milano, Nota Prelim.

Dowler, W.M. and King, F.D. (1966). Proc. Amer. Soc. Hort. Sci., 89: 80-84.

Dudzinski, F. (1985). Roczniki Akademii Polniczej W Paznonniu agrodnic, 156: 75-87.

Durner, E.F. and Gianfagna, T.J. (1988). J. Amer. Soc. Hort. Sci., 113: 404-406.

Eaton, F.M. (1944). Agr. Res., 69: 237-47.

Eckert, J.W. (1969). World Review Pest Control, 8: 116-37.

Edgerton, L.J. (1966). Proc. Amer. Soc. Hort. Sci., 88: 197-203.

Edin, M. and Garcin, A. (1986). Compte Rendud' Experimentation CTIFL No.87007, p.11.

Edin, M. and Garcin, A. (1994) L'Arboriculture Fruitiere, 475:20-23

Edwards, G.R. (1987). Acta Hort., 199: 61-62.

Edwards, J. H., Horton, B.D. and Kirkpatrick, H.C. (1976). J. Amer. Soc. Hort. Sci., 101: 139-42.

Edwards, J.H. and Horton, B.D. (1979). J. Amer. Soc. Hort. Sci., 104: 97-99.

EI-Banna, G.L., Hassan, A.H. and Abdel Naby, H.M. (1981). Egyptian J. Hort., 8: 65-76.

Eksteen, G.J., Visagie, T.R. and Laszlo, J.C. (1986). Dec. Fr. Grower, 36: 128-32.

El-Badawy, H.E.M. (2012). Res. J. Agric. Biol. Sci., 8: 272-81.

Ende, B. van den and Chalmers, D.J. (1982). HortSci., 17: 218-20.

Ende, B. van den, Chalmers, D.J. and Jene, P.H. (1987). HortSci., 22: 561-67.

Ende, B. van den, Godley, G. and Chapman, B. (1985). Compact Fruit Tree, 18: 135-38.

Erez, A. (1975). HortSci., 10: 251-53.

Erez, A. (1986). Acta Hort., 179: 513-20.

Erez, A. and Couvillon, G.A. (1983). HortSci., 18: 480-81.

Errani, A. and Baroni, D. (2000). Rivista di frutticoltura, 11: 58-59.

Evert, D.R. (1988). HortSci., 23: 700-701.

Fallahi, E., Lee, R.R. and Lee, G.A. (1988). Hort. Tech., 8: 556-60.

Faust, M. and Timon, B. (2011). Scripta Hort, 11: 11-54

Feldstein, J. and Childers, N.F. (1957). Proc. Amer. Soc. Hort. Sci., 69: 126-30.

Fernandez-Trujillo, J.P. and Artes, F. (1997). Food Res, Intl., 30: 441-50.

Filite, N. and Costa, G. (1986). Acta Hort., 179: 689-92.

Filti, N., Costa, G., Daghia, U. and Mazza, A. (1982). Riv. Ortoflorafruttic. Ital, 66: 277-95.

Fisher, D.V. and Britton, J.E. (1940). Sci. Agr. (Canada). 21: 103.

Foster, W.R. and Moisey, J.A. (1959). Proc. British Columbia Dept. Agr. PI. Path. Circ.No. 1.

Gambrell, C.E. J. (1975). In: The Peach-Varieties, Culture, Pest control, Storage, Mrketing. Hort. Pub. New Brunswick, New Jersey, pp. 241-43.

Gambrell, C.E.Jr., Coston. D.C. and Sims, E.T. J. (1983). J. Amer. Soc. Hort. Sci., 108: 605-608.

Gambrell, C.F. and Rhodes, W.E.H. (1967). S. Carolina Agr. Exp. St. Res. Serv., 102, 1-11.

Gammon, N. J. and Shoemaker, J.S. (1963). Proc. Fla. St. Hort. Sci., 76: 380-84.

Garg, R. C., Ram, H.B., Srivastava, R.K. and Singh, S.K. (1975). Progressive Hort., 6: 57-66.

Garg, R.C., Joshi. B.P., Ram, H.B., Km. Jain, S.P. and Triparhi, U.K. (1978). Progressive Hort., 10: 77-84.

Gathercole, F.J. (1982). Aust. J.Exp. Agric and Animal Husb., 21: 354-56.

Gautam, D.R. and Chauhan, J.S. (1984). J. Tree Sci., 3: 99-104.

Gautarn, D.R., Jindal, K.K. and Chauhan J.S. (1981). Haryana J.Hort. Sci., 10: 17-19.

Genard, M. (1992). Canadian J. Pl. Sci., 72: 517-25

George, A.P., Nissen, R.J. and Baker, J.A. (1986). Qd. Agric. J., 112: 27-33.

Ghosh, S.P. (1976). Fruit Res. Workshop Res. Rept. held at Ranikhet, India, pp. 13-19.

Ghosh, S.P. and Chand, S.(1984). Indian J. Agric. Sci., 54: 450-51.

Ghrab, M., Mimoun, M.B., Masmoudi, M.M. and Mechlia, N.B. (2014). Int. J. Environ. Studies, 71: 3-14

Gianfagna, T.J. and Rachmiel, S. (1986). Physiol. Plantarum, 66: 154-58.

Gianfagna, T.J., Marini, R. and Rachrniel, S. (1986). HortSci., 21: 69-70.

Girona, J., Gelly, M., Mata, M., Arbones, A., Rufat, J. and Maral, J. (2005). Agric. Water Management, 72: 97-108

Girona, J., Mata, M., Arbones, A., Alegre, S., Rufat, J. and Marsal, J. (2003). J. Amer. Soc. Hort Sci., 128: 432-40.

Goldhamer, D.A., Salinas, M., Crisosto, C., Day, K.R., Soler, M. and Moriana, A. (2002). Acta Hort., 582: 343-50.

Gonzalez Dugo, V., Ruz, C., Soriano, M.A., Suarez, L., Berni, J.A.J. Savco-Tejada, P.J. and Feres, E. (2011). Acta Hort., 922: 217-20.

Gopalan, C., Ramasastri, B.V. and Balasubramanian, S.C. (1987). Nutritive Value of Indian Foods, NIN, ICMR, Hyderabad, India, p.90.

Grappadelli, L.C. and Sansavini, S. (1998). Fruitticoltura ad alta densita, Edagricole, Bolgona, pp. 191-216

Griffin, E.P. (1959). Ph.D. Thesis. Rutgers Univ., USA.

Guerriero, R. and Scalabrelli, C. (1985). Acta Hort., 173: 185-90.

Guerriero, R., Loreti, F. and Massai, R. (1981). HortSci., 23: 117-18.

Gupta, B.P. and Joshi, N.K. (1976). Fruit Res. Workship Rept. held at Ranikhet, India, pp. 414-24.

Gupta, B.P., Joshi, R. and Tripathi, G.M. (1986). Progressive Hort., 18: 129-31.

Gupta, O.P., Singh, B.P., Singh, S.P. and Chauhan, K.S. (1984). Punjab Hort. J., 24: 105-10.

Gur, A., Altman, A., Stern, R., and Wolowitz, B. (1986). Scientia Hort., 30: 97-108.

Guyot, A.L. (1948). Prog. Agr. et Vitic; 48: 406-408.

Gyuro, F. and David, M. (1978). Acta Hort., 80: 393-95.

HaiChun, G., Fengying, Y. and Zheng, Z. (2010). Acta Hort Sinica, 37: 673-74.

Haji, T., Yaegaki, H. and Yamaguchi, M. (2003). J. Jap. Soc. Hort. Sci., 72: 212-17.

Hansche, P.E. and Beres, V. (1979). Calif Agric., 33: 4-6.

Hansche, P.E., Hesse, C.O. and Beres, V. (I972). J. Amer. Soc. Hort. Sci., 97: 76-79.

Harter, W.Da.R., Grutzmacher, A.D., Nava, D.E., Goncalves, R.Da, S. and Protton, M. (2010). Pesquisa Agropecuaria Brasileira, 45 : 229-35.

Hartley, M.J. (1987). In: Proc. New Zealand Weed and Pest control Conf., pp. 140-143

Hartmann, F.O. and Hill, R.G.Jr. (1983). Fruit Var. J., 37: 108-12.

Harvey, J.M., Smith, L.W.Jr. and Kaufman, J. (1972). USDA. Agric. Handbook No. 414, p.64.

Hatzehariese, I.A. (1981). Epistemonike Epetcries, Geoponike Kai Dasologike Shole Aristoteleion Panepistemion Thessalonikes. 24: 339-46.

Hatzetheodorou, I. and Hatzehariese, L.A. (1983). Georgike Ereuna, 7: 5-15.

Havis, A.L. (1962). Proc. Amer. Soc. Hort. Sci., 80: 172-75.

Havis, L. and Gourley, J.H. (1937). Ohio Agr. Exp. St. Bull., No.581, p.41.

Hayden, R.A. and Emerson. F.H. (1975). In: The Peach-Varieties, Culture, Pest control, Storage, Marketing, New Brunswick, New Jersey, pp. 255-62.

Hedrick, U.P. (1917). Rep. N.Y. Agr. Exp. Sta., 1916.

Heighway, N. (1984). Comb. Proc. Intern.PI. Prop. Soc., 34: 68-70.

Hesse, C.O. (1975). In: Advances in Fruit Breeding, (eds. J. Janick, and J.N.Moore). Purdue Universty Press, West Lafayette, Indiana. pp. 285-335.

Hirai, J.S., Nakagawa, Y.J.I. and Hirato, N. (1961). Osaku, U.B.Ser.B., Agr. and BioI., 11: 85-98.

Hoffman, M.B. (1975). The Peach Varieties, Culture, Pest control, Storage, Marketing, New Brunswick, New Jersey, pp. 135-38.

Hooker, H.D. (1946). Flora of British India.

Horton, B.D. (1985). HortSci., 20: 244-46.

Hough, L.F. and Weaver, G.M. (1959). J. Hered., 50: 59-62.

Hough, L.F., Moore., J.N., and Bailey, C.H.(1964). In: The Use of Induced Mutations in Plant Breeding, Pergamon Press, New York, pp. 679-86.

Humbert, R.P. (1968). Better Crops, 52: 10-11.

Hutchins, L.M. (1936). Proc. Amer. Soc. Hort. Sci., 34: 330-38.

Hutton, R.J., McFadyen, L.M. and Lill, W.J. (1987). HortSci., 22: 552-60.

Issell, L.J. and Chalmers, D.J. (1979). J. Hort. Sci., 54: 33-38.

Janie, V. (1979). Jugoslovensko Vocarstvo, 13: 3-14.

Jawanda, J.S., Josan, J.S. and Singh, S.N. (1979). J. Res. India, 16: 408-12.

JianBo, Z., Jiying, G., Quan, J., QingHua, C., XinYue, L., Fei, R. and XiaoHong, W. (2011). Acta Hort. Sinica, 38: 2225-26.

JianLin, D., WenGui, L. and Quan Jun, Z. (2011). Acta Hort. Sinica, 38: 2223-24.

Johnson, C.E., Pollet, O.K. and Lancaster, D.M. (1985). Louisiana Agric., 28: 18-20.

Johnston, S. (1952). Mich. Agr. Exp. St. Cir., 177: 37.

Jona, R. and Vigliocco, R. (1985). Acta Hort., 173: 223-28.

Joshan, J.S. and Chauhan, G.S. (1982). Haryana J. Hort. Sci., 11: 1-4.

Kabu, K.L. and Dalal, M.A. (1976). Fruit Improvement Workshop. Ranikhet, India Tech. Doc., 9.

Kader, A.A. (1983). 21st. Int. Hort. Congr. Proc. Vol 1; Abst. No. 1232.

Kader, A.A. and Morris, L.L. (1977). Food Tech., 34: 51-54.

Kajura, T. (1975). Scientia Hort., 3: 179-87.

Kamali, A.R. and Childers, N.F. (1970). J. Amer. Soc. Hort. Sci., 95: 656-66.

Kanwar, J.S. and Kahlon, G.S. (1983). Punjab Hort. J., 23: 162-63.

Kappel, H. (1985). Mitteitungen Klosterneuburg Rebe Wein, Obstbau and Fruchteverwertung. 35 : 54-60.

Kaul, G.L. and Mitra, V. (1976). Fr. Res. workshop Res. Rept. held at Ranikhct. India. pp. 177-78.

Kavvadias, V., Daggas, T., Paschalidis, C., Varoulidou, E. and Theocharopoulos, S. (2012). Comm. Soil Sci. Pl. Analysis, 43: 134-48

Kenworthy, A.L., Larsen, R.P. and Gamble, S.J. (1959). Paper presented at Ann. Mtg. Amer. Soc.Hort.Sci.,

Khan, M.S., Wazir, F.K. and Muhammad, A. (2000). Sarhad J. Agric., 16: 25-28.

Khera A.P., Makhija, Moo, Chitkara, S.D. and Chauhan K.S. (1981). Haryana Agric. Univ. J. Res.,11: 181-84.

Kim, K.R., Byun, J.K. and Kim. C.C. (1969). Res. Rep. Rur. Dev. Korea, 12: 20-34.

Kinmann, C.E. (1941). Proc. Amer. Soc. Hort. Sci., 38: 191-99.

Kirkpatrick, H.C., Thompson. J.M. and Edwards, J.H. (1975). HortSci., 10: 132-34.

Kluge, R.A. and Jacomino, A.P. (2002). Scientia Agricola, 59: 69-72.

Kobashi, K., Gemma, H. and Iwahori, S. (2000). J. Amer, Soc. Hort. Sci., 125: 425-28.

Kochba, J. and Samish, R.M. (1972). J. Amer. Soc. Hort. Sci., 97: 115-19.

Kochba, J., and Samish, R.M. (1971). J. Amer. Soc. Hort. Sci., 96: 458-61.

Kondo, S.; Hayata, Y. and Iwasaki, N. (2000a) Acta Hort., 514:75-82

Kondo, S.; Tomiyama, A. and Seto, H. (2000b) Jr. Amer. Soc. Hort. Sci., 125: 282-87

Koutinas, N., Sotiropoulos, T., Antonopoulou, C., Therios, I. and Stavrou, C. (2010). Agrochimica 54: 41-51.

Kumar, A. (1982). Thesis Abstracts, 8: 48-49.

Kumar, A. and Chitkara, S.D. (1983). Haryana J. Hort. Sci., 12: 48-51.

Kumar, M., Rawat, V., Rawat, J.M.S. and Tomar, Y.K. (2010). Scientia Hort., 125: 218-21.

Ladurner, E., Benuzzi, M., Fiorentini, F. and Lucchi, A. (2012). 10BC/WPRS Bull., 74: 259-63.

Larson, K.D., Dejong, T.M. and Johnson, R.S. (1988). J. Amer. Soc. Hort. Sci., 113: 296-300.

Lavattil, L., Spada, G.L. and Zerbini, P.E. (1986). Rivista de Frutticoltura e di Ortofloricoltura, 48: 49-52.

Layane, R.E.C. (1987). In: Rootstocks for fruit crops, (eds, R.C. Rom. and R.F. Carlson.). John Wiley and Sons, New York, pp. 185-216.

Layne, R.E.C. and Tan, C.S. (1984). J. Amer. Soc. Hort. Sci., 109: 795-99.

Layne, R.E.C. and Tan, C.S. (1988). J. Amer. Soc. Hort. Sci., 113: 518-25.

Lekson, M. (1981). Jugoslovensko Vocarstvo, 15: 515-23.

Lelievre, J.M., Latche, A., Jones, B., Bouzayen, M. and Pech, J.C. (1998). Food Technol., 15: 492-94.

Lesley, J.W. (1975). Proc. Amer. Soc. Hort. Sci., 70: 93-103.

Li, H. and Yu, T. (2000). J. Sci. Food. Agric., 81: 269-74.

Li, S.H., Bussi, C. and Atger, P. (1988). Agronomico., 8: 657-62.

Li, Z. (1984). HortSci., 19: 348-51.

LiBin, Z. (2000). China Fruits, 1: 1-2.

Ligouori, G., Weksler, A., Zutahi, Y., Lurie, S. and Kosto, I. (2004). Postharvest Biol. Technol., 31: 263-68.

Lilleland, O. (1935). Proc. Amer. Soc. Hort. Sci., 33: 114-19.

Lilleland, O. (1946). Blue Anchor., 23: 14-16.

Lilleland, O. and Brown. J.G. (1942). Proc. Amer. Soc. Hort. Sci., 41: 1: 10.

Liwerant, J. (1959). C. R. Acad. Agr. Fr., 45: 818-26.

Liwerant, J. (1960). France Inst. Nat. de la Rech.Agron. Ann.Ser. A., Ann. Agron., 11: 93-111.

Lombardo, V.A., Osorio, C.S., Borsani, J., Lauxmann, M.A., Bustamante, C.A., Budde, C.O., Andreo, C.S., Lara, M.V., Fernie, A.R. and Drincovich, M.F. (2011). Pl. Physiol., 157: 1696-710.

Loreti, F., Morini, S. and Grilli, A. (1985). Acta Hort., 173: 261-69.

Lott, R.V. and Simons, R.K. (1966). Proc. Amer. Soc. Hort. Sci., 85: 141-53.

Lou, R.V. (1931). Proc. Amer. Soc. Hort. Sci., 28: 23-27.

Lu, Ming-Te, Weisong, C., Chia Huang, C. and Ou, S.K. (2008). Acta Hort., 768: 153-59.

Lurie, S. (1992). Int. J. Food Sci. Technol., 27: 407-15.

Lurie, S. and Crisosto, C.H. (2005). Postharvest Biol. Technol., 31: 195-208.

Magnani, M., Freire, C.J. Das and Moraes, E.C. De (1984). In Anais do VII Congress Brasileiro de 'Frutticoltura. Florianopolis. Brazil Empreso Catorinense de Pesqeura Agropecuraria S.A., 4: 1105-15.

Mahhou, A. and Elghazzali, M. (2000). Fruits, Paris, 55: 153-61.

Mansour, N.M., Fathi, M., Elbeheri, M. and Ibrahim, N. (I981). Agric. Res. Rev., 59: 137-56.

Marangoni, B., Antonelli, M., Scudellari, D. Cobianchi, D. and Liverani, A. (1985). Acta Hort., 173: 389-94.

Maria, C.B., Raseira and Moore, J.N. (1987). HortSci., 22: 216-18.

Marini, R.P. (1984). Acta Hort., 146: 287-92.

Marini, R.P. (1985). J. Amer. Soc. Hort. Sci., 110: 133-39.

Marini, R.P. and Rossi, D. (1985). HortSci., 20: 242-43.

Marini. R.P. (I986). J. Amer. Soc. Hort. Sci., 111: 391-94.

Martin, G.C., Yashikawa, F. and La Rue, J.H. (1987). J. Amer. Soc. Hort. Sci., 112: 915-21.

Matsumoto, T.T. and Sommer, N.F. (1967). Phytopathology, 57: 881-84.

Matta, F.B., Rashberry, S. and Little, S. (1987). Res. Rep. Mississipi Agri. and Forestry Exp. St., 12: p.3.

McClung, A.C. (1953). Proc. Amer. Soc. Hort. Sci., 62: 123-30.

McClung, A.C. (1954). Proc. Amer. Soc. Hort. Sci., 64: 75-80.

McClung, A.C. and Lott, W.L. (1956). Proc. Amer. Soc. Hort. Sci., 67: 113-20.

McClung, N.C. and Clayton, C.N. (1956). Pl. Dis. Reptr., 40: 542-48.

McGeorge, W.T. (1949). Univ. Ariz. Agr. Exp. St. Tech. BuII. No.117, p.48.

McKenry, M.V. and Kretsch. J. (1987). Nematologica, 33: 343-54.

McLarty, H.R. and Woodbridge. C.G. (1950). Sci. Agr., 30: 393-95.

Mehanna, H.T. and Martin, G.C. (1985). Scientia Hort., 25: 247-54.

Mehanna, H.T., Martin, G.C. and Nishijima, C. (1985). Scientia Hort., 27: 63-73.

Mika, A. (1986). Hort. Reviews, 8: 339-78.

Minguzzi, A. (1993). Allevamento e potatura. II divulgatore, 3: 28-29.

Mishra, K.A., Sud, G. and Arora, I.K. (1984). Indian J. Hort., 4: 85-87.

Mishra, R.S. and Srivastava, R.P. (1972). Progressive Hort., 4: 127-36.

Mitchell, P.D. and Chalmers, D.J. (1983a). HortSci., 18: 472-24.

Mitchell, P.D. and Chalmers, D.J. (1983b). Aust. Hort. Res. Newsletter, No.55, p.51.

Mitra, S.K. (1988). In: Mineral Nutrition of Fruit Crops. (eds. T.K. Bose, M.K. Sadhu, and S.K. Mitra)., Naya Prakash, Calcutta, India, pp: 512-49.

Mizutani, F. (2011). Adv. Hort. Sci., 25: 179-86.

Mizutani, F. and others (1985). J. Jap. Soc. Hort. Sci., 54: 327-35.

Moharan, F.A. and El-Deen, E.M.A.Z. (2011). Res. J. Agric. Biol. Sci., 7: 434-44.

Moing, A., Rothan, C., Svanella, L., Just, D., Diakou, P., Raymond, P., Gaudillere, J.P. and Monet, R. (2000). Physiologia Plantarum, 108: 1-10.

Mondedji, A.D., Ketoh, G.K., Amevoin, K., Ameline, A., Giordanengo, P. and Glitho, I.A. (2014). African J. Agric. Res., 9: 1344-52.

Monet, R. (1989) Acta Hort., 254:49-57

Morini, S. and Concetti, S. (1985). Acta Hort., 173: 205-10.

Morris, 1.R., Roy, L.D. and Cowthoru, D.L. (1978). J. Amer. Soc. Hort. Sci., 103: 716-22.

Morris, J.R., Khattan, A.A. and Arrington, E.H. (1962). Proc. Amer. Soc. Hort. Sci., 80: 177-89.

Mowry, 1.B. (1963). Amer. Fr. Grower, 83: 56-57.

Mrvic, V. (1998). Jugoslovensko Vocavstvo, 32: 111-18.

Murray, R., Lucangeli, C., Polenta, G. and Budde, C. (2007). Postharvest Biol. Technol., 44: 116-21.

Myers, S.C. (1986). Abst. 22nd Int. Hort. Congr. California, No. 148.

MyongDong, C., HeeSeung, P. and YongKoo, K. (2000). Korean J. Hort. Sci. Tech., 18: 353-59.

Nakasu, B. H. and Raseira, M.Do., C.B. (1999). Ciencia e Agrotechnologica, 23: 234-36.

Narn, K.U. and Kim C.C. (1985). J. Korean Soc. Hort. Sci., 26: 342-49.

Nasir, F.R. and Mohamed, A.R.S. (1987). Iraqi J. Agric. Sci., 'Znco', 5: 55-63.

Natali, S., Kiloyannis, C. and Pezzarossa, B. (1985). Acta Hort., 171: 167-80.

Nauriyal, J.P. (1967). Indian Hort., 11: 75- 79.

Nauriyal, M.C. and Misra, R. (1982). Progressive Hort., 14: 19-42.

Nerri, D. and Massetani, F. (2011). Adv. Hort. Sci., 25: 170-78.

Nightingale, G.T., Addorns, R.M. and Blaske, M.A.(1930). N. J. Agr. Exp. St. Bull., No.494, p.16.

Nitrantsky, S. (1984). Vedeeke Prace Vyskumnehq Vstavn Ovoncnyeh a Okrasnych Orevin V Bogniciach, 5: 209-20.

Nitrantsky. S. (1987). Sbornik UVTIZ, Zahradnictvi, 14: 28-37.

Nurstem, H.E. and Williams, A.H. (1967). Chem.lnd.London, 12: 486-97.

Okie, W.R., Ramming, D.W. and Scorza, R. (1985). HortSci., 20: 633-41.

Olmo, H.P. (1965). Proc. Int. Peach Congr. (Verona). pp. 317-79.

Ortiz, A., Sopena, P., Comabella, E. and Lara, I. (2012). Acta Hort., 940: 475-81.

Ortiz, A., Vendrell, M. and Lara, I. (2011) J. Hort. Sci. Biotech., 86: 175-81.

Overholser, E.L. and others (1941). Warsh. Agr. Exp. St. Bull., No.162.

Padlog, N. and others (1983). Rivista de Ie Faculted de Agronomic Univ. Nacional de La Plata, 59: 49-63.

Paltineanu, Cr., Septar, L., Moale, Cr., Oprita, A. and Lamureanu, G. (2011). Acta Hort., 922: 195-202.

Pandey, D. and Upadhya, S.N. (1981). Progressive Hort., 13: 71-72.

Pandey, V.S. (1977). Ph.D. thesis. Kanpur University, India.

Pandey, V.S. and Awasthi, D.N. (1976). Proc. All India Coordinated Fr. Improv. Res. Workshop on Pome stone and Nut fruits held at Ranikhet India, pp. 175-78.

Parnia, P., Mladin, G.H., Dutu, I. and Stanciu, N. (1988). HortSci., 23: 107-109.

Pascal, T.; Pfeiffer, F. and Kervella, J. (2010) HortSci., 45:150-52

Pathak, R.K. and Pandey, V.S. (1976). Fruit Improvement Workshop, Ranikhet, India, Tech. Doc., 9.

Pathak, R.K., Pandey, D. and Pandey, V.S. (1976). Indian Fmg. Dig., 9: 31-34.

Pathak, V.N. (1980). Diseases of Fruit Crops. Oxford and IBH Publishing Co. New Delhi.

Pickel, C., Bentley, W.J., Hasey, J.K. and Day, K.R. (2006). Pest Management Guidelines: Peach. UC Davis, INR Pub. No: 3454

Powell, A.A. and Howell, J.C. (1985). Amer. Fr. Grower, 105: 12-14.

Powell, C. A. (1984). Plant Diseases, 68: 908-909.

Powell, D. (1975). The Peach-Varieties, Culture, Pest Control, Storage, Marketing, New Brunswick, New Jersey, pp. 358-59.

Proebsting, E.L. and Mills, H.H. (1978). J. Amer. Soc. Hort. Sci., 103: 192-98.

Proebsting, E.L.Jr. (1975). The Peach-Varieties, Culture, Pest control, Storage, Maturity. New Brunswick, New Jersey, pp. 120-22.

Qrtiz, A., Vendrell, M. and Lara, I. (2011). J. Hort. Sci. Biotech., 86: 175-81.

Quamme, H.A., Layne, R.E.C. and Ronald, W.G. (1982). Canad. J.Plant Sci., 62: 137-48.

Raff, J.W. and Hallipike, A.L. (1983). Aust. Hort. Res. Newsletter, 55: 332-33.

Ramming, D.W. (1985). HortSci., 20: 419-20.

Randhawa, S.S. (1987). Wild Germlasm of Pome and Stone Fruits. IARI Reg. Stn. Shimla. pp. 51.

Raseira, M. do. C.B. and Nakasu, B.H. (1998). Agropecuaria Clima Temperado, 1: 271-73.

Raseira, M.C.B. and Moore, J.N. (1987). HortSci., 22: 216-18.

Raseira, M.do.C.B., Nakasu, B.H., Grellman, E.O. and Simonetto, P.R. (1998). Agropecuaria Clima Temperado, 1: 275-77.

Rathore, D.S. (1984). Punjab Hort., 24: 111-15.

Reeves, D.W., Horton, B.D. and Couvillion, G.A. (1983). Scientia Hort., 21: 353-57.

Regland, C.H. (1934). Proc. Amer. Soc. Hort. Sci., 31: 1-21.

Rester, D. E. and Assay, R.N. (1986) HortSci., 21:331-32

Ridley, J.D. and Sims, E.T. (1966). S.Carolina Agr. Exp. St. Res. Serv., 90: 1-13.

Riggoti, R. (1942). Ital. Agr., 21: 664-67.

Ritter, C.M. (1957). Proc. Amer. Soc. Hort. Sci., 68: 48-55.

Robertson, J.A. and Meredith, F.J. (1988). HortSci., 23: 1032-34.

Robitailie, H.A. and Yu, K.S. (1979). Compact Fruit Tree, 12: 95-67.

Robitailie, H.A. and Yu, K.S. (1980). HortSci., 15: 579-80.

Rom, R.C. (1975). The Peach-Varieties, Culture, Pest Control, Storage, Marketing. New Brunswick, New Jersey, pp. 67-71.

Rom, R.C. (1983). Fruit Var. J., 37: 3-14.

Rom, R.C. (1984). Proc. 43rd. Nat. Peach Council Ann. Convention, pp. 59-68.

Rom, R.C. and Ferree, D.C. (1984). HortSci., 19 : 543-45.

Rood, P. (1957). Proc. Amer. Soc. Hort. Sci., 70: 104-12.

Rudikova, N.D. (1988). Refcrativnyi Zhurnal, 1,55: 618.

Rufato, L., Rossi, A.D., Picolotto, L., Parizoto, E. and Fanchinello, J.C. (2004). Acta Hort., 636: 711-15.

Ruzic, D., Rasati, P. and Marino, E. (1984). Rivista della Ortoflorofrutticoltura Italiana, 68: 413-22.

Ryugo, K. and Davis, L.D. (1958). Proc. Amer. Soc. Hort. Sci., 72: 106-12.

Salesses, G. and Juste, C. (1971). An. Amelior Plantes, 21: 265-80.

Samani, R.B. (2014). Agricultural Communication, 2: 37-42.

Sandhu, A.S. and Singh, Z. (1984). Punjab Hort. J., 24: 101-104.

Sandhu, B.S., Chahil, B.S., Brar, M.S. and Athwal, B.S.(1980). Haryana J. Hort. Sci., 9: 47-49.

Sandhu, S.S. and Dhillion, B.S. (1981). J. Res. PAU. India, 18: 6-12.

Sandhu, S.S. and Dhillion, B.S. and Singh, S. (1983). J. Hort. Sci., 58: 203-207.

Sandhu, S.S. and Dhillon, B.S. (1983). J. Res. PAU. India 20: 452-66.

Sandooja, J.K. Sharma, R.K., Singhrot, R.S. and Singh, J.P. (1986). Haryana J. Hort. Sci., 15: 210-17.

Sansavini, S. (1983). Dec. Fr. Grower, April-May, p. 3

Sansavini, S., Corelli, L. and Giunchi, L. (1984). Acta Hort., 173: 139-58.

Sansavini, S., Corelli, L. and Giunchi, L. (1985). Rivista di Frutticoltura e di Ortofloricoltura, 47: 23-31.

Sarkadi, B. and Iordanescu, O. (2012). Sci. Papers Series B. Hort., 56: 161-65.

Sato, K. (1955). Bult. Nat'l. Inst. Agr. Sci.Series E, 4: 146-68.

Sato, K., Inhihara, M. and Harada, R. (1953). J. Jap. Soc. Hort. Sci., 23: 159-64.

Scalabrelli, G. and Couvillon, G.A. (1985). Acta Hort., 173: 103-12.

Scalabrelli, G. and Couvillon, G.A. (1986). Acta Hort., 179: 855-62.

Scalabrelli, G., Couvillon, G.A. and Pokorny, F.A. (1984). Acta Hort., 150: 459-63.

Schiaparelli, A., Ristori, G.F. and Ferrera, A. (1982). Frutticoltura, 44: 69-76.

Schneider, G.W. (1956). Amer. Fr. Grower, Aug., pp. 38-40.

Schneider, G.W., Jones, J.D. and McClung, A.C. (1958). Proc. Amer. Soc. Hort. Sci., 71: 110-15.

Schwartz, E., Fachinello, J.C., Carvalho, F., J.C. and Silva, J.G.C. da (1999). Rivista Brasileira de Fruticultura, 21: 304-307.

Scorza, R. (1984). J. Amer. Soc. Hort. sci., 109: 455-57.

Scorza, R., Shawn, A. M and Lightner, G.W. (1985). J. Amer. Soc. Hort. Sci., 110: 547-52.

Semash, D.P. and Storchous, V.N. (1987). Melioratsiya Vodmoe Khozyai Stvd. 66: 8-11.

Sen, S.M. and Couvillon, G.A. (1983). HortSci., 18: 304-305.

Serradilla, M.J.; Lopez-Corrales, M.; Lozano, M.; Ayuso, M.C. and Bernalte, M.J. (2014) Acta Hort., 1020: 157-63

Seyit, M.S. and Couvillon, G.A. (1983). HortSci., 18: 324-25.

Sharma, A.K. (1981). Progressive Hort., 12: 231-33.

Sharma, A.K. (1982). Progressive Hort., 14: 211-13.

Sharma, C.M. (1984). Indian J. Hort., 41: 225-29.

Sharma, H.C. and Singh, R.N. (1978). Scientia Hort., 9: 47-53.

Sharma, H.C. and Singh, R.N. (1979). Punjab Hort., 19: 53-55.

Sharma, H.C. and Singh, R.N. (1980). Indian J. Plant Physiol., 23: 26-32.

Sharma, H.C. and Singh, R.N. (1981). Indian J. Plant Physiol., 24: 218-23.

Sharma, J.P. (1986). M.Sc. Thesis. HPKVV., Palampur India.

Sharma, M.R. (1976). Plant Sci., 8: 82-83.

Sharma, M.R. and Gautam, D.R. (1981). Indian J. Agric. Sci., 51: 362-64.

Sharma, M.R. and Singh, R.N. (1982). Punjab Hort. J., 22: 93-98.

Sharpe, R.H. and Sherman, W.B. (1975). In: The Peach (ed. N.F. Childers). Hort. Publ, New Brunswick, NJ.

Sharpe, R.H., Hesse, C.O., Lowsnbessy, B.F., Perry, V.G. and Hansen, C.J. (1970). J. Amer. Soc. Hort. Sci., 94 : 209-12.

Sharples, R.O. (1980). Rept.East Malling Res. Sat., for 1979. p.108.

Shen, T. and Hsiang, T. (1949). Proc. Amer. Soc. Hort. Sci., 53: 11-12.

Sherman, W.J., Lyrene, P.M. and Andrews. C.P. (1986). Proc. of the Trop. Region. Amer. Soc. Hort. Sci., 23: 25-28.

Shyikuan, O.U., MingFwu, H., Chingpin, Li, ShihJen, L.U. and Chiawei, S. (2000). J. Ag. Res, China, 49: 46-53.

Singh, B.B. and Srivastava, R.P. (1978). XX Int. Hort. Congr. held at Sydney. Australia.

Singh, B.P., Gupta, O.P. and Chauhan, K.S. (1982). Indian J. Agric. Sci., 52: 235-39.

Singh, D. (1983). Ph.D. Thesis. H.A.U., Hissar, India.

Singh, G., Shrama, K.K. and Jawanda, J.S. (1984). Punjab Hort. J., 24: 92-95.

Singh, H. (1963). Himachal Hort., 5: 8-14.

Singh, I.D. and Sirohi, S.C. (1976). Fruit Improvement Workshop, Ranikhet. India. Tech Doc., No. 9.

Singh, K., Grewal, S.S. and Dhatt, S.S. (1986). Haryana J. Hort. Sci., 15: 4-8.

Singh, L.B. (1960). Ann. Rep. Hort. Res. Inst., Saharanpur, India, pp. 47-50.

Singh, S. (1969). Hand Book of Agriculture, I.C.A.R. New Delhi.

Singh, Z., Grewal, G.P.S. and Singh, H. (2000). Acta Hort., 525: 467-69.

Sinha, M.M., Tripathi, S.P., Tiwari, J.P. and Misra, R.S. (1983). Punjab Hort. J., 23: 43-46.

Slack, J.M. (1978). Farmers Newsletter, 139: 26-27.

Smith, R.B. (1985). J. Amer. Soc. Hort. Sci., 110: 650-54.

Smith, W.L. J. (1962). Pl. Dis. Reptr., 46: 861-65.

Sommer, N.F. (1983). Plant Science Short Course, 196: 14-21.

Souza, A., Das, G., Chalfun, N.N.J., Musser, R., Dos, S., Fachinello, J.C., Campos, V.P. and Souza, A.A. de. (2014). Amazonian J. Agric. Environ. Sci., 57: 108-13.

Souza, A.P.de., Leonel, S. and Silva, A.C.da. (2011). Pesquisa Agropecuaria Brasileira, 46: 1588-96.

Spencer, M.D., Pandborn, R.M. and Jennings A. (1978). J. Agr. Food. Chem., 26: 725-32.

Srivastava, R.P. (1965). Indian Hort., 9: 11-2.

Srivastava, R.P. (1966). Indian Hort., 10: 9-11.

Srivastava, R.P. (1974). Temperate Fruit Culture in India. ICAR., New Delhi.

Stan, S. and others. (1985). Acta Hort., 173: 395-403.

Stassen, P.J.C. (1987). Dec. Fr. Grower, 37: 245-49.

Stembridge, C.E. and Lorue, J.H. (1969). J. Amer. Soc. Hort. Sci., 94: 492-95.

Stembridge, G.E. and Gambrell, C.E.Jr. (1969). J. Amer. Soc. Hort. Sci., 94: 570-73.

Stojkovska, A. and others (1972). Jugoslov. Vocarstvo, 6: 525-35.

Storchous, V.N. (1984). Sadovodstvo i Vinogrdarstvo Moldavii, 3: 15-17.

Storchous, V.N. (1986). Sadovodstvo i Vinogradarstvo Moldavii, 4: 52-54.

Sud, G., Dhuria, H.S. and Kar, P.L. (1978). Udyanika, 2: 57-59.

Svoboda, A. and Bazani, Z. (1983). Vedecke Prace Ovocnarste, 9: 71-77.

Taheri, A., Cline, J.A., Jayasankar, S. and Pauls, P.K. (2012). Amer. J. Pl. Sci., 3: 295-301.

Tate, K.C. and Mespel, G.J. vander (1985). Orchardist of New Zealand, 58: 94-95.

Teskey, B.J.E. and Shoemaker, J.S. (1972). Tree Fruit Production. The AVI Publishing Co. Inc., Westport Connecticut, pp. 149-237.

Thapar, A.R. (1960). Horticulture in the hill regions of North India; Directorate of Extension, Ministry (Food and Agric. New Delhi).

Thorne, D.W. and Wan, F.B. (1950). Utah Agr. Exp. Sta. Bull., 338: 29.

Tonutti, P., Ramina, A., Baroni, G. and Costa, G. (1986). Acta Hort., 179: 571-72.

Tressler, D.K. and Joslyn, M.A. (1961). Fruit and Vegetable Juice Processing Technology. The AVI Publishing Co., Inc. Westport, Connecticut.

Ulrich, R. (1979). Harvesting, Storing and Packing of Fruits. HR., Paris.

Upshall, W.H. (1943). Sci. Agr. (Canada). 23: 12.

Veerhoff, O. (1948). Proc. Amer. Soc. Hort. Sci., 50: 209-18.

Venig, A. (2008). Med. Vet. Bucuresti Seria-B. Hort., 52: 72-74.

Visai, C., Marro, M. and Poma, C.T. (1985). Acta Hort., 173: 177-82.

Vitagliano, C., Testolin, R. and Messina, R. (1985). Rivista della Ortoflorofrutticolture Italiana, 69: 171-83.

Wang, S., Zhang, H., Pi, Y. and Gao, H. (2003). Acta Hort., 628: 643-52.

Weaver, G.M. (1968). HortSci., 3: 260-62.

Webster, A.D.; Spencer, J.E.; Dover, C. and Atkinson, C.J. (2006) Acta Hort., 727: 467-72

Weinbaum, S.A. and Erez, A. (1983). Fruit Var. J., 37: 113-14.

Weinberger, J.H. (1929). Proc. Amer. Soc. Hort. Sci., 26: 174-79.

Weinberger, J.H. (1934). Md. Agr. Exp. Sta. Bull., 336: 69-99.

Weinberger, J.H. (1954). Proc. Amer. Soc. Hort. Sci., 63: 157-62.

Weinberger, J.H. (1955). Handbook der Pflanzenzrechtung, 6: 624-36.

Weinberger, J.H. (1956). Proc. Amer. Soc. Hort. Sci., 67: 107-12.

Weinberger, J.H. and Cullinan, F.P. (1936). Proc. Amer. Soc. Hort. Sci., 34: 249-54.

Weinhold, A.R. (1961). Phytopathology, 51: 699-703.

Welker, W.V.Jr. (1983). Proc. 37th Ann. Meeting Northeastern Weed Sci. Soc., p. 249.

Welker, W.V.Jr. and Glenn, D.M. (1988). J. Amer. Soc. Hort. Sci., 113: 652-56.

Wells, J.M. (1970). U. S. Dept. Agr. Mkg. Res. Rept., 908. pp. 10.

Wells, J.M. and Bennett, A.H. (1976). Phytopath., 66: 801-805.

Wells, J.M. and Gerodits. H.G. (1972). Plant Dis. Rept., 55: 69-72.

Werner, D.J. and Young, E. (1987). Fruit Var. J., 41: 140-41.

Werner, D.J., Mowrey, B.D. and Young, E. (1988). J. Amer. Soc. Hort. Sci., 113: 775-78.

West, L.D., Hilderth, R.C. and Schlesschmann (1983). Proc. West. Soc. Weed Sci., 36: 189-90.

Westwood, M.N. (1978). Temperate Zone Pomology. W.H. Freeman and Co.,

Williams, C.F. and Veerhoff, O. (1948). Proc. Amer. Soc. Hort. Sci., 58: 88-96.

Wills, R.B.H., Serven, F.M. and Greenfield, H. (1983). J. Sci. Food Agric., 34: 1383-89.

Wills, R.P., McGlasson, B., Graham, D. and Joyce, D. (2007). In: Postharvest an Introduction to the physiology and handling of fruit, vegetables and ornamental, (eds. R. Wills, B. McGlasson, D. Graham and D. Joyce), CABI, UK. pp. 202

Wittwer, S.H. and Hibbardand, A.D. (1947). Proc. Amer. Soc. Hort. Sci., 49: 116-20.

Woodbridge, C.G. (1954). Candian J. Agr. Sci., 34: 542-51.

Woodroof, J.G., Heaton, E.K., Boggers, T.S.Jr. and Li, K.C. (1975). The Peach-Varieties. Culture. Pest. Control, Storage. Marketing. New Brunswick, N.J., pp. 484-494.

Xiao, X., LiBin, Z., JianZhen, L., LiQin, S., YuMin, F. and FengMing, Y (2011). Acta Hort. Sinica, 38: 2411-12.

XuiSen, C., MaoPai, L., PeiGang, X., Qun, L., FuTian, P., Chengxiang, J., Jingli, W., Xiaoliu, C. and Yan Min, Z. (2010). Acta Hort. Sinica, 37: 1019-20.

Yadav, V., Yadav, P. and Singh, P.N. (2013). Asian J. Hort., 8: 690-95.

YaJun, G., ZeHua, W., Baocai, S., ZongJiang, K., Liang, Z., XiaoJun, G., Jianhua, L. and Shujun, W. (2011). Scientia Agri. Sinica, 44: 4385-94.

Yakovleva, L.V. (1985). Byulleten Gosudarstvennogo Nikitskogo Botanicheskogo Sada, 56: 41-48.

Yamamoto, T.; Mochida, K. and Hayashi, T. (2003). J. Jap. Soc. Hort. Sci., 72:116-21

Yamdagni, R. and Jindal, P.C. (1979). Progressive Hort., 10: 41-44.

Yoshida, M. (1970). Bull. Hort. Res. Sta. (Hiratsuka). 9: 15.

Yoshikawa, F.T. (1988). J. PI. Nutrition, 11: 1387-96.

Young, C.S. and Jcse, P.H. (1985). Transactions of ASAE., 28: 75-78.

Young, M.J. and Crocker, T.E. (1982). Amer. Fr. Grower, 102: 26-28.

Yu, K.S. (1985). Compact Fruit Tree, 18: 109-110.

Zanandrea, I., Raseira, M. do, C.B., Santos, J. dos. and Silva, J. B. da (2011). Ciencia Rural, 41: 2066-72.

Zeven, A.C. and Zhukovsky, P.M. (1975). Dictionary of Cultivated Plants and their Centres of Diversity. Centre for Agricultural Publishing and Documentation, Wagenigen.

Zhang-Bo, Jiyuan, S., WenWen, W., WengPeng, Xi, ChangJie, Xu, Ferguson, I. and KunSong, C. (2010). J. Agric. Fd., Chem., 58: 6157-65.

ZhengWen, Ye, MingShen, Su, YuLiang, Wu, JiHong, Du, XueYing, Z., QingHua, Gao and EnJi, Z. (2011). J. Fr. Sci., 28: 932-33.

ZhiHong, F., Yingli, Z., GenZhu, Y., JianHua, L., LiAng, W., JunFeng, S., HuaRui, W. and ChunSheng, W. (2010). Acta Hort. Sinica, 37: 207-12.

ZhongKui, Gao, XunSheng, Hu, ShenFu, Li., QingTang, Wu, SiQing, Z. and ZhongDong, Z. (1999). China Fruits, 4: 9-11.

Ziliotto, F., Begheldo, M., Rosari, A., Bonghi, C. and Tonutti, P. (2008). J. Exp. Bot., 59: 2781-91.

Zilokh, S., Klein, I. and David, (1980). J. Hort. Sci., 63: 209-16.

Zini, A. (1984). Informatore Agrario, 40: 73-76.

Zucconi, F. (1986). In: CRC Handbook of Fruitset and Development (ed. S.P. Monselise). CRC Press, Inc., Florida. pp. 303-21.

Chapter 5

Pears

Ancient Greek poet Homer praised pears as one of the 'gifts of god'. This prehistoric fruit has been under cultivation both in Europe and Asia for about 3000 years (Hedrick *et al.*, 1921; Kikuchi, 1946). Early development in its cultivation took place in Greece and Rome around 200-300 BC. The description of cultivars and methods of vegetative propagation were well documented by the contemporary writers. The major improvement in the cultivars, however, occurred in Belgium, France and England in the 16th, 17th and 18th centuries. Asia, China and Japan were the leading pear growing countries where the pear culture remained in a semi-wild state for a very long time. Advances in pear culture and varietal improvement were brought only in the 18th century. In India, the improved pear cultivars were introduced in the later part of the 19th century and the cultivation got momentum with the success of Bartlett and Gala cultivars in hills and Patharnakh in the submountaneous region of Punjab.

Amongst the delectable temperate fruits, the pear is next only to apple in importance, acreage, production and varietal diversity. Because of its tolerance to a wide range of soil and.climatic conditions, it is grown both in temperate and subtropical conditions. However, the lack of attractive shape and colour, and perishable nature of the fruit are serious impediments in expanding its cultivation on a very large scale.

Composition and Uses

The pear fruits available in the market belong either to *Pyrus communis* or *P. pyrifolia* or to the hybrid group of these two species. Their chemical composition is given in Table 5.1.

Table 5.1: Composition of Edible Portion of Important Types of Pears Grown in India*

	Bagugosha	*Kashmiri Nakh*	*Country*	*Kieffer*
	(1)	*(2)*	*(3)*	*(4)*
Moistureg /100g	86.5	83.6	86.0	86.3
Protein g /100g	0.4	0.2	0.2	0.2
Fat g /100g	0.1	0.3	0.1	0.2
Minerals, g /100*g*	0.3	0.4	0.3	0.2
Fibre g /100g	2.1	0.6	1.0	1.4
Other carbohydrates g /100g	10.6	14.9	12.4	11.7
Calcium mg/100*g*	20	20	6	10
Phosphorus. mg /100g	20	20	10	10
Iron mg /100g	1.5	1.0	1.0	0.4
Vitamin A IU /100g	0	0	14	9
Thiamine mg /100g	-	-	0.02	0.03
Riboflavin, mg /100g	-	-	0.03	0.02
Nicotinic acid mg /100g	0.2	0	0.2	0
Vitamin C mg /100g	1	3	0	7

Note-Nos 1 and 2 are types of *P. communis;* No.3 is a type of *P. pyrifolia;* No.4 is a hybrid .

* Anon. (1969).

Carbohydrates

Sugars, starch and cellulose are the main carbohydrates found in the pear fruit. In ripe fruits, reducing sugars constitute about 70 per cent of the total carbohydrates present. Fructose is the predominant sugar (6.5 per cent to 11.2 per cent) followed by glucose (0.5 per cent to 3.5 per cent) and sucrose (0.1 per cent to 2.4 per cent) in the fruit juice of 26 types of ripe pear fruits Round starch granules of pear resemble very much in shape and amylase content to that of apple starch (Anon., 1969).

Protein

A protein content of 0.6 percent recorded in pear fruit is rather low (Gopalan *et al.*, 1987). However, all the essential amino acids except tryptophan have been identified in the fruits (Anon., 1969). According to Matsuoka *et al.* (1979) the main amino acids found in Bartlett pear juice are asparagine (57 per cent), aspartic acid (17 per cent) and glutamic acid (8 per cent). They also reported the reduction in glutamic acid on advancing maturity and its correlation with fruit hardness. Fruit size had no effect on the variation in amino-nitrogen but fruits harvested late or overripened were low in amino-nitrogen content.

Organic Acids

Rathore (1982) reported 0.12 to 0.22 per cent titratable acidity as malic acid in the fruits of five pear cultivars. The organic acids detected in pear fruits are malic, citric, quinic, α-ketoglutaric, succinic, lactic, glycolic, shikimic, glyceric and mucic, but malic and citric acids are predominant (Anon., 1969).

Vitamins

The pear fruit is not rich in any particular vitamin. The range of values of various vitamins are I-50 IU. vitamin A, 2-50 μg thiamine, 7-30 μg riboflavin and 1-11 mg/100g ascorbic acid (Anon., 1969). The other vitamins present are niacin, biotin, pantothenic acid, folic acid and vitamin B_{12}. The fruit peel is richer in ascorbic acid and biotin content than the pulp.

Minerals

Pear fruits are comparatively a rich source of minerals. The values of different minerals found in mg per 100g fruit pulp are calcium 8, phosphorus 15, copper 0.4, sulphur 14, and chlorine 1 (Gopalan *et al.*, 1987).

Pectic Substances

Pectin is found in abundance in pear. The exact amount, however, depends on the cultivar, stage of ripening and storage conditions. Fruit ripening is accompanied by a rise in water and EDTA-soluble pectic fractions and in free galaturonic acid (Ben-Arie *et al.*, 1979). At harvest, in unripe Bartlett fruits, 0.07 per cent soluble pectin was recorded which rose to 0.7-0.8 per cent on ripening of fruit (Anon., 1969).

Aroma Constituents

Volatile emitted by pear fruits are primarily esters, alcohols, hydrocarbons, and aldehydes, and many are high moleculear weights. Studied with different European pear varieties showed, esters accounted for as little as 60 percent to as high as 99 percent. The characteristic pear aroma is attributed to the presence of ethyl and methyl esters of trans-2, cis-4-decadienoic acid in association with n-hexylacetate which is also an indicator of picking maturity (Strandzhev, 1974). According to Polesella (1980), more than ninety volatile compounds (esters) have been separated and identified but the precise quantity of these compounds is determined by the cultivar, growing conditions, ripeness, postharvest treatment and processing. The compounds do not increase steadily during the course of fruit development, yet they are substantially higher at the harvest than at initial stages of the development (Strandzhev, 1974).

Tannins

The main constituents of tannins are complex leucoanthocyanins composed of units similar to cyanidin. The fruits contain about 0.1 per cent tannin (total catechol derivatives) which increases during early fruit growth period but declines afterwards (Anon 1969). A 100g serving of pear fruit with peel provides between 27.2 and 40.7 mg of phenolics (Galvis-Sanchez *et al.*, 2003). Arbutin and chlorogenic acid are the major phenolic compounds found in Oriental pears (Cui *et al.*, 2005).

Enzymes

Amylase, catalase, peroxidase, xanthine oxidase, pectin polyga lacturonase, protopectinase and pectase are the enzymes reported in pear fruits (Anon., 1969). Polygalacturonase is active in the ripe Bartlett fruits. Browning of cut pear fruits is considered due to polyphenolase.

Composition of Seed

The various constituents of pear seeds are crude protein 33.4 per cent, fat 24.8 per cent, lecithin 1.2 per cent, sugars 5.1 per cent, pentosans 6.7 per cent, fibre 10.9 per cent and ash 3.8 per cent (Anon., 1969). Pear seed oil is clear, yellow with mild odour and taste resembling very much to apple seed oil.

The pear is a highly delicious fruit which helps in maintaining a desirable acid-base balance in human body. The Occidental (European and West Asian) pear is relished for its buttery, juicy fine texture and flavour, while Oriental (East and North Asian) pear for its crisp and sweet taste. The fruit can be consumed even by diabetic patients because of the low sugar content. Like majority of other fruits, pear fruits are mostly used for table purposes. In the Western countries, a substantial portion of the crop is subjected to juice and wine making, canning, drying and cooking. With the treatment of pectic enzyme, a clear pear juice of excellent flavour is prepared from Bartlett fruits which can be utilized in jellies and sherbets, and after acidification as a beverage (Anon., 1969). Tressler and Joslyn (1961) reported the preparation of pear nectar from Bartlett in California.

Bartlett fruits are mostly canned because of their uniform shape, white flesh and comparative freedom from grit cells. Lee *et al.* (1975) evaluated 8 Oriental and 10 European cultivars for canning and found Paldal and Josaeng Paldal of the former and Bartlett, La France and Duchess'd Angouleme of the latter group as promising. The harvesting of fruits at proper maturity and subsequent ripening at optimum temperature determine the quality of canned pears. A 30-35 per cent pear waste left after canning is good for making vinegar, brandy or alcohol (Anon., 1969).

Both sun and oven-drying are employed for preserving the pear fruits. Oven-dried product is better than sun-dried one which is tender, pliable, translucent and light in colour. Pears can also be candied and sweet pickled.

Origin and Distribution

The genus *Pyrus* has probably originated in the mountainous regions of western China wherefrom it moved both in east and west directions and evolution took place in different regions under geographical isolation, ecological tolerance, introgression and interspecific hybridization (Bell and Hough, 1986; Layne and Quamme, 1975). Speciation has occurred mainly in eastern and central Asia, in the Himalayas, Caucasus, Asia Minor and Eastern Europe. Vavilov (1951) proposed three centres of origion (centres of diversity) for the cultivated pears.

Chinese Centre

It is a primary gene centre comprising the regions of North and Central China, Japan and Korea. The important species originated in this tract are *P.pyrifolia, P. ussuriensis, P. betulifolia,* and *P. calleryana.*

Central Asiatic Centre

Western Tian-shan, Pamir-Alai, Tadjikistan, Uzbekistan, Northwest India and Afghanistan are included in this centre. The indigenous species are *P. communis, P. salicifolia, P. regeli,* and *P. pashia.*

Near Eastern Centre

Asia Minor and Causasus mountains are the tracts where great variability exists in *P. communis, P. syricaca* and *P. caucasica* are the other important species found in this centre.

Table 5.2: Pyrus Species and Hybrids from Asia[1]

Species	*Site of Origin*	*Crop*
Pyrus alnifolia (S. and Z.) Franch and Sav.	Russian Far East, China, Japan, Korea, Taiwan	*
Pyrus armeniacifolia T.T. Yu	China	*
P. aucuparia var. *randaiensis* Hayata	Taiwan	*
Pyrus baccata L.	Russia, Mongolia, China, Korea	*
Pyrus baccata var. *aurantiaca* ***Ragel***	Russia, Mongolia, China, Korea	*
Pyrus baccata var. *himalaica* Maxim.	China, Bhutan, India, Nepal	*
Pyrus baccata var. *mandshurica* Maxim.	Russia, China, Japan, Korea	*
Pyrus betulifolia Bunge	China, Laos	*
Pyrus x bretschneideri Rehder	China	*
Pyrus calleryana Decne.	China, Korea, Taiwan, Vietnam	USA, Canada
Pyrus calleryana var. *dimorphophylla* (Makino) Koidz.	Japan	*
Pyrus calleryana var. *fauriei* (C.K. Schneid.) Rehder	Korea	*
Pyrus calleryana var. *koehnei* (C.K. Schneid.) T.T. Yu	China	*
Pyrus cathayensis Hemsl.	China	*
Pyrus delavayi Franch.	China	*
Pyrus discolor Maxim.	China	*
Pyrus doumeri Bois	China, Taiwan, Laos, Vietnam	*
Pyrus folgner (C.K. Schneid.) Bean	China	*
Pyrus foliolosa Wall.	Burma, Bhutan, India, Nepal, China	*
Pyrus glabra Boiss.	Iran	*
Pyrus gracilis Siebold and Zucc.	Japan	*
Pyrus harrowiana Balf. F. and W.W. Sm.	China, India, Nepal, Burma	*
Pyrus heterophylla Regel and Schmalh.	Kyrgyzstan, Tajikistan, china	*
Pyrus hondoensis Nakai and Kikuchi	Japan	*
Pyrus x hopeiensis T.T. Yu	China	*
Pyrus hupehensis Pamp.	China, Taiwan	*
Pyrus indica Wall.	South Asia and Far East Asia	*

Species	*Site of Origin*	*Crop*
Pyrus japonica Thunb.	Japan	*
Pyrus keissleri (C.K. Schneid.) H. Lev.	China, Mayanmar	*
Pyrus kansuensis Batalin	China	*
Pyrus lanata D. Don	Afganistan, India, Nepal, Pakistan	*
Pyrus matsumurana Makino	Japan	*
Pyrus nussia Buch. Ham. Ex D. Don	Far East, South Asia	*
Pyrus x phaeocarpa Rehder	China	*
Pyrus pohuashanensis Hance	Russia, China, Korea	*
Pyrus prattii Hemsl.	China	*
Pyrus prunifolia Willd.	China	*
Pyrus pseudopashia T.T. Yu	China	*
Pyrus pyrifolia var. *pyrifolia*	China, Laos, Vietnam	*
Pyrus ringo Wenz.	China, Korea	*
Pyrus ringo var. *kaido* Wenz	China	*
Pyrus scabrifolia Franch.	China	*
Pyrus scalaris (Koehne) Bean	China	*
Pyrus x serrulata Rehder	China	*
Pyrus sieboldii Regel	China, Japan	*
Pyrus sikkimensis Hook L.	China, Bhutan, India	*
Pyrus sinensis var. *maximowicziana* H. Lev.	Korea	*
Pyrus x sinkiangensis T.T. Yu	China	*
Pyrus spectabilis Aiton	China	*
Pyrus taiwanensis Iketani and H. Ohashi	Taiwan	*
Pyrus ussuriensis Maxim.	Russia, China, Japan, Korea, Brazil	Brazil
Pyrus x uyematsuana Makino	Japan, Korea	*
Pyrus vestita Wall. Ex G. Don	China, Bhutan, India, Nepal, Mayanmar	*
Pyrus vilmorinii (C.K. Schneid.) Asch. and Graebn.	China	*
Pyrus xerophila T.T. Yu	China	*
Pyrus yunnanensis Franch.	China, Myanmar	*
Pyrus zahilbruckneri (C.K. Schneid.) Cardot	China	*
Pyrus tschonoskii Maxim	Japan	*
Pyrus cydonia L.	Iran, Armenia, Azerbaijan, Russia, Turkmenistan	*
Pyrus germanica (L.) Hook. f.	Middle East and Northern Asia	*
Pyrus korshinskyi Litv.	Afganistan, Tajikistan, Uzbekistan	*
Pyrus kumaoni Decne.	Middle East, Far East and South Asia	*
Pyrus salicifolia Pall.	Iran, Armenia, Turkey, Arzebaijao	*
Pyrus trilobata (Poir.) DC.	Israel, Lebanon, Turkey, Bulgaria, Greece	*
Pyrus turkestanica Franch.	Kyrgyzstan,Tajikistan, Turkmenistan, Afganisistan	*

*The same origin. 1: *Source*: USDA (2012).

In addition to the above, Zeven and Zhukovsky (1975) have proposed the fourth centre of diversity as European Siberian Centre. *P. communis, P. nivalis, P. salvifolia,* and *P. cordata* are endemic to this centre. *P. communis* (syn. *P. domestica*) has been divided into two species *P. pyraster* and *P. nivalis*. The former grows in central Europe and West Asia and the latter in West Switzerland and France.

The commercially important Oriental pear cultivars have been derived mainly from the species of *P. pyrifolia, P. ussuriensis,* and their natural hybrids (Shen, 1980). *P. pyrifolia,* undoubtedly, is the first Pyrus species which was domestricated followed by *P. ussuriensis* (Lombard and Westwood, 1987). Selections of *P. pashia* are also under cultivation in Southern China and Northern India (Layne and Quamme, 1975). *P. pyrifolia* is resistant to blight and drought while *P. ussuriensis* is winter-hardy.

Almost all the commercial cultivars of European pear predominating throughout the world except China and Japan are descendents of *P. communis*. The ancestors of this species are considered to be *P. communis* and probably *P. novalis, P. nivalis,* snow pear, is mainly utilized for making pear cider in Europe (Lombard and Westwood, 1987). The *Pyrus* species and hybrids from Asia, Europe, South Africa and Americas are prescuted in Tables 5.2–5.4.

Table 5.3: Pyrus Species and Hybrids Originating in Europe and Southern Africa[1]

Species	*Geographic Distribution-Site of Origin*	*Crop*
Pyrus aria (L.) Ehrh.	Canary Islands, North Africa, All of Europe	*
Pyrus aria (L.) Ehrh. var. *cretica* L. indl.	North Africa, Middle East, central Europe Oriental and Southern and Turkemenistan	*
Pyrus aucuparia var. *dulcis* (K.) A. and G.	All Europe	North America
Pyrus boissieriana Buhse	Azerbaijan, Turkmenistan, Iran	*
Pyrus korshinskyi Litv. subsp. *Bucharica* (Litv.) B.K.	Former Soviet Union	*
Pyrus bulgarica Kuth. and Sachokia (*Pyrus x nivalis* Jacq.)	Western Europe, Central Eastern and Southern	*
Pyrus caucasica Fed.	Eastern Europe and Central Greece	*
Pyrus chamaemespilus (L.) Ehrh.	Western Europe, central eastern and Southern	*
Pyrus communis L.	All Europe	Eastern Europe Central, South and West and South America
P. communis var. *cordata* (Desv.) H.f.	UK., Portugal, Spain, France	*
P. communis subsp. *gharbiana* (T.) Maire	Algeria, Morocco	*
P. communis subsp. *marmorensis* (Trab.) Maire	Morocco	*
P. communis subsp. *pyraster* (L.) Ehrh.	Western Europe, central Eastern, and Southern	*
Pyrus x complexa Rubtzov	Former Soviet Union	*

Species	*Geographic Distribution-Site of Origin*	*Crop*
Pyrus cossonii Rehder	Algeria	*
Pyrus crataegifolia Savi	Turkey, Albania, Serbia, Greece, Italy, Macedonia	*
Pyrus cuneifolia Guss.	Central Eastern Europe, South and Central	*
Pyrus decipiens Bechst.	All Europe and North Africa	*
Pyrus domestica (L.) Sm.	Algeria, Cyprus, Eastern Europe Central West and Meridional	*
Pyrus elaeagrifolia Pall.	Turkey, Ukraine, Albania, Bulgaria, Greece, Romania	*
Pyrus elaeagrifolia subsp. *Kotschyana*	Turkey	*
Pyrus germanica (L.) Hook f.	Middle East, Eastern Europe, central Southern and Northern Asia	*
Pyrus gharbiana Trab.	Algeria, Morocco	*
Pyrus intermedia Ehrh.	All Europe	*
Pyrus malus subsp. *Paradisiacal* (L.) schubl. and G. Martens	Western, Eastern and Central Europe and Greece	*
Pyrus minima Ley	UK	*
Pyrus nebrodensis Guss.	Italy-Sicily	*
Pyrus pinnatifida Ehrh.	All Europe	*
Pyrus praemorsa Guss	South of Italy, France	*
Pyrus sachokiana Kuth.	Georgia	*
Pyrus spinosa Forssk.	Central Eastern Europe, South, and Central	*
Pyrus sudetica Tausch	Western Europe, Central eastern, and Southern	*
Pyrus syriaca Boiss.	Caucasus and Middle East Region	*
Pyrus torminalis (L.) Ehrh.	North Africa, Middle East, South Caucasus, whole Europe	*
Pyrus trilobata (Poir.) DC.	Turkey, Bulgaria, Greece, Israel, Lebanon	*

* The same origin.

1: Source: USDA (2012).

Production

The pear had its dispersion from northern Italy, Switzerland, former Ygoslavia, Germany, Greece, Moldova, and Ukraine to East, in countries such as Iran, Uzbekistan, India, China, Japan, Korea and Bhutan.

Pear is the ninth most important cultivated fruit and second most important temperate fruit (after apple) in the world. World production of pear in 2012 was 23, 580, 845 tonnes. China is the world's largest producer (Asian pear) and the United states is the second largest producer being the first producer of European type. Together, the top ten producers occupy an area of 1,360,230 ha annually (Table 5.5). China's pear production has increased steadily during the 1980's and early

1990's because of the extensive planting. Data show that China produces more than twice production by other countries in the world production, making the crop a commodity of great importance to this country.

Table 5.4: *Pyrus* Species and Hybrids Originating in the Americas[1]

Species	*Site of Origin*	*Crop*
Pyrus americana DC.	Greenland, USA, Canada	*
Pyrus angustifolia Aiton	USA, Canada	*
Pyrus arbutifolia (L.) L.f.	USA	*
Pyrus arbutifolia (L.) L. f. var. *nigra* Willd.	USA	Northern and Eastern Europe Center
Pyrus coronaria L.	Canada, USA	*
P. coronaria var. *ioensis* Alph. Wood	USA	*
Pyrus diversifolia Bong.	USA, Canada	*
Pyrus floribunda Lindl.	USA, Canada	Korea, Russia, Sweden, Czech Republic, Slovakia, Germany, Latvia, Bulgaria
Pyrus fusca (Raf.) c.K. Schneid.	USA, Canada	*
Pyrus sanguina Pursh	Canada, USA	*

The same origin.

1: Source: USDA (2012).

Table 5.5: World Production of Pear in 2012 (tones) and the Area (ha) Harvested in the Ten Most Productive Countries

Position	*Country*	*Production (tonnes)*	*Area Harvested (ha)*
1.	China	16,266,000	1,136,700
2.	USA	778,582	22,015
3.	Argentina	700,000	26,500
4.	Italy	645,540	35,195
5.	Turkey	439,656	34,067
6.	Spain	400,600	25,000
7.	Republic of Korea	394,596	14,353
8.	India	340,000	38,500
9.	South Africa	338,584	13,000
10	Japan	299,000	14,900

FAOSTAT, 2012.

In India, high chilling good quality pear cultivars are grown in the states of Jammu and Kashmir, Himachal Pradesh and Uttarakhand, while low chilling (mostly Asian pears) and grown in northern plains, north-eastern states and Darjeeling district of West Bengal. In 2012-13, the total cultivated area of pear in

India was 42.30 thousand hectare with a total production of 74.12 thousand tonnes, which was 34.0 thousnd hectares and 38.5 thousand tonnes in 2011-12.

Taxonomy

All the pear cultivars belong to the genus *Pyrus* which is a member of subfamily Pomoideae, family Rosaceae and order Rosales. The basic chromosome number is 17 and somatic number is 34, rarely 51 or 68 (Westwood, 1978). Most cultivated pears are diploid (2n = 34), but there are a few ployploid cultivars of *P. communis* and *Pyrus* x *bretschneideri* (Silva *et al.*, 2014). The number of cataloged species varies greatly according to the interpretation of each author, 20 to 75 species (Erhardt, 2002). Zeven and Zhukovsky (1975) mentioned about 60 species but there appears to be a general agreement on 22 primary species and the rest as non-primary species which may be botanical varieties, subspecies or interspecific hybrids (Westwood, 1982; Bell and Hough, 1986). Challice and Westwood (1973) made detailed taxonomic studies of genus *Pyrus* on the basis of both chemical and botanical characters and categorized 22 primary species into four geographical groups. Morphological characters of these species are given in Table 5.6.

Some of the primary species of Bell and Hough (1986) differ from that of Challice and Westwood (1973). *Pyrus pashia, P. psuedopashia,* and *P. kawakamii* were considered as primary species by Challice and Westwood (1973) but did not find a place in the list of primary species of Bell and Hough (1986), instead they included *P. glabra, P. kansuensis* and *P. kohnei* in their list and classified all the primary species into five groups on the basis of geographical distribution as under:

1. European Group: *P. communis, P. nivalis, P. cordata, P. caucasica.*
2. North African Group: *P. longipes, P. gharbiana, P. memorensis.*
3. West Asian Group: *P. syriaca, P. elaeagrifolia, P. amygdaliformis, P. salicifolia, P. glabra, P. regelii.*
4. East Asian Group: *P. pyrifolia, P. kansuensis, P. ussuriensis, P. hodoensis.*
5. Asian Group: *P. calleryana, P. betulifolia, P. fauriei, P. dimorphophylla, P. koehnei.*

A grouping of the subspecies has also been done by Terpo (1985) on the basis of the origin after preparing a new classification scheme. Salient features of very important species are given below:

P. communis

The trees are pyramidical in shape, medium tall and spiny when young. The leaves are orbicular-ovate to elliptic, crenate-serrate and glabrous when old but new growth and inflorescence are pubescent. White flowers are borne mostly in corymbs. Fruits vary in shape but mostly pyriform, calyx persistent, fleshy pedicels and pulp with gritty concretions. The pulp is melting and buttery in texture.

P. pyrifolia (syn. *P. serotina* var. *culta*)

The tree is tall, vigorous, spreading and leaves are ovate-oblong, dark green, pubescent, elliptical and comparatively large in size. White flowers appear prior to

Table 5.6: Morphological Characteristics of *Pyrus* Species*

Species	Tree Size (Height in m)	Adult Leaf Margins	Juvenile Leaves	Leaf Length (cm)	Fruit			
					Skin	Calyx Type	Fruit Dia (cm)	Carpel No.
P. caucasica	5-6	Crenate	Simple	4.81	Smooth	Persistent	4.70	5
P. communis	5-6	Crenate	Simple	4.95	Smooth (some russet)	Persistent	3.80	5
P. cordata	2-3	Crenate	Simple	7.87	Russet	Deciduous	1.50	2-3
P. nivalis	3-4	Crenate	Simple	4.87	Smooth	Persistent	5.60	5
P. amygdaliformis	1-2	Entire or slight crenate	I-lobed	6.52	Smooth	Persistent	2.60	5
P. elaeagrifolia	3-4	Entire or slight crenate	Simple	5.70	Smooth	Persistent	2.40	5
P. gharbiana		Crenate	Simple	5.15				
P. longipes	3-4	Crenate	Simple	3.77	Russet	Deciduous (some persistent)	1.70	3-4
P. mamorensis		Crenate	Simple	4.58				
P. syriaca	1-2	Crenate	Simple	4.78		Persistent		5
P. pashia	3-4	Crenate	2-lobed	7.99	Russet	Deciduous	2.39	4-5
P. regelii	1-2	Crenate	Laciniate	5.80	Smooth	Persistent	2.0-3.0	5
P. salicifolia	1-2	Entire	Simple	5.10	Smooth	Persistent	1.90	5
P. betulifolia	5-6	Coarse Serrate	I-lobed	5.74	Russet	Deciduous	0.88	2 (3)
P. calleryana	3-5	Crenate	Simple	6.60	Russet	Deciduous	1.06	2
P. dimorphophylla	3-4	Crenate, Serrate	2-lobed	7.16	Russet	Deciduous	1.24	2
P. fauriei	1-2	Crenate	Simple	4.35	Russet	Deciduous	1.34	2

Species	*Tree Size (Height in m)*	*Adult Leaf Margins*	*Juvenile Leaves*	*Leaf Length (cm)*		*Fruit*		
					Skin	*Calyx Type*	*Fruit Dia (cm)*	*Carpel No.*
P. hondoensis	3-4	Fine serrate, Setose	Simple	7.57	Smooth	Persistent	2.80	5
P. psuedopashia	3-4	Crenate	1-lobed	6.42	Smooth	Persistent	4.26	5
P. Kawakamii	1-3	Crenate	3-lobed	6.14	Russet	Deciduous	1.04	4(3)
P. pyrifolia	3-5	Fine serrate, setose	Simple	10.65	Russet	Deciduous	4.10	5
P. ussuriensis	1-3	Coarse serrate Setose	Siple	8.23	Smooth	Persistent	3.75	5

* Modified after Challiee and Westwood (1973).

the emergence of leaves. Fruits are mostly round (apple shaped) with a depression at stem end, deciduous calyx, russeted, very hard and full of grit cells. They have crisp texture, a refreshing sweet taste but lack aroma.

P. pashia

It is comparatively medium-sized, open-headed tree, locally known as *Mehal, Kainth* or *Shiara.* Sometimes stem has stout spines when young. Simple leaves are lanceolate, acuminate, crenate, wooly when young but glabrous and shining above when old. Flowers are white and fragrant. Fruits globose, small, dark brown in colour, scurfy and covered with raised white dots. They mature during November-December, being gritty and unacceptabale for table purpose due to high tannin content (Srivastava *et al.*,1977). This species is commonly used as a rootstock for commercial cultivars in India. Wild trees are also topworked with superior cultivars which commence fruiting within 3-4 years of operation.

Cultivars

The cultivars of pear mainly belong to three groups, *viz.*, European, Asian and their hybrids. In general, the European cultivars are predominent in cultivation as they are of superior fruit quality but susceptible to fire blight. The cultivars can also be classified on the basis of chilling requirement. The cultivars requiring high chilling are Anjou, Bartlett, Bose, Beurre Hardy, Doyenne du Cornice, Conference, Flemish Beauty, Fertility, Starking Delicious, Winter Nelis and Packham's Triumph. In contrast, low chilling cultivars are Keiffer, Le Conte, Gola, and Patharnakh. Summer pears are early cultivars of moderate keeping quality while winter pears ripen late and store well for many months. The adaptability to climate and soil, market needs, fruit quality, productivity and resistance to pests and diseases are to be given prime consideration in choosing cultivars for any particular locality.

Vulnerability and impact of climate change on pear production are now a cause of concern by many pear growing countries. Pears are sensitive to the risks posed by extremes in rainfall and temperature and to gradual warming in an already warm production area. Impacts are expected to be felt in both yields and fruit quality, including reduced chilling units and disrupted reproductive processes, increase incidence of sunburn, poor colour development in some blushed cultivars and higher risk of drought stress in years when water availability is insufficient. Adaptation options, which could include breeding, microclimatic amelioration, water conservation methods and climate monitoring systems (Wand *et al.*, 2008).

European Types

Bartlett

It is also known as William's or William's Bartlett or William's Bon Chretien and has its origin in England. Except China and Japan, Bartlett is the most popular commercial cultivar all over the world because the tree is adapted widely to soil, climatic conditions and geographical ranges (Childers, 1976) whereas the fruit is well suited to dessert, drying and canning. The plant is a prolific and regular bearer. Bartlett is generally considered as the standard of excellence which is used in describing other pear cultivars.

The fruit is medium-large, ovate-pyriform, green when picked, turning bright yellow on ripening. White flesh is melting, firm, fine and juicy. The flavour is mildly subacid and good with a fine aroma. It is picked in midseason and can be held up to two months in cold storage. Some undesirable traits of this cultivar are its lack of precocity, upright growing habit and susceptibility to fire blight. Guerriero *et al.* (1984) after assembling 19 clones of Bartlett from different countries studied for 14 years, found variations in shape, size and yield.

Bartlett Improved-I, Verona 25, Ferrera and Ravenna clones were selected for high yield efficiency with large fruit size.

Max Red Bartlett

This cultivar is a bud mutant of Bartlett introduced in 1945. The plant and fruit characters resemble the parent except that the fruit skin colour is dark cranberry red. Tree is tall, spreading and productive and shoots and leaves have a reddish tinge (Rathore, 1982). Fruits mature about a week later than Bartlett. It is gaining popularity as a fresh-fruit cultivar.

Anjou (Beurre'd Anjou)

It is thee most important winter pear of French origin after Bartlett. The tree is vigorous, regular producer and slow in coming to bearing but fairly resistant to cold and fire blight. It requires pollination. The fruit is large, bright green, turning greenish-yellow when ripe. The flesh is fine, mildly acidic to sweet with high dessert qualtity. The fruit has an excellent keeping quality as it can be stored for about eight month at -1°C and is free from storage disorders. The cultivar is suitable for fresh fruit market.

Bosc (Beurre Bosc). (Kaiser Alexander)

Open pollinated seedlings, either from Van Mons, Belgium or Haute Saone, France in 1807. It is widely cultivated now in many countries as an early winter pear of excellent fruit quality. It is a vigorous, precocious, productive and cold hardy cultivar but difficult to give desired shape when the plant is young. The fruit is medium-large, oblong-ovate, pyriform with yellow skin which is almost covered with a brown russet. The flesh is very sweet, juicy and has grit cells in the core. Fruit does not suffer from core or flesh breakdown but requires very careful handling. It ripens in September about a fortnight later than Bartlett. Bosc is seriously affected by fire blight, blossom blight and scab.

Comice (Doyenne du Comice)

This cultivar has originated in France and is called Europe's queen of dessert pear (McAlpin, 1980). The tree is upright, open and inconsistent in bearing. The fruit is large green to lemon yellow with a flesh of outstanding texture and flavour. It is a late cultivar the fruit of which can be kept in storage up to January. The tender skin is readily bruised by the rubbing of leaves and branches. Of late, the cultivar

Concorde

D'Anjou-Pear-450w

Corck Spot

Conference

Le Conte

Packham

Eldorado

Pear Scab

Pear Decline

Pear in Flowering

Pear Scald

Abate Fetel 1

is not being commercially planted owing to its susceptibility to handling damage, russeting and unsatisfactory productivity.

Flemish Beauty

It has originated as chance seedling in Belgium. The tree is big, spreading due to more branching, and a high yielder, but susceptible to blight. Fruit is large, obovate, smooth and attractive. The flesh is pure white, very juicy and almost free of grit cells. It has a sweet aromatic flavour and high keeping quality after ripening (Rathore, 1982). Flemish Beauty is a self-fertile as well as a good pollenizing cultivar.

Hardy (Beurre Hardy)

This heavy bearing and strong upright growing cultivar was evolved in France. The plant is easily manageable in pruning as it does not branch freely and remains open. Since it unites well with quince, it is commonly used as an interstock. The fruit is of good size and shape but not very attractive. The flesh is white, juicy and of fair to good flavour. It has good keeping quality.

Eldorado

It originated as a chance seedling of probably of Winter Nelis x Williams' in California. Eldorado is a winter type, late-season cultivar which is considered as a possible replacement of Anjou. Medium vigour and productive, however highly susceptible to fire blight.

The fruit is large, oblong pyriform with a narrow neck, clear greenish-yellow skin when picked. Moderately firm flesh and juicy, smooth and sweet with a good blend of sugar and acid with very good flavour. The fruit is resistant to pressure and friction bruising and can be stored at - 1.1°C until April and May.

Packham's Triumph

The cultivar has been bred in Australia (thought to be a seedling from Uvedale's St. Germain x Williams) and has become popular elsewhere also. The tree is upright, irregular in growth, good yielder but susceptible to blight, stony pit virus and cold. The fruit is large, resembling Bartlett in shape but the skin surface is a bit uneven. The flesh is fine, juicy and free of grits with a rich and pleasing flavour. Harvesting is done after 30 days of Bartlett and the fruits are easily preserved for about 6 months in cold storage. Very susceptible to blossom blast, moderately susceptible to fire blight and scab.

Magness

Magness has been evolved as a fire blight resistant cultivar in the USA. Plant is very vigorous, upright, spreading and male-sterile. The fruit is of medium size, oval and partly covered with russet. Its flesh is soft, very juicy and almost free of grit cells. The ripe fruits are extremely sweet with exceedingly pleasant aroma. Picking is done after a week of Bartlett and the fruit stores for three months.

Corella

Probably a seedling of Forelle, Australia, tree moderately vigorous and is moderately resistant to scab, blast and mites, fairly low chilling requirement, fruit small, obtuse-pyriform, skin green-yellow base with red over half the surface or more. Flesh white soft and juicy, good storage life.

Concorde

This is a hybrid between conference and Doyenne du Comice selected at East Malling, UK in 1977. Tree medium vigour, upright, high chilling requirement. Fruit pyriform, similar to Conference. Skin green, developing an attractive yellow ground colour with excellent skin finish, flesh firm, juicy, sweet with flavour. It is a late season variety, precocious and heavy cropper.

Aurora (NY 7620)

Aurora is a hybrid of Marguerite Marillat X Williams' from New York Agricultural Experiment Station, Geneva, USA. Tree vigorous, spreading, susceptible to fire blight, mid-season variety, fruit attractive, large, pyriform. Skin bright yellow, slightly russeted, Sometimes blushed. Smooth flesh, juicy, sweet and aromatic with good eating quality.

Moonglow

Moonglow is a spur-type tree, vigorous, late blooming, and has good fire blight resistance. Fruit quality is good with a mild flavour, few grit cells and soft flesh, good for canning.

Conference

The fruits are medium to large, long-pyriform in shape with a smooth, uneventy russeted green skin. It is a late-maturing pear with a very long storage life. It can produce parthenocarpic fruits and is a heavy regular cropper but is susceptible to fire blight.

Abate Fetel

Originated in France and is the number one pear variety of Italy. Produced large fruit, long pyriform in shape, russeted, sometimes blushed. Melting, smooth flesh, Juicy, sweet and aromatic with good eating quality. Highly productive, susceptible to fire blight, moderately susceptible to scab.

Asian Types and Hybrids

Nijisseiki (Twentieth Century)

This was the commercial cultivar of Japan in the past was introduced in 1898. Tree is medium in vigour, spreading and heavy yielder. The fruits are medium in size, round, crisp and juicy. The skin colour is yellow while flesh colour is creamy. It is harvested in August and the fruits can be stored at 0°C for several months.

Kieffer

It is a widely grown pear because the tree is well adapted to different geographical, climatic and soil conditions and moderately resistant to fire blight. Tree is vigorous and very productive but self-unfruitful. The fruit is medium to large, brownish, gritty and hard. Flavour is somewhat inferior but it is a suitable cultivar for processing. The fruits mature in September.

Le Conte

Le Conte can be grown both at low and midhills because its chilling requirement is low. Tree growth is medium and spreading but bearing is quite heavy. The fruit is small, round, yellowish green. The flesh is white, juicy and acidic sweet. The crop is ready to harvest in August and plant is susceptible to blight.

Gola

It is another prolific bearing cultivar which has become very popular at lower altitude (Singh *et al.,* 1983). Roundish fruits are not very large in size but have an excellent keeping quality and can be transported to distant markets for fresh consumption.

Patharnakh

This is a low chilling cultivar which is commercially grown in submountainous and adjoining regions of India. The plant can withstand very high temperatures, hot winds, waterlogging and drought conditions. The tree is very vigorous, well branched and heavy yielder. Medium-sized fruits are round with prominent dots. The flesh is juicy, crispy and gritty but flavour is not so good. Its tough skin allows rough handling, distant transport and long shelf life.

New Varieties

China

Yan Xiangli

The variety was selected from the seedlings of the natural hybrid seeds of fine strain 78-14 derived from the cross combination of Ping Gouli x Nan Gouli (Piao-Yu *et al.,* 2012). The fruit is oblate, weighing 180g on average with the maximum weight of 280g. Skin yellow-green at maturity which turns to yellow after storage. Flesh is white, fine texture and few stone cells, crisp, sweet (TSS 12.5-14.0 per cent), juicy. The variety is productive with relatively resistance to pear scab.

Hongyue

This new red pear was bred by the cross Red Clapp's x Apple-pear (Li-Juncai *et al.,* 2011). Fruit is concical, average weight 245g, red skin, Flesh white and juicy. Flavour is sour-sweet with a little aroma and less stone cells. The TSS content is 14.4 per cent and vitamin C is 24.8 mg. An early variety, highly productive and hardy.

Yulu

This is an early maturing and high quality variety derived from the cross between Laiyangcili (*P. bretsechneideri* cv. Laiyangcili) and Taibai (*P. pyrifolia* cv. Taihaku) was bred by the Hubei Academy of Agricultural Science, China (ZhongQi *et al.*, 2011). Fruits mature at the end of July to early August in Wuhan area. Average fruit weight 270g, but reaching to about 434g, with a green skin and a near round shape. The flesh is white fine, tender, crisp, juicy with a balanced sweet-acid taste. (TSS 10.5 per cent to 12.2 per cent, acidity 0.28 per cent). The variety is precocious and productive, high and stable yield and high resistance to diseases.

Selection 29-1

This was selected from the cross between Cili and Taihuku (ZhongQi *et al.*, 2008). It is a mid-season selection, mature in early August, fruits large, average weight 329g, smooth green skin, flesh white, crisp, juicy, TSS 12.0 per cent, acidity 0.24 per cent. Trees are resistant to black spot, tolerant of drought, waterlogging and infertile soil. It is precocious and productive.

Korea

Supergold

The variety supergold (*Pyrus pyrifolia* var. Culta Nakai) was originated from the cross between Chuwhangbae and Manpungbae with the aim of improving the fruit quality of Chuwhangbae. It was selected in 2002 and named in 2008 (Samseok *et al.*, 2011). Tree vigorous, resistant to black leaf spot (*Alternaria kikuchiana*). Fruit shape oblate, skin colour greenish-white at maturity. Average fruit weight 570g and TSS 13.6 per cent. Flesh soft and juicy, good eating quality and good shelf life.

The National Horticultural research Insitute (NHRI) of Republic of Korea released the following varieties between 1984 and 2001. (See Table 5.7 for important characters)

i) Shinecheon - released in 1999
ii) Minibae - released in 1996
iii) Hanareum - released in 2001
iv) Gamro - released in 1996
v) Sunhwang - released in 1996
vi) Wonhwang - released in 1994
vii) Josengwhangkeum - released in 1998
viii) Shinil - released in 1995
ix) Whangkeumbae - released in 1984
x) Soowhangbae - released in 1988
xi) Yeongsanbae - released in 1984
xii) Whasan - released in 1992

xiii) Manpoong - released in 1997
xiv) Gamcheonbae - released in 1990
xv) Chuwhangbae - released in 1985
xvi) Mihwang - released in 1995
xvii) Mansoo - released in 1995

Table 5.7: Characteristics of Pear Varieties Released by NHRI, Korea*

Varieties	*Fruit Shape*	*Skin Color*	*Weight (g)*	*Soluble Solid (per cent)*	*Storability*	*Pollen*
Shincheon	Round oblate	Light yellowish brown	350	13.2	Weak	Little
Minibae	Round	Brighe yellowish brown	240	10.5	Weak	Plenty
Hanareum	Round ablate	Lighe yellowish brown	482	13.8	Weak	Little
Gamro	Round oblate	Bright yellowish brown	315	13.8	Weak	No
Sunhwang	Round oblate	Lighe yellowish brown	390	13.2	Medium	Plenty
Wonghwang	Oblate	Lighe yellowish brown	570	13.4	Medium	Plenty
Josengwhangkeum	Round	Yellowish green	410	12.8	Medium	Plenty
Shinil	Round oblate	Lighe yellowish brown	377	13.8	Medium	Plenty
Whangkeumbae	Round	Yellowish green	430	14.9	Medium	No
Soowhangbae	Round	Lighe yellowish brown	490	12.0	Medium	Plenty
Yeongsanbae	Round	Lighe yellowish brown	540	13.2	Medium	little
Whasan	Round oblate	Lighe yellowish brown	580	13.2	Strong	Plenty
Manpoong	Oblate	Light yellowish brown	662	13.8	Medium	Plenty
Gamcheonbae	Round oblate	Lighe yellowish brown	590	13.3	Strong	Plenty
Chuwhangbae	Oblate	Brighe yellowish brown	395	12.0	Strong	Plenty
Mihwang	Round	Light yellowish brown	480	12.7	Strong	Plenty
Mansoo	Round oblate	Light yellowish brown	663	12.4	Strong	plenty

* Adopted from leading, Deciduous Fruit cultivars in Korea, National Harticultural Research Institute, R.D.A. Rep. of Korea.

United States of America

Gem

Gem (Sheldon x US62563-004) is a new variety that combines exceelent appearance, fruit quality, and long storage potential with precocious and high yields. It can be eaten without ripening as result of a crisp, juicy texture as well as ripened to a soft texture (Bell *et al.*, 2014).

Sunrise

Sunrise is composed of *P. communis* and *P. pyrifolia* Nakai parentage. It is an early maturing pear variety that combines exceelent appearance and fruit quality. It has also good storage potential for an early-maturing pear with resistance to fire blight. The fruit is obviate-acute-pyriform to oblong-pyriform in shape. Skin colour at harvest is light green and most fruits are blushed. The skin turns yellow when ripe. The variety is moderately resistant to pear scab but susceptible to powdery mildew (Bell and Van der Zwet, 2011)

Japan

Almost all redcent Japanese pear (*Pyrus pyrifolia* Nakai) varieties bear round fruit weighing 300 to 500g, but some late-maturing varieties bear fruit more than 1000g. The fruit skin of Japanese pear cultivars is of two types, russet or smoth and yellowish-green, called 'Akanashi' and 'Aonashi' in Japanese, respectively. Japanese pear has a soften, smother texture. Japanese pear is mainly consumed fresh and most harvested fruit is marketed with no storage. The two major varieties 'Kosui' and 'Hosui' make up more than 65 per cent of the area under pear. The other important varieties are 'Nijisseki' and 'Niitaka'. Many new cultivars have been released since 1990. 'Gold-Nijisseki' -a mutant of Nijisseki is tolerant to black spot disease. 'Osa-Gold' and 'Kotobuki-Shinsui' were also bred as mutant varieties and are tolerant to black spot. Several other varieties were released.

The Japanese pear exhibits gametophytic self-incompatibility (GSI), which is controlled by a single S-locus with multiple S-haplotype. Allthough, GSI is a genetic mechanism to prevent inbreeding and promote outcrossing to mechanism to prevent inbreeding and promote outcrossing to maintain genetic diversity, it be problematic in fruit trees because it causes unstable fruit set. Therefore, orchardist interplant Japanese pear with other lines as pollinizers or conduct artificial pollination to ensure fruit set (Okada, 2015).

Climate

Pear is found growing in a wide range of climatic conditions as it can tolerate as low as -26°C in dormancy and as high as 45°C in growing period. Different pear cultivars have inherited this wide adaptability from their progenitor, *i.e.*, *P. communis*, *P. ussuriensis* and *P. pyrifolia* which have originated in extremely varied climatic conditions. A large number of pear cultivars require about 1200 hours below 7°C during winter to complete their chilling requirement in order to flower and fruit satisfactorily. 'Bartlett' needs slightly more chilling hours (about 1500) while an inferior cultivar Patharnakh fruits heavily even at about 150 chilling hours and can withstand high temperatures and hot winds during the summers. In between these two cultivars, there are many medium- chilling requiring cultivars such as Kieffer, Le Conte, Gola, Pineapple and Hood, which perform well in places with mild winters.

Spring frosts are detrimental to pear production and temperatures at -3.3°C or lower kill the open blossoms. Air circulation is therefore, necessary in the vicinity of a pear orchard which is achieved by avoiding low lands. Likewise, hail prone localities must be avoided because pear will not give a satisfactory crop as hails damage both plants and fruits. The fruits of cultivars Bartlett and Bosc attain the best dessert and keeping qualities where temperatures are higher for two months preceding the harvest (Teskey and Shoemaker, 1972). In warm areas, fruits ripen more slowly after harvest, remain in prime eating or canning condition longer and have less tendency towards core break-down. But higher spring and early summer temperatures enhance the incidence of fire blight. If lower temperatures (7.1°C during night and 21°C during day) prevail for several days before harvest, premature ripening occurs impairing quality and storage life of fruits.

Arid and moderately hot climate is favourable for pears which are cultivated under both rainfed and irrigated situations. A well-distributed annual precipitation of about 100 cm is desirable for the successful pear growing if irrigation facilities are not available. Dry summers are, however, required for an effective control of fire blight.

Selection of proper site is important for regular cropping. The land which is at a slightly higher level than the surrounding area and on northern aspect needs to be selected. Ample moisture is available for plant growth and fruit development on northern aspect as compared with southern aspect where fruit size remains small but crops mature earlier.

Soil

Pears thrive on a variety of soils but deep, well-drained, fertile, medium-textured and relatively more clayey soil is the best. The trees are short lived on sandy and loam soils.When compared with apple, pear is less tolerant to drought but more tolerant to wet soils.

Deckers *et al.* (1978) compared the tree dimension and fruit growth of Doyenne du Comice cultivar in a wet sandy loam soil with a dry sandy loam and found that the tree growth was greater in wet soil. Pears even do well on poorly aerated heavy soil with high water table which is too heavy in texture for most of the deciduous fruits. Pear trees growing on deeper soil (182 cm deep) give about twice yield than in shallow soil (50 cm deep) and similarly trunk girth, cross sectional area and pruning weight of trees are more in deeper soil (Alberto *et al.*, 1982). A deeper soil holds high reserve of nutrients and water than a shallow soil. A soil depth of about 180cm is, therefore, requires for proper root development and heavy fruit production. A neutral pH range of 6.0 to 7.5 is desirable because iron deficiency is encountered on highly alkaline soils. Highly fertile soils very rich in nitrogen are not suitable for pear growing as they increase the incidence of psylla and fire blight.

Rootstocks

In the past, the propagation by cuttings in majority of fruit trees was not successful making it imperative to resort to grafting and budding and in that case the seed of commercial cultivars was used for raising the rootstock. Since such seedling rootstocks do not impart any specific capability to the scion and efficient techniques of propagation through cuttings have developed, there is general trend in modern fruit culture toward growing cultivars on their own roots. In pear also, the cultivars can be regenerated by rooting of cuttings and such self-rooted plants perform equally well or better than on seedlings or clonal rootstocks (Hartmann *et al.*, 1963; Thibault and Hermann, 1982). But the stress conditions of climate and soil, incidence of pests and diseases and demands of high density culture coupled with the availability of all types of genotypes in pear to combat these problems make the use of rootstock in pear cultivation indispensable (Layne and Quamme,

Table 5.8: Influence of Important Rootstocks on the Performance of Pear Trees*

Rootstocks	*Tree Size (Per cent of standard)*	*Yield*		*Fruit Quality*		*Pests and Diseases*		
		Preco-city	*Effi-ciency*	*Black End*	*Over all*	*Pear Decline*	*Fire Blight*	*Wooly Aphid*
***P. communis* seedling**	90	3	4	No	4	4	1	2
OH x F 69, 230, 333	60	3	4	No	4	5	5	2
OH x F 217, 267, 361	70	3	5	No	4	5	5	2
OH x F 18, 112, 198	80	3	4	No	4	5	5	2
P. nivalis	100	3	4	No	4	5	5	2
P. pashia	90	3	3	No	3	4	2	4
P. betulifolia	90	3	3			4	2	3
Oregon 260, 261, 264								
***P. calleryana* seedling**	110	3	3		4	5	3	4
Oregon 211	90	3	4	No	4	4	5	5
D-6	15	3	4			4	5	5
P. pyrifolia	70	3	4	No	5	3	5	
P. ussuriensis	70	3	4	Yes	3	1	3	
Quince A	100	3	4	Yes	3	1	5	1
QC	50	4	5	No	5	5	1	4
PQ Lepage C	50	4	5	No	5	5	1	4
	55	4	5	No	5	5	1	4

Precocity-time required to come into bearing: above 10 year= 1, below 5 years=5.

Efficiency-yield per surface area of mature tree, based on trunk cross-sectional area.

Overall quality-flavour, texture, storage and shelf life of fruit.

Rating for pear decline, fire blight and woolly aphid -l=susceptible, 2=low tolerance, 3=moderate tolerance, 4=high tolerance, 5=very tolerant

*Modified after Lombard and Westwood (1987).

1975; Lombard and Westwood, 1987). Since many *Pyrus* species are cultivated in the world, they are also employed as rootstock. Lombard and Westwood (1987) have reviewed the literature on pear rootstocks and have given the performance of various rootstocks used in the different countries (Table 5.8).

Chacteristics of Rootstocks

Pyrus communis

The seedling rootstocks of *P. communis* are compatible with all pear cultivars. They impart vigour to the scion and adapt to a wide range of soil and climatic conditions. They are mostly resistant to oak root fungus (*Armillaria mellea*) and pear decline but are susceptible to fire blight and woolly aphid. Old Home and Farmingdale of European origin and several species from Asia resistant to blight have been collected (Reimer, 1925). A series of clonal rootstocks (Old Home x Farmingdale) resistant to blight (except 361) with high yield potential has been developed (Westwood, 1976). The cultivar Anjou gave the highest yield on an unnumbered OH x F among the eight pear rootstocks (Fallahi and Larsen, 1986). On the basis of their control on the size of grafted trees, OH x F rootstocks are grouped as under (Westwood and Lombard, 1977).

Vigorous	*Semi-vigorous*	*Semi-dwarf*	*Dwarf*
OH x F 18	OH x F217	OH x F34	OH x F51
OH x F97	OH x F267	OH x F69	
OH x F 112	OH x F361	OH x F87	
OH x F 198		OH x F230	
		OH x F333	

P. pyrifolia and *P. ussuriensis*

The susceptibility of these species to pear decline and physiological disorder 'black end' has rendered them unsuitable for using as a rootstock (Batjer *et al.*, 1967). *Pyrus ussuriensis* is used as a vigorous and winter hardy rootstock in northern Asia and North America.

P. calleryana

This blight resistant rootstock is very satisfactory except for its lack of winter-hardiness. It produces vigorous trees which are almost free from the problem of 'black end'. D-6 rootstock developed from Australia is resistant to drought (Cole, 1966), fire blight, leaf spot (quince fleck) and woolly aphid, moderately resistant to crown gall, collar rot, mildew, D-6 is compatible with European pears and nashi. It is the most commonly used rootstock for commercial pear production in Australia.

P. betulifolia

This vigorous rootstock is free from woolly aphid, leaf spot, and tolerant to alkaline soils and adaptable to wide range of climatic conditions. In a comparison of eight different pear rootstocks, it has given the highest yield of Bartlett (Larsen and Fritts, 1986). It is, however, not resistant to fire blight and is incompatible with the cultivar Anjou. Oregon 260, 261 and 264 are the clonal rootstocks of this species.

P. nivalis

In France ten clonal rootstocks have been evolved from snow pear which have greater yield efficiency than QA and clone P 2267 is 25 per cent of the tree size of it (Michelen, 1980).

Cydonia oblonga

Quince is a widely accepted dwarfing rootstock of the pear. It is resistant to woolly aphids and nematodes but susceptible to oak root fungus, fire blight, excess lime and cold (Hartmann and Kester, 1959; Tukey, 1935). The propagation of quince rootstocks is rather easy making it possible to select and maintain as clones. Therefore, many clonal rootstocks such as QA, QB, QC, PQBA 29, PQ Lepage C, PQLepage BC 132, *etc.*, have been developed (Parry, 1981; Hatten, 1933). Quince B is semi-vigorous while Quince C is very dwarfing. C 132, a quince of Caucasian origin is not as precocious as Quince C but produces the tree of same vigour (Parry, 1981). Many commercial cultivars *viz.*, Bartlett, Bose, Seckel, Winter Nelis, Eldorado, Clapp's Favouritte and Farmingdale need an interstock (Old Home or Hardy) to form a compatible union with quince. The compatibility of scion on quince rootstock is affected by virus infection (Sansavini *et al.*, 1980).

Quince A (Angers Quince)

Selection made at East Malling Research Station, UK for producing bush and smaller trees and high density planting. This has high tolerance to woolly aphid, resistant to root lesion nematode, mildew and crown gall, medium-resistant to pear blast, susceptible to quince fleck and fire blight. It is not compatible for cvs. Beurre Bose, Packham's Triumph, Winter Cole, Josephine or nashi.

Quince C

Another selection from East Malling Research Station, UK for bush and smaller trees and high density planting. This has also high tolerance to woolly aphid, root lesion nematode, mildew and crown gall like Quince A. It is also not compatible with all the varieties mentioned for Quince A.

Quince MH

This quince stock has a vigour similar to MC. Though initial bearing is not as early as MC, MH induces better fruit size.

Quince Adams

Released by a Belgiam nursery in the 1970s, it features a vigour intermediate to MA and MC and induces good cropping and good fruit size.

EMH (QR 193-16)

This was introduced by East Malling in 2001. It is more dwarfing than Quince A and produced larger fruits of pear scion cultivars than both Quince A and Quince C.

BA 29

Developed at France, more invigorating than Quince A and is suitable for poor soils.

Pyriam

A *P. communis* selection bred by the Angers INRI station in France. It has higher vigour than quince BA 29.

Pyrodwarf

A clonal stock from the cross Old Home x Buona Luisa D'Avranches bred at Geisenheim in Germany. Its vigour is similar to or slightly less than BA 29 and is slower to begin cropping than quince.

Angelini *et al.* (1986) grafted 32 pear cultivars on Quince BA 29 and pear seedlings and reported that the trees on BA 29 were much smaller gave higher yields and the fruit on it matured 6 days earlier than on seedling rootstocks. The South African pear industry used pear (*P. communis*) seedling rootstock until the development of the BP range of *P. communis* clonal rootstock. The major commercial pear rootstocks in South Africa, BPI and BP3, which are stuitable for the relatively poor soils and limited water supply. 'Angers' Quince A (QA), 'Provence' Quince C51(QC51), *P. calleryana, P. betulaefolia* and 'Old Home' x 'Framingdale' (OH x F) selections have been evaluated in South Africa since the early 1970s and BPI was found to be the favoured rootstock, although trees on BPI, BP3 and QA showed similar yield efficiencies (Jacobs and Cook, 2003), Internationally, QC 51 is classified as dwarfing, BP1 and QA as semi-dwarfing, OH x F97 as semi-vigorous and BP3 as vigorous (Wertheim, 1998) under South African conditions, QC 51 proved to be less precocious than BPI and more precocious than QA, both the former having higher cumulative yields and being more production efficient than the latter (van Huyssteen and vanZyl, 1992).

Major structural root differences between pear and Quince stocks have been reported (Lombard and Westwood, 1987) with *P. communis* providing excellent anchorage while quince generally being poorly anchored and requiring some support. North and Cook (2008) evaluated the commercially available rootstocks in South Africa using the popular commercial blushed pear 'Forelle', planted at 4.2 x 0.5 x 0.5m and trained on a modified Tatura trellis. They reported that after 5 seasons, trees on BP3 were excessively vigorous while trees on Bl 1 and BA 29 had filled their allocated space however, trees on OH x F97, QA and QC 51 were less vigorus, Trees on pear rootstocks had higher levels of N and P and lower Mg levels than trees on Quince (Table 5.9) which may influence storage life.

Rufato *et al.* (2011) in an experiment with European pear scions cultivars in combination with quince rootstocks in Urupema Santa Catarina State, Brazil observed that the 'Packam's Triumph/EMA and 'Abate Fetal'/Adams combinations were good for high vegetative growth while cvs. William's, Conference, Rocha, Abate Fetal, all grafted on EMC, showed morphological incompatibility.Sugar and Basile (2011) reported that in a sandy loam soil, Comice, pear trees on Quince

H (also known as Quince 193-16) rootstock were similar in fruit set, tree size, and yield efficiency to trees on Quince C while yield and fruit size were similar to trees on Quince BA 29C. The cumulative average (for 10 years) individual fruit weight was recorded 231g on Quince BA-29C, 214g on Quince H and 167g on Quince C. The cumulative yield efficiency (kg/cm^2) was 0.56 on Quince BA – 29C, 1.04 on Quince H and 0.98 on Quince C.

Table 5.9: Effect of Rootstock on Leaf Mineral Content (Per cent dry mass) of 'Forelle'*

Rootstock	*N per cent*	*P per cent*	*K per cent*	*Ca per cent*	*Mg per cent*
BP1	2.27a	0.182a	1.20a	2.24a	0.44cd
BP3	2.34a	0.162b	1.12a	2.33a	0.52c
OH x F97	2.35a	0.152bc	1.13a	2.42a	0.42d
QA	2.15b	0.135d	1.29a	2.38a	0.66ab
QC 51	2.17b	0.135d	1.17a	2.19a	0.63b
BA29	2.15b	0.145cd	1.07a	2.46a	0.73a
LSD ($P \leq 0.05$)	0.084	0.0173	0.323	0.389	0.093
P	<0.0001	0.0002	0.775	0.654	<0.0001

Means in the same column, which are followed by different letters, differ at P<0.05.

* Adopted from North and Cook (2008).

Rootstock Relationship to Yield and Quality

In a number of rootstock trials, high yield efficiency has been obtained in various cultivars on quince, Old Home x Farmingdale (OHxF) *and P. betulifolia* rootstocks (Wertheim, 1987; Larsen and Fritts, 1986; Trachev, 1984; Westerlaken, 1981; Patzold, 1981). Being vigorous, *P. betulifolia* is particularly a good rootstock for less vigorous scions such as Seckel and not for Anjou (Westwood, 1978). Old Home as an interstock is more productive than Hardy. *P. pyrifolia* and *P. ussuriensis* rootstocks cause 'hard end' and *P. betulifoli caused* 'cork spot' fruit disorder in Anjou. The fruit of both Bartlett and Anjou were higher in soluble solids than several other rootstocks (Fallahi and Larsen, 1981). Among various quince rootstocks, EMA induced the lowest fruit acidity (Rivalta *et al.*, 1986). Generally, high soluble solids and acidity increase the fruit quality. Rootstocks exert more influence on acidity and firmness than on soluble solids (Westwood, 1978).

Kosina (2003) evaluated new pear rootstocks (OH x F.69, OH x F87, OH x F230, OH x F333 and BA 29) with scion cvs. Red Barttlet, Conference and Lucas for six years in Czeck Republic. The cv.Red Barttlet showed the maximum cumulative yield on OH x F69 and the lowest on OH x F 333. The cvs Conference and Lucas produced highest total yield on OH x F87. Fruit weight was however found higher on BA 29 than on OH x F rootstocks (Table 5.10).

Rootstock Adaptation

According to Lombard and Westwood (1987), a rootstock may be good for a particular set of conditions but no single rootstock can be the best for all the conditions. Therefore, soil, climate, cultivar, prevalence of pests and diseases and planting density are the various factors which are taken into consideration while making the choice of a suitable rootstock. *Pyrus amygdaliformis* and *P. elaeagrifolia* rootstocks perform well in soils with pH 7.5 to 8.0, but *P. pashia, P. betulifolia and P. calleryana* will produce chlorotic plants in such soils. Zyl and Etsebeth (1981) recommended BPI for rich soils and BP3 for poor soils. Likewise, Patzold (1984) is in favour of using Quince A rootstock on good soils for precoicity and improving fruit quality instead of pear seedlings which are desirable on poor soils.

Table 5.10: Yield and Fruit Weight of Pear cvs. Red Barttlet, Conference and Lucas on different Rootstock*

Rootstock	*Cumulative Yield per Tree in kg. (1992-2002)*			*Mean Fruit Weight (g) (2001-2002)*		
	Red Barttled	*Conference*	*Lucas*	*Red Barttlet*	*Conference*	*Lucas*
BA 29	49.9	56.5	73.3	191	146	246
OHxF69	52.7	56.3	70.4	172	117	188
OHxF87	46.7	68.5	96.6	181	135	228
OHxF230	47.7	46.1	79.4	155	110	176
OHxF333	37.2	34.8	56.5	177	111	169
LSD (0.05)	3.2	4.3	4.9			

* Compiled from Kosina (2003).

Pyrus ussuriensis rootstock will be preferred for severe climate. 'Cork spot' develops in Anjou fruits when this cultivar is grown on *P. betulifolia.* However, this disorder is not observed when *P. betulifolia* rootstock is used for other cultivars (Westwood, 1978). In the localities having pear decline problem, the least susceptible rootstock *P. betulifolia* is a right choice over the most susceptible *P. ussuriensis* and *P. pyrifolia* rootstocks. For high density planting, early cropping and fruit maturity, quince rootstocks (MC. BA29, Lepage B) have been found better than other stocks (Angelini *et al.*, 1986; Trachev, 1984; Loreti and Natali, 1975).

Propagation

Pear cultivars are commercially propagated by budding or grafting on seedling rootstocks because of easy availability. Seedlings of commercial cultivars serve the purpose of rootstock in most of the countries while seedlings of several *Pyrus* species are also used in some countries. The clonal rootstocks are used only in Europe, USA, South Africa and Israel (Lombard and Westwood, 1987).

Seed Propagation

For raising rootstocks, seeds are obtained either from processing factories or from ripe fruits. Pear seeds need stratification (chilling treatment) for proper germination. The growth inhibitors present in the seed-coat and embryo decrease and auxin-like substances increase due to the physiological action of chilling (Shawky *et al.*, 1978). The temperature and duration of chilling vary with the species and related genera and their optima are given in Table 5.11.

Table 5.11: Chilling Requirements of Several *Pyrus* and Related Genera for Seedling Rootstock Propagation*

Species	*Seed Size (numbers per g)*	*Chilling Duration (Days)*	*Optimal Chilling Temperature (°C)*
Pyrus amygdaliformis	24	25	7
P. betulifolia	90	55	5
P. calleryana	55	30	7
P. communis	22	90	5
P. dimorphophylla	77	65	5
P. elaeagrifolia	22	90	5
P. fauriei	57	35	7
P. gharbiana	99	60	5
P. mamorensis	11	50	5
P. nivalis	18	110	5
P. pashia	55	15	10
P. pyrifolia	26	120	5
P. ussuriensis	20	80	7
Amlanchier alnifolia	112-250	180	3
Crataegus mollis	-	90	5
Malus sylvestris	21-35	100	4
Sorbus americana	185-520	90	5
S. aucuparia	230-375	90	1
S. decora	280	90	5

* Lombard and Westwood (1987).

The stratification can be accomplished either by sowing the seeds in open nursery in November or placing the seeds in alternate layers with the sand, perlite or vermiculite at optimum low temperature for the required duration. Among the typical species of rootstock for grafting pear trees *P. calleryana* is one of most used for producing plants of Asian cultivars. The use of seedlings from seeds has been preferred by fruit growers. Seeds of Taiwan Nashi-C, stratified for 30 days at 5°C showed high germination indices (Maeda *et al.*, 1997). Treatment of pear seeds with the plant regulators has been reported to increase the percentage of germination and reduce the period of stratification. Dhillon and Sharma (1978) recorded an increase in germination from 3 to 93 per cent after treating the seeds with GA_3 at 150 ppm or thiourea at 5000 ppm followed by 14 days stratification. According to Lin *et al.* (1987), the effect of thiourea (2 per cent) was equivalent to 360 hours of chilling and that of potassium nitrate (3 per cent) and hot water (37°C) was 240 hours in the commercial cultivars of *P. pyrifolia*. A temperature of 12°C for a period of 17-20 days is required at the time of seed germination for production of normal plants (Brossier *et al.*, 1975). On the contrary, germination at 18° or 20°C produces seedlings with short internodes. Vasilenko (1986) planted seedling rootstocks at 90 x 10,15, 20 or 25 cm in nursery and budded. For the production of high quality transplants. He recommended planting at 90 x 15 cm, giving about 75000 rootstocks per hectare.

The rootstocks can be budded or grafted *in situ* or after transplanting. Seedlings *in situ* were ready for budding in six months and gave 100 per cent success while 73 per cent success was recorded in transplanted seedlings (Singh and Jawanda, 1978).

Vegetative Propagation

Various techniques of vegetative propagation such as budding, grafting, topworking, stooling, cutting can be used in the multiplication of pears.

Budding and Grafting

Shield or T budding is the most popular method practised in the propagation of pear. Other methods like patch, flute, ring and chip budding are not so common. However, chip budding may be more successful as in apple (Howard, 1987). When an interstock is required to be inserted between scion and rootstock to overcome the incompatibility, double shield budding can be done in a single operation saving one full year (Westwood, 1978). Satisfactory growth and cropping of important pear cultivars on Quince A with an interstock bark shield of Hardy were recorded for eleven years by Allison and Erasmus (1985). Budding is usually done in May-June or in September.

Grafting in pear is accomplished by whip and tongue method on young rootstocks and cleft on older ones. Maric (1975) suggested bench grafting (whip and tongue) in January-February in order to reduce the graft producing period from 3 to 1 or 2 years.

Topworking

When comparatively older rootstocks are grafted or budded at higher level, the process is termed topworking. It is usually adopted for the conversion of wild large seedlings into productive commercial plants or inferier cultivars to improved ones or to provide pollenizers. A large number of the wild seedling trees of *P. pashia* found in the Himalayas have been topworked with improved cultivars. In China, *P. ussuriensis* have been transplanted and topworked with an average survival of 90 per cent (Cao *et al.*, 1981). The topworked plants began fruiting in two years after the operation and gave high yields in the fifth year.

Disease-free young trees should be selected for topworking. Three to five branches with wide angles and projecting in all the directions are retained for this purpose removing the other branches in the month of November. Thompson (1975) used whip and tongue method when the scion and stock were of equal thickness but cleft grafting was done on the stocks thicker than the scion. According to him, thick plastic wrapping hastened the tissue union and encouraged callus formation. In cleft grafting, nailing and grafting wax sealing are required to be attended properly.

Early spring is the appropriate period for topworking. No shoot other than the scion should be allowed to grow on the plant.

Stooling

Pear rootstocks can be multiplied in the same way as Malling apple stocks but it is not adopted commercially (Hartmann and Kester, 1959). Encouraging results have been reported in the stooling of *P. pashia* by using IBA (Rana and Singh, 1966; Yadava *et al.*, 1971; Pathak *et al.*, 1982). Sharma and Dhillon (1985) have advocated the use of root suckers as rootstocks for cultivar Patharnakh. It reduces the incidence of tree falling due to horizontal spread of roots.

Cutting

Both scion and rootstock can be propagated by hardwood, semi-hardwood and softwood cuttings employing IBA, mist and bottom heat with varying degrees of success. Juvenile hardwood cuttings of Patharnakh gave 61.2 per cent rooting in comparison with 22.5 per cent in non-juvenile cuttings (Kahlon and Singh, 1981). They found 100 ppm IBA to be the best both for rooting and shoot growth but Pathak *et al.* (1982) reported better rooting in hardwood cuttings of cv. Gola by using 2500 ppm IBA and propagating at 21 ± 1°C. Hardwood cuttings of Old Home are planted in nursery after being taken in October and treated with 100 ppm IBA for 24 hours and stored at 21°C for 3 weeks (Lombard and Westwood, 1987). Hartmann *et al.* (1963) were successful in rooting the softwood cuttings of Old Home and Bartlett by treatment with the high concentrations of IBA (6000 to 10000 ppm). Propagation of Old Home and OH x F rootstock is successful through hardwood stem cuttings (Lombard and Westwood, 1987). They have also suggested of propagation through root cuttings for quick multiplication of a rootstock. Up to 71-77 per cent rooting was obtained in 15 cm long *P. pashia* root cuttings when planted in August (Randhawa and Kishore, 1981).

Micropropagation

Micropropagation technique has been useful in rapid multiplication of healthy plant material, retrieval of virus-free stocks and in hastening the pace of breeding. Successful regeneration of both scions and rootstocks from shoot tip and cotyledon culture has been accomplished in pear by using modified Murashige and Skoog's (MS) medium with addition of benzyladenine and sugar for callusing and NAA or IBA for rooting (Lane, 1979; Marino, 1984; Bhojwani *et al.*, 1984; Turovskaya, 1987; Browning *et al.*, 1987; Zhao, 1982).

The procedure of *in vitro* multiplication of pear cultivar Williams Bon Chretien has been described in detail by Marino (1984). He got the proliferation rate of 10-12:1 when shoots without their tips were cultured on a modified MS medium with 3 per cent sucrose, 0.7 per cent agar and 1.0-1.5 mg BA/l with or without 0.1 mg IBA or NAA/l. Highest percentage of rooting (90-100) and number of roots (4-5) were obtained by providing a 10-day period of darkness. The survival of rooted plants was as high as 90-100 per cent and their growth improved further with the spray of 100-200 ppm GA_3 at 4-5 days interval. Bhojwani *et al.* (1984) reported the enhancement of percentage of shoots forming roots and survival of plantlets after transplanting in *P. pyrifolia* with the addition of phloroglucinol to the rooting medium. According to Maarri *et al.* (1986), meristems from adult tree *in vitro* grew weakly, proliferate poorly and the resultant shoots elongated slowly than the meristems from seedlings. In Egypt, *Pyrus betulaefolia* is largely used as a rootstock for pear cultivation. Seeds are used to develop rootstock which takes long time to have the desired size for grafting. Hassanen and Gabr (2012) developed an effective *in vitro* culture system from stem segments and shoot tips of *P. betulaefolia*. They successfully induced culture on MS basal medium supplemented with 2.0 mg/l 6BAP and 0.05 mg/l βNAA, 30g/l sucrose and 3g/l phytagel. The optimal shoot multiplication was achieved on MS medium supplemented with 2.5 mg/l BAP plus 2.0 mg/l isopentenyl adenine (2ip). They also observed that IBA was more effective for rooting than NAA. Eighty five percent of rooting was obtained on half-strength MS medium containing 2.0 mg/l sucrose. Regenerated shoots were successfully acclimatized to green house conditions and grew vigorously with no apparent phenotypic aberrations. Kainth (*P. pashia*) is used as a rootstock of pear in northern subtropical region of India. Rehman *et al.* (2014) developed protocol for micropropagation of Kainth using nodal explants. During proliferation stage, Woody Plant Medium (WPM) supplemented with BAP at 3.0 mg/l resulted in highest proliferated cultures (83.19 per cent). The highest shoots per explant were obtained using MS mediun supplemented with BAP at 3.0 mg/l. The rooting was found maximum (27.46 per cent) by using half-strength MS medium supplemented with NAA at 1.0 mg/l. Cosac *et al.* (2008) developed regeneration protocol for European pear cvs. Cure, Argessis, Euras, Carpica, Republica, Daciana, Getica, Ervina and Conference. The highest multiplication rate was obtained on MS medium supplemented with 0.1mg/l BAP. High rooting was achieved with 2.5 mg/l IBA and 2.5 mg/l NAA. Decreasing mineral concentration (50 per cent) was found to increase rooting. At the beginning of the rooting stage, the explants benefitted from a light exclusion treatment for 8 days. Browning *et al.* (1987) recommend cotyledon culture for rapid multiplication of desired genotypes and for the generation of somaclonal variation.

Embryo culture is useful in rescuing abortive embryos derived from interspecific and intergeneric hybridization. Hiratsuka and Katagiri (1988) cultured the various parts of immature seeds of Japanese pear at different stages of seed development but they could get shoot development only 8-13 weeks after pollination.

On the basis of a 6-years trial. Sansavini *et al.* (1986b) reported that micropropagated self-rooted William's trees were more vigorous than on Quince A and gave similar yield excepting that they ware less precocious.

Planting

The land for establishing a pear orchard needs to be prepared at least one year in advance so that the plants can make a good start instead of struggling for survival in their most important period of life. It should be cleared of all the old stems and roots of previously growing trees and shurbs and be levelled properly giving a gentle slope for water drainage. After cultivating the soil thoroughly, green manure or other agricultural crops are grown in order to improve the tilth.

A planting plan is prepared adopting a particular layout system before actual planting. The layout system depends on the plant density to be adopted and topography of the land. In the conventional planting density, rectangular system is preferred over square system because it utilizes ground and air space in a better way by accommodating more number of trees per unit area without reducing the tree to tree spacing. The latest trend towards high density planting has increased the adoption of hedgerow system. The plants form a continuous hedge in the row because of closer spacing between the plants. The distance between the rows is also kept just the minimum to provide adequate room for light and orchard operations.

In hilly terrain, the contour system is followed on a rolling land. Sometimes on a very steep land, bench terraces are made which are, however, very expensive both in construction and maintenance. The retaining walls are frequently damaged by heavy rains. In the contour layout, the first row is drawn at the highest elevation with the help of a surveyor's level and rod. All trees in a row come at the same elevation. The distance between the rows depends on the slope, being closer on steeper slope and distant otherwise.

Planting Distance

Soil fertility, cultivar, rootstock, training, implements and climate are decisive factors in plant spacing. Keeping these factors in view, the spacing between plants and rows is so decided that the maximum bearing surface per unit area develops in a minimum possible time and overcrowding does not occur when the trees attain full size. For standard plants, 6 x 4 m permanent spacing is considered adequate which should be kept initially 3 x 2 m under filler removal plan. Using Quince rootstocks, high density plantings have been successfully attempted in pear.

A planting distance of 3-3.5x1.2m on Quince MA has been recommended for cultivars William's, Beurre Clairgeau and Beurre Bosc by Ristevski (1982). He also found that planting distance and yield per unit area are negatively correlated. In contrast, planting distance has positive correlation with vegetative growth, flowering intensity and yield per tree.

Pepelyankov and Granevski (1986) recommended 3.5 x 1.1m spacing for William's Bon Chretien on rootstock Quince MA. Povar and Adaskalitsii (1986) compared the crown parameters and productivity of four cultivars on wild pear rootstock at 4.5 x2.0, 2.5, 3.0, 3.5, 4.0 and 4.5m spacings with the control (7x4m). They obtained the best crown parameters and yields by free palmette trained trees of the cultivars Beurred' Hardenpont, Clapp's Favourite and Beurre Bosc spaced at 3.0-3.5m and William's at 2.5-3.0 m in the row. Wagenakers (1987) planted the cultivars Conference and Doyenne du Comice at 2000, 2667 or 4000 tree/ha in arrangements of 1, 3 or 5 rows. Maximum yield/ha of Conference was recorded in the highest planting density in third and fourth year but with Comice, highest planting density gave the highest yield in the third year in one trial only and thereafter the yield differences between various planting densities narrowed down. Fruit size usually decreased with the increasing planting density (Wagenakers, 1987; Wertheim, 1984). Extra planting cost involved in dense planting is, however, soon recovered by early high yield and low production cost per tree.

The European pear industry is undergoing a steady progression towards more intensive plantings, a phenomenon largely linked to the use of Quince rootstocks to constrain tree vigour. This trend is even more marked in Southern Europe and especially in Italy, where dwarfing rootstock like Quince MC have led to planting densities ranging over 5,000 up to a peak 13,000 trees/ha (Musacchi, 2008). Italy's Emilia-Romagna region, Europe's leading pear district, produces an average yield of 24 t/ha. The high density plantings use high inputs to optimize mutrition (on daily scheduling), growth regulators to control fruit set and vigour, vertical – axis training systems and are trained in the double-hedge V-shape and use exclusively Quince MC as stock.

After determining the planting system and spacing, the field layout is done indicating the exact location of fruit trees by stakes or lime. A pit of 1 x 1 x 1 m dimension is dug at each point and filled again with a mixture of soil and compost. The plants obtained from a reliable nursery must have good root and shoot system, strong graft union and freedom from pests and diseases.

Late fall and early spring are the common planting periods. In the regions where the winter is mild and the soil has enough moisture, late fall planting is desirable but in the contradictory conditions early spring planting is the rule. The hole is made just big enough to accommodate all the roots. Very long roots can be shortened. The plant should remain straight in its position when the roots are being covered with the soil firmly. The roots should never be allowed to remain exposed to the sun and wind for more than twenty minutes before or during planting. If a delay is anticipated in this operation, the plants can be heeled in a shallow trench, roots being covered with moist soil.

Immediately after planting, the basin should be prepared and irrigation be given. If soil settles down exposing the roots, more soil needs to be added in order to bring the floor of the basin to original level. Water is not allowed to come in direct contact of the trunk. To avoid sun scald, trunk is painted white. Staking is also useful. In order to enhance the growth, a very small quantity of urea may be applied to the soil. Weeds should be removed as often as necessary as they rob the plants of moisture and nutrients.

Training and Pruning

Training

A number of training systems have been evolved currently to increase the orchard efficiency and secure early and high returns from the investment. The systems used are free standing (pine-shaped tree), pyramid, spindle (bush, arched and Fourger), palmette (early and oblique) and Tatura trellis which have been illustrated by Evequoz and Evequoz (1986). They concluded that the systems requiring much basal framework involved excessive pruning, thus reduced the yield. The systems which are easy to prune, do not impede the growth and quickly occupy the allotted space, are palmette for vigorous cultivars, arched spindle for medium vigorous cultivars, and Fourger spindle for less vigorous cultivars.

The oblique arm palmette system has given highest yield in the cultivars William's Bon Chretien, Clapp's Favourite, Santa Maria Morettini and Beurre Bosc but it is the next best in the cultivar Beurre d' Hardenpont (Kuzmenko *et al.*, 1982; Burlak, 1980; Talice, 1980). The production cost of fruits per kilogram was about 30 per cent lower for Fourger's spindles and palmettes than for bush trees (Darbellay and Dessemoz, 1987). Bellini (1987) has reviewed the literature on training and pruning systems of the pear trees illustrating the various systems diagrammatically. According to him, pyramid and bush training systems are being replaced by palmette systems because the latter has been better adapted to modern mechanization. The advantage of early production from the early palmette system over the palmette one was apparent with yield of 26, 37 and 46 kg fruits per tree for cultivars Buttira Precoce Morettini, Abbe Fetal and Passe Crassane, respectively, with the early system for the period 1964-67 compared with 8, 18 and 26 kg per tree, respectively, for the same cultivars with the palmette system which came into production a year later. Reviewing the research work on changes in planting and training of pear trees in Italy, Sansavini (1986) favoured the early palmette, spindle bush, vertical column and transverse Y training systems because they gave the best results.

Tatura trellis is the recommended system for high density planting of pear trees (Lichou *et al.*, 1987; Ende and Chalmers, 1983). In this system, V-shaped trees are planted closely together in single rows. Each tree has two limbs trained at 60° from the horizontal across the inter-row alley way. It facilitates hand picking and simple mechanization but does not require dwarfing rootstock. The last two decades has seen pronounced innovations in orchard design with the new layout being functionally linked to planting density. The low density orchards have low inputs of energy and technology, are trained in hedgerow or palmette systems using vigorous stocks like seedlings or self-rooted trees. Medium density orchards, which pioneered fertigation, are trained to the spindle system and its offshoots with medium vigour Quince stocks like BA29 and Sydo. The high density planting use high inputs, vertical-axis training systems and are trained in the double-hedge V-shape and use exclusively Quince MC as stock.

Spindle, currently the most widely used system in mid-to-high density orchards, makes it possible to intensify spacing without excessive costs for trellising and labour. One of the prerequisites for this system is the use of feathered scions and Quince MC rootstock so as to achieve yields of 7-8 t/ha even by year 2 of orchard life. Various alterations have also been developed to enhance tree control. These include the S-shape to bend the tree and reduce its height, the more recent Y shape, or Bibuan® system, which utilizes preformed split feather trees in the nursery and controls vigour and notably reduce pruning costs and the V-shapes. The V system, which tilt alternate trees to a 15° angle from vertical are particularly suited to pear's growth-yield habit and have the key advantage of exploring a double cropping curtain so as to maximise light interceptions, although the higher costs for trellising and labour for pruning are notably drawbacks (Musacchi, 2008)

The vertical axis, a recently developed system for ultra high densities, *i.e.* no more than 30-35 cm in row spacing and 2.5-2.75m between row spacings, features a vertical axis and short limbs with spurs that are periodically renewed (Sansavini *et al.*, 1997). This kind of planting has a high initial outlay and must be used with Quince MC rootstock because the other stocks are two vigorous. 'Abbe Fetal' in a V system at 5,555 trees/ha produced an average yield of 45.5 t/ha and average fruit weight of 253 g (Musacchi *et al.*, 2005)

In Korea, Asian pear trees used to be trained mostly by either pergola system or open-center natural form since 1906, but after the introduction of Y-trellis system in 1985, the Y-system is prevailing throughout Korea. However, the V-system has been tried in commercial orchards since 1996. Results of the experiment comparing V-system and Y-system for Asian pear of Niitaka/seedling trees showed wider tree width for Y-trellis. Fruit number per tree and yield were higher in V-trellis than the Y-trellis (Table 5.12).

Table 5.12: Tree Growth and Fruiting of 'Niitaka'/Seedling Asian Pear Trees Planted at 6 x 1 m Spacing and Trained to a V-trellis or Y-trellis Training Systems in the 5th Leaf (2003)*

Orchard System	*Main Stem Length (cm)*	*Tree Width (cm)*	*TCA (cm^2)*	*Fruit No./ Tree*	*Yield/Tree (kg)*	*Yield (t/ha)*	*Fruit Size (g)*
V-trellis	320	150	137	56.3	48.2	80.6	857
Y-trellis	338	199	174	51.6	40.3	67.1	782
t-Test	NS	**	NS	*	**	**	**

TCA: Trunk cross sectional area; *: Significantly different at P=0.05; **: Significantly different at P=0.01; NS: Non significant.

* Adopted from Koo Kim *et al.*, 2008.

The average fruit weight was also significantly higher in V-trellis than for Y-trellis (Koo Kim *et al.*, 2008). To determine how different training systems affected early growth and fruiting (first 5 years), Kappel and Brownlee (2001) tried four different training systems (angled trellis, slender spindle, vertical axe and Y-trellis) of 'Conference' pear on Quince A rootstock at Summerland, Canada. They observed that trees of the Y-trellis had the greatest spread after 5 years and the vertical axe and slender spindle trees were the tallest. The Y-trellis trees had the highest light interception and had significantly higher yields than the other training systems.

Pruning

Pruning is one of the most important cultural practices requiring considerable expertise. The objectives of pruning are to obtain good annual crop of quality fruits, to control tree size for the convenience of various horticultural operations like interculture, spraying, pruning, thinning, picking, *etc.*, to avoid limb breakage due to heavy crop load and to remove undesirable interfering and diseased branches. The characteristics of a cultivar, rootstock, training system, planting density, soil and climatic conditions play an important role in determining the kind and degree of pruning. Pruning reduces plant size, delays fruiting and decreases the total yield but enhances the fruit size and maintains the regularity in cropping. Instead of heavy pruning, moderate pruning is advised to maintain a balance between fruit quality and yield. Pruning of young plants differs considerably from that of bearing ones.

Pruning of Young Trees

The aim of pruning of young plants is to develop proper tree structure which can carry heavy crop loads. Since the branches are chosen and forced to grow in proper direction so as to build a strong framework, it is known as training. Depending on the growth habit of a cultivar, anyone of the three training systems, *i.e.*, central leader, modified central leader or vase can be adopted in pear but the modified central leader system is very common. Gordon and Richens (1980) found little difference in yield in the eighth year of the cultivar William's Bon Chretien trees on Quince rootstock when trained as palmette, central leader and vase systems but they came into bearing in 3rd, 4th and 5th year, respectively.

An unbranched plant should be headed back at 60 to 90 cm above the ground level at planting time which will form relatively low head profitable tree. In the first dormant pruning, three or four wide angled, 15 to 20 cm apart branches arising in all directions on the stem need to be selected to serve as scaffold limbs in future. If only narrow angled and closely growing branches are available, they may also be retained because pear limbs are not easily breakable as apple limbs by the crop load. Bending of such branches may, however, be necessary with the help of ropes and pegs. The selected branches are headed back at varying heights at lower levels than the central leader while other branches are thinned out. Each selected branch gives rise to several shoots but only two shoots are retained on it in opposite direction at the second dormant pruning. These shoots are headed back at a level where further shoots are desired to form third set of branches. In the third and fourth year, the pruning is mainly carried out to maintain a proper relationship among the branches forming scaffold system. Efforts are, however, made to correct the imbalance, if any. If the trees have grown well, the framework gets established by the fifth year when the leader requires to be cut back to a well-placed outward growing lateral. The leader may be shortened earlier as well if the plant has attained a good height. Heavy pruning of young plants results in smaller trunks and branches, fruiting and lower yields. The principle of minimum pruning comprising only light thinning out of branches, rare heading back of strong growth is applied at this stage till the tree commences bearing.

Pruning of Bearing Trees

In pruning bearing trees, a certain amount of thinning out and heading back of outward growing laterals are considered adequate. In fact, a balance is required to be maintained between fruit production and vegetative growth. Pruning intensity, however, varies with the bearing habit of a cultivar and vigour of a tree. Pear bears fruits on spurs arising on two-year-old wood and a spur continues to bear for more than six years. Some cultivars such as Manning Elizabeth, Hardy, Bosc, Flemish Beauty and to a lesser extent Anjou and Comice produce more spurs curtailing annual vegetative growth and in a period of 15-20 years, the older branches become a mass of spurs which produce few small fruits or none. In such situation, rather heavy and detailed annual pruning is required. The limbs with spurs over 6 to 8 year-old need to be removed in a phased manner. The branches and new shoots are headed back to induce new growth, and old fruiting branches and spurs are thinned out to maintain the vigour of the retained ones.

The cultivars Bartlett and Winter Nelis produce a large number of side shoots along with the terminal growth but comparatively less spurs. The side shoots should be thinned out to open the tree for sunlight and to impart the vigour in new growth to encourage the spur formation. Some moderate heading back of vigorous shoots is also performed. Thus Bartlett requires lighter pruning than Anjou.

Vigorous growth is more susceptible to fire blight. Therefore, in areas of heavy blight infection, pruning should be carried out in such a way that the trees make only a thrifty growth. Light pruning reduces cork spot and increases yield of desirable fruit size, Mechanical pruning is much quicker than manual pruning and it also gives higher yield (Bellini, 1987). Sansavini and Guinchi (1980) report that mechanical pruning takes only one-tenth of hand pruning time. Montevecchi *et al.* (1986) compared many mechanical devices of pruning and found Nobili pruner with 5 discs of 50 cm diameter with a total cutting length of 230 cm as the best.

Dormant season, when the danger of heavy snowfall is over, has been found to be the best time of pruning. Early pruning may result in severe cold injury because the cells near the cut get activated losing the hardiness (Westwood, 1978). Bliek (1985) reported higher yield in cultivar Doyenne du Comice with winter pruning alone than winter plus summer pruning. But in case of mechanical system, pruning in summer or soon after harvest was faster and easier than in winter (Montevecchi *et al.*, 1986 Sansavini and Giunchi, 1980).

Pruning regimes for high density planting (HDP) and V-shape training system under HDP need to consider the growth and cropping habit of the cultivar and incorporate pruning strategies specific for that cultivar. In case 'Abbe Fetal', its growth-yield habit along with fruit prices in the Italian market make it well suited to planting over 4000 trees/ha (Musacchi, 2008). 'Abbe Fetal's habit is usually marked by its copping on 2-year wood, though in some cases even on 3-year branches (Sansavini, 2002). The spring growth is maintained by retaining at winter pruning a certain number of one year old shoots, which will not be bent the following year so as to promote flower bud differentiation. Pruning is best applied late and, ideally timed to coinside with bud-swelling. Back cutting of the 1-year wood in both the apical area and at branch ends is also becoming more widely used today. The apical cuts ensure a strong regrowth and promotes vigour in the shoot system and the branch cuts, while also promoting the former, prevent limb ageing and consequent exhaustion (Musacchi, 2008).

Nutrition

Nutrition of any fruit is the most complicated cultural practice because it varies a great deal with the climate, soil, cultivar, rootstock, orchard management, planting density, available moisture and season.

Optimum supply of nutrient elements is essential for a good yield of quality fruits. The availability of one element can, however, determine the utilization of another element to a large extent through antagonism or synergism of ion interactions. N absorption is inversely proportional to P uptake, while high content of K in soil decreased the availability of Mg and Ca (Westwood, 1978). Mota *et al.* (2011) working with peer cv. Rocha reported that nitrogen was preferentially absorbed as NH_4+ than as NO_3 and about 62-66 per cent of the total labeled N recovered in the trees. The fertilize N use efficiency was also higher for ammonium nitrate treatment than for nitrate treatment.

Iron deficiency is mostly lime-induced but it may also be caused by potassium deficiency by an excess of cobalt, on neutral soils by an excess of phosphorus, or on acid soils by an excess of zinc. Indeed, soil pH plays a predominant role in the availability of various elements to the plant. Increasing the supply of one element may enhance the demand for another element which then becomes deficient and a limiting factor for fruit production. In order to maintain the adequate availability of essential elements for the best plant performance, it is important to know their deficient, optimum and excess levels in the tissue and deficiency and toxicity symptoms as well.

Tissue Analysis

In conjunction with soil analysis, tree condition, yield and quality of fruits, the leaf analysis serves as a good guide for fertilizer application in pear. It helps in determining the need of a nutrient and mode of its application before the occurrence of a nutrient disorder and also the nutrient element when any disorder is noticed. Kenworthy (1950) suggested the following optimum levels (per cent dry matter) for Williams' pear on the basis of surveys of highly productive orchards in the USA:

N-2.13-2.75	Mn-0.0068-0.0220
P-0.110-0.161	Fe-0.0028-0.0240
K-0.80-2.16	Cu-0.005-0.010
Mg-0.317-0.419	B-0.0010-0.0043

Vang-Peterson and Hansen (1973) reported the following range of the optimum values in Dermark:

N-2.0-2.5 per cent	Mg-0.20-0.30n
P-0.15-0.30 per cent	B-20-50 ppm
K-1.2-1.6 per cent	Mn-30-150 PPm
Ca-1.2-1.8 per cent	

On the basis of the review of literature and high yielding young and old orchards in Australia. Ende and Leece (1975) established the pear leaf composition standards related to major and minor elements (Table 5.13). They found close agreement between the standards of most elements and the orchard's evaluations except Ca, Na and Cl standards which needed amendments.Use of multiple regression analysis enabled them to define the contribution of each element to growth and fruit production.

Table 5.13: Standard Values for Nutrient Contents of Pear Leaves from the Middle of the New Shoot Sampled in Mid Summer*

Nutrient per cent of Dry Matter	*Deficient*	*Low*	*Optimum*	*High*	*Excessive*
N	1.8	1.8-2.2	2.3-2.7	2.8-3.5	3.5
P	0.10	0.10-0.13	0.14-0.20	0.21-0.30	0.30
K	0.7	0.7-1.1	1.2-2.0	2.0	
Ca	0.8	0.8-1.3	1.4-2.1	2.2-3.7	3.7
Mg	0.13	0.13-0.29	0.30-0.50	0.51-0.90	0.90
S	0.10	0.10-0.16	0.17-0.26	0.26	
Na			0.01	0.01-0.02	0.20
Cl			0.05	0.05-0.10	0.10
		ppm in dry matter			
Fe		60	60-100	200	
Zn	10	10-19	20-50	750	
Mn	25	25-59	60-120	121-220	220
Cu	5	5-8	9-20	21-50	50
B	10	10-19	20-40	40	

* Ende and Leece (1975).

Leaf nutrient levels, however, vary with the cultivars and rootstocks. Georgiev and Koleva (1977) analyzed the leaves of pear cultivars William's Bon Chretien, Beurre Bosc, Passe Crassane and Clapp's Favourite grafted on Quince A, B12 and Provence Quince for three years. They found that B12 reduced leaf N content by 4.35-8.7 per cent than the other rootstocks. The rootstocks had no effect on leaf P content but Provence increased K content by 6.5-11.2 per cent. The differences in N. P and Fe levels were affected by the cultivars. Leaves of Passe Crassane contained 8.7 per cent less N, 9.9 per cent less P, and 15 per cent less Fe than those of William's, Clapp's Favourite had 22.54 per cent more P than the other cultivars. Dotti and Fontana de Gradi (1976) reported that Passe Crassane pear on Quince rootstocks had a higher N requirement than self-rooted pear which might be due to the restricted movement of nitrogenous compounds through the graft union. From a study of nutritional status of William's pear on different rootstocks with or without double grafting, Chaplin and Westwood (1980) mentioned only few differences which were mainly in Mg, Mn, and Fe. Double grafting did not affect nutrient uptake by William's on Quince rootstock, pear takes up less N and B, and more Mg than on pear stocks.

Leaf Sampling

Leaf nutrient standards are, however, valid only when proper procedures of leaf sampling, handling and analysis are followed. Lilleland and McCollam (1961) and Chapman (1964) recommended sampling of entire leaves from spur in mid-summer. Sampling of entire leaf from mid-terminal shoots, 8-12 weeks after full bloom has been suggested by Kenworthy (1964).

N and K deficiencies were, however, best diagnozed by taking rosette leaves from non-fruiting spurs in late July or early August (Filippov and Pilipenko, 1971). Sampling of 20 leaves from the middle of current year's shoot of average vigour and length from each of 10 selected trees is the best procedure (Ende and Leece, 1975). Davis and Moore (1934) noted considerable changes throughout the season in mineral content of basal leaves from current season's growth of Bartlett pear. Kumar and Grewal (1977) collected the leaf samples of Patharnakh pear shoots at monthly interval from the time of emergence to fall and analyzed NPK contents. They found that leaf N content declined from 2.36 per cent in 2 months old leaves to 1.94 per cent during 9 month period, but remained constant for 5 to 8 months old leaves (July to October). Similar trend was shown by leaf P and K. Casero and Ferran (1985) estimated the foliar nutrient levels throughout the vegetative growth cycle in three consecutive years. According to them, throughout the vegetative cycle, N, P, K and Cu contents decreased, Ca, Fe and Mn contents increased, while Mg, Zn and B contents varied. On the basis of analysis of data, they recommended the most suitable period for sampling between 125 and 150 days after flowering.

Fertilizer Recommendations

Macronutrients

Nitrogen is the only element required for proper nutrition of pear in many orchards but there is a general tendency to supply P and K along with it. An element should, however, be only given to the plant when it is actually needed. In a study of 15 Bartlett pear orchards in California, Proebsting (1961) obtained higher yields in only 6 orchards with N application which led him to conclude that vigour, leaf colour and production combined with leaf analysis should give definite indications of the possible response to N. Total shoot and root growth and number of flower buds per tree increased with increasing N concentration in nutrient solution of pot-grown trees of the *P. serotina* cvs. Shinsui, Nijiseiki and Hosui (Banno *et al.*, 1984). Lazarov (1979) evaluated 13 fertilizer treatments using N at 12 to 36 g/m^2 and P and K at 6 to 18 g/m^2 to William's trees on Quince A rootstock. There was no marked effect on tree growth or fruit quality though the best result was found by 36g N+18g P+18g K/m^2. Later on, he (1984) reported the best yield, fruit quality and returns by using 360 kg N+180 kg P_20_s+180 kg K_20/ha. Drobota and Drobota (1979), however, got maximum yield of 180 kg fruits from 10-year-old tree of cultivar Cure when 20 tonnes of farmyard manure per hectare was applied. Urdukhanov *et al.* (1985) treated the trees of cultivar Cure annualy in spring with 60 kg N, 45 Kg P_2O_5 and/or 60 kg K_2O/ha and got 40.5 per cent increase in yield over control during three years with full NPK treatment.

Darfeld and Lenz (1985) gave 8 different NPK treatments to Alexander Lucas pear on Quince. A rootstock and Gellerts interstocks over a 16-year period. Without N, the trunk girth was smallest and only 77 per cent of the yield of fully fertilized trees was produced. Fruit size and quality were rarely affected by the treatments. N at 50 or 100g, P_2O_5 at 25 or 50g and K_2O at 50 or 100g/tree per year of age were supplied until 6th year and thereafter the rates were stabilized (Henry *et al.*, 1984). The highest yield (67.1 kg/tree) of good quality fruits was obtained from trees receiving the fertilizers at 600:150:300g/tree. The yield in non-fertilized control was 37.3 kg/tree. Fruits from pear trees low in N (1.65-2.16 per cent) had lower respiration rate, a higher content of soluble solids and titratable acidity and a longer shelf life than those from trees high in N (2.14-2.72 per cent). High N level had however, a much lesser effect on the fruit quality of pears than the fruit quality of apples (Vangdal and Ystaas, 1980).But high N content of pear foliage makes it highly susceptible to psylla and fire blight attack. Pfieffer and Burts (1983) found that the density of eggs and nymphs of psylla increased at a much rapid rate on the trees which received higher rate of nitrogen fertilizer, and the rate of N application had a greater effect in that context than the time of application.

For the pear orchards under clean cultivation, 1.5 1 b N per tree per year was considered sufficient while an additional 0.5 to 1.0 Ib N is needed under sod culture. The use of herbicides has, therefore, been useful in economizing N application (Quast, 1982). From the results of pot culture experiments, Banno *et al.* (1984) reported the greatest increase in the number of flower buds for all cultivars on supplying N from June to July and the least on supplying in May. Jankovic (1986) fertilized William's Bon Chretien and Passe Crassane trees with NPK (124 kg N:84 kg P_2O_5:182 kg K_20/ha) in four modes, *i.e.* (i) in a single dose in November-December (ii) split dose in November-December and February (iii) single dose in February or (iv) in April, at flowering. He obtained the highest yield by a single dose application in November-December. Dormant urea sprays of 2 and 6 per cent increased the bud burst and enhanced the number of flowers per spur but not the fruit set (Khattab *et al.*, 1981).

Raese and Burts (1983) found that high rate of urea in combination with soil applied paclobutrazol at 20g/tree increased fruit number, leaf N and green leaf colour but reduced shoot growth, fruit weight and leaf injury from transpiration burn. Foliar sprays of urea are not utilized by pear (Westwood, 1978). Testoni and Granelli (1986) reported increased yield and fruit size with foliar spray of NPK (26-6-12) and micronutrients applied several times during the season in comparison to a standard soil applied programme of 200kg N/ha as $(NH_4)_2S0_4$. Foliar treatment however, had no influence on the fruit quality, flesh firmness, soluble solids and acidity.

Pear trees normally do not respond to P and K fertilizer application and can perform satisfactorily under their low availability conditions (Gayford, 1945; Overholser and Claypool, 1936; Proebsting, 1934). K application increased fruit size in Bartlett pear only when its content was lower than 0.96 per cent in the dry matter of mid-summer leaves (Fisher *et al.*, 1959).

Phosphorus deficiency occurs under high pH and is exhibited by leaf margin and tip burn, short terminal growth, die back of new growth and failure of fruits to mature normally (Bryant and Gardner, 1943). Purphlish-red leaf margins indicating high anthocyanin production due to P-deficiency was observed in leaves of the trees grown in low–P soils (Raese, 2002). Stefan *et al.* (1964) reported greater uptake of P fertilizer when applied during the growing season. Marginal leaf scorching appearing first on older leaves is a well-known K deficiency symptom in all fruits which is noticed in pear also if the mature Bartlett leaves contain less than 0.5 per cent K (Reuther, 1941). The other K deficiency symptoms are upward and inward rolling of leaf blades, reduction in fruit size and acceleration of maturity. The deficiency can be accurately diagnozed by leaf analysis and corrected by the application of K fertilizers to the soil.

Calcium deficiency symptoms in pear have been reported on the basis of studies carried out in sand and water culture and affected orchards (Raese, 1984; Yokornizo, 1974). In Ca-deficient solution, the leaves and fruits had low Ca content but not in the seeds. Likewise, fruit disorders caused by Ca deficiency are black end, green stain, alfalfa greening cork spot and reduction in fruit firmness and enzyme activity. Raese (1984) obtained beneficial effects of $CaCl_2$ spray either once at 40 Ib/acre or 3 or 4 times at 7.5 lb/acre in July or August or as soil application.

Magnesium deficiency symptoms usually appear late in the season. The leaves begin yellowing and browning at the margins and show distinct interveinal chlorosis and browning. The leaves, particularly on spurs, are small and in severe cases partial defoliation occurs before harvest (Wallace, 1961).

Micronutrients

The condition known as 'little leaf or 'rosette' is caused by zinc deficiency. Small, very narrow and light coloured leaves develop in small clusters or 'rosette'. Bould *et al.* (1949) found spraying of zinc sulphate at 0.1 to 0.2 per cent early in the season as effective in correcting the condition, but sometimes it caused foliage injury. Soil application of zinc is not useful because it gets fixed in the surface soil.

Boron deficiency is generally expressed by the development of shallow depressions on the fruit surface mostly around the calyx end and corky tissues (Keinholz, 1942). Severely affected fruits generally crack without showing noticeable cork in early season but well-defined pitting develops in late season. 'Blossom blast' another disorder is also caused by boron shortage. Blossoms wilt, become dry and fail to abscise persisting until the following year. Both in the orchard and on a particular tree blast attacks randomly affecting only 10 to 12 per cent of trees. In solution culture Yokomizo (1977) found depressed tree and fruit growth due to lack of B.

Boron requirements of pear trees can effectively be supplied as foliar sprays, preferably in late summer or early fall. Batjer *et al.* (1953) reported that foliar spray of B in the fall increased its concentration in leaves, shoots, flowers and fruits in the following season. Foliar treatments with 0.4 per cent borax solution before, during and after flowering showed good control of deficiency and enhanced tree growth and productivity (Delcheva and Makariev, 1982). Soil applications are also effective in alleviating B deficiency symptoms provided the soil is not dry and its temperature is not low (Whiting *et al.*, 1978). Soil application of boron at 400 mg/l/tree after harvest resulted highest fruit set, fruit retention and yield in the following year in 'Le-Conte' pear. The treatment also improved the physio-chemical properties of fruit and leaf contents of N, P, K, Ca, Mg and B (El-Shazly and Kotb, 2011).

The symptoms of B toxicity, such as, shoot die back, chlorosis of leaf midribs and distorted apical leaves have been reported by Choe *et al.* (1986), in soils having higher B content (0.56 ppm) than the normal soils (0.14 ppm). As high as 120 ppm B was recorded in the leaves from affected trees compared with only 39 ppm in the leaves from normal trees.

When soil pH is higher than 7, deficiencies of Fe and Mn occur. The leaves turn yellow with a fine network of green viens when Fe is in shortage as a result of lime-induced chlorosis. Aocording to Boynton and Oberly (1966), the situation of slow availability of Fe in soil and immobilization of Fe in the trees cause the lime-induced chlorosis in pears. Soil injection of an iron chelate solution is the most effective means of correcting Fe deficiency because foliar sprays damage the leaves (Benson, 1963). Though trunk injections of both Fe 330 and Fe 504 at 1 per cent or 2 per cent are equally effective in reducing chlorosis, yet Fe 504 is cheaper and easier to use and less phytotoxic (Reil and Beutel, 1978). Raese and Parish (1984) also reported that trunk injection of 1 per cent ferrous sulphate in the fall and spring improved tree performance and increased the level of Fe and green colour in the leaves and fruits of Anjou and Bartlett pears. Foliar sprays of various iron compounds were, however, effective only in rectifying the chlorosis for one year. In a study of Fe, Mn and Ca status of Bartlett and Le Conte leaves, El-Gazzar *et al.* (1974) obtained a higher level of Fe in Bartlett leaves than Le Conte leaves. The range of Fe content in Le Conte leaves was 60-193 ppm (Taha *et al.*, 1974).

Symptoms of Mn deficiency are observed on the older leaves only. The interveinal chlorosis pattern is similar to apple (Wallace, 1961). The leaves become yellow or golden without any reduction in size. The leaf tissues adjacent to the major veins remain green while the tissues between the veins turn yellow. Taha *et al.* (1974) reported 8-38 ppm range of Mn in Le Conte pear leaves. If the values are lower than 25 ppm, it indicates Mn deficiency which can easily be corrected by manganese sulphate sprays, though seldom required.

Irrigation

If there is not enough moisture in the soil and rains do not occur during the summer season when shoots and young fruits grow actively irrigation of pear trees becomes essential. Water deficiency at this stage restricts the growth of shoots and fruits, ultimately reducing the yield. In pear an accurate response to daily water demand of the trees is necessary for an optimal fruit size and production. Water stress caused smaller fruit size and yield decline (O'Connel and Goodwin, 2007). However, during cetain stages, *e.g.* during shoot growth period, it is possible to reduce the irrigation supply without any great loss (Anconelli and Mannini, 2002). In Belgium, pear trees are mainly grown on loam and sandy loam soils with high water storage capacity. Sensitivity to water stress and limitations for deficit irrigation under these conditions was shown by Janssens *et al.* (2011a). The core breakdown of fruit upon ripening after cold storage was, however, less when the tree received reduced water supply in the summer (Ryall and Aldrich, 1937). On the basis of his studies on Japanese pear, Endo (1975) reported that optional watering in the hot dry conditions gave better size, appearance and quality but slightly later fruit maturation than no watering, and unwatered fruits were higher in free acid content. He further pointed out that fruits in excessive irrigation were intermediate in size, poorest in total soluble solids and often displayed a skin disorder. Irrigation of cultivars William's and Passe Crassane increased the total shoot length, trunk girth, crown size, fruit yield and fruit size (Radojevic *et al.*, 1979). Likewise, the irrigation at the shortest interval of 13 days than either of 26 or 39 days interval enhanced vegetative growth and the production of the largest fruits in Le Conte pear trees (Tawfik *et al.*, 1978).

Mitchell *et al.* (1984) investigated the effects of regulated water deficits on growth and bearing of Bartlett trees and found that shoot and frame growth declined in proportion to the water deficit. The yield was marginally increased in the regulated deficit irrigation treatments. Higher level of irrigation to bearing trees of cultivar Beurre'd Anjou though produced larger and firmer fruit yet increased the incidence of alfalfa greening and cork spot (Raese *et al.*, 1982). Maximum fruit size distribution and high productivity in 'Conference' pear was noted by applying regulated deficit irrigation (Janssens *et al.*, 2011b). However, unpredictable summer rains and drought periods make it difficult to establish the optimal irrigation thresholds. Janssens *et al.* (2011b) developed a soil water balance model to monitor the evolution of soil water content in the rootzone on a daily basis. The input parameters for the model are soil water retention characteristics, rain, irrigation events and evapotraanspiration. They developed a specific algorithm for drip irrigation and tree transpiration in an orchard with weed free strips under the trees and grass strips between tree rows. A significant correleation ($R>0.80$) was observed between predicted observed soil water content.

Brun *et al.* (1985) applied different irrigation regimes and thereafter concluded that fruit disorders, especially, cork spot, were positively correlated with increased irrigation, shoot growth, fruit N and K concentration, fruit N: Ca and K: Ca, and early levels of fruit B, Fe, Mn, and Zn, but negatively correlated with Ca. The trees subjected to dry irrigation regime had a lower fruit set. In laboratory and field trials, Victoria, Vicar of Winkfield, Clapp's Favourite, Lousie Bonne and Lesnaya Krassavitsa have been found to be the most drought resistant and productive cultivars (Maksimova, 1986).

Irrigation Systems

Irrigation systems are designed and operated to obtain optimum crop yield and efficient water use, but other considerations, such as, type of soil, quantity of water available, the cost of initial installation, operation and maintenance help a great deal in making a final choice of a system. Good water management not only influences the yield and quality of the fruit, but also the mineral nutrition of the trees, incidence of pests and diseases, weed control and soil erosion.

Four irrigation systems, *viz.*, flood, furrow, sprinkler and drip, are generally used in pear orchards. Flood systems with some variations, *i.e.*, border check, ring basin, square basin and contour basin, are widely prevalent on relatively level and of medium or heavier texture. It is the least expensive system but uneven percolation and excessive water losses are its serious drawbacks. Sometimes 5-8 equally spaced furrows are formed between each two rows of trees. Water from a head ditch or pipeline runs in these rows at a speed that will not cause soil erosion. Skill and care are essential for uniform application of water from furrows because an unirrigated area of soil remains in the tree row.

Sprinkler system using light weight aluminium pipes and quick coupling devices is practised on uneven lands or very light textured soils. Sprinklers are either fixed head or rotating head or perforated pipe. Land grading and preparation are not required for this system but cost of installation is very high. Uniform water distribution, elimination of deep percolation and above all tree response have encouraged the use of this system. Describing the multiple uses of overhead sprinkler systems with pesticide and fungicide sprays and in application of liquid fertilizers. Akiyama *et al.* (1977) also reported an increase in fruit yield and quality. Furthermore, sprinkler irrigation of pear decreases orchard temperature but increases relative humidity, concentration of CO_2 in crown canopy, absorbing roots/cm^2, stomatal opening and photosynthesis (Gu *et al.*,1982). Consequently, fruit weight and total yield also increase. By this system, soil puddling, salt accumulation and frost injury to blossom can also be avoided. Applying water through micro-jets influenced the orchard temperature decisively. At higher temperature, (about 20°C), temperature reduction by 5-7°C is possible. The beginning of bloom in 'Bosc' pear could be delayed by about 10 days by micro-jet irrigation. The main effect is the diminition of the frost damage in the spring that assure yields (Lakatos *et al.*, 2011). However, orchard cooling with overtree mircro sprinkler irrigation increases fruit russet on 'Conference' pear (Asin *et al.*, 2011).

Drip/trickle or daily-flow irrigation is continuous or frequent application of water to a limited soil volume through small tubes or mechanical devices called emitters. In the situations of water shortage, excessive soil permeability, steep land, and scarcity of labour, this system is particularly useful. Clogging of emitter orifices is a very serious limitation of drip irrigation. Feyen *et al.* (1983) on evaluating this system, concluded that it is economical for 5 to 7 years out of 10 in cv. Commice and 3 to 8 years out of 10 in Durondeau. In young pear trees, the shoot growth was 40-72 per cent greater under drip irrigation than control but the early cultivar William's Bon Chretien responded best to irrigation at 70 per cent field capacity whereas the midseason cultivar Beurre Bosc and the late cultivar Beure'd Hardenpont at 20 per cent field capacity (Semash and Panasenko, 1984).

Cheng *et al.* (2012) reported that withholding of irigation from 120 days after full bloom (DAFB) to 150 DAFB increases the fruit TSS, soluble sugars and dry matter content in cv. Yali. The treatment also resulted in an increase in water use efficiency. Withholding of irrigation in this period of fruit growth can be used to save irrigation water without adverse effects on the quality of fruits. Marshal *et al.* (2012) recommended postharvest deficit irrigation for 'Conference' pear but also suggested that midday stem water potential (psi) should not be lower than-1.2 MPa. Pacheco *et al.* (2011) in an irrigation experiment under deficit irrigation (irrigation was applied when the pre-dawn leaf water potential (Psi PD) reached a pre-established threshold depending on the development stage of the crop at -0.5 to -0.8 MPA). The deficit irrigation (DI) caused a superior ratio of fruits above 65mm diameter, 10 per cent higher TSS when compared with Maximum Water availability treatment (MWA) in which daily irrigation of 265 mm water was applied. Water use was also recorded 63 per cent lower in DI compared with MWA. This irrigation treatment (MWA) caused an economic benefit of 2.73 Euro/m^3 of applied water, while this value was 7.51 Euro/m^3 in DI. Crop coefficients (KC) of pear cv. Abate Fetel was determined during the second growing season after planting. (Reguena *et al.*, 2011). Maximum monthly crop water evapotranspiration (ETC) was also determined in January with a value of 2.1 mm/day. This was found equivalent to 16.8 l/tree/day. The crop coefficients increased from 0.22 at full bloom to 0.36 at the middle of December and from their, they remained almost constant until the middle of February when they decreased up to 0.28 at the end of April.

Soil Management

Cover Cropping

In order to prevent soil erosion, improve soil texture and fertility and to obtain some economic return, cover crops are raised many a time. Fidanovski and Ristevski (1982) recorded 9.87 per cent increase in the yield of William's pear by growing a mixture ot grasses (*Lolium perenne* + *Festuca pratensis* + *Dactylis glomerata*) as cover crops over clean cultivation. Maidebura and Cherbinka (1983) assessed the effect of seven soil management systems on the growth: (i) clean cultivation (control), (ii) winter-sown green manure (rye) in the inter rows, (iii) spring-sown green manure (mustard) in the inter rows, (iv) summer-sown green manure (unspecified) in the inter rows, (v) natural sward (mown periodically), (vi) sward in the inter rows, clean cultivation in tree rows, and (vii) sward in the tree rows, clean cultivation in inter rows. The average yield was the highest in treatment (iv), followed by the control (i), and lowest in treatment (v). The commonly grown cover crops are legumes (sunn-hemp, soybean, pea, clover, vetch), mustard and cereals (rye, barley and oats). The choice of a cover crop will, however, depend on the soil fertlity, water supply and prevailing temperature. The tree basins should, however, be kept clean. Intercropping of radish did not affect the yield of a commercial pear orchard (Newman, 1986).

Mulching

Under rainfed conditions, where water scarcity restricts severely the growth and yield of the trees, mulching is often used to conserve moisture which also helps in controlling the weeds, erosion and temperature fluctuations. As regards to the yield, straw mulching is better than clean cultivation (Selimia and Keatley, 1970). Black polyethylene increased the yield of pear trees by 14 per cent over the clean cultivation and also enhanced trunk circumference, shoot length and number of bearing branches (Rahovic and Petrovic, 1977). Young pear trees grow more vigorously and bear more flowers in black plastic covered soil than in the grass covered soil (Frimanslund, 1984). Littua *et al.* (2009) in their experiment grows alfalfa (*Medicago sativa*) in the space between rows of pear orchards for five years. Results of 5 years showed that the production, fruit weight and soluble solids content of the alfalfa planted plot was increased by about 25 per cent, fruit weight by 30.8g, soluble solids content by 0.8 per cent. Ground cover management by using pink bark mulch, straw or straw plus cow manure was reported as an useful tool to satisfy the demand of trees because it improves the availability of soil nutrients, reduce weed competition and results harmonized nutrient ratios in leaves (Nagy *et al.*, 2011). Application of more nitrogen fertilizers has been advocated by Teskey and Shoemaker (1972) with mulching than sod culture.

Weed Control

In view of the many disadvantages of repeated cultivation, herbicides have been used instead for controlling weeds. The labour has also become very expensive. Besides controlling the weeds, simazine application in pear orchards increased the tree growth, yield and fruit quality (Jazbec, 1972). Ristevski (1977) compared the different soil management systems in pear orchards. Hoeing gave the highest yield while paraquat and mowing gave the next higher and similar yields but the herbicide was more expensive. In fact, this system is expensive to establish during the first few years. Once weed control is established, the costs per hectare are substantially lesser than with other systems. The tree basin should be kept free from weeds by spraying 6 ml paraquat per litre of water in April and thereafter in July.

Flowering and Pollination

Flowering

Pear trees begin flowering at the age of 5-8 years mostly on spurs and terminal shoots. Bartlett commences fruit production about 2-3 years earlier than Bosc whereas Seckel and Cayuga are rather late (Teskey and Shoemaker, 1972). Bearing habit of pear cultivars is influenced by the rootstocks, mainly for reproductive spur formation. Pasa *et al.* (2011) in their experiment with three cultvars (Carrick, Packham and Williams) on different Quince and *P. calleryana* rootstocks at Brazil, observed an inverse relationship between the percentage of vegetative and reproductive spurs, and between productive efficiency and vigour for all scion and rootstock combination. Floral bud differentiation in pear takes place in July while bloom occurs next year in spring (Jaumien, 1983a). Flower initiation depends on the levels of organic and inorganic storage compounds and endogenous growth regulators, notably cytokinins and gibberellins (Banno *et al.*, 1985; Jaumien, 1983b) but it has been hastened and increased by the application of SADH and paclobutrazol (Banno *et al.*, 1985; Badwai *et al.*, 1978; Dheim and Browning, 1988). Paclobutrazol is the most effective plant regulator in enhancing the flower bud numbers by suppressing the growth of l-year-old shoots (Krieken and Lindenberg, 1984) it increases floral initiation by decreasing apical dominance (Dhiem and Browning, 1988). Crisosto *et al.* (1988) showed that application of putrescine at anthesis increased yield efficiency in Comice pear along with improvement in fruit set, crop density, fruit size and seed number. In the indeterminate inflorescence of pear, lateral flowers open first and the terminal ones the last (Westwood, 1978).

Pollination

Pear fruit set is affected by a number of genetic and environmental factors and one of the most important genetic factors is incompatibility (Haring *et al.*, 1990). Most pear cultivars have been classified as completely or almost completely self-incompatible, while others are partially self-compatible, depending on the environment in which they are grown (Nyeki, 1996). Self-incompatibility (SI) is the mechanism that prevents self-fertilization in flowering plants. SI in rosaceous species such as pear, apple, cherry is gametophytically controlled by a single, multi-allelic locus (S locus) (Sanzol and Herrero, 2002). When a pollen grain lands on a stigma, a discrimination process takes place as to whether the S-alleles of the pollen matches one of the S-allels of the pistil. The pollen grain germinates on the stigma and grows toward the embryo but, if its S-alleles match one of the two S-alleles of the pistil, pollen tube growth is stopped in the style, and no fertilization takes piace. Therefore, the provision of pollinizers inter-planted in an orchard is required to produce an economic crop from most cultivars. Since, the pollinizers must be cross compatible with the commercial cultivar, the self-incompatible allelic constitution of both cultivars must be different in at least one of the alleles. Thus pollinators and pollinizers must be carefully chosen to guarantee optimal fruit set on any given cultivars, Gametophytic self-incompatibility (GSI) system in pears show the ability to molecularly S-Genotype cultivars of the Rosaceae family based on their S-RNase genes and ontheir SFB (S-haplotype-specific F-box) genes (indentified in Japanese pear but not in European pear). In many cases the cultivar couples in the orchards are genetically semi compatible and not fully compatible with each other. Semi-compatibility was shown to correlate with low yield. In Israel, the main European pear cultivar 'Spadona' is semi-compatible with its three pollenizers 'Coscia' 'Gentile' and 'Spadnochina' and provides low yield probably because of pollen deficiency. The wild pear *P. syriaca* is expected to carry different S-genotypes than the European pear cultivars and therefore could be a "Universal Pollenizer". Goldway *et al.* (2008) examined the fruit set of 'Spadona' pollenized by *P. syriaca*. The fruit set of 'Spadona' was similar to that when it was pollinated by a fully compatible European pear cultivar, and significantly higher than when it was pollinated by semi-compatible cultivars. Beurre Hardy, Doyenne du Comice and Flemish Beauty are self-fruitful, whereas Le Conte, Magness and Winter Nelis are self-unfruitful and majority of other commercial cultivars are partially self-fruitful. Bartlett and Anjou are virtually self-sterile but they behave as self-fruitful under certain conditions because of the production of parthenocarpic fruits. The extent of self-fruitfulness is determined by the location, season and tree vigour. Even under optimum plant vigour and weather conditions, cross-pollination results in a higher fruit set. Furthermore, parthenocarpic fruits are more prone to pre-harvest drop than the seeded fruits.

In a study on self-incompatibility reaction of Japanese pears, Hiratsuka *et al.* (1985) obtained 20-26 per cent fruit set by self-pollination and 90-100 per cent by cross pollination. The fruits produced from self and cross pollination had similar characteristics except that the fruits from self pollination were more malformed and slightly smaller in size. Cross pollination has been found advantageous in various commercial pear cultivars and best pollenizers for them have also been indentified (Seth and Kukshal, 1981). Almost all important cultivars producing viable pollen are compatible with each other. The cultivars Waite and Magness are completely male sterile and obviously cannot function as pollenizers. Bartlett and Seckel, which are genetically cross-incompatible: neither can be used as a pollen source for other. Kieffer is a poor pollen producer and blooms too early to pollenize Bartlett and most other common cultivars.

Apart from the compatibility with the commercial cultivar, a pollenizing cultivar should have sufficiently overlapping bloom period, bear fruits of commercial value and be blight resistant. When more than two cross-compatible cultivars are grown together, they act among themselves as pollenizers without posing a problem of placement. But when only one commercial cultivar is under cultivation, the placement of pollenizer becomes crucial. In such situation, planting of every fourth tree in every fourth row as pollenizer is adequate. It is further desirable to place two colonies of bees per hectare on opening of first blossom for ensuring transfer of pollen from one cultivar to another (Teskey and Shoemaker, 1972). Pear flowers produce less nectar than apple flowers. Bees often work pear bloom only during the first day or even the first few hours after hives are placed in the orchard. Faoro and Orth (2011) stated that pear flowers produce nectar with low amounts of sugar (10-15° Brix). This makes the flowers less attractive to pollinating insects and it is necessary to use a high percentage of polinizer cultivars (from 10-20 per cent).

Pollen viability and its germmation both *in vivo* and *in vitro* are reduced by fungicidal sprays at bloom (Marcucci *et al.*, 1980). The best medium for pollen germination is 15 per cent sucrose, 20 ppm boric acid, and 1 per cent agar (Filiti and Marcucci, 1982). Herrero and Gascon (1987) found that the life span of embryo sac in pear flowers was prolonged by about 10 days with cross-pollination or GA_3 treatment in comparison to self-pollination. It indicates that pollination stimulus can be mediated through GA_3. Seeley *et al.* (1987) accelerated the bloom of pollenizer more effectively with polyethylene hot caps than with reflector boards. The pollenizers also influence the cell size of the fruit flesh in pears (Vujanic-Varga, 1985). The largest flesh cells were obtained when William's was pollinated by Doyenne d Hiver and Passe Crassane by William's. The larger flesh cells make the fruit longer but less firm.

Fruit Set

Generally, fruit set is not a problem in commercial pear cultivars, being as high as 73.2 per cent in cross-pollination and ranges from 17.3 per cent to 32.6 per cent in open-pollination (Dhaliwal *et al.*, 1982; Seth and Kukshal, 1981). Endogenous plant regulators, *viz.*, auxin, gibberellin, cytokinin, abscisic acid and ethylene play an important role in the fruit set which, however, is not clearly understood. Martin *et al.* (1982) investigated the changes in hormone content of pear receptacles from anthesis to shortly after fertilization as affected by pollination and GA_3 treatment. They concluded that IAA and GA levels in pollinated and GA_3 treated flowers appeared to signal subsequent rapid receptacle growth, but otherwise no consistent correlation between receptacle growth and endogenous growth regulator concentration was found.

Among the various growth regulators, the role of gebberellin is definitely more important in the setting of pear fruits because many experiments with gibberellin sprays on the flowers have shown an enhancement in fruit set. Several gibberellins have been detected in the immature seeds of cultivar Shinseiki and out of which GA_4 is the principal gibberellin while GA_3 could not be traced (Nakagawa *et al.*, 1979). Gibberellin application increased the yield, particularly after flower damage by spring frost, and should be used as a curative rather than a preventive measure (Flick and Hermann, 1984; Buban and Borka, 1979; Modic and Kodric, 1978). GA_{4+7} or GA_3 at 10-20 ppm concentration was applied 10-14 days after full bloom or 28-36 hours after frost. Sansavini *et al.* (1986a) found an increased fruit retention with gibberellin treatment but it reduced flower bud differentiation in the following year. Gibberellin treatments enhanced the proportion of parthenocarpic fruits, especially in cross-pollinated trees and, interestingly, GA_{4+7} showed more pronounced effect on self-pollinated trees and GA_3 on cross-pollinated trees. Application of exogenous GA_{4+7} is most effective GA combination in inducing parthenocarpic fruits or maintaining the fruits with excised embryos on the tree until harvest (Nakagawa *et al.*, 1979; Yudil *et al.*, 1981). Exogenous GA_3 was not effective in this respect. In addition, the biologically active GA-like substances extracted from immature seeds of *P. serotina* has been tentatively identified as GA_4 (Nakagawa *et al.*, 1979) while Martin *et al.* (1977) have identified GA_{17}, GA_{25} and GA_{45} in *P. communis* and indicated that GA_{45} was the only biologically active GA among them. Yuda *et al.* (1983) were able to induce parthenocarpy in two cultivars of Japanese pear when GA_{4+7} (500 ppm) in combination with or without other plant growth substances were applied on the emasculated and decapitated (stigma removed) flowers in full bloom and 3 weeks after. Parthenocarpic fruits were usually much smaller and showed a somewhat elongated obovoid shape compared to open-pollinated fruits (Table 5.14). A similar tendency was observed in both *P. communis* and *P. ussuriensis* and the parthenocarpic fruit of *P. ussuriensis* became extremely elongated obovoid shape as compared with open-pollinated fruit (Table 5.15). Dheim and Browning (1987) stated that the embryo sac development during the first 45 days after full bloom appeared to depend upon gibberellin synthesized within the ovular region, without which embryo sacs degenerated and rendered the flower incapable of setting in response to exogenous gibberellin.

Beside the gibberellin, fruit set has also been augmented by the application of SADH, CCC, 2, 4, 5-TP, and boron sprays (Degman, 1953; Griggs, 1968; Watts and Villiers, 1979; Stahly and Williams, 1976.) The cultural practices, such as, pruning, girdling, nutrition and limb positioning also promote fruit set. Spreading limbs to the 45° position with the trunk is the most effective practice. Fruit drop of parthenocarpically develop Bartlett pears before harvest can be prevented by spraying 5-10 ppm NAA. Torres *et al.* (2011) evaluated the effect of Maxim Reg (3, 5, 6-TPA) in preventing pre-harvest fruit drop in pear cvs. Conference and Blanquilla. Maxim Reg at 10 to 30 g/hl was applied at 3, 4 and 5 weeks before harvest (wbh). MaximReg, effectively reduced fruit drop in both the cultivars and the efficacy showed a tendency to increase with dose application and with application times closer to the harvest period. Spraying of AVG at 250 mg/l applied two weeks after full bloom can increase fruit set and yield in European pear cvs. Abate Fetal and Packham's Triumph (Sanchez *et al.*, 2011)

Table 5.14: Fruit Characteristics of Parthcnocarpic *Pyrus serotina* at Maturity (GA_{4+7}, GA_3: 500 ppm, NAA: 10 ppm, BA: 20 ppm Applied at Bloom (1st) and 3 Weeks after (2nd)*

CV	Treatment		Fruit Set (per cent)	Width (mm)	Length (mm)	W/L	Fresh Weight (g)	Sugar (per cent)	Acid (per cent)
	1st	2nd							
Shinseiki	GA_{4+7}+NAA+BA	,	66	50.82	52.58	1.03	78.08	12.52	2.78
	GA_{4+7}	NAA	60	49.43	48.75	0.99	69.61	13.04	2.69
	GA_{4+7}+NAA	BA	75	46.67	48.38	1.04	66.63	12.94	3.05
	GA_3	-	0	-	-	-	-	-	-
	NAA	-	0	-	-	-	-	-	-
	Control		0	-	-	-	-	-	-
	Open		-	65.51	53.34	0.81	131.71	14.00	2.35
Chojuro	GA_{4+7}+NAA+BA	,	39	61.13	61.53	1.01	136.36	12.38	3.08
	GA_{4+7}	NAA	29	63.65	62.35	0.98	141.87	12.90	2.33
	GA_{4+7}+NAA	BA	21	50.70	57.80	1.02	94.38	11.80	1.62
	GA_3	,	0	-	-	-	-	-	-
	NAA	,	0	-	-	-	-	-	-
	Control		0	-	-	-	-	-	-
	Open		-	74.34	60.84	0.82	190.60	12.58	3.78

* Yuda *et al.* (1983).

Table 5.15: Fruit Characteristics of Parthenocarpic *P. ussuriensis* and *P. communis* at Maturity ($GA_{3,}GA_{4+7+}$,GA_{45}: 500ppm applied at bloom and 3 weeks after)*

CV	*Treatment*	*Fruit Set (per cent)*	*Width (mm)*	*Length (mm)*	*W/L*	*Fresh weight (g)*	*Sugar (per cent)*	*Acid (per cent)*
Pyrus ussuriensis	GA_3	87	66.27	92.48	1.40	175.12	14.93	4.49
	GA_{4+7}	73	64.08	92.96	1.45	172.24	15.29	4.62
Yali	Control	0	-	-	-	-	-	-
	Open	-	78.66	88.50	1.13	281.52	15.03	2.82
Pyrus communis	GA_3	42	89.76	91.15	1.02	402.33	13.56	3.72
	GA_{4+7}	45	90.52	87.71	0.97	404.56	13.23	4.11
Passe crassane	GA_{45}	46	88.28	87.19	0.99	393.56	13.73	5.52
	Control	0	-	-	-	-	-	-
	Open	-	97.04	89.10	0.92	472.38	14.38	3.98

* Yuda *et al.* (1983).

Bagging of Fruit

Fruit bagging to improve fruit appearance and reduce pesticide residues is rapidly being adopted in pear production, particularly in China (Shaomin *et al.*, 2002). Various kinds of bags are used and the time of bagging also effect quality of fruits. Bagging of fruits from an early stage of fruit growth to until harvest by triple layer paper bags showed no negative effects on fruit size, weight, firmness and soluble solids concent of 'Concorde' pear. However, the appearance of bagged fruit was found less attractive. Their colour was muddy yellow, and their extent of russeting was much higher than unbagged fruit (Hudina and Stampar, 2011). DongSheng *et al.* (2011) evaluated different types of bags on cv. Cuiguan. They observed that using bags with yellow outer and white inner the fruits at harvest keep intrinsic green colour of peels, and flavour. However, the fruit firmness by using this bags decreased (6.30 kg/cm^2) compared with yellow outer and black inner (7.34 kg/cm^2) and control (6.65 kg/cm^2). Fruit bagging however, affected accumulation of polysaccharide and lipid, and a close relationship was also observed between the decrease contents of polysaccharide and lipid and occurrence of browing spot (YingTao *et al.*, 2011). Hudina *et al.* (2012) reported that bagging European pear cv. Concorde after June drop until harvest decreased the content of phenolic compounds in skin (catechin, chlorogenic acid, epicatechin, *p*-coumaric acid, quercetin 3-0-galactoside, quercetin 3-0-glucoside, quercetin 3-0-rhamnoside). The removal of bags 7 days before harvest significantly increased glucose, shikimic and fumaric acids.

Thinning of Fruits

Thinning is done to obtain the fruits of good size and quality and to stimulate floral bud formation for the next year's crop. Some pear cultivars, such as, Bartlett, Hardy, Bosc, Winter Nelis and Anjou often set very heavy crops requiring thinning to avoid limb breakage and to improve fruit size and quality. The degree of thinning, however, depends upon the tree vigour, growing conditions and the fruit set. Teskey and Shoemaker (1972) suggested retention of 30-40 leaves per fruit in all cultivars for better fruit growth. High fruit densities often result in poor light distribution within the cluster and low assimilate import to fruit. Thinning of fruit therefore result in better light distribution and higher nutrient import by the remaining fruit, thus improving colour development in fully coloured and blushed cultivars. Thinning is done by hand, mechanical and chemical methods. All small fruits grow at about the same rate as the large ones, thus remain smaller even at harvest. Therefore, selective removal of smaller and deformed fruits is done regardless of spacing in hand thinning about 6 weeks after bloom. Mechanical thinning is performed either by a direct blast of high pressure water or a tree shaker but this method is not so useful. Chemical thinning is mostly followed on account of lower cost. Ammonium Thiosulphate (ATS) which is used as a foliar fertilizer, is also used for flower thinning (Bertelsen, 2002). ATS was found effective in flower thinning by desiccating vital female organs in 'Conference' 'Winter Cole' 'Clara Frijs' and 'Packham's Triumph' (Wertheim, 2000, Bertelsen, 2002, Bound and Mitchell, 2002, Bound and Jones, 2004). ATS is usually used at 2-3 per cent spray and was found more effective on flowers that have reached anthesis at the time of application (Bertelsen, 2002). In organic fruit production system, lime sulpher (LS) is used as a bloom thinning agent. Garriz *et al.* (2007) reported that in 'Abate Fetel' pear, spraying of LS at 7 per cent at 30 per cent bloom stage reduced fruit set and increased final fruit weight by 17 per cent, without affecting fruit quality. Ethephon is used in many countries as a flower thining chemical. Spraying of ethephon at 400 mg/l at full bloom stage adequately thinned flowers of 'Winter Cole' pear. (Bound *et al.*, 1991). Bonghi *et al.* (2002) working with 'Conference' and 'Rosada' pear, observed that spraying of naphthylacetamide (NAD) at 15mg/l applied at 5 days after full bloom reduced crop load and increase fruit size. Benzyladenine (BA) is an effective thinning agent for 'Williams' pear, the most important variety of Argentina. Spraying of BA at 150 or 200 mg/l with silicone surfactant (Silwet Reg), when the fruit diameter were 10-12 mm; reduced fruit set by 20-40 per cent and fruit number by 10-35 per cent compared with control. At harvest fruit weight were improved by 10-20 per cent, yield of small fruit (diameter < 65mm) was reduced by 40-65 per cent and the yield of large fruit (diameter > 70mm) increased by 60-100 per cent (Curetti, *et al.*, 2011). Similar effect of BA in fruit thinning was observed in South Africa on pear cv. Early Bon Chretien (Theron *et al.*, 2011). Spraying of 150 ml/l BA significantly reduced fruit set and improved return bloom in the following season.

Fruit Growth and Maturity

Cumulative fruit growth in pears, measured as fresh weight, usually follows a sigmoid curve. Fruit growth and development can be divided into two distinct, but overlapping, phases; cell division and cell expansion. The first phase, *i.e.*, cell division in 'Bartlett' pear lasts for between 6 and 8 weeks after full bloom (Sterling, 1954). Length, diameter, and weight of pear fruits increase from fruit set to maturity. (Mann and Singh, 1988). Since the fruits are usually harvested at mature green stage, the third phase of growth does not occur (Westwood, 1978). Fruit weight and volume advance with a corresponding decline in specific gravity during fruit growth. The growth rate of fruit is determined by the cultivar, soil moisture, heat stress, fruit set, thinning and growth regulator sprays. Biochemical changes in the growing pear fruits have been investigated by Mann and Singh (1986). Sugar content increases throughout the growth period, being maximum nearing the maturity and then drop in ripe fruits (Fazekas and Graeza, 1970). Total phenolic contents are high up to 20 days after fruit set gradual decline occurs subsequently. Acidity shows a declining trend during most of the stages of fruit growth (Mann *et al.*, 1978).

Determination of fruit maturity is very important for successful pear growing which is a definite period between the final stages of fruit growth and commencement of ripening. The fruits are picked in this period depending upon their ultimate use. Immature fruits on picking do not attain full size, lack sweetness and flavour, shrivel quickly in storage and are susceptible to stroage scald. In contrast, overmature fruits on picking ripen rather quickly, are gritty in texture, and have a short storage life, and are prone to core breakdown and scald during storage (Avelar, 1984). Therefore, a number of maturity indices have been assessed to fix the standards for the optimal maturity.

The factors evaluated for determining the maturity of pear fruits are flesh firmness, skin colour, sugar content, titratable acidity, sugar/acid ratio, total soluble solids, starch content, appearance of 2-4 decadienic acid, corking of the lenticels, fruit size, occurrence of fruit drop, case of separation from spur, number of days from full bloom and heat unit accumulation during specific periods of growing season (Ozelkok *et al.*, 1988; Kvale, I 986; Palazyan *et al.*, 1983; Kajiura and Omura, 1982; Westwood, 1978; Strandzhev, 1975; Teskey and Shoemaker, 1972). The colour changes occurring as European pears develop can be used as a guide for picking, but are not considered a reliable enough index to be used alone (Wang, 1982). In Asian pears, ground colour is correlated positively with soluble solids concentration and negatively with fruit firmess (Kajiura *et al.*, 1975).

Many European pears will only commence ripening after a preconditioning treatment, often cool storage (Wang, 1982). Anumber of Asian pear cultivars are non-climaeteric (Kitamura *et al.*, 1981). For these fruit, ethylene measurements are of no use as a maturity index.

Flesh firmness as measured by a pressure tester is the most satisfactory single index of pear maturity. Its realiability can further be improved if total soluble solids, starch content and heat units are also taken into consideration simultaneously. A plunger of 0.8 cm head is forced into the flesh at two or three locations on the circumference of a pear fruit recording (the reading on scale). The firmness range for optimum maturity varies with the cultivar, climate, rootstock and the utilization of the fruit. For Bartlett pear, it is 10-7.7 kg in USA, but 6.7-5.3 kg in Yugoslavia (Westwood, 1978; Rudie *et al.*, 1977). The maturity indexes recommended for harvest of 'Packham's Triumph' pears firmness between 68 and 78 N and soluble solids from 10-11° Brix (Benitez, 2001), where as the recommendation for 'Rocha' pears is flesh firmness between 54 and 64 N and soluble solids 11 to 13° Brix (Silva *et al.*, 2005). Antoniolli and Czermainski (2012) working with 'Packham's Triumph' and 'Rocha' pears at the south region of Brazil reported that fruits of 'Packham's

Triumph' when harvested between 67 and 76 N flesh firmness showed firmness loss after 30 days of clod storage, whereas fruit harvested at 58N flesh firmness, retained the initial values, resulting in firmer fruit after 60 days of cold storage. Both the varieties of pears when harvested at flesh firmness of 58 N, showed better quality attributes after 60 days of cold storage (1 ± 1° C, 90-95 per cent RH) plus 5 days of shelf-life than fruit harvested at flesh firmness of 67N and 76N. Hot climate and Oriental rootstocks increase the fruit firmness at maturity. Total soluble solids can easily be recorded by a hand refractometer, and thus, is a useful index of maturity in conjuction with pressure test. Although fruit firmness is often used to quantify optimum harvest maturity of pears, quantitying optimum harvest maturity of 'Forelle' pears cannot rely on firmness only. Firmness has been found to be lower in a season with low summer temperatures, thus climate has to be considered when this parameter is used for prediction of harvest maturity. Bartlett pears should be considered mature to harvest starting at firmness of 8.6 Kg for CA storage and continuing to 7.7 kg for short term storage. Anjou pears start at 6.8 kg firmness for CA storage and continue to 5.9 kg for short-term storage. 'Bosc' pears start at 7.3 kg firmess for CA storage and continue to 6.4 kg for short-term storage. In all pears, it is important to monitor the starch level, because in certain seasons firmness will remain high for a time while the starch content changes. This starch change indicates a movement in maturity. Lotze *et al.* (2011) developed prediction model based on historical maturity indexing and hourly climate data.

DeEll and Slingerland (2012) from Ministry of Agriculture, Food and Rural Affairs, Ontario, Canada suggested the followings:

i) Harvest all pears destined to be stored for any duration at proper maturity, and handle them with care during harvest. Cool pears as quickly as possible and store then at proper temperature with high humidity

ii) Harvest Bartlett pears when firmness reaches 81-85.5 N (18-19 lb) and Bosc at 72-81N (16-18 lb). The later the pears are harvested and the softer the pear, the more likelihood this is of storage disorders developing.

iii) Harvest Harrow Crisp, Harovin Sundown and HW620 pears at around 58.5-67.5 N (13-15 lb) pressure.

iv) Using all best management practices to reduce bruishing and fruit injury in the field and during the packing process will help ensure good quality pears going into the storage. Maintain proper temperature and high relative humidity during storage to reduce disorders and diseases, and permit pears to be stored for longer periods of time.

Sugar/acid ratio is another good index for maturity (Palazyan *et al.*, 1983). When Bartlett fruits were harvested on different dates, optimum picking maturity was indicated by 3.9 to 3.5 values of starch reaction to iodine and the index proved practicable (Kvale, 1986; Ozelkok *et al.*, 1988; Rudic *et al.*, 1977). Kajiura *et al.* (1975) prepared a fruit ground colour chart of Japanese pears for studying the maturity and they have pointed out the calyx end to be the best position for evaluating the ground colour. Lenticels of immature fruits are white which turn brown on the development of cork cells giving an indication that the fruit, if picked, will ripe without shrivelling. Heat unit accumulation for 5 to 9 weeks following bloom can be calculated for each cultivar to find out the degree days required for harvest maturity (Westwood, 1978). Pear cultivars are also classified as early, mid-season and late, indicating the probable time of their harvest maturity.

In recent years non-destructive techniques are being evaluated for determining the maturity. Gomila *et al.* (2011) in studies with 'Williams' pear used spectroscopy in the Vis-NIR range, and established the AD index (absorbance difference) as the difference between the average value of absorbance between two points near the peak of chlorophyll absorbtion: 677 and 722 nm. They harvested fruits between 93 and 131 days after full bloom (DAFB) and ethylene production, traditional maturity indexes and AD index were determined on each date. Four significantly different maturity groups were characterized by their ethylene production pattern: immture (<99 DAFB), transition (102-105 DAFB), intermediate (107-124 DAFB) and late (>126 DAFB). The AD value decreased with advanced harvest and had a high correlation with ethylene production. They established that AD values between 1.45 and 0.85 correspond to the pre-climacteric, between 0.85 and 0.50 to the rising phase, and values below 0.5 for climacteric. It was concluded that AD index can be useful to identify the physiological change that occur during 'Williams' repening, with the advantage of being an instant and non-destructive determination.

The maturity of pear fruits can be hastened by the application of growth regulators. Treatment with SADH on Le Conte pear accelerated maturation by one week at 500 and 1000 ppm and by two weeks at 2500 ppm (Badawi *et al.*, 1978). Likewise, the maturity of the fruits of *P. serotina* cv. Kosui was advanced by 4 to 5 days with GA treatment alone, 7 to 8 days with ethephon spray alone and 9 to 12 days by the combined treatment (Takase *et al.*, 1982).

Harvesting

European pears are usually harvested at pre-climacteric stage and the maturity for harvest depends on their final destination. However, harvesting too early results in off-flavour development and poor coloured fruits where as late harvested fruit is more susceptible to decay than fruit harvested at optimum maturity. The dessert quality of 'Anjou' fruits harvested at optimum maturity remains good up to 150 days in storage but deteriorate after 90-day storage in late harvested fruits (Chen and Mellenthin, 1981).

Pear fruits are plucked individually by giving a gentle twist rather than direct pull. Two or three pickings at an interval of 3 to 4 days are better than single picking. In order to avoid bruising and stalk damage, the fruits need careful handling during the harvest. Mechanical harvesting of pear is still in experimental stage. A self-propelled platform (Argiles AF-8) was tested in four mature. 'Bartlett' pear orchards in Lake County, California in 2006 and 2007 to address productivity, post harvest quality, ergonomics and worker satisfaction, (Elkins *et al.*, 2011). Productivity was found much higher using platform-harvester compared with ladder crew. Platform harvested fruit had 57 per cent fewer stem punctures. Worker ergonomics and satistaction favoured the platform.

To secure better price and goodwill, proper grading of fruits is desirable. 'Bartlett' pears are graded as extra large, large, medium, and small with diameters of 8, 7, 6.5 and 5 cm respectively (Nauriyal and Bajwa, 1968). Sometimes, these grades are also known as Extra Class., Class I, Class II and Class III. All damaged and misshapen fruits should be excluded from grading. The fruits after wrapping need to be arranged in layers in the containers. Both bottom and top portions of the container should be cushioned. Label indicating the grade, cultivar and name of the orchard should be pasted, printed or stamped on the container.

Yield

The yield of fruits varies greatly with the cultivar, age of the plant, orchard management and the climatic conditions. A well-managed orchard of Bartlett pear gives about 30 to 40 tonnes of fruit per hectare.

Storage

Pears are stored in cold storage or controlled atmosphere storage to prevent the decay before consumption because the entire crop cannot be put to use immediately after the harvest. In fact, all efforts should be made to control the losses between harvest and utilization which is usually achieved by reducing the fruit temperature. It may be realized by the fact that the amount of ripening taking place in three days at 21°C and in one month at -1°C is the same (Westwood, 1978). Therefore, the fruits, immediately after the harvest are cooled to a core temperature of -0.6 to -1.6°C to arrest ripening. Precooling, however, becomes unnecessary if the fruits are to be consumed within few weeks after harvest. Hydrocooling reduces the incidence of shrivelling and brown core without affecting the weight loss or incidence of rot (Zerbini *et al.*, 1980). Very fast cooling rates are necessary to increase storage potential of 'Conference' fruits when harvested late (Stow, 1988). Forced-air cooling is the most common method of rapid cooling for pears, especially in areas of large production. This can cool pears five to eight times faster than traditional room cooling.

Mellingthin and Chen (1981) investigated the influence of simulated transit temperatures on softening and ripening of 'Anjou' pears. The fruits held for 3 days at 1.7 or 7.2°C had not shown significant softening for 4 weeks at a transit temperature of -1.1°C but fruits held for 3 days at 12.8°C softened from 6.0 to 4.0 kg after 4 weeks at -1.1°C. They further stated that higher the simulated transit temperatures, more the softening occurred.

The storage life of pears varies with the cultivar, picking maturity, orchard management, climatic conditions, time lapse between picking and precooling and temperature in storage. Storage life of the fruits of different cultivars ranges from 120 to 245 days at -1°C (Prusse and Sopalova, 1970). The fruits of cultivars Anjou, Winter Nelis and Easter Beurre can be stored for a period of 6 to 7 months. Fruits of early cultivars can be kept at 1-4°C for 18 to 33 days, mid-season cultivars for 32 to 70 days and late season for 52 to 158 days (Gherghi *et al.*, 1972). Likewise, Mukherjee *et al.* (1980) were able to store fruits of cultivars Le Conte, China and Nakh at 0-1.7°C and 85 per cent relative humidity for 10, 10 and 14 weeks, respectively, but cultivar Smith showed internal browning only after 6 weeks of storage. 'Blanquilla' fruits stored better and had the best quality when picked on 9 and 16 September than harvested on 8 August (Alique *et al.*, 1983; Canet *et al.*, 1983). Teixeira and others (1972-73) reported that the pears could be stored for 3 to 6 months at 0°C, 6 to 7 months at -1°C and only for 2-3 months at 4°C.Seedling rootstock imparts more storage life than Quince rootstock whereas high nitrogen application causes more loss of fruit weight through respiration during storage (Firuzeh and Ludders, 1979). The fruits grown in cooler climate have generally a shorter storing capability (Teskey and Shoemaker, 1972). Sharples and Stow (1986) suggested storage temperature of -1.0°C to - 0.5°C for air storage as well as for CA storage of commercial pear cultivars in the UK (Table 5.16).

Table 5.16: Recommended Storage Conditions for some Cultivars of Pear *

<table>
<tr><th rowspan="2">Cultivar</th><th colspan="2">Air Storage</th><th colspan="5">Controlled Atmosphere (CA) using a Scrubber</th></tr>
<tr><th>Tempe-rature</th><th>Terminate (Month)</th><th>Tempe-rature</th><th>Per cent Co₂</th><th>Per cent O₂</th><th>Terminate (month)</th><th>Notes</th></tr>
<tr><td>Beurre Hardy</td><td rowspan="5">-1.0°C to -0.5°C</td><td>December</td><td>-</td><td>-</td><td>-</td><td>-</td><td>1-4</td></tr>
<tr><td>Confe-rence</td><td>April</td><td>-1°C to -0.5°C</td><td><1</td><td>2</td><td>May</td><td>1-6</td></tr>
<tr><td>Doyenne du Comice</td><td>March</td><td>-</td><td>-</td><td>-</td><td>-</td><td>1-4</td></tr>
<tr><td>Williams Bon</td><td>November</td><td>-</td><td>-</td><td>-</td><td>-</td><td>1-4</td></tr>
<tr><td>Chretien</td><td></td><td></td><td></td><td></td><td></td><td></td></tr>
</table>

1. Set a second (over-riding) thermostat to cut out the refrigeration plant when temperature of air leaving cooler falls to -2ºC.
2. Ensure field heat is removed rapidly. Aim to cool to 4ºC within 2-3 days of loading and obtain recommended storage temperature within a further 7-10 days. If fruit cannot be maintained below 0ºC, terminate storage earlier.
3. The minimum storage temperature should be raised by 0.5ºC when soluble solids content of fruit is below 11 per cent.
4. Pears should be conditioned at a higher temperature after storage and before distribution and sale.
5. Picking dates are very important for pears when storing in CA. Use starch iodine test to assist judgement of maturity.
6. (a) In CA ensure that CO_2 level remains below 1 per cent and O_2 does not fall below 1.8 per cent. (b) Remove field heat before establishing low O_2 regime. (c) Unscrubbed CA storage is no longer recommended for pears.

* Sharples and Stow (1986).

Details of controlled atmosphere storage conditions and period of storage life of fruits of important pear cultivars are given in Table 5.17. The storage life of

'Anjou' pears could be further prolonged by 2 months in CA storage (Wang and Mellingthin, 1975). The fruits are treated with 12 per cent CO_2 immediately after the harvest for 2 or 4 weeks and then brought to normal condition (1.5 per cent O_2 + 1.0 per cent CO_2). In another experiment, Alique *et al.* (1983) increased the storage period of 'Blanquilla' fruits by one month with changing controlling atmospheric conditions over the constant ones. Wax emulson and polymeric coating of fruits help in maintaining the firmness, titratable acidity and green colour but adversely affect the ripening of 'Bartlett' fruits by causing blotchy apperance when held at ripening temperature (Meheruik and Lau, 1988). Storing of some pear cultivars has usually included a reliance on postharvest chemical treatments, *e.g.*, ethoxquin or diphenylamine, to control superficial scald. The current trend towards elimination of pre- and postharvest synthetic chemical treatments in Europe and other markets necessitates a change to reliance on other, preferably non-chemical, methods to achieve scald and quality control, which can be achieved with storage at very low O_2 levels. Wide-spread adoption, however, has not taken place due to concerns of anaerobic damage when the pear furit are held below the lower oxygen limit (LOL) for long periods of time. Dynamic Controlled Atmosphere (DCA) storage is able to identify the LOL of pears using chlorophyll fluorescence. It is termed 'Dyanamic' CA since it allows the storage manager to customize the O_2 concentration at the beginning of storage and change it during the storage as LOL changes. DCA is appealing because it: (1) is non-chemical, (2) uses existing CA technology, (3) can be monitored electronically in real-time, (4) extends the storage time of fruits, (5) control superficial scald and other storage disorders, and (6) can warn the operator of equipment malfunctions, *e.g.*, refrigeration and CA equipment failure. The principal of DCA is based on the discovery that a sudden increase in the fruits fluorescence occurs at the LOL. If the O_2 concentration is increased, by as little as 0.2 per cent above the LOL, the fruit can be safely stored. This has allowed fruits to be stored below 1.0 per cent O_2 and as low as 0.4 per cent O_2, thereby achieving both longer storage life and superficial scald control in scald-prone fruit. Its (DCA) use is increasing at more than 40 per cent per year. It is useful for extending the marketing window of pears and as an alternative to ethoxyquin for controlling superficial scald in pear fruit (Prarge *et al.*, 2011)

I-Methylcycloropene

I-Methylcycloropene (1-MCP) is a competitive inhibitor of ethylene action, as it binds irrereversibly to ethylene receptors to regulate plant tissue responses to ethylene. Therefore, treatment with 1-MCP can delay fruit ripening and senescence, as well as associated processes such as respiration, ethylene production, colour changes, softening, acidity less, and volatile production (Watkins, 2006). 1-MCP has been shown to inhibit ethylene production, retard respiration rate and delay or prevent softening in various pear cultivars (Kubo *et al.*, 2003; Ekman *et al.*, 2004, Trinchero *et al.*, 2004, Mwaniki *et al.*, 2005, DeEll and Ehsani Moghaddam, 2011, Villalobos-Acuna *et al.*, 2010).

Table 5.17: Controlled Atmosphere Storage Conditions of Important Pear Cultivars.

Cultivar	*Temperature (°C)*	*Oxygen (per cent)*	*Carbon-dioxide (per cent)*	*Storage Period (months)*	*Reference*
Anjou	11	1.5	1.0	8	Mellingthin *et al.* (1980)
Bartlett	-0.5	(.3	1.5	2	Truter (1987)
Blanquilla	0.5	3.0	5.0	8	Alique *et al.* (1983)
Comice	0	5.0	5.0	6	Leblond (1986)
Conference	-0.5-0	2-3	0.5	8	Schaik (1986)
Kaiser	0-1	4-5	2-3	5	Zerbini *et al.* (1980)
Krystalli	0	3.0	5.0	6	Karaoulanis (1984)
Packham's Triumph	0	5.0	3.0	7	Plestenjak *et al.* (1986)
Passe Crassane	4.5-5.0	3.0	4.0	5	Artes (1980)

The effect of 1-MCP, however varies with the cultivars. In 'Blanquilla' and 'Conference' pears treated with 1-MCP, often remain 'evergreen' meaning that their ripening process is blocked and does not resume after removal from cold storage (Chiriboga *et al.*, 2010). In 'd'Anjou pears, 1-MCP treatment slowed down the whole metabolic system of fruit including inhibititing the biosynthesis of alpha – farnesene and the treated fruit lost ripening ability (Bai, 2010) 'Bartlett' pear is a climacteric fruit that is harvested at 80-89 N when the fruit is still green and firm but physiologically mature. Villalobos-Acuna *et al.* (2010) tested prehervest application of 1-MCP on California 'Bartlett' pears at 80N maturity at 0-106 mg/l. They found that 1-MCP was more effective in retarding colour, softening, and ethylene production during ripening than delaying fruit maturation on the tree (loss of firmness). The effect of 1-MCP on ripening was lost if fruit remained on the tree 21 days after 1-MCP application or after the fruit were stored for 3.5 months in cold storage. DeEll and Ehsani-Moghaddam (2011) investigated the effect of postharvest 1-MCP treatment timing on the ripening and physiological disorders of 'Bartlett' pears during cold storage and shelf life. Pears were held for 1, 3 or 7 days at 3°C after harvest and then treated with 0.3µl/l 1-MCP for 24 hours at 3°C. The most substantial benefit of 1-MCP observed was the marked reduction in disorders, especially senescent scald and internal break down. The result suggested that 1-MCP treatment 3 days after harvest provided the best balance of reduced disorder development during storage and the ability of 'Bartlett' pears to soften adequately thereafter. However, postharvest application of 1-MCP may not always respond same which was found more effective in relatively cooler growing regions.

Several fruit disorders cause decay in the storage. Brown heart in CA-stored pears is accelerated by early harvest, larger fruit size and low O_2 and high CO_2 levels which can be reduced considerably by delaying the harvest, increasing the levels of O_2 with concomitant decreasing CO_2 level in the storage (Yoshida *et al.*, 1986; Chen and Borgic, 1984; Pratella and Bertolini, 1979). To control soft scald during CA storage, pear fruits need a preharvest treatment with 4-5 per cent $CaCl_2$ solution because there is an inverse relationship between the extent of soft rot and calcium content of the fruits (Pratella and Bertolini, 1979).

Ripening

Ripening pear fruits exhibit a typical climacteric pattern of respiration (Kitamura *et al.*, 1981). Among the various cultivars studied, Bartlett had the highest respiration and ethylene evolution. Ethephon application (100-300 ppm), therefore, hastens ripening by about two weeks, increase, colouring but reduces frimness (Mann and Singh, 1986; Kimura *et al.*, 1977; Nicotra *et al., 1974-1975*).

Most pear cultivars ripen and develop the best dessert quality at 15.6° to 21°C and relative humidity of 80 to 85 per cent. Pekmezci and Kaska (1975) reported better ripening of 'Ankara' pears at 25°C and 85 per cent to 90 per cent relative humidity in 3-6 days as compared with 12° to 15°C and 50 per cent to 60 per cent relative humidity in 9 to 15 days. Prange *et al.* (1988) compared the consumers acceptance and weight loss of the fruits of many cultivars stored in simulated shelf conditions (5°, 18°, 22° and 26°C). The quality of fruits was the poorest at 5°C. 'Anjou' attained higher quality at 18° than at 22° and 26°C. Weight loss increased with rising temperatures in all the cultivars. The fastest ripening temperature for 'Barlett' fruits was 23.8°C but excellent quality developed at 15° to 20°C. High ripening temperatures (above 26°C) may impair the flavour and texture of the fruits as they become mealy and fail to ripen. Species within the genus *Pyrus* exhibit different softening characteristics during ripening. The European pear 'La France' softens dramatically during ripening; the Chinese pear (*P. bretschneideri*) 'Yali' produces ethylene but does not soften to the same extent and remains crisp, while firmness in the non-climaeteric Japanese pear (*P. pyrifolia* Nakai) does not change markedly (Hiwasa *et al.*, 2004).

Interestingly, many pear cultivars (Anjou, Bosc, Comice, Eldorado, Packham's Triumph), even on picking at optimal maturity, did not ripe without being stored at -1° and -2°C for weeks (Westwood, 1978). According to Lawes (1977), Bosc fruits, given low temperature pretreatment alone, attained optimal eating quality in 5-6 days whereas fruits not so treated failed to reach acceptable dessert quality during the holding period of 19 days. 'Anjou' fruits harvested at right maturity needed 60 days of postharvest chilling at -1.1°C for proper ripening whereas, late harvested fruits ripened with only 30 days of chilling (Chen and Mellingthin, 1981). When cold-stored fruits were compared with the fruits which remained continuously at 15°C, the latter had diminished intensity of respiration and very slow ethylene emission (Morin *et al.*, 1985). Instead of cold treatment, ethylene application caused normal ripening along with polyribosome and m-RNA synthesis and increased in galacturonase activity (Attibayeba and Hartmann, 1985; Drouet and Hartmann, 1979). Endogenous levels of l-aminocyclopropane-1- carboxylic acid (ACC) and ethylene in Eldorado fruits were low at harvest but increased rapidly on storing for 4 weeks at 0°C (Wang *et al.*, 1985). Thus, cold treatment was responsible for increasing ethylene synthesis thereby triggering the ripening process. Blankenship and Richardson (1985) investigated the biosynthesis of ethylene in d' Anjou pears during the cold requirement for ripening and they established that the capacity of converting ACC into ethylene was developed first during the 46 days at -1°C and the production of ACC and ethylene followed subsequently.

Ripening process has been delayed by the treatment of 'Bartlett' pears with aminoethoxyvinylglycine (AVG) and mannose. The ripening was postponed up to 112 days by giving a preharvest treatment to the Bartlett fruits, kept at 20°C, with 400 ppm AVG, but prolonged non-ripening storage might adversely affect the fruit quality (Romani *et al.*, 1983). Watkins and Frankel (1987) inhibited pear fruit ripening and the concomitant processes like softening of the flesh, ethylene evolution, respiration by vacuum infiltration of mannose into intact fruit of cultivar Bartlett.

Pear fruits have a definite period of storage life. If kept in storage for a longer period, they will not ripe normally even on removal to high temperatures.

Pests

Increasing period and intensity of pear culture have enhanced the pest infestation, warranting the use of costly pesticides. Injudicious application of pesticides not only aggravates the cost of production but also destroys the useful insects. Therefore, a number of pest control programmes, *viz.*, selective chemical control programme (Westigard *et al.*, 1986), integrated control (Mori and Sancassani, 1984), soft pesticide programme (Burts, 1983), natural control (Westigard *et ai.*, 1984) and sex pheromone baits have been under investigation from time to time. The applicability of these programmes in any locality, however, depends on the type of pests, extent and nature of damage, surrounding vegetation and useful parasites and predators found there. Since control programmes may change rapidly, consultation with the local plant protection agencies is always desirable. Some common pests of temperate fruits such as San Jose scale, defoliating beetles, root and stem borer, aphids and tent caterpillar also attack pears. Their control measures given on apple, peach and plum are applicable on pears as well. Information on specific pear pest is, however, given here.

Pear Psylla (*Cacopsylla pyri* L.)

All commercial varieties in Europe belong to the species *Pyrus communis* L. and they are all susceptible to *C. pyri*. The susceptibility increases when orchard techniques are aimed to maximize fruit production, such as high density orcharding large use of fertilizers and intense irrigation. Psylla is a serious pest of pear because, besides damaging the foliage and fruit, it is also associated with the development and spread of pear decline disease. Honey-dew secreted by these insects results into an accumulation of sooty mould which causes blackening of fruits and foliage. The other damages are prevention of fruit bud formation, premature leaf drop, injection of phytotoxins and transmission of pathogen (Adams *et al.*, 1981). Prolonged attacks and intense weaking by *C. pyri* may lead to plant death. The adults of *C. pyri* are responsible for transmission of the phtoplasma of 'pear decline'. The phtoplasma is acquired by the psylla upon feeding on an infected plant and transmitted to a healthy one by salivation during feeding on phloem (Carravo *et al.*, 1998). Among the European pear cultivars developed in the last 25 year, cvs. Cornia, Getica, Monica, Ina-Estival. Haydeea (all from Romania) were reported as resistance/

tolerance to psylla. The cultivar Spina Carpi is, however, known to be non-preferred for oviposition by psylla which may be associated with morphological and or environmental factors as this character is non-genetical (Quarta and Puggioni, 1985).

The hibernating adults become active as soon as the temperature rises above 4.4°C and start laying eggs in late February or March in small crevices above the buds and hatch in 10-30 days. The young nymphs feed on the opening leaves and soon cover themselves with the honey-dew they secrete. After going through five stages they moult into adults. Three to five generations are born in a year.

A great deal of research work has been done in various countries for controlling this pest. According to Riedl and Hoying (1980), two sprays of fenvalerate, prebloom as well as the foliar period at 6.7 and 13.5 g/100 I provide commercial control of psylla. The other pesticides which have been used for the same purpose are mineral oil diflu-benzuron, SN-72129, avermectin BI, amitraz, oftanel and fenoxycarb (Westigard *et al.*, 1986; Staubli, 1985; Burts, 1984; Badiali, 1978). The use of pesticides on large scale is however, detrimental to the useful insects which help in controlling the population of pear pests. Presently the defence against *C. pyri* is mainly based on integrated pest management (IPM), supported by natural control aimed to equilibrate the complex biological relationship of the field community. Autum treatment (at leaf fall) with mineral oil alone or associated to imidacloprid showed good control of adult winter forms (Civolani, 2000). Late winter treatment, to break down the population of females emerging from winter shelters and about to lay eggs, is presently based on broad spectrum pyrethroids. Pasqualini *et al.* (2003) however, reported good reduction of the number of eggs laid by spraying with pure mineral oil (dormant oil). In spring-summer, treatments against summer generation can be made with diflubenzuron and teflubenzuron. According to Westigard *et al.* (1984), parasites and predators in some unsprayed orchards were able to control the population of psylla to a desirable extent. But the successful biological control depends on many other factors such as the development of a non-disruptive control programme for *Cydia pomonella,* a reduction in tree vigour by the use of growth retardants reduction in nitrogenous fertilizer, summer pruning of water sprouts and terminal shoots, the use of overhead tree sprinklers to reduce honey-dew accumulation, manipulation of ground cover in the orchards to provide refuge for beneficial species, food spray to increase populations of predators and parasites and the availability of a monitoring services for orchard pests and useful species (Burts, 1984; Westigard *et al.*, 1984).

Codling Moth (*Laspeyresia (Cydia) pomonella*)

This is another very important pest of the pear which is injurious only in the larval stage. The pinkish-brown-headed larva is about 2 cm long which burrows into the fruit either through the stem or calyx ends or in the side where another fruit or leaf is touching it. After feeding on the seeds and core, the full grown larva again tunnels to the surface and leaves the fruit. Mature larvae overwinter in cocoons under the loose bark of the tree or in the debris on the ground. They pupate in the spring and adult moths appear in May. Moths mate during warm evenings (15.6°C) and lay eggs singly on furits, leaves and twigs. On hatching in one to two weeks, the larvae begin feeding in the fruit. There are one to three generations in a year.

Fenvalerate provides commercial control against both pear psylla and codling moth. According to Riedl and Hoying (1980), even a single well-timed first cover spray at 3.4 to 13.5g/100 litres gives sufficient protection. But Westigard and Gut (1986) advocate the use of insect growth regulator, diflubenzuron, which gives equal or better results than that achieved with standard phosmet. Diflubenzuron does not disrupt the non-target pests, the psyllid and mite, and can be sprayed at the rate of 27.2 to 54.5g/ha in late spring as the first summer treatment to coincide with the first flight of moths (Westigard, 1979). When the initial population of codling moth was low, it was significantly controlled through mating disruption with sex pheromone (Moffit and Westigard, 1984).

Oriental Fruit Moth (*Grapholitha molesta*)

The biology, nature of damage and control of the Oriental fruit moth are very much similar to that of codling moth. The fully grown worms hibernate in cocoons and several generations are encountered in a season. The pinkish white larva burrows into the fruit and its presence is evident from the brown frass that plugs the holes. The damaged fruit becomes unmarketable. Agrialoba (phosalone) at 0.2 per cent and Gordona (tetrachlorvinphos) at 0.1 per cent have been found as the most effective insecticides against this pest (Zapryanov, 1984). Sex pheromone baited traps can also be used for the Oriental fruit moth.

Pear Leaf Tortrex (*Rhopobota (Acroclita) raevana*)

Although leaf tortrix is not a major pest of pear, newly hatched caterpillars sometimes attack at bud burst stage causing considerable damage. Within a month, one generation takes place on pear. The pest can be controlled by spraying with 0.05 per cent of endosulfan or dichlorovos.

Orchard Mites

Many mites, such as, the European red mite (*Panonychus ulmi*), the brown mite (*Byrobia rubrioculus*), the two spotted spider mite (*Tetranychus urticae*) and the pear rust mite (*Epitrimerus piri*) are important pests of pear often causing serious damage. The mites usually feed in spring on new plant growth, *i.e.*, leaves, blossoms and fruits. Infested foliage bronzes and fruit skin russets. European red and brown mites pass the winter in egg stage on trees and hatching of eggs usually happens before the opening of blossoms. The young mites feed on the developing foliage. The spider and rust mites hide in winter under bark scales or in litter on the ground and migrate to new foliage in spring.

The mites multiply rather rapidly in warm weather producing eight or more generations in a year. This makes their control more difficult requiring measures to be taken before serious damage occurs. Pesticides are usually applied at delayed dormant, prepink, summer or at the postharvest periods. Gupta *et al.* (1975) got almost complete control on brown mite with a single high volume spray of carbophenothion or dimethoate. In controlling a cyhexatin resistant population of two spotted spider mites, clofentezine and flubenzimine proved superior to cyhexatin and oxythioquinox (Stone and Edge, 1984). The pear rust mite was controlled effectively with application of DNOC at 600g/hl at the bud stage, followed at petal fall 15 days later by one of 42 per cent dicofol at 150g/hl, or 2 applications of dicofol applied at the bud and petal fall stages (Laffi, 1983).

Diseases

Many diseases such as pink disease, stem black, stem brown, powdery mildew, *etc.*, are common to both pear and apple. Fire blight, leaf spot and decline are, however, specific to pear only.

Fire Blight (*Erwinia amylovora*)

This is the most serious disease which is a limiting factor for commercial growing of pears in many localities. The disease appears initialJy as blossom blight and later spreads to shoots, developing fruits and in severe cases to almost all parts of the tree. Trunk infection sometimes causes girdling which may kill the tree. The extent of damage depends on varietal susceptibility, weather conditions, cultural practices and control measures.

Infected blossoms suddenly wilt, and as the disease progresses down the pedicel, the tissues become water soaked and dark green in colour. Affected parts turn brown, then black and usually remain firmly attached to the tree, giving a fire burnt appearance. The bacteria exist either in the form of ooze or exudate usually associated with blighting blossoms and shoots during the growing season and may overwinter in cankers. Ooze and cells are spread by insects and rain splashes. Temperature above 19°C combined with high moisture in the form of rain, dew or high humidity (70-100 per cent) at least for 48 hours are most conducive for infection and multiplication of bacteria. Predictive models based on these factors have been developed (Reil *et al.,* 1979).

An integrated approach comprising orchard management, preventive measures and chemical sprays is required to control the disease. A good orchard sanitation is essential. Overstimulation of growth either by fertilizer application or pruning or irrigation is to be avoided. Drip irrigation can be preferred over sprinkler irrigation as the latter helps in disease spread. Infected spurs, crotches and trunks showing cankers should be removed during pruning by making cuts 20 to 30 cm below the visible blight symptoms. Water sprouts and root suckers need to be cut down routinely. Pruning tools should be disinfected between each cut to prevent possible transmission of the causal organism. Instead of growing extremely susceptible cultivars in the disease prone localities, blight resistant or tolerant cultivars should be planted (Janick. 1976).The Asian pear cvs. Chojuro Kosui, Olympic (Korean Giant) Seuri, Shinko, Shinsui, Singo and TseLi are considered resistant while Hosui, Kiusui, Okusankichi, Seigyoku, Nijisseki and Shinseiki are susceptible. The highly resistant European pear cvs are Honey sweet, Kieffer, Laconte, Magness, Moonglow and Old Home; cvs. Seckel and Maxine are resistant while cvs. D'Anjou, Aurora, Bartlett, Bosc, Comice, Clapp's Favourite and Dutchess are susceptible. The pear rootstocks Old Home (OH), Old Home X Farmingdale (except OH X F 51) *P. calleryana* and

P. betulifolaefolia seedlings are resistant and both Bartlett and Quince seedlings are susceptible.

A thorough spray of Bordeaux mixture (8:8: 100) plus 2-3 per cent oil should be given at green tip in delayed dormant stage to reduce the overwintering inoculum, crawling insects and eggs. First application of streptomycin should be made when blossoms are starting to open. Fire blight has been significantly reduced by foliar application (5-12) of streptomycin (100µg/ml) following periods of rain throughout the late spring and early summer (Bonn and Morand, 1980). The aminoglycoside antibiotic kasugamycin (KS) targets the bacterial ribosome and is particularly active against *E. amylovora*. Blossom blight control was found statistically equivalent to the Industry standard streptomycin (McGhee and Sundin, 2011).

Pseudomonas Blight (*Pseudomonas syringae*)

The disease is also known as blossom blast, lilac blight, bacterial canker, false fire blight and bacterial blossom blight and resembles to a large extent to fire blight, particularly in the blossom stages of infection. Pseudomonas blight is primarily a disease of blossom and rarely affects other parts. One or more minute black lesions appear first on the floral parts which enlarge subsequently forming necrotic areas. Bacterial ooze does not occur usually by the infection of this bacterium. In contrast to fire blight, pseudomonas blight develops rapidly at lower temperatures and frost increases the incidence markedly.

Bordeaux sprays have been successful in controlling this blight but not the antibiotic sprays. At the onset of leaf shedding, the first spray is given which is followed by the second spray when most of the leaves have gone.

Scab (*Venturia pinna*)

Pear scab is important in some regions of the world and resembles apple scab in many respects. The two, however, differ in morphological characters and mode of overwintering. Unlike the apple scab, this pathogen overwinters in imperfect stage (conidial state) on infected twigs in addition to perithecia on fallen leaves and infected twig lesions. The pustules on the infected twigs rupture during moist weather in the spring and liberate conidia which cause the primary infection. A combination of suitable temperature (23.9°C) and continual wet period of 5 hours is the most ideal for the infection process.

Losses occur due to reduction in yield, devaluation in quality, grade, loss in twigs and foliage as well as increased cost of production. Disease symptoms on leaves and fruits are found as dark velvety patches, spots are generally more prevalent on the lower leaf surface and lesions are often found along the mid-rib causing puckering and twisting of infected leaves. Leaves infected early in the season usually drop and severe defoliation adversely affects the tree growth and decreases the fruit bud formation which ultimately reduces the fruit yield in the following season. Fruits infected early usually drop whereas those infected later are deformed. Almost all commercial pear cultivars are susceptible to scab but few cultivars have been identified as resistant in the USSR (Matsura *et al.*, 1980).

For the control of disease, infected twigs should be pruned off and diseased leaves should be buried by deep ploughing. Pear scab control strategy is more complicated because perithecia are formed on infected twigs in addition to fallen leaves. Therefore, pre-bud-burst fungicide sprays are crucial to suppress the development of ascospores. Later, the best inhibition of ascospore production was obtained with the application of fanarimol at 30 ml/100 l followed by bitertanol at 60g/100 l and 2 per cent urea (Latorre and Martin, 1982). Dodine has been more effective and less injurious than other fungicides. Systematic fungicides like carbendazim (bavistin or agrozim), Thiophenate methyl (topsin-M), *etc.*, give protection for longer duration. Fungicidal sprays are repeated at regular interval during disease favouring weather conditions. Sugar and Hilton (2011) stated that when scab infection is very high, potassium bicarbonate 1 per cent, and copper hydroxide (Kocid 3000 1.0g/l) treatments reduced fruit scab infections by 94.3 and 87.5 per cent respectively. While there were no apparent negative effects of these two chemicals on naturally russet cv. Golden Russet Bosc, their use may not be appropriate in some green–skinned cultivars where a russet-free finish is desired.

Leaf Spots

Early spot is caused by *Fabraea maculata* which appears in early spring. The spots are circular, dark brown to black with purplish margins. Fruits of susceptible cultivars develop cankers and cracks after infection. Ascospores formed on the fallen leaves are responsible for primary infection while secondary spread of conidia occurs by wind and rain splashes. Leaf spot caused by *Cercospora minima* appears late in the season. Indefinite brown to greyish spots between major veins are its symptoms. Another leaf spot has been reported from India and the identified fungus is *Mycosphaerella chaubattiensis* (Bose and Roy, 1970). In early stages the spots are small brownish areas of lighter colour which on maturation have greyish white centres and defined margins.

Fungicidal sprays (Bordeaux mixture, lime sulphur) should be given at dormant, open cluster and petal fall stages (Singh, 1943). Shah *et al.* (1985) obtained the best control of *M. chaubattiensis* with several fungicide combinations, *i.e.*, carbendazim at 0.1 per cent applied in the third week of June, dithane Z-78 at 0.3 per cent applied 20 days later and carbendazim at 0.1 per cent again 20 days after dithane Z-78. Pruning of infested twigs and burial of leaves by deep ploughing help in disease control.

Blossom-End Rot of Fruits

Blossom-end rot of fruits is caused by *Alternaria, Botrytis,* and *Penicillium* spp. in storage. Stamens and styles get infected by fungi and thereafter retention of colonized fragments within the floral tube of the mature pear is associated with blossom-end rot of fruit in storage (Sommer *et al.*, 1985). Rotting starts rather late in storage period when fruits are in post-climacteric stage.

To control the rot, first blossom should be sprayed with benomyl to reduce colonization of floral parts, and secondly, harvest treatment of fruits should be followed with carbendazim (Mori and Sozzi, 1980).

Root Diseases

Several species of *Phytophthora* have been found to be associated with root diseases which are commonly called as collar rot, crown rot or root rot. Armillaria root rot (*Armillaria mellea*) is termed as oak root fungus or shoestring root rot. Losses from root disorders vary from poor root growth to death of trees. The prominant *Phytophthora* species attacking pear roots are *P. cactorum*, *P.cryptogea* and *P. cinnamoni.* Trees infected by *P. cactorum* may eventually die due to girdling of trunk at or below the ground level. *P. cinnamoni* kills feeder rootlets but does not attack larger roots and crown tissues.

Resistant rootstocks offer the best solution for root diseases but little information on the subject is available. *Pyrus calleryana,* and Old Home and Old Home x Farmingdale seedling rootstocks have lower susceptibility to root diseases than Bartlett and Winter Nelis seedling rootstocks.

Effective control of the diseases can be obtained by various preventive and curative measures. There should be good provision for water drainage. The graft union should be kept about 30-40 cm above the ground level. In phytophthora infested soil, new pits should be dusted with a mixture of zinc sulphate, copper sulphate and lime in the ratio of 5:1:4. The diseased portions need to be scrapped out along with healthy portion with a sharp knife without injuring the wood and be protected with wound dressing paste like Bordeaux paint. *Armillaria* root rot is a serious disease in only some areas where fungal rhizomorphs penetrate the roots of pear plants of all the ages. The rhizomorphs grow rapidly throughout the root between bark and wood forming a finger-like growth. The roots are killed, crown becomes girdled, the tree declines and eventually dies. Fungus has a wide range of hosts and live saprophytically in the soil. Infested soils should be avoided for new plantings or they should be disinfected properly.

Viral Diseases

Viral diseases often pose a threat to pear industry because virus infection reduces tree vigour, fruit quality and yield. Post *et al.* (1980) has reported better bearing and fruit size grades from virus-free 'Conference' and 'Doyenne du Comice' trees than from virus affected trees. Four virus diseases, *viz.*, pear vein yellow, stony pit, rough bark, and ring mosaic have been found responsible for causing yield reduction in pears (Thomsen, 1975).

Stony pit produces hard areas of dead tissues in the fruit skin usually at the base of depressions distorting the shape of fruits and making them useless (Thomsen, 1975). Bosc and Anjou cultivars are more susceptible than other cultivars. Pear vein yellow virus is comparatively more harmful than other viruses. It is characterized by the formation of chlorotic bands along the smaller veins of infected leaves which are more prnounced during the early growth of scion. The growth of trees affected by rough bark is stunted. Light green to greenish yellow rings and lines are the symptoms of ring pattern mosaic virus which is induced by apple chlorotic leaf spot.

Viruses are transmitted by vectors (insect, mite and nematode) or by mechanical means during grafting and pruning. The virus affected plants should be removed from the orchard as soon as symptoms are first noticed. The scion-wood for grafting should only be used from virus-free indexed trees. Virus-free clones may be regenerated by heat treatment. Pear vein yellows and other viruses can be eliminated from the tips by heat treatment at 38°C after 32 days and from the axillary buds after 67 days (Minoiu, 1972).

Pear Decline

Three basic syndromes, *i.e.*, quick decline or tree collapse, slow decline and leaf curl are associated with the disease which is caused by mycoplasma and is transmitted by pear psylla. Mycoplasma damages the phloem tissues at graft union depriving the roots of food material which ultimately result into quick or slow decline. It happens particularly when commercial cultivars are grafted on *Pyrus serotina* or *P. ussuriensis* rootstocks. Leaf curl symptoms, however, develop when pear cultivars growing on more tolerant *P. communis, P. betulifolia* and *P. calleryana* rootstocks get infected with the mycoplasma. Other symptoms include reddened leathery leaves curled upward along the longitudinal axes. McIntyre *et al.* (1978) studied the pear decline in detail and described the symptoms, distribution, aetiology and symptom remission by oxytetracycline. They found 28.6 per cent trees had symptoms. Orchard surveys further revealed that 2 to 10 years old trees were more susceptible than 40-year-old trees (McIntyre *et al.*, 1978). Oxytetracycline hydrochloride is administered by transfusing it through small holes into the trunk even to alleviate severely infected trees. Another rapid and inexpensive method using pipettes has been suggested for the infusion of antibiotic solutions into the trunks after harvest by Lacy *et al.* (1980).

Physiological Disorders

A) Preharvest

Pitting or Cracking

The disorder is due to boron deficiency. At the initial stage fruits showing russeted skin with cracks which at later stage showed pitting with deep corky areas (Williams *et al.*, 1978). The cvs Bartlett, Anjou and Bosc are susceptible with the disorders. Spraying of boron as sodium pentaborate (Solubor 20.5 per cent WP) in late fall or early spring is suggested to prevent the boron deficiency symptoms and the disorder (Beers *et al.*, 1987).

Corck Spot

The disorder is also known as Bitter pit, Cork and Drought Spot. The symptoms appears on the fruit as a bumpy uneven appearance with sunken areas on the skin. It was reported in 'Anjou' pear. The disorder is believed to be due to unbalanced water relation, excessive shoot growth, high N and K and low Ca supply to the fruit. Increasing the Ca content of fruit by spraying of $CaCl_2$ was found to decrease the disorder in 'Anjou' pear (Richardson and Lombard, 1979). The other suggestions to reduce the incidence of Corck spot are (a) maintaining adequate soil moisture, (b) summer pruning, (c) limitations of N and K fertilizers and (d) avoid late harvesting.

Water Core

Water core is a serious problem of certain cultivars of Japanese pears. The symptoms appear as water-soaked areas on the skin and the affected portion is sweeter than healthy fruit. The incidence of water core was found to increase by bagging and gibberelin treatments during the early period of fuit development and induced by exposure to lower temperature during summer.

Gemma *et al.* (2002) reported that a signal of water core might be found in the extraordinary shaped cells in which less starch grains and different colour appearance under the light microscope were observed in July in the flesh of 'Hosui' pear grown in volcanic soil, but not in alluvial soil. It is believed that there may be a possibility of structural failure during cell enlargement which is possibly due to the weak accumulation of cell wall components (Lee *et al.*, 2008). Cultural practices that improve fruit-Ca concentrations may reduce the potential development of Water core.

B. Postharvest

Core Breakdown

The disorder is characterized by a softening and browning of the tissue in the core area of the fruit which sometimes extended in the surrounding cortex (Pierson *et al.*, 1971). Affected tissues becomes soft, mushy and watery and at later stages, the colour of the affected tissue forms brownish or black. Most cultivars of pears are subject to this disorder. Delayed harvesting, keeping too long in storage or relatively high storage temperature may develop core breakdown. Controlled atmoshphere storage (Porritt *et al.*, 1982) and spraying of $CaCl_2$ in the orchard or dipping the fruits immediately after harvest were found effective in reducing core/internal breakdown/browning of furit.

Scald

Many pear cultivars develop brown discolouration of the skin when kept in storage too long (Hall and Scott, 1977). Attached fruit show yellowing of skin, core breakdown and various rots while still in cold storage. Hall and Scott (1977) stated that over storage scald may be caused by delayed harvesting (over- maturity at harvest), delayed storage, slow cooling or high storage temperature. It is suggested to remove the fruits from storage while they are still green and firm to avoid this disorder.

Black Speck

Black Speck or skin speackling of d' Anjou pears is a concern to the pear industry after long storage. The distinct dark brown specks scatter randomly on the skin tissue of affected fruit. The symptoms appears on the fuit if they are stored in CA storage with oxygen concentration at 1 per cent or lower for longer than four months. Occasionally, fruit stored in regular CA storage with 1.5 per cent to 2 per cent plus 0.8 per cent to 2.0 per cent CO_2, for seven months or longer may also develop black speck disorder (Chen and Varga, 1996).

Breeding and Improvement

A very high level of improvement in pear has been made since early days and as a result more than 5000 cultivars have been evolved. Pear improvement passed through various stages such as speciation, natural selection, domestication, selection from open-pollinated seedlings, controlled hybridization within the species and interspecific hybridization.

History of Improvement

Early developments in pear began with the speciation under geographical isolation and adaptation to extreme ecological and edaphic factors. Interspecific hybridization, introgression and selection played pivotal role subsequently. The forms of Oriental pears developed in China, Japan, Korea and Himalayan region are the descendents of *P. pyrifolia, P. ussuriensis, P. bretschneideri, P. pashia* and their hybrids (Shen, 1980). Improvement mainly occurred through domestication of superior chance seedlings.

The Chinese or Japanese sand pear (*P. pyrifolia*) is the most widely cultivated form in the whole of Asia. Beside the sweet and juicy fruits, the sand pear is very heavy yielder and adapts to warm climate. It is also resistant to fire blight. Twentieth Century of Japanese origin is the most important cultivar in Japan and China while Patharnakh and Gala are performing superbly in India (Jawanda, 1981; Layne and Quamme, 1975; Srivastava, 1976).

The Ussurian pear (*P. ussuriensis*) is the most cold-hardy form being grown in northeastern China. Its important commercial cultivars are Sion-sui-li and An-Ii (Pieniazek, 1966). Small to medium-sized globose fruits though not too gritty, yet are poor in taste. The Chinese white pears (*P. bretschneideri*) are medium cold-hardy and Yar-li and Pin go-li are the two popularly grown cultivars. The medium-sized fruits have the best texture and flavour among the Oriental pears (Pieniazek, 1966). *P. pashia* is grown on a limited scale in India and southwestern China and the medium-sized fruits happen to be of poor quality. The fruits have dark brown flesh with high tannin content. The plants, however, grow well even in hot and humid climate. U-Ii is one of the popular cultivars of *P.pashia* in China. Interspecific hybrids of *P.pyrifolia* and *P.ussuriensis* appear to be the most improved cultivars of China. Likewise, Kikuchi (1946) feels that the cultivated forms of *P. bretsehneideri* may be a hybrid between *P. betulifolia* and the cultivated form of *P. pyrifolia.*

Improvements in only two species, *P. communis* (the common European pear) and *P. nivalis* (the perry or snow pear), were made in Europe (Layne and Quamme, 1975). Prehistoric evolutionary trends in *P. communis* have not been fully investigated. According to Rubzov (1944), at least three species, *viz., P. elaeagrifolia, P. salicifolia* and *P. syriaca*, are involved in the evolution of modern cultivars of *P. communis.* Intercrossing of *P. communis* with *P. heterophylla. P. korshinskyi* (*P.regeli*), *P. boisseriana* and *P: korshinskyi* has been suggested by Bell and Hough (1986) during the course of evolution of cultivated forms in the central Asiatic centre of diversity.

Although improvement in the European and Oriental pear began simultaneously, the number of cultivars evolved in the former were much more than the latter. Romans were first to describe the cultivated forms as is apparent from the description of 35 cultivars by Pliny. Improvement in the cultivars is not well documented up to 15th century but apparently France, Belgium and Germany had developed a large number of seedling selections as cultivars. However, during the golden age (1750-1850) of pear improvement in Europe, many cultivars were selected from the open-pollinated seedlings of common cultivars (Layne and Quamme, 1975). Beurre Bosc, Beurre 'd Anjou, Flemish Beauty and Winter Nelis were selected by Belgian breeders in the eighteenth century. The most popular cultivar William's Bon Chretien (Bartlett) was a chance seedling and its plants were distributed in 1816 (Hooker, 1818).

Pear was introduced to North America by early French and English settlers in 1629 (Hedrik *et al.*, 1921). In the beginning, *P. communis* cultivars were badly damaged by fire blight. Since *P. pyrifolia* is blight resistant, its introduction in 1800 from Europe is considered an important event in the history of pear improvement. Le Conte, the first hybrid of *P. communis* x *P. pyrifolia*, was released for cultivation in 1846. Likewise, Keiffer appeared in 1873 and Garber in 1880. The cultivars were probably the result of outcrosses rather than controlled crosses (Layne and Quamme, 1975). Another significant event in pear improvement in the USA was the introduction of cold-hardy Russian cultivars of *P. communis* in 1879 and *P. ussuriensis* in 1867. Later, more superior blight-resistant material was introduced from the Orient which became the basis for breeding blight-resistant cultivars in the USA and Canada. Comparatively little work has been done on the improvement of pear in Australia, New Zealand and Africa. Packham's Truimph cultivar evolved in Australia is, however, next only to Bartlett in Australia and New Zealand (Cole, 1966).

Breeding Objectives for Scion Cultivars

Breeding objectives depend on the nature of breeding programme, *i.e.*, general or specialized. In a general breeding programme, the improvements are desired mostly in pomological characteristics relating to productivity and fruit quality. Sometimes, in a region, the cultivation of a fruit gets restricted due to one serious problem; in that case specialized breeding programme is implemented to overcome that constraint. In the past, mostly general breeding programmes were undertaken in pear and a large number of good quality cultivars were evolved. Since pear is a heterozygous perennial plant, the breeding of further superior cultivars over the existing ones becomes difficult subsequently. Attention was, therefore, diverted to solve serious regional problems of pear cultivation and thus specialized breeding programmes came into existence. In such programmes, the cultivars resistant to fire blight disease or psylla pest are to be evolved or they should be tolerant to cold, high temperature, wet or drought conditions. The important common objectives of pear breeding are discussed here. Important pear growing countries of Europe like France, Germany, Italy, Romania, Spain, Switzerland and UK focused the breeding programme mainly on fruit anality and fruit type diversification while in North America, the focus was mainly on pests and disease resistance. Between 99 and 2007, about 300 pear cultivars have been released of which about 200 are of European pear and about 100 of Asian pear. In these period China alone has developed > 60 cultivars. US 309 is a natural dwarf which can be donor for compact growth (Layne and Quamme, 1975). Dominance of some fruit characters has been established. Juiciness of the fruit is dominant over dryness, grit over its absence, yellow skin colour over green and white flesh over green or creamy (Zielinski *et al.*, 1965). But many characters, *viz.*, fruit size, russeting, ripening season, and keeping quality are under complex polygenic control (Crane, 1949; Sielinski *et al.*, 1965).

Higher Yield

In order to get higher profit, the grower wants to plant only a high yielding cultivar which has to be precocious, annual bearer, and more productive per unit of land it occupies. All other attributes of a cultivar such as fruit quality, climatic adaptability, and resistance against pests and diseases are overlooked if it is not productive. Interestingly, in India, the poor fruit quality cultivars Patharnakh and Gala have attained commercial status due to the high productivity. Beside the inherent high yield potential, the environmental yield contributing factors are ecological tolerance, pest and disease resistance and adaptability to high planting density system.

Improvement in Fruit Quality

The fruit quality is another very improtant objective in all pear breeding programmes. It includes size, shape, skin colour, flesh colour, texture, flavour, storage life and processing quality of fruits. European and Oriental pears differ considerably in determining the quality attributes of their fruits. In a new pear cultivar, the fruit should be between 5 and 10 cm in diameter and pyriform in shape, but other shapes are also acceptable (Layne and Quamme, 1975). The fruit colour on ripening is required to be bright yellow with a bright red blush. Max Red Bartlett is preferred over its parent Bartlett in new plantings indicating that red fruited bud sports or hybrids need to be selected in future. Russeting and corking of the skin mark the shine and attractiveness of the skin colour and, therefore, are undesirable. The texture of Oriental pear fruits is desired to be crisp, breaking and juicy while that of European pears should be soft and buttery. Absence of fibre and grit in the flesh is wanted but grittiness is a character of sand pears. Fruit flavour is a very significant marketing attribute, being decided by a balance of sugars, acids, aromatic compounds and tannin. Generally for good flavour, sugar and aromatic compund contents should be high, and acid and tannin contents low. Long storage life without internal breakdown is another improtant attribute which should be ascertained in the fruits of any new cultivar because it enhances the profitability of grower by avoiding the glut in market. In European pear breeding programme, cvs, William, Kaiser Alexander, Dr. J. Guyot, President Heron *etc.* are used for source of quality. Some of the good quality European pear cultivar includes Isolda®, Tistan®, Armida®, Elektra®, Hortensia®, Manon®, Agata®, David® and Uta® (Fischartand Mildenberger, 2002). Red skin colour in pear cultivars is under single-gene control. Most of the red-skin cultivars is a chimeric mutation and are frequently unstable, however in cvs. Homored and Rosired, the red-skin colour is stable.

Prolonging Season of Maturity

In order to cater the needs of fresh market and processing industries, pear breeders will have to evolve early, mid and late season cultivars to prolong the picking season. To develop the cultivars for ripening in a particular season (early or late), both parents should be selected for the same season in order to have a larger progeny ripening at the period (Layne and Quamme, 1975). The Italian bred hybrids Etrusca (Coscia x Gentile), Sabina (Santa Maria Morettini x Doyenne du Comic), and Tosca (Coscia x William), are early ripening cultivars. The hybrids, Angelys (Doyenne d' Hiver x Doyenne du Commice), Delmoip and Bauroutard developed from France are the late ripening cultivars.

Ecological Tolerance

Generally, other more profitable crops are grown under favourable ecological and edaphic conditions whereas pear cultivation is expanded to more adverse situations like severe cold, high temperatures, high or low humidity, drought, waterlogging, saline and acid soils, *etc.* Since one or few commercial cultivars cannot meet the requirements of all three situations, fortunately, pear has enough genepool to meet all these requirements. Many years will, however, elapse in incorporating the genes tolerant to different adverse conditions from distant parents in the commercial cultivars.

Resistance against Diseases and Pests

The objective of breeding cultivars resistant to serious diseases and pests is of paramount importance. Even after resorting to a well-planned spraying schedule of fungicides, it becomes very difficult to control fire blight and scab diseases. Furthermore, controlling insect pests through pesticides is harmful to man and creates an ecological imbalance as many beneficial insects get killed. Spraying is expensive, laborious and risky. Evolving cultivars resistant to diseases and pests, therefore, is an integral part of pear breeding programmes all over the world. A newly developed pear cultivar must be resistant or tolerant to fire blight, psylla, leaf blight, scab disease, leaf blister mite and nematodes. Breeding programme to develop fire blight resistant cultivars were initiated in 1920's in Canada and United States and later on to many European countries. Many fire blight resistant cultivars were released which include cvs. Harrow Sweet and Harrow Delight from Canada; cvs. Monnglow, Potomac, Magness and Blake's Pride from USA; and cvs. Aida and Boheme from Italy. Resistance to scab has been reported in cultivars like Navara, Winter Nelis, Muscat, Abbe Fetel and Wilder.

Compact and Efficient Plant Type

High density culture is gaining ground in the cultivation of all the fruits, meaning thereby that newly released pear cultivars must have compact nature of vegetative growth being very efficient in utilizing solar energy and other nutrients for increasing fruit production. Pruning, spraying and harvesting of small plants are much easier, amenability to mechanization. Machines will be increasingly used for fruit harvest and probably for pruning as well. It is, therefore, imperative to evolve pear cultivars which can meet the needs of mechanization. Such cultivars should mature at a time, develop uniform fruit size and be easily separable on shaking but their fruit skin needs to be resistant to rupture and bruising during bulk handling (Layne and Quamme, 1975).

Breeding Systems

A breeding strategy in pear has to be planned on the type of donor parents, character and inheritance.

Hybridization

In the past, selections were made from the open-pollinated seedlings and many outstanding cultivars in pear were evolved. However, the pace of improvement was very slow and its direction was also uncertain. Subsequently, when the desirable characters found in two or more cultivars were required to be combined in one cultivar, the process of controlled hybridization was practised. This has now become a conventional method of varietal improvement. In controlled hibridization, the genetic base should not be narrowed down making the progeny vulnerable to serious pest or disease. Further, it reduces the variability in the progeny and thereby the chances of selection and genetic progress (Anderson, 1970).

Intraspecific, interspecific, and intergeneric hybridization are possible in pears because no major crossing barrier exists (Bell and Hough, 1986). The work on interspecific and intergeneric hybridizations in pears have also been reviewed. When wild seedlings are employed as better sources of disease resistance or cold-hardiness, back-crossing is resorted to improve the quality of progeny (Layne and Quamme, 1975). Li-JuCai *et al.* (2002) used 342 seedlings of 12 pear cross combinations to study the inheritance of some characters of pear fruit. They reported that fruit size was a quantitative character and the fruit of the progenics had a regressive tendency to smaller size. Fruit size of the progenies was similar to the correlation level of mid-parent value, *i.e.* the size of the female and male parent. The soluble solids content of pear fruits was shown to have a quantitative character, and the sexual progenies exhihited a wide segregation. Andreies and Erculescu (2011) from Romania reported the important gene sources for fruit quality were Williams, TN 30-44, Angers, Oliver de Serres, Josephine de Malines and Untosa Hardy. The late fruit maturation trait was transferred to the progenics of Oliver de Serres. Passe Crassane, Rosior Pietros and Winter Decana. Increased resistance to pear scab, fire blight and Psylla sp was also obtained through intraspecific hybridization with genitors (cultivars) from *Pyrus communis*.

Clonal Selection

Variation in a renowned cultivar is often observed which may be due to gene mutation, Since these cultivars are already popular, their improvement through clonal selection is desirable. Guerriero *et al.* (1984) studied 19 clones of Bartlett from different countries for fourteen years and recommended Verona 25, Ferrara and Ravenna for cultivation.

Polyploidy

Zielinski and Thompson (1967) have found all *Pyrus* species as normal diploids (2n=34, x= 17). Most of the cultivars are diploid but a few triploid, tetraploid and hexaploid cultivars have also been recorded in *P. communis* (Layne and Quamme, 1975). According to them, polyploids may have greater vigour and larger fruit size but they suffer from generation sterility. The need for more pollenizers is another handicap. Polyploidy may be utilized in breeding for disease and pest resistance because sometime a triploid hybrid may produce more and larger fruits through parthenocarpy.

Mutation

It is considered a short-cut system of breeding. When the improvement is desired only in one or few characters, such as skin colour, disease resistance, compact growth, self-fertility and maturity season, mutation, either through ionizing radiation or chemical mutagenic agents, becomes handy (Bishop, 1966). Mutation induction may be useful in breaking up lined genes and creating greater variability but it is not a substitute for conventional systems of selection and hybridization. Spontaneous mutations (bud sports) have given red coloured cultivar Starkrimson from Clapp's Favourite and Red Anjou from Anjou. Spotting of coloured fruit mutants is easier than others. Mutations have been induced that effected disease resistance in *P. pyrifolia* (Masuda *et al.*, 1997) and self- compatibility (Hirata, 1989). Japanese pear cv. Gold Nijisseiki which was developed through mutation breeding has a great impact on the Asian pear industry.

Breeding for Rootstocks

All pear cultivars are propagated by grafting (budding) on rootstocks which influence their performance to a very great extent. Rootstock determines plant vigour, adaptability to climate and soil, and resistance to pests and diseass. Mostly seedlings of commercial cultivars serve as rootstocks in pear barring few cases of clonal Quince rootstocks. High density culture has, however, made the use of dwarfing rootstocks imperative. The objective of rootstock breeding is, therefore, to evolve pear rootstocks which have adaptability to varying climates and soils, resistance to major pests and diseases, good anchorage and compatibility with scion cultivars, dwarfing ability and capacity to induce precocious fruiting (Layne and Quamme, 1975)

Old Home x Farmingdale (OH x F) rootstock is the results of successful rootstock breeding (Brooks, 1984). These rootstocks give a good range of scion vigour, control from dwarf to vigorous sizes (Lombard and Westwood, 1987). Brossier *et al.* (1975) developed twelve clonal rootstocks from the seedlings of Perry pear cultivars. According to Michelen (1980) one clone, P 2267, is only 25 per cent of the tree size of Quince A and all the ten dwarfing rootstocks have better yield efficiency than the trees on Quince A. Therefore, these rootstocks can serve as a good donor for both dwarfing and yield efficiency (Lombard and Westwood, 1987). *Pyrus* rootstocks have a more promising future than *Cydonia* rootstocks because the former have more sources of desirable characters than the latter which have incompatibility problem with some pear cultivars (Layne and Quamme, 1975). Tamura (2012) reviewed the research on Japanese pear rootstocks. In Japan, several clones of *P. betulaefolia* and *P. calleryana* seedlings have been indentified based on their ability to induce dwarfing scion growth. Among the various rootstock species, *P. betulaefola* exhibits good adaptability to different envioronmental conditions, with high tolerance to drought, cold and salt. *P. calleryana* exhibits not only a relatively high drought tolerance but also the highest flood tolerance. *P. xerophila* exhibits superior Fe^{3+} - chelate reductace activity in roots, endowing the species with tolerance to alkaline soils. *Pyrus calleryana* has been used in breeding in USA, China and Australia because of its fire blight, pear decline and wooly aphid resistance (Lombard and Weswood, 1987). *P. amygdaliformis* has high tolerance to salinity stress (Matsumoto *et al.*, 2006).

Molecular Markers

Most of the information on molecular markers in pear are from 2000 onwards and still very little information is available. Yamamoto *et al.* (2001, 2002a) was the first to identify a number of markers suitable for the analysis of the genetic diversity among *Pyrus* spp by SSR (Simple Sequence Repeats) for both Japanese and European pears. They successfully confirm the synteny amongh *Malus* and *Pyrus* genomes. Wunch and Hormaza (2007) described the genetic relationships among 63 European pear cultivars by using SSR. Several maps based on SSRs and MFLPs (Microsatellite-anchored length polymorphism) have been constructed. More than 100 SSR markers have been developed for pear and most of the available apple SSSRs can be be used in pear because of high synteny between apple and pear species.

Iketani *et al.* (1998) reported four cp DNA (Chloroplast DNA) haplotypes in Asian pears, but only one in Occidental pears. They believed that the two groups have evolved separately, and that considerable hybridiszation and introgression has occurred between species. Many studies have been conducted in China to identify genetic variations and clustering of population of cultivated pear. Zhengwang (2009) determined the level of genetic diversity of 233 landraces of *P. pyrifolia* (Chinese sand pear) by 14 SSR markers. A genetic linkage map of an interspecific cross between European (*P. communis* cv. Bartlett) and Asian (*P. pyrifolia* cv. Housui) pears using isozymes, AFLPs, SSRs, morphological traits from pear apple, peach and cherry has been developed by Yamamoto *et al.* (2002b). In the map of the female parent of cv. Bartlett, 226 loci were identified on 18 linkage groups over a total length of 926cM. In the male parent, Housui, 154 loci were represented on 17 linkage groups encompassing a genetic distance on 17 linkage groups encompassing a genetic distance of 926cM. They opined that the position of 14 SSRs from apple could be placed on the pear map, along with few SSR from *Prunus*. A RAPD marker linked to fruit skin colour in Japanese pear was reported by Inoue *et al.* (2006).

Genomics

The S-locus in pear is similar to the one in apple. Sassa *et al.* (2007) identified two F-boxgenes inside the S. locus in apple (*Malus* x *domestica*) and three in Japanese pear (*P. pyrifolia*) called SFBB, or S-locus F-Box Brothers. De Franceschi *et al.* (2011a, b) reported that S-locus region of *P. communis* contains no less than six SFBB members surrounding S-RNases and that its structure seems to be rather conserved between apple and pear species. Chinese sand pear (*P. pyrifolia* Nakai) is widely cultivated throughout China. Molecular mechanism underlying anthocyanin accumulation in red Chinese sand pears, gene encoding the anthocyanin biosynthesis enzymes including phenylalanine ammonia-lyase (PAL), chalcone synthase (CHS), chalcone isomerase (CHI), flavones 3-hydroxylase (F_3H), dihydroflavonol 4-reductace (DFR), anthocyanin synthase (ANS) and UDP-glucose: flavonoid 3-0 glucosyltransferase (UFGT) were isolated from the mature fruit of 'Meirensu' pears and were designated as P_pPAL P_pCHI, P_pCHS, P_pDFR, P_pF_3H, P_pANS and P_pUFGT, respectively (Bo *et al.*, 2012). The expression of PyMYBIO and all seven anthocyanin biosynthetic genes were detected. The transcription level of all these genes was markedly up-regulated and peaked at IDABR (days after the removal of bagging of fruit) followed by a decline. Bo *et al.* (2012) also observed higher mRNA abundance of all seven anthocyanin biosynthetic genes and PyMBYBIO in the skin of the sun side of cv. Yunhongli No. 1 than in Meirensu. The results suggested that (a) anthocyanin accumulation in fruit of red Chinese sand pears was closely related to the coordinated expression of PyMYBIO and multiple biosynthetic genes and the level was modified by the genetic background of cultivars and (b) light exposure after bag removal enhanced anthocyanin accumulation by inducing the expression of PyMYBIO and the anthocyanin biosynthetic genes.

Identifying scab resistance genes, developing molecular markers linked to these genes and establishing marker assisted selection would be an effective way to improve European pear breeding for scab resistance. Most of the European pear cultivars are sensitive. The pear cv. Navara was shown to carry a major scab resistance gene whose phenotypic expression in seedling progenies was a typical stellate necrosis symptom. The resistance gene was called Rvp1, for resistance to *V. pirina*, and was mapped on linkage group 2 of the pear genome close to microsatellite marker CHO2b10. This genomic region is known to carry a cluster of scab resistance genes in apple indicating a first functional synteny for scab resistance between apple and pear (Bouvier *et al.*, 2012).

Transgenic

Development of transgenic to introduce new traits in pear was initiated by researchers in late 1990's. Fire blight resistance gene D5C1 (Puterka *et al.*, 2002), AttE (Reynoird *et al.*, 1999), HrpN (Malnoy *et al.*, 2005) have been inserted into pear and reported to cause significant reduction in the cultivar's susceptibility to fire blight. The gene of supersweet protein thaumatin IL from *Thumatococcus danielli* has been transferred to pear to improve fruit taste and phytopathogen resistance. Based on the coding sequence, cloned at Unilever and plasmid PBI121, Dolgov *et al.* (2011) constructed the binary vector for its expression in plants. The successful transformation of the pear cv. Burakovka has been obtained by using the disarmed supervirulent agrobacterical strain CBE21. The introduction of thaumatin gene was confirmed by PCR analysis and its expression in different tissues by Western blotting. Fruit from the transgenic line showed taste modification. In developing dwarfing rootstock, some success was reported by Bell *et al.* (1999) and Zhu *et al.* (2003) by intergrating rolC and rolB gene from *A. rhizogenes*. Freiman *et al.* (2012) developed transgenic early flowering European pear genotype by RNAi silencing of PcTFLI-1 and PcTFLI-2. A transgenic line of cv. Spadona name Early Flowering spadona (EF-Spa) was produced using MdTFLI RNAi cassette targeting the native pear genes PcTFLI-1 and PcTFLI-2. The developed transgenic line had three T-DNA insertions, one assigned to chromosome 2 and two to Chromosome 14. PcTFLI-1 and PcTFLI-2 were completely silenced and EF-Spa displayed an early fowering phenotype. EF-Spa developed solitary flowers from apical or lateral buds, reducing vegetative growth vigour.

The Genome of Pear (*Pyrus bretschneideri* Rehd.)

Wu *et al.* (2013) reported on a high-quality draft genome sequence of the diploid *P. bretschneideri* Rehd. cv. Dangshansuli (also known as 'Suli'), the most important commercial Asiatic pear cultivar grown in the world (more than of million tons per year) and having more than 500 years of cultivated history in China. In this study, a novel combination of BAC- by -BAC strategy, with illumina sequence technology, is used for the first time for *de novo* assembly of a highly heterozygus genome of this size with highly repetitive DNA sequences. A 512.0 Mb sequence corresponding to 97.1 per cent of the estimated genome size of this highly heterozygous species is assembled with 194 x coverage. The pear genome sequencing project concluded that the average density of genes is one per 12 Kb in at least 42, 812 gene loci, a similar number comparing to other plants, and that the pear and apple genome are almost equal in gene numbers. (for further reading download from genome.cship.org – published by Cold Spring Harbor Laboratory Press).

References

Adams, R.G., Kollas, D.A. and Domeisen, H. (1981). Res. Rept. Storrs Agric. Exp. Stn. Connecticut Univ., 60 : 14.

Akiyama, T., Okayama, I., Aono, N., Takahashi, E. and Yasunobu, Y. (1977). Bull. Kanagawa Hort. Exp. Stn., 2: 105-13.

Alberto, F., Arrue, J.L., Machin, J., Combra, M., and Herrero, J. (1982). Annales dela Estacion. Experimentale de Aula Dei., 16: 111-38.

Alique, R., Calro, M.L. and Plaza, J.L. Dela. (1983). Soc. Espanola de Cienncias Horticolas Primer Congresso Nacional II, pp. 887-94.

Allison, E. and Erasmus, S.P. (1985). Dec. Fruit Grower 35: 171-15.

Anconelli, S. and Mannini, P. (2002). Acta Hort., 596: 687-90.

Anderson, R.L. (1970). Ph.D. Dissertation. Univ. of Minnesota.

Andreies, N. and Erculescu, M. (2011). Sci. Paper, Res. Inst. Fruit Growing, Pitesti, Romania, 27 unpaginated.

Angelini, S., Bergamini, S., Nogare, L. Dalle and Migliorni, C. (1986). Rivista di Fruitticoltura e di Ortofloricoltura., 48: 31-33

Anonyrnous (1969).The wealth of India, VIII. Publications and Information Directorate, CSIR, New Delhi, pp. 327-34.

Antoniolli, L.R. and Czermainski, A.B.C. (2012). Acta Hort., 934: 865-70.

Artes, F. (1980). Information Technica Economica Agraria., 11: 23-32.

Asin, L., Torres, E. and Vilardell, P. (2011). Acta Hort., 909: 557-64.

Attibayeba, A. and Hartmann. C. (1985). Fruits, 40: 665-68.

Avelai, M.L. (1984). Acta Hort., 16l: 25-29.

Badawi, A.M., Stino, G.R., Sweidan, A.M. and Al-Doori, A.H.(1978). Egyptian J. Hort., 5: l43-52.

Badiali, G. (1978). Frutticoltura,40: 3l-34.

Bai, J. (2010). Acta Hort., 857: 37-44

Banno, K., Hayashi, S. and Tanabe, K. (1984). J. Jap. Soc. Hort. Sci., 53: 265-70.

Banno, K., Hayashi, S. and Tanabe, K. (1985). J. Jap. Soc. Hort. Sci., 54: 15-25.

Batjer, L.P., Rogers, B.L. and Thompson, A.H. (1953). Proc. Amer. Soc. Hort. Sci., 62: 119-22.

Batjer, L.P., Schomer, H.A., Newcomer, E.J. and Coyien, D.L. (1967). Commercial Pear Growing, USDA, Handbook pp. 1-330.

Beers, E.H., Covey, R.P., Rom, C.R., Park er R., Askham, L. and Maxwell, R.C. (1987). Wash. State. Coop. Exten, E B0419

Bell, R. and Vander Zwet, T. (2011). HortSci., 46: 118-20

Bell, R., Vander Zwet, T., Caslagnoli, S., Einhorn, T., Turner, J.D., Spotts, R., Moulton, G.A., Reighard, G.L. and Shane, W.W. (2014). HortSci. 49: 361-63.

Bell, R.L. and Hough, L.F. (1986). HortSci., 2l: 62-64.

Bell, R.L., Scorza, R., Srinivasan, C. and Webb, K. (1999). J. Amer. Soc. Hort. Sci., 124: 570-74.

Bellini, E. (1987). lnformatore Agrario., 43: 91-100.

Ben-Arie, R., Sonego, L. and Frankel, C. (1979). J. Amer. Soc. Hort. Sci., l04: 500-505.

Benitez, C.E. (2001). INTA. EEA, Alto Valle

Benson, N.R. (1963) Proc. Washington State Hort. Assoc., pp. 23-26.

Bertelsen, M.G. (2002). J. Amer. Pomol. Soc., 56: 149-55.

Bhojwani, S.S., Mullins, K. and Cohen, D. (1984) Scientia Hort., 23: 247-54.

Bishop, C.J. (1966). Proc. 17th Int. Hort. Congr., 3: 15-25.

Blankenship, S.M. and Richardson, D.G. (1985). J. Amer. Soc Hort. Sci., 110: 520-23.

Bliek, W.P.V.D. (1985). Fruitteelt., 75: 52-53.

Bo, Y., Dong, Z., ChunHui, H., MinJie, Q. Xiaoyan, Z. YuanWen, T., Jun, S. and Qun, S. (2012 Scientia Hort., 136: 29-37.

Bonghi, C., Poja, M. and Ramina, A. (2002). Acta Hort., 596: 717-22.

Bonn,W.G. and Morand, J.B. (1980). Canad J. Pl. Pathol., 2: 39-41.

Bose, S.K. and Roy, A.J. (1970). Progressive Hort., l: 57-62.

Bould, C., Nicholas, D.J.D., Potter, J.M.S., Tolhurst, J.A.H. and Wallace, T. (1949). Ann. Rep. Agri. and Hort. Res. Stn. Long Ashton, UK pp. 45-49.

Bound, S.A. and Jones, K.M. (2004). Aust. J. Exp. Agr., 44: 931-37.

Bound, S.A. and Mitchell, L. (2002). Acta Hort., 596: 729-33.

Bound, S.A., Jones, K.M. and Koen, T.B. (1991). Aust. J. Exp. Agr., 31: 133-36.

Bouvier, L., Bourcy, M., Boulay, M., Tellier, M., Guerif, P., Denanu, C., Durel, C. E. and Lespinasse, Y. (2012). Tree Genet. Genomes, 8: 53-60

Boynron, D. and Oberly, G.H. (1966). In: Fruit Nutrition (ed. N.F Childers) Horticultural Publications, Rutgers State Univ., New Brunswick, New Jersey.

Brooks, L. (1984). Fruit Var. J., 38: 126-28.

Browning, G., Ognjanclv,V., Passey, A.J. and James, D.J. (1987). J. Hort. Sci., 62: 305-11.

Brun, C.A., Raese, J.T. and Siahly, E.A. (l985). J. Amer. Soc. Hort. Sci., 110: 835-40.

Buban, T. and Borka, G. (1979). Kertgazdasag, 11: 2l-27.

Burlak, V.A. (1980). Referativnyi Zhurnal, 11: 55.506.

Burrs, E.C. (1983). J. Econ. Entom., 16: 936-41.

Burts, E.C.(1984). Bull. SROP., 7: 325-29.

Canet, W., Arique, R. and Plaza, J.L. Deha (1983). Soc. Espanola de Horticolas Primer Congresso Nacional II. pp. 943-52.

Cao,Y.Z., Chen, R.D. and Chen, D.Y.(1981). LianingNongye Kexue, l: 29-30.

Carraro, L., Loi, N., Ermacora, N., Gregories, A., Oster, R. and Hadidi, A. (1998). Acta Hort., 472: 665-68.

Casero, T. and Ferran, J. (1985). ITEA. 16: 39-51.

Challicc, J.S. and Westwoocl, M.N. (1973). Bot. J. Linn. Soc., 67: 121-48.

Chaplin, M.H. and Westwood, M.N. (1980). J. Amer, Soc. Hort. Sci., 105: 60-63.

Chapman, H.D. (1964). World Crops, 16: 36-46.

Chen Paul, M. and Warga Diana, M. (1996). Proc. Storage Challenge with Winter Pears, Washington 12-13 March, 1996 pp. 1-5

Chen, P.N. and Mellinthin, W.M. (1981). J. Amer. Soc. Hort. Sci., 106: 38-42.

Chen. P.M. and Borgic, D.M. (1984). Good Fruit Grower, 35: 14-15.

Cheng, F., Sun, H., Shi, H., Zhao, Z., Wang, Q and Zhang, J. (2012). J. Agric. Sci. Technol., 14: 183-94.

Childers, N.F. (1976). Modern Fruit Science Hort. Publ. Rutgers Univ., New Brunswick, New Jersey,

Chiriboga, M.A., Schotsmans, W. and Larrigaudiere, C. (2010). Acta Hort., 858: 149-54.

Choe, J.S., Lee, J.G., Kim, S.B. and Moon, J.Y. (1986). J. Korean Soc. Hot. Sci., 27: 149-56.

Civolani, S. (2000). Informatore Fitopato Logico, 50: 27-34.

Cole, C.E. (1966). Proc. 17th Int. Hort. Congr., 3: 321-68.

Cosac, A.C., Frasin, L.B. and Isac, M. (2008). Acta Hort., 800: 447-52.

Crane, M.B. (1949). Heredity, 3: 85-97.

Crisosto, C.H., Suear. D. and Lombard, P.B. (1988). Adv. Hort. Sci., 2: 27-29.

Cui, T., Nakamura, K., Ma, L., Li, J.Z. and Kayahara, H. (2005). J. Agric. Food Chem., 53: 3882-87.

Curetti, M., Rodriguez, R., Magdalena, C. and Rodriguez, A. (2011). Acta Hort., 909: 395-401.

Darbellay, C. and Dessemoz, H. (1987). Revue Suisse d Viticulture d' Arboriculture et d' Hort., 19: 225-29.

Darfeld, H.J. and Lenz, F. (1985). Erwerbsobstbau, 27: 236-39.

Davis, L.D. and Moore, N.D. (1934). Proc. Amer. Soc. Hort. Sci., 32: 131-38.

De Franceschi, P., Pierantoni, L., Dondini, L., Grandi, M., Sanzol, J. and Sansavini, S. (2011a). Tree Genet. and Genomes, 2: 231-40.

De Franceschi, P., Pierantoni, L., Dondini, L., Grandi, M., Sansavini, S. and Sanzol, J. (2011b). Tree Genet. and Genomics, 4: 663-83.

Deckers, J.C., Sweldens, V. and Smeets, L. (1978). Fruit Belge, 462: 291-303.

DeEll, J.R. and Ehsani-Moghaddam, B. (2011). Can. J. Plant Sci., 91: 853-58.

DeEll, J.R. and Slingerland, K. (2012). Factsheet, AGDEX 215/50, July 2012

Degman, E.S. (1953). Proc. Amer. Soc. Hort. Sci., 62: 167-72.

Delcheva, S. and Makariev, Z. (1982). Gradinarska i Lozarska Nauka, 19: 43-49.

Dhaliwal, G.S., Aulakh, P.S. and Jawanda, J.S. (1982). Punjab Hort. J., 22: 164-68.

Dheim, M.A. and Browning, G. (1987). J. Hort. Sci., 62: 313-27.

Dhillon, B.S. and Sharma, M.R. (1978). Indian J. Agric. Sci., 48: 370-72.

Dolgov, S.V., Lebedev, V.G. and Firsov, F.P. (2011). Acta Hort., 909: 67-73.

DongSheng, W., SiPu, Z., ZhongYing, W., ChaoFeng, G. and Lubo (2011). J. Henan Agric. Sci., 40: 105-107.

Dotti, F. and Fonrana De Gradi, C. (1976). Frutticoltura, 38: .27-32.

Drobota, G. and Drobota, M. (1979). Iasi Horticultura, 23: 25-27.

Drouet, A. and Hartntann, C. (1979). Plant Physiol., 64: l104-08.

Ekman, J.H., Clayton, M., Biasi, W.V. and Mitcham, E.J. (2004). Postharvest Biol. Technol., 31: 127-36.

El-Gazzar, A.M., Taha, M.W., Minessy, F.A. and Gajjar, S.A. (1974). Egyptian J. Hort., l: 103-12.

Elkins, R.B., Meyers, J.M., Duraj, V., Miles, J.A., Tejeda, D.J., Mithenam, E.J., Biasi, W.V., Asim, L. and Abreu, J. (2011). Acta Hort., 909: 241-49.

Ende, B. Van Den and Chalmers, D. (1983). HortSci., l8: 946-47.

Ende, B. Van Den and Leece, D.R. (1975) Aust. J. Agric. and Anim. Husb., 15: 129-35.

Endo, M. (1975). J. Jap. Soc. Hort. Sci., 44: 248-59.

Erhardt, W., Gotz, E., Bodeker, N. and Seybold, S. (2002). Zander, Handworterbuch. der Pflanzennamen. Vol.1. Stuttgart, Germany.

Erossier, J., Michelesi, C. and Flick, J.D. (1975). Physiologia Vegetale, 13: 853-67.

Evequoz, M. and Evequoz, N. (1986). Revue Suisse de Viticulture d' Arboriculture et d' Hort., l8: 355-65.

Fallahi, E. and Larsen, F.E. (1981). HortSci., 16: 650-51.

Fallahi, E. and Larsen, F.E. (1986). Good Fruit Grower, 37: 50-51.

FAOSTAT, Food and Agricultural Organization of the United Nations, http: // faostat.fap.org/site/342

Faoro, I.D. and Orth, A.I. (2011). Acta Hort., 909: 409-14.

Fazekas, S. and Graeza, P. (1970). Acta Agron. Nuj., 19: 184-86.

Feyen. J., Veken, L. Van Der, Bosschaerts, L. and Deckess, J.C. (1983). Revue del. Agric., 36: 164-166.

Fidanovski, F. and Ristevski, B. (1981). Jugoslovensko Vocarstvo, 15: 435-39.

Fidanovski, F. and Ristevski, B. (1982). Godisen Zbornik na Zemjodelskiot Fakultet na Univerzitctot vo Skopje, 30: 31-39.

Filippov, L.A. and Pilipenko, V.G. (1971). Agrokhymia, 11: 139-43.

Filiti, N. and Marcucci, M.C. (1982). Rivista della Ortoflorofrutticoltura Italiana, 66: 201-16.

Firuzeh, P. and Ludders, P. (1979). Erwerbsobstbau, 2l: 12-15.

Fisher, *E.G.*, Parker, K.G., Luepschen, N.S. and Kwang, S.S. (1959). Proc. Amer. Soc. Hort. Sci.,73 : 78-90.

Fisher, M. and Mildenberger, G. (2002). Acta Hort., 596: 225-31.

Flick, J.D. and Hermann, L. (1984). Fruit Belge, 52: 251-65.

Freiman, A., Shlizerman, L., Golobovitch, S., Yablovitz, Z, Korchinsky, R., Cohen, Y., Samach, A., Chevreeau, E., Le Roux, P.M., Patocchi, A. and Fisherman, M.A. (2012). Planta, 235: 1239-51.

Frimanslund, E. (1984). Farskning og Forsok i Landbruket, 35: 35-40.

Galvis-Sanchez, A.G., Gil-Izquierdo, A. and Gil, M.I. (2003). J. Sci. Food Agric., 83: 995-1003.

Garriz, P.I., Alvarez, It.L., Colavita, G.M. and Gajdos, M.S. (2007). J. Appl. Hort., 9: 46-49.

Gayford, C.W. (1945). J. Dept. Agr. Victoria, 43: 459-63.

Gemma, H., Oomori, S., Sugaya, S., Peng, S.A. and Iwahori, S. (2002). Acta Hort., 596: 845-50.

Geoffrion, R. (1984). Bull SROP, 7: 13-15.

Georgiev, S. and Koleva, A. (1977). Ovoshcharstvo, 56: 31-33.

Gherghi, A., Milliam, K. and Panait, E. (1972). Hortiviticole, 3: 69-87.

Goldway, M., Zisovich, A., Raz, A. and Stern, R.A. (2008). Acta Hort., 800: 109-18.

Gomita, T., Calvo, G. and Candan, A.P. (2011). Acta Hort., 909: 295-302

Gopalan, C., Ramasastri, B.V. and Balasubramaniam, S.C. (1987). Nutritive value of Indian Foods. NIN, ICMR, Hyderabad, India, p. 90.

Gordon, R.J. and Richens, K.F. (1980). Farmer's Newsletter, 149: 2-4.

Griggs, W.H. (1968). Proc. Amer. Soc. Hort. Sci., 92: 155-66.

Gu, O.F., Yang, Z.X., Fang, J.P. and Shou, S.W. (1982). Sprinkling Techno. (Penguin Jishu), 3: 28-30.

Guerriero, R., Scalabrelli, G. and Biagioni, E. (1984). Rivista della Ortoflorofruittcoltura Italiana, 68: 123-38.

Gupta, S.K., Dhooria. M.S. and Sandhu, G.S. (1975). Sci. and Cult., 4l: 126-27.

Hall, E.C. and Scott, K. J. (1977). CSIRO, Fwd Res. Lab., North Ryde, N.S.W. 2113.

Haring, V.H., Gray, J.E., McClure, B.A., Anderson, M.A. and Clarke, A.E. (1990). Science, 250: 937-71.

Harrtmann, H.T., Griggs, W.H. and Hatten, R.C. (1963). Proc. Amer. Soc. Hort. Sci., 82: 92-102.

Hartamnn, H.T, and Kester, D.E. (l959). Plant Propagation- Principles and Practices. Pretice-Hall. Inc. Englewood Cliffs. N.J.

Hassanen, S.A. and Gabr, M.F. (2012). American-Eursian J. Agric. and Environ. Sci., 12: 484-89.

Hatten, R.C. (1933). Rept. East. Malling Res. Stn. for1932, pp. 75-86.

Hedrik, U.P., Howe, G.H., Taylor, O.M., Francis, E.H. and Tukey, H.B. (1921). The Pears of New York, N.Y. Dept. Agr. 29th Ann. Rept., 2. part 2.

Henry, V.R., Mani, A.K. and Sampath, V. (1984). South Indian Hort., 32: 119-21.

Herrero, M. and Gascon, M. (1987). Ann. Bot., 60: 287-93.

Hirata, N. (1989) Gamma Field Symp., 28: 71-80.

Hiratsuka, S. and Katagiri, M. (1988). Scientia Hort., 34: 193-99.

Hiratsuka, S., Hirata, N., Tezuka, T. and Yamamoto, Y. (1985). J. Jap. Soc. Hort. Sci., 54: 9-14.

Hiwasa, K., Nakano, R., Hashimoto, A., Matsuzaki, M., Murayama, H., Inaba, A. and Kubo, Y. (2004). J. Exp. Bot., 55: 2281-90.

Hooker, W. (1818) Trans. Hort. Soc. London, 2: 250-51.

Howard, B.H. (1987) In: Rootstocks for Fruit Crops (eds. R.C. Rom and R.F. Carlson.) John Wiley and Sons, Inc. New York, pp. 29-77.

Hudina, M. and Stampar, F. (2011). Acta Hort., 909: 625-30.

Hudina, M., Stampar, F., Orazem, P., Petkovsek, M.M. and Veberic, R. (2012). Can. J.Pl. Sci., 92 : 67-75.

Iketani, H., Manabe, T., Matsuta, N., Akihama, T. and Hayashi, T. (1998). Genent Res. Crop. Evol., 45: 533-39.

Inoue, E., Kasumi, M., Sakuma, F., Anzai, H., Amano, K. and Hara, H (2006). Sci. Hort., 107: 254-58.

Jacobs, J.N. and Cook, N.C. (2003). S. Afr. J. Plant Soil, 17: 57-59.

Janick, J. (1976). Amer. Fr. Grower, 96, p. 21.

Jankovic, R. (1986). Agrohemia, l: 43-51.

Janssens, F., Elsen, F., Elsen, A., Deckers, T. and Vandendriessche, H. (2011b) Acta Hort., 919: 39-46.

Janssens, P., Elsen, F., Vandendriess che, H., Deckers, T., Schoofs, H. and Verjans, W. (2011a). Acta Hort., 889: 281-89.

Jaumien, F. (1983a). Acta Agrobotanica, 36: 83-94.

Jaumien, F. (1983b). Acta Agrobotanica, 36: 103-33.

Jazbec, M. (1972). Jugoslovensko Vocarstvo, 6: 659-66.

Kahlon, P.S. and Singh, S. (1981). Haryana J. Hort. Sci., l0: 36-41.

Kajiura, F., Suzuki, K. and Yamazaki, T. (1975). HortSci., l0: 257-58.

Kajiura, I. and Omura, M. (1982). Bull. Fr. Tree Res. Stn., Japan, A (Yatabe), 9: 61-114.

Kappel, F. and Brownlee, R. (2001). HortSci., 36: 69-71.

Karaoulanis, G. (1984). Acta Hort., 161: 47-56.

Keinholz, J.R. (1942). Phytopathology, 32: 1982-86.

Kenworthy, A.D. (1964). Rept.Hort. Dept. Michigan State Univ., pp. 1-39.

Kenworthy, A.L. (1950). Proc. Amer. Soc. Hort. Sci., 55 : 4l-46.

Khattab, M.M., Stino, G.R. and Bakr, S.I. (1981). Annals Agri. Sci., Moshtohor, 15: 265-71.

Kikuchi, A. (1946). Col. Records of Hort. Res., No. 3.

Kimura, N., Okada, N., Aoki, M. and Kawabuchi, A. (1977) Res. Bull. Aichi-Ken Agric. Res. Centre, B. 9: 112-18.

Kitamura, T., Iwata, T.,Fukushima, T., Furukawa, T. and Ishiguro, T. (1981). J. Jap. Soc. Hort. Sci., 49: 608-16.

Kookim, Y., Woong Jung, H., DaeJin, J. and YoungLee, J. (2008) Acta Hort. 772: 459-66.

Kosina, J. (2003). Hort. Sci. (Prague), 30: 56-58

Krieken, W. Van and Lindenberg, C.M.J. (1984). Fruitteelt, 74: 594-96.

Kubo, Y., Hiwasa, K., Owino, W.O., Nakano, R. and Inaba, A. (2003). HortSci., 38: 1414-16.

Kumar, S. and Grewal, S.S. (1977). Haryana J. Hort. Sci., 6: 112-14.

Kuzmenko, M.S., Shenshina, S.V., Untilova, A.E. and Kobal'skaya, V.M. (1982). Sadavodstvo Vinogradarstvo i Vinodelie Moldavii, 10: 51-52.

Kvale, A. (1986). Acta Agriculturae Scandinavica, 36: 301-04.

Lacy, G.H., Mclntyre, J,L., Walton, G.S. and Dodds. J.A. (1980). Canad. J. Pl. Pathol., 2: 96-101.

Laffi, F. (1983). Informatore Fitopatologico, 33: 49-54.

Lakatos, L., Dussi, M.C., Gonda, I., Soltesz, M., Nyeki, J., Szabo, Z. and Zhongfu, S. (2011). Acta Hort., 909: 631-38.

Lane, W.D. (1979). Pl. Sci. Letters, l6: 337-42.

Larsen, F.E. and Fritts, R.Jr. (1986). Good Fruit Crower, 37: 38-39.

Latorre, B.A. and Martin, G. (1982). Pl. Disease, 66: 585-86.

Lawes, G. S. (1977). Orchadist of New Zealand, 50 : 15-17.

Layne, R.E.C. and Quamme, H.A. (1975). In : Advances in Fruit Breeding (eds. J. Janick, and J.N. Moore). Purdue Univ. Press, West Lafayette, Indiana, pp. 38-70.

Lazarov, I. (1979). Gradinarska i Lozarska Nauka, 16 : 28-35.

Lazarov, I. (1984). Gradinarska i Lozarska Nauka, 2l : 28-32.

Lebland, C. (1986). Arboriculture Fruitiere, 33: 34-38.

Lee, C.S., Kim, Y.H. and Kong, S.J. (1975). Res. Rept. Rural Dev. Hort. Agric.No.17, pp. 1-5

Lee, S.H., Gemma, H., Sugaya, S., Sekozawa, Y. and Kim, W.S. (2008). Acta Hort., 772: 327-32.

Lichou, I., Garcin, A, and Arnaud, F. (1987). Infos. Centre Tech. Interprofessional des Fruits et Legumes, France, 32: 9-14.

Li-JuanCai, Cheng, L., Ziazhen, W., ZhongMin, C. and ShouFeng, S. (2011) J. Fruit Sci., 28: 1122-23.

Li-JunCai, Kai, Y., Cheng, L., HongTao, S. and JiaZhen, W. (2002). J. Fruit Sci., 19: 87-93.

Lilleland, O. and McCollam, M.E. (1961). Better Crops, 45: 1-7.

Lin, H.S., Lin, C.H. and Lee, C. L. (1987). Acta Hort., 199: 118.

Littua, Lu, Ying, Han and Haisen, Song (2009). China Fruits, 5: 26-28.

Lombard, P.B. and Westwood, M.N. (1987). In: Rootstocks for Fruit Crops (eds. R.C. Rom and R.F. Carlson). John Wiley and Sons, New York, pp, 145-83.

Loreti, F. and Natali, S. (1975). Italia Agricola, 112: 113-16.

Lotze, E., Sadie, A. and Theron, K.I. (2011). Acta Hort., 909: 693-98.

Maarri, K.Al., Duron, M., Arnaud, Y. and Miginiac, E. (1986). C.R. Agric. France,72: 413-21.

Maeda, J.A., Barbosa, W., Lago, A.A. do, Medina, P.F., Dall'-Orto, R.A.C. and Ojima, M. (1997). Revista Brasileira de Sementes, 19: 271-75

Maidebura, V.I. and Cherbinka, N.K. (1983). Sadovodstvo Vinogradarstvo i Venodelie Moldavii, 11: 21-24.

Maksimova, I.N. (1986). Sbornik Nauchnykh Trudov, Vsesoyuznzi Nauchno Issledovatelskii Istitut Sadovod-stvo imeni I.V. Michurin, 46: 52-54.

Malony, M., Faize, M., Venippe, J.S., Geider, K., and Chevreau, E. (2005). Plant Cell Rep., 23: 632-38.

Mann, S.S. and Singh, K. (1986). Proc. Nat. Symp. Temp. Fruits, 1984, Solan, pp. 311-14.

Mann, S.S., and Singh, B. (1988). Acta Hort., 218: 175-79.

Mann, S.S., Chand, D. and Dhillon, R.S. (1978). Punjab Hort. J., l8: 87-91.

Marcucci, M.C., Camorani, F. and Fiorentino, M. (1980). Societa Orticola Italiana, pp. 147-60.

Maric, Z. (1975). Nauka U Praksi., 5: 365-72.

Marino, G. (1984). Rivista della Ortoflorofrutticoltura Italiana, 68 : 95-106.

Marsal, J., Lopez, G., Mata, M and Girona, J. (2012). Agricultural Water Management, 103: 1-7.

Martin, G.C., Dennis, F.G.Jr., Mac Millan, J. and Gaskin, P. (1977). J. Amer. Soc. Hort, Sci., 102: 16-19.

Martin, G.C., Horgan, R. and Nishijima, C. (1982). J. Amer. Soc. Hort. Sci., 107: 419-82.

Masuda, T., Yoshioka, T., Inoue, K., Murata, K, Kitagawa, K., Tabira, H., Yoshida, A., Kotobuki, K. and Sanda, T. (1997). J. Jap. Soc. Hort. Sci., 66: 85-92.

Matasumoto, K., Tamura, E., Chun, J.P. and Tanabe, K. (2006). J. Jap. Soc. Hort. Sci., 75: 450-57.

Matsuoka, T., Tamura, T., Satake, H. and Sawada, Y. (1979). J. Jap. Soc. Fd. Sci. and Tech., 26: 487-92.

Matsura, T.R., Rylov, G.P. and Adeshchick, A.G. (1980). Plodovodstvo, 4: 89-95.

McAlpin, M. (1980). Fruit World and Market Grower, 8l : 21-23.

McGhee, G.C. and Sundian, G.W. (2011). Acta Hort., 896: 519-20.

McIntyre, J.L. and others (1978). Plant Dis. Reptr., 62: 503-07.

Meheruik, M. and Lau, O.L. (1988). J. Amer. Soc. Hort. Sci., 113: 222-26.

Mellingthin, W.M. and Chen, P.M. (1981). J. Amer. Soc. Hort. Sci., 113: 35-38.

Mellingthin, W.M., Chen, P.M. and Kelly, S.B. (1980). J. Amer. Soc. Hort. Sci., 105: 522-27.

Michelen, J.C. (1980). Rivista della Ortoflorofrutticoltura Italiana, 64: 93-104.

Minoiu, N. (1972). Analale Instituttalui de Ceretari pentru Protectia Plantelor, 10: 11-19.

Mitchell, P.D., Jerie, P.H. and Cholmess, D.J. (1984). J. Amer. Sco. Hort. Sci., 109: 604-06.

Modic, D. and Kodric, I. (1978). Jugoslovensko Vocarstvo, 12: 3-14,

Moffit, M.R. and Westigard, P.H. (1984). J. Econ, Entom., 77: 1513-19.

Montevecchi, P., Sansavini, S. and Guinchi, L. (1986) Rivista di Frutticoltura e di Ortofloricoltura, 48: 34-43.

Mori, P. and Sancassani, G.P. (1984). Bull. SROP, 7 : 354-57.

Mori, P. and Sozzi, A. (1980). Rivista della Ortoflorofrutticoltura Italiana, 64: 623-29.

Morin, F., Rigault, C. and Hartmann, C. (1985). Physiologia Vegetale, 23 : 353-59.

Mota, M., Neto, C.B. and Oliveira, C.M. (2011). Acta Hort., 909: 317-22.

Mukherjee, P.K., Srivastava, R.B. and Dayal, K. (1980). Progressive Hort., 12: 77-80.

Musacchi, S. (2008). Acta Hort., 772: 447-58.

Musacchi, S., Ancarani, V., Gamberini, A., Gaddoni, M., Grandi, M. and Sansavini, S. (2005). Acta Hort., 671: 463-69.

Mwaniki, M.W., Mathooko, F.M., Matsuzakai, M., Hiwasa, K. Tateishi, A. and Lishijima, K. (2005). Postharvest Biol. Technol., 36: 253-63.

Nagy, T., Nyeki, J., Szabo, Z. and Kocsisne, G.M. (2011). Acta Hort., 909: 323-29.

Nakagarva, S., Mitsui, H., Yuda, E., Takaheshi, N., Akimorl, N. and Hishida, S. (1979). Phytochem., 18: 1695-99.

Naurial, J.P. and Bajawa, M.S. (1968). Pear Cultivation. Farm Bull. PAU. Ludhiana, India. 18: 1-28.

Newman, S.M. (1986). Exp. Agric., 22: 383-92.

Nicotra, A., Manzo, P. and Damiano, C. (1974-75). Annali dell' Instituto Sperimentale per la Frutticoltura, 6: 139-47.

North, M.S. and Cook, N.C. (2008). Acta Hort., 772: 97-103.

Nyeki, J. (1996). In: Floral biology of temperate zone fruit trees and samall fruit (eds. Nyeki, J. and Soltesz, M). Akademiai Kiado Pub. pp. 185-265.

O'Connell, M.G. and Goodwin, I. (2007). Aust. J. Agr. Res., 58: 670-79.

Okada, K. (2015). The Hort. J., 84: 183-94.

Overholser, E.L. and Claypool, L.L. (1936). Proc. Amer. Soc, Hort. Sci., 33: 299-303.

Ozelkok, S., Ertan, U. and Buyukyilomoz, M. (1988). Bahce, 12: 43-54.

Pacheco, A., Azevedo, J. and Oliveira, C.M. (2011). Acta Hort., 909: 331-37.

Palazyan, T.N., Snapyan, G.G. and Melkolnyan, I.T. (1983). Biologicheskii Zhurnal Armenii, 36: 863-67.

Parry, M.S. (1981). J. Hort. Sci., 56: 139-43.

Pasa, M. da. S., Fachinello, J.C., Schmitz, J.D., Souza, A.L.K. de and Herter, A.G. (2011). Pesquisa Agropecuraria Brasileira, 46: 998-1005.

Pasqualini, E., Civolani, S. and Corelli Grappadelli, L. (2003). Bull Insectology, 55: 39-42.

Pathak, R.K., Pandey, D. and Pandey, V.S. (1982). Nat. Sem. Plant Propagation, Calcutta, p. 35.

Patzold, G. (1984). Gartenbau, 3l: 118-20.

Pauold, G. (1981). Gartenbau, 28: 304-05.

Pekmezci, M. and Kaska, N. (1975). Colloques Internationaux du Centre National de la Recherche Scientifique, 238: 31-34.

Pepelyankov, G. and C'ranevski, V. (1986). Rasteniev' dni Nauki, 13: 9l-96.

Pfeiffer, D.G. and Burts, E.C. (1983). Environ. Entom., 12: 895-901.

Piao-Yu, BingYan, L., YongZhen, P., Xiangjun, Q., Lan, Z., JingRun, X and LinXian, Y. (2012). J. Fruit. Sci., 29: 149-50.

Pieniazek, S.A. (1955). Proc. 17th Int. Hort. Congr., 4: 427-56.

Pierson, C.F., Ceponis, M.J. and McCulloh, L.P. (1971). US. Dept. Agr. Husb. 376.

Plestenjak, A., Hribar, J., Pitako, D. and Levicnik, M. (1986). Jugoslovensko Vocarstvo, 20: 653-57.

Polesella, A. (1980). Rivista della Ortoflorofrutticoltura ltaliana, 64: 631-42.

Porritt, S.W., Meheriuk, M. and Lidster, P.D. (1982). Agriculture Canada Pub. No. 1737/E

Post, J.J., Stem, J.C., Janse, A.J.G. and Spruit, G. (1980). Fruitteelt, 70: 426-64.

Povar, F.F. and Adaskalitsii, M.M. (1986). Sadovodstvo i Virrogradarstvo Moldavii, 5: 16-18.

Prange, R.K., Delong, J.M. and Wright, A.H. (2011). Acta Hort., 909: 707-17.

Prange, R.K., Embree, C.G. and Ju, H.Y. (1988). Fruit Var. J., 42: 76-79.

Pratella, G.C. and Bertolini, P. (1979). Frutticoltura, 41: 65-67.

Proebsting, E.L. (1934). Proc. Amer. Soc. Hort. Sci., 30: 27-30.

Proebsting, E.L. (1961). Calif. Agric., 15: 14-15.

Prusse, A.G. and Sopalova, E.W. (1970). Sadovodstvo, No, 10 p. 18.

Puterka, C.J., Bocchetti, C., Dang, P., Bell, R.L. and Scorza, R. (2002). J. Econ. Entomol., 95: 797-802.

Quast, P. (1982). Mitteilungen des Qbstbauversuchssringes des Alten Landes, 36: 159-68.

Qurata, R. and Puggioni, D. (1985). Acta Hort., 159: 77-86.

Radojevic, R., Nenic, P., Kedic, M., Radojkovic, D. and Milankovic, N. (1979). Jugoslovensko Vocarstvo, 13: 65-80.

Raese, J.T. (1984). Good Fruit Grower, 35: 16-19.

Raese, J.T. (2002). Comm. Soil Sci. Pl Anal., 33: 461-77.

Raese, J.T. and Burts, E.C. (1983). HortSci., 18: 212-14.

Raese, J.T. and Parish, C.L. (1984). J. Pl. Nutrition, 7: 243-49.

Raese., J.T., Brun, C.A. and Seeley, E.J. (1982). HortSci., l7: 666-68.

Rahovic, D. and Petrovic, M. (1977). Jugoslovensko Vocarstvo, 10: 553-59.

Rana, R.S. and Singh, A.N. (1966). Himachal Hort., 7: 1-6.

Randhawa, S.S. and Kishore, D.K. (1983). Indian J. Hort., 40: 213-14.

Rathore, D.S. (1982). Progressive Hort., 14: 202-24.

Rehman, H.U., Gill, M.I.S., Sidhu, G.S. and Dhaliwal, H.S. (2014). J. Exp. Biol. Agric. Sci., 2: 188-96.

Reil, W.O. and Beutel, J.A. (1978). Calif. Agric., 32: 22-24.

Reil, W.O., Moller, W.J. and Thomson, S.V. (1979). Plant Dis. Reptr., 63: 545-49.

Reimer, F.C. (1925). Oregon Agri. Exp. Stn. Bull.,No. 485.

Requena, A., Nordenstron, G. and Castillo, E. (2011). Acta Hort., 909: 339-43.

Reuther, W. (1941). Cornell Univ. Agri. Exp. Stn. Mem., p. 241.

Reynoird, J.P., Mourgues, F., Norelli, J., Aldwinckle, H.S., Bassi, D. and Marangoni, B. (2008). Italus Hortus,6: 27-35.

Richardson, D.G. and Lombard, P.B. (1979). Prol. Int. Symp. On Calcium Nutrition of Economic Crops. (ed. C.B. Shear) Vol. 10. Comm. SoilSci. Plant Anal.

Riedl, H. and Hoying, S. A. (1980). J. Econ. Entom., 73: 117-22.

Ristevski, B. (1977). Jugoslovensko Vocarstvo, l0: 561-68.

Ristevski, B. (1982). Godisen Zbornik na Zemjodelskiot Fakultet na Univerzitetot vo Skopje, 30: 19-30.

Rivalta, L., Cobienchi, D. and Bogirara, G.L. (1986). Rivista di Frutticoltura e di Ortofloricoltura, 48: 47-51.

Romani, R., Labaritch, J., Yamashita, T., Hess, B. and Rae, H. (1983). J. Amer. Soc. Hort. Sci., 108: 1046-49.

Rubzov, G.A. (1944). Amer. Nat., 78: 358-66.

Rudic, M., Gvozdenrrvic, D., Rodojkovic, D., Mijajlovic, M., Kedic, M., Milenkovic, N. and Milovanovic, M. (1977). Jugoslovensko Vocarstvo, 10: 675-79.

Rufato, L., Kretzschmar, A.A., Bogo, A., MacHado, B.D., Marcon-Filho, J.L., Luz, A.R. and Marchi, T. (2011). Acta Hort., 909: 207-13.

Ryall, A.L. and Aldrich, W.W. (1937). Proc. Amer. Soc. Hort. Sci., 35: 283-88.

Samseok, K., Yoonkyeong, K., Kwangsik, C., SangBouk, J., HeaSeong, H., MyungSu, K., Ilsheob, S., YongUK, S., KyeongHo, W. and Jangjeon, C. (2011). Korean J. Hort. Sci. Technol., 29: 645-50.

Sanchez, E., Curetti, M. and Retamalos, J. (2011). Acta Hort., 909: 435-40.

Sansavini, S. (1986). Rivista di Frutticoltura della Ortotflorfrutticoltura Italiana, 48: 21-30.

Sansavini, S. (2002). Acta Hort., 596: 627-33.

Sansavini, S. and Giunchi, L. (1980). Rivista della ortoflorofrutticoltura Italiana, 64: 591-602.

Sansavini, S., Corclli, L. and Ragazzini, D. (1986a). Rivista della Ortoflorofrutticoltura Italiana, 70: 65-81.

Sansavini, S., Costa, G., Credi, R., Grani, M., Bindi, V. and Monti, C. (1980). Rivista della Ortoflorofrutticolura, Itaiiana. 64: 573-78.

Sansavini, S., Grandi, M., Farina, M. and Tagliani, F. (1997). Rivistadi. Frutticoltura, 3: 11-17.

Sansavini, S., Neri, D., Grandi, M. and Lane, W.D. (l986b). Rivista di Frutticortura e di Ortofloricoltura, 48: 23-30.

SanZol, J. and Herrero, M. (2002). Euphytica, 128: 325-31.

Sassa, It., Kakui, It., Miyamoto, M., Suzuki, Y., Hanada, T., Ushijima, K., Kusaba, M., Hirano, H. and Koba, T. (2007). Genetics, 175: 1869-81.

Schaik, A.C.R. van (1986). Fruitteelt, 76: 694-96.

Seeley, S.D., Seeley, E.J. and Richardson, E.A. (1987). HortSci., 22: 51-52.

Selimia, A. and Keatley, J.C. (1970). Aust. J. Exp. Agri. Ani. Husb., 9: 553-59.

Semash, D.P. and Panasenko, I.N. (1984). Sadovodstvo Vinograderstvo i Vinodalie Moldavii, 6: l2-14.

Seth, J.N. and Kukshal, R.P. (1981). Progressive Hort., 13: 23-25.

Shah, A., Bor, S.S. and Roy, A.J. (1985). Progressive Hort., 17: 363-65.

Sharma, K.K. and Dhillon, B.D. (1985). Indian Hort.,30 : 14.

Sharples, R.P. and Stow, J.R. (1986). Rept. East Malling. Res. Stn. for 1985, pp. 155-70.

Shawky, I., Tomi, A. El., Rawash, M.A. and Makarem, A. (1978). Res. Bull. Aim Shams Univ., No. 826. p. 12.

Shen, T. (1980). HortSci., 15: 13-17.

Silva, G.J., Souza, T.M., Barbieri, R.L. and Oliveir, A.C. (2014). Adv. In Agric. (online Article ID S41097, 8 Pages) http: //dx.doi.org/10.1155/2014/541097

Silva, J. M., Barba, N.G., Barrors, M.T. and Torres-Paulo, A. (2005). Acta Hort., 671: 219-22.

Singh, R.P., Sriuuriuuu, R.P., Verma, V.K. and Phogat, K.P.S. (1983). Punjab Hort. J., 22: 193-96.

Singh, S. and Jawanda, J.S. (1978). Punjab Hort. J., 8: 58-59.

Singh, U.B. (1943). Indian Fmg., 4 : 411-12.

Sommer, N.F., Buchanan, J.R., Forttage, R.J. and Bearden, B.E. (1985). Pl. Diseases, 696: 340-43.

Spieget-Roy and Alston, F.H. (1979). J. Hort. Sci., 3l: 115-20.

Srivastava, R.P., Pathak, R.K., Bana, D.S. and Pandey, V.S. (1977). In: Fruit Breeding in India (ed. G.S. Nijjar) Oxford and IBH Pub. Co. New Delhi, pp. 1-8.

Sryun, E.R. and Gardner, R. (1943). Proc. Amer. Soc. Hort. Sci., 42: 101-03.

Stanly, E.A. and Williams, M.W. (1976). HortSci., 11: 502-04.

Staubll, A. (1985). Obstbau Weinbau, 22: 299-301.

Stefan, V. and others (1964). Lucr. Sti. Inst. Agron. No. Balcescu, Ser., 7: 75-85.

Sterling, C. (1954). J. Food Sci., 19: 433-43.

Stone, C. and Edge, V.E. (1984). South Australian Govt. Printer, pp. 28l-85.

Stow, J.R. (1988). J. Hort. Sci., 63: 59-67.

Strandzhev, A. (1974). Ovoshcharstvo, 53 : 40-43.

Strandzhev, A. (1975). Gradinarska i Lozarska Nauka, 12: 3-10.

Sugar, D. and Basile, S.R. (2011a). Acta Hort., 909: 215-18.

Sugar, D. and Hilton, R.J. (2011b). Acta Hort., 909: 527-30.

Suirturu, R.P. (1916). Fruit Res. Workshop, Tech. Doc., 9: 45.

Taha, M.W., El-Gazzar, A.M., Menessy, F.A. and Gaffor, S. (1974). Egyptian J. Hort., l: 93-101.

Takase, S., Kawabuchi, A. and Okada, N. (1982). Res. Bull. Aichi-Ken, Agri. Res. Centre, 14: 205- 10.

Talice, B.R. (1980). Investigacious Agronomicas, 1: 14-18.

Tamura, F. (2012). J. Jap Soc. Hort. Sci., 8: 1-10.

Tawfik, M., Adbel-Maksoud, M.Z., Khatil, M.A.F. and Hamid, Z.A. (1978). Agri. Res. Review, 56: 1-4.

Teixeira, A.R. and others (l972-1973) Janra Nonional Das Fruitas, pp. 63-68.

Terpo, A. (1985). Feddes Repertorium, 96: 73-77.

Teskey, J.E. and Shoemaker, J.S. (1972). Tree Fruit production. The AVI Publishing Co. Inc. Westport, Connecticut.

Testoni, A. and Granelli, C. (1986). Rivista di Frutticoltura e di Ortofloricoltura, 48: 44-46.

Theron, K.I., Chabikwa, T.G. and Lotze, G.F.A (2011). Acta Hort., 909: 387-93.

Thibault, B. and Hermann, L. (1982). Arboricurture Fruitiere, 29: 35-36.

Thomsen, A. (1975). Acta Hort., 44: 119-22.

Torres, E., Asin, L. and Alegre, S. (2011). Acta Hort., 909: 441-50.

Trachev, D. (1984). Gradinarska i Lozorska Nauka, 2l: 3-12.

Tressler, D.K. and Joslyn, M.A. (1961). Fruit and vegetable Juice Processing Technology, The AVI Publishing Co., Westport, pp. 881-22.

Trinchero, G.D., Sozzi, G.O., Covatta, F. and Fraschina, A.A. (2004). Postharvest Biol. Technol., 32: 193-204.

Truter, A.B. (1987). Deciduous Fr. Crower, 37: 136-39.

Tukey, H.B. (1935). N.Y. Agri. Exp. Stn. Bull., 657.

Turovskaya, N.I. (1987). Sadovodstvo, 6: 22-24.

Urdukhanov, A.S., Kardashev, A.T. and Adamov, M.G. (1985). Sadovodstvo, 6: 16.

USD A, National Agriculture Library, "National Nutrient Database for Standard References" 2012, http: //ndb.nal.usda.gov/

Vallalobos-Acuna, M.G., Biasi, W.V. and Flores, S. (2010). HortSci., 45: 610-16.

Van Huyssteen, P. and van Zyl, H. J. (1992). FFTRI Bull., 618.

Vangdal, E. and Ystaas, J. (1984). Forskning oj Forsok i Landbruket, 35 : 137-43.

Varig-Peterson, O. and Hansen, P. (1973). Tidslk. Plantevl, 77 : 241-54.

Vasilenko, R.K. (1986). Sadovodstvo i Vinogradarstvo Mordavii, 9: 39-42.

Vavilov, N.I. (1951). The Origin, Variation, Immunity and Breeding of Cultivated plants. Ronald press, New York.

Vujanic-Varga, D. (1985). Jugoslovensko Vocarstvo, 19 : 97-102.

Wagenakcrs, P.S. (19g7) Fruitteelt, 77 : 17-19.

Wallace, T. (1961). The Diagnosis of Mineral Deficiencies in Plants by Visual Symptoms. H.M. Stationery, Office, London.

Wand, S.J.E., Steyn, W.J. and Teron, K.I. (2008). Acta Hort., 800: 263-72.

Wang, C.Y. (1982). In: The Pear (Eds. T. Vander Zwet and N.F. Childers) Horticultural Publishers, Gainesville, Fl, USA, pp. 431-43.

Wang, C.Y. and Mellingthin, W.M. (1975) J. Amer. Soc. Hort. Sci., 100: 492-95.

Wang, C.Y., Sains, C.E. and Gross, K.C. (1985). J. Amer. Soc. Hort, Sci., 100: 687-91.

Watkins. C.B. and Frankel, C. (1987). Plant Physiol., 85: 56-61.

Watts, J.E. and Viiliers, O.T.De (1979). Agroplantae, 11: 75-77.

Wertheim, S.J. (1984). Fruitteelt, 74: 1306-308.

Wertheim, S.J. (1987). Fruitteelt, 77: 10-11.

Wertheim, S.J. (1998). Rootstock Guide, FRS, Wilhelminadrop, The Netherlands, 25: 61-84.

Wertheim, S.J. (2000). Pl. Growth Regul., 31: 85-100.

Westerlaken, J. (1981). Fruitteelt. 71 : 479-81.

Westigard, P.H. (1979). J. Econ. Entom., 72: 552-54.

Westigard, P.H. and Gur, L.J. (1936). J. Econ. Entom., 79: 247-49.

Westigard, P.H., Gut, L.J. and Liss, W.J. (1986). J. Econ. Entom., 79: 250-51.

Westigard, P.H., Willett, M.J. and Gut, L.J. (1984). Bull. SROP, 7: 198-205.

Westwood, M.N. (1976). Fruit Var. J., 30: 63-64.

Westwood, M.N. (1978). Temperate Zone Pomology. W.H. Freman and Co, San Francisco.

Westwood, M.N. (1982). Acta Hort., 124: 57-65.

Westwood, M.N. and Lombard, P.B. (1977). Acta Hort., 69: 117-22.

Whiting, D.E., Crandall, P.C. and Woodbridge, C.G. (1978). J. Amer. Soc. Hort. Sci., 102 : 641-45.

Williams, M.W., Couey, H.M., Mofitt, H. and Coyier, D.L. (1978). Pear Production. U.S. Dept. Agr. Hanb. 526.

Wu, J.Z., Wang, Z., Shi *et al.* (2013). GenomeRes., 23: 396-408

Wunsch, A. and Hormaza, J.I. (2007). Sci. Hort., 113: 37-43.

Yadav, I.S., Pandey, S.N. and Singh, M.P. (1971). Indian J. Hort., 28: 210-30.

Yamamoto, T., Kimura, T., Sawamura, Y., Kotobuki, K., Ban, Y., Hayashi, T. and Matsuta, N. (2001). Theor. Appl. Genet., 102: 865-70

Yamamoto, T., Kimura, T., Sawamura, Y., Manabe, T., Kotobuki, K., Hayashi, T., Ban, Y. and Matsuta, N. (2002a). Euphytica, 124: 137

Yamamoto, T., Kimura, T., Shoda, M., Imai, T., Saito, T., Sawamura, Y., Kotobuki, K., Hayashi, T. and Matsuta, M. (2002b). Theor. Appl. Genet., 106: 9-18

YingTao, W., Xiao, L., Yong, L., Liulin, L. and Shaoling, Z. (2011). J. Fruit Sci., 28: 959-63.

Yokomizo, H. (1974). Bull. Fruit Tree Res. Stn., A. (Hiratsuka), l: 79-92.

Yokomizo, H. (1977). Bull. Fruit Tree Res. Stn., A. (Hiratsuka), 4: 93-103.

Yoshida, T., Borgic, D.M., Chen, P.M. and Mielke, E.A. (1986). HortSci., 21: 472-74.

Yuda, E., Hirakawa, M., Nakagawa, S., Yamaguchi, L., Marofushi, N. and Takahashi, N. (1983). Acta Hort., 137: 277-84.

Yuda, E., Nakagawa, S., Matsui, H., Wada, K. and Imai, T. (l981). HortSci., 16: 436.

Zapryanov, A. (1984). Rastitelna Zashchita, 32: 27-29.

Zerbini, P.E., Sozzi, A. and Testoni, A. (1980). Societa Orticola Italiana, pp. 269-79.

Zeven, A.C. and Zhukovsky. P.M. (1975). Dictionary of Cultivated Plants and their Centres of Diversity, Centre for Agricultural Publishing and Documentation, Wageningen.

Zhao, H.X. (1982). Acta Botanica Sinica, 24: 392-94.

Zhengwang, J., Feiyan, T., Hongwen, H., Hongju, H. and Quilang, C. (2009). HortSci., 44: 619-26.

ZhongQi, Q., XianMing, Li, XianQin, Liu, Juanfan, Tu, FuChen, Yang and HongFan, Zhee (2011). J. Fruit. Sci., 28: 928-29.

Zhu, L.H., Li, X.Y., Ahlman, A. and Welander, M. (2003). Plant Sci., 165: 829-35.

Zielinski, G.B. and Thompson, M.M. (1967) Bot. Gaz., 128: 109-12.

Zielinski. G,B., Reimer, G.C. and Ouackenbush, V.L. (1965). Proc. Amer. Soc. Hort. Sci., 86 : 8l-87.

Chapter 6

Plums

Plum is an important temperate fruit crop. It ranks next to peaches so far as its economic importance is concerned. The plum tree thrives well where the high chilling requiring fruits like apple and cherry cannot be grown profitably. Plum is a delicious juicy fruit prized both for its exquisite fresh fruit flavour and in the fruit preservation industry.

Composition and Uses

Plums are rich in sugars and carotenes. The range of various plum fruit components as reported by Wills *et al.* (1983) are presented in Table 6.1. Plum pits which constitute about 2 to 7 per cent of the fruit, yield up to 26.7 per cent of kernels. The kernels are bitter in taste and their composition is given in Table 6.2. A fatty oil can be extracted from the kernels which resembles the expressed almond oil and can be used for lubricating, cooking, as an illuminant and for hair dressing. Fructose, glucose, sorbitol and sucrose are the major soluble sugars in Japanese plums. (Singh *et al.*, 2009). Malice acid is the principal organic acid in Japanese plums and the other organic acids present in minor concentrations include shikimic, succinic, citric and fumaric acid (Robertson *et al.*, 1992; Singh and Singh, 2008). Nutraceutical value of yellow- and red-fleshed South African plums (*P. salicina*) was evaluated by Venter *et al.* (2014). They observed wide variation both qualitatively and quantitatively and cultivars in phenolic composition. Neochlorogenic acid, cynadin-3-0-glucoside among quercetin-3-0-glucoside are the predominant hydroxycinnamic acid,anthocyanin and flavonol, respectively, although not present in all cultivars. The flavon-3-ols, (+) ¯Catechion, (-) ¯epicatechin and procyanidins B_1 and B_2, were found in all ten plum cultivars studied. Red-fleshed plum showed higher total antioxidant capacity (TAC) and total polyphenol content (TPC) than yellow fleshed plums. Os *et al.* (2000) working with European plum (*P. domestica*) cultivars at Poland reported

Table 6.1: Proximate Composition of Plum Fruit*

Fruit Component	*Range*
Fresh weight	38-126g
Edible weight	93-98 per cent
Water	86-88g (100^{-1}g of edible portion)
Protein	0.4-0.8
Fat	0.1
Fructose	0.9-2.1
Glucose	1.7-4.9
Sucrose	1.0-3.4
Dietary fibre	1.3-2.4
Organic acid (total)	1.4-2.7
Malic acid	0.14-2.54
Citric acid	0.03-0.04
Quinic acid	0.12-0.41
Ash	0.3(100^{-1}g of edible portion)
Energy	125.0-187 kJ
Titratable acidity	16.3-30.5 meq $H^{+}100^{-1}$
Soluble solids	9.5-16.3^{0} Brix
Minerals	
Potassium	120.0-190.0 (mg 100^{-1}g edible portion)
Sodium	0.0-3.0
Calcium	6.0-8.0
Magnesium	4.0-7.0
Iron	0.1-0.4
Zinc	0.1
Vitamins	
Vitamin – C	4.0-6.0,
Thiamine	0.02-0.05
Riboflavin	0.04-0.05
Niacin	0.2-0.9
Carotene (total)	260.0-778.0 (µg 100^{-1}g edible portion)

* Wills *et al.* (1983).

that the total polyphenols as well as anthocyanins and flavanols were rather low (160-300 mg/100g, 1833 mg/100 and 914 mg/100g of fruits, respectively). (+) catechin, occurring in quantities from 1.3 to 3.9 mg/100g, was reported to be a dominant compound representing biologically active monomers and dimmers of flavanols. The highest quantities of hydroxyl-cinnamic acids, especially neochlorogenic acid (46-85 mg/100g) were found in all cultivars studied. The main anthocyanin pigments of plums were identified as cyanidin-3-glucoside and cyanidin-3-rutinoside; they make up 42-62 per cent of the total anthocyanins. All investigated plum cultivars show high polyphenoloxidase activity but there are substantial differences between particular cultivars ranging from 3200 to 17200 U/g. Jaisawal *et al.* (2012) used LC-MSn to detect and characterize the phenolic compounds in plum cultivars. Forty-one phenolics were detected comprising caffeoylquinic acids, feruloylquinic acid, p-coumaroylqunic acids, methyl caffeoylquinates, methyl p-coumaroylquinate, caffeoylshikimic acids, catechin, epicatechin, rutin, esculin, quercetin, quercetin-3-0-hexosides, dimeric proanthocyanidins, trimeric proanthocyanidins, caffeoyl-glucoside, feruloy-glucoside, p-coumaroyl-glucoside, vanillic acid-glucosides, 3, 4-dihydroxybenzoyl-glucoside, quercetin-3-0-pentosides, quercetin-3-0-rhamnoside, quercetin-pentoside-rhamnosides, and 3-p-methoxycinnamoylquinic acid. For first time they detected methoxycinnamoyliquinic acid in nature. Chlorogenic acids and proanthocyanidins were the major phenolics present in plums. Furthermore, HPLC with DAD and chemical reaction detection was used to generate quantitative phenolic fingerprints from the fruit flesh of 33 plum cultivars. The predominant compound was found as 3-caffeylquinic acid in nearly all cultivars studied.

Table 6.2: Proximate Composition of Plum Kernels*

Kernel Component	*Per cent by Weight*
Water	9.6
Protein	20.2
Fat	39.5
Lecithin	0.15
Carbohydrates	15.0
Fibre	12.2
Ash	3.4
Amygdalin	0.3
Phosphorus	0.3

* Anon. (1969).

The aroma of the plum blossoms is due to benzaldehyde (Anon., 1969). The gum present on the fruit surface and that exuding from the bark is composed of D-galactose, D-mannose, L-arabinose, D-xylose, L-rhamnose, glucuronic acid and hexuronic acid (Anon., 1969).

Plums are used either as fresh dessert fruit or cooked. They are also canned. dried and made into jam. Plums containing a high proportion of solids, particularly sugars are preferred for drying and are known as prunes. Prunes are demulcent, mildly laxative and refrigerant and often added to cathartic decoctions to improve their flavour and make them more effective (Anon., 1969). Prunes are also canned and used for the preparation of prune pulp and prune beverage. The prune pulp is used in ice-cream mix, confectionery products and meat sauces.

Origin and Distribution

Watkins (1976) identified five centres of origin for plums. These include: Europe for *Primus domestica* (European plum). western Asia for *P insititia* (Damson plum) western and central Asia for *P. cerasifera* (cherry plum), China for *P. salicina* (Japanese plum) and North America for *P. americana* (American plum).

Plum species evolved in the North American centre of origin have been placed in the section *Prunocerasus;* while the Asian and European plums have been placed in the section Euprunus (Knight, 1969). Watkins (1976) could not identify any genetical basis for justifying such a classification.

Area and Production

Plums are cultivated in all the temperate regions of the world. According to FAOSTAT (2012), China is the largest single producer of plums producing about 60,00,000 MT (Table 6.3). The second largest producer is Serbia followed by Romania and Chile. Other major plum producing countries are Turkey, Iran, USA, India, France and Spain.

Table 6.3: Top Ten Plum Producing Countries*

Country	*Area (ha)*	*Production (MT)*
China	3580608	6000000
Serbia	233625	391485
Romania	191416	424068
Chile	179030	300000
Turkey	177255	297026
Iran	176048	295000
USA	137096	229731
India	128305	215000
France	124904	209302
Spain	122516	205300

* FAOSTAT (2012).

In India plum is grown on a commercial scale in Jammu and Kashmir, Himachal Pradesh, Uttarakhand, Punjab and North-Eastern states. They are also grown to some extent in the Nilgiri hills of South India. India is the eighth largest producer of plum in the world.

Taxonomy

Plum belongs to the genus *Prunus* of the subfamily *Prunoidae* of the family Rosaceae (Weinberger, 1975; Watkins, 1976). Plums have a basic chromosome number of X = 8. Rehder (1967) identified 77 species in the genus *Prunus*. Plums and related species have been placed in the subgenus *Prunophora*, sections *Euprunus* and *Prunocerasus* (Knight, 1969). Six species have been placed in the section *Euprunus*. Their chromosome number range from diploid to hexaploid and are native to the land extending from Europe to China. Cherry plum (*P. cerasifera*) and Japanese plum (*P. salicina*) are diploid, blackthorn (*P. spinosa*) is tetraploid and European plum(*P. domestica*) are hexaploid (Okie and Weinberger, 1996). According to this derivation, there is a wide diversity among plums both at the inter-and intra-species level. This diversity is particularly evident in the characteristics of the fruit (size, shape, colour, texture, aroma) and plant (shrubs to large trees, spreading to upright, thick to thin leaves, and early to late blooming) (Ramming and Cociu, 1990). Thirteen species native to the North American continent are all diploid in nature and have been placed in the sections *Prunocerasus.* Separation of the plums into the two sections does not appear to have any genetical basis since the species appear to be closely related. In fact Watkins (1976) views *Prunus salicina* of China as the link between the sections *Prunocerasus* and *Euprunus. Prunus domestica,* the European plum, is believed to be the most recent crop species to have arisen within the section *Prunophora. Prunus insititia, Prunus cerasifera* and *Prunus spinosa* are thought to be involved in its ancestry (Endlich and Murawski, 1962: Watkins, 1976, Salesses, 1977).

The cultivated plums belong to two species, *viz., Prunus domestica* L. (the European plum) and *Prunus salicina* (the Japanese plum). Other species like *Prunus spinosa* (the black thorn sloe), *Prunus cerasifera* (the myrobalan plum), *Prunus ussuriensis* (the ussurian plum), *Prunus nigra* (the Canadian plum) and *Prunus americana* (the American plum) are of interest to the plum breeders for imparting winter-hardiness to the cultivated plums or for use as rootstocks. Vitkovskii (1986) could classify more than one thousand plum cultivars into five basic types based on floral polymorphism. He observed petal shape to be the most polymorphic trait. Cultivars could be classified on the basis of: (i) temporal relationship between flower bud burst and leaf burst; (ii) the number of flowers/bud; (iii) the diameter of the corolla at the beginning of the peak flowering stage; (iv) the spacing of petals in the corolla; (v) petal shape; (vi) calyx shape; (vii) sepal shape (viii) pedicel length; and (ix) degree of pedicel pubescence.

European Plum

The wild form of the European plum, which comprises the bulk of the cultivated plums, has never been found (Kolesnikov, 1966). Many botanists and horticulturists believe it to be a hybrid of the diploid myrobalan plum (*Prunus cerasifera*) and the tetraploid black thorn (*Prunus spinosa*) both of which occur wild in Europe and western Asia. The triploid hybrid so generated could have double its chromosome number resulting in the hexaploid *Prunus domestica* (Hedrick, 1911). However, Salesses (1977) demonstrated that both *Prunus domestica* and *Prunus spinosa* carry the *Prunus cerasifera* genome suggesting thereby a more complex origin for *Prunus domestica.* This origin is supported by the fact that in *Prunus cerasifera* the fruit has a yellow ground colour and red anthocyanin. *Prunus spinosa* fruits have a green ground colour and blue anthocyanin with very limited variation, *Prunus domestica* fruits exhibit both yellow and green ground colours and also both red and blue skin clours, with an unlimited range of variation. Fruit characteristics of size, shape and flavour also lend support to the above origin. *P. domestica* is considered a relatively young species. Werneck (1959), using archeological evidence, concluded that *P. domestica* is indigenous to middle Europe. Population of *P. domestica* was reported growing in Hungary-Serbia-Bosnia and the neighboring countries Romania and Bulgaria and Gavrilvic and Paunovic (1962) placed the origin of this species in the Balkans. Eryomin (1991) also considered Trans-Caucasia or Iran as the place of origin for *P. domestica*. Endlich and Murawski (1962) provided a categorical evidence from their artificial crossing experiments between *Prunus cerasifera* and *Prunus spinosa.* The F_1 were all nearly sterile. The majority of F_2, were hexaploid (57.69 per cent, 2n = 48). Some of these hexaploids closely resembled the *Prunus domestica.*

Prunes

This is commercially the most important group in the Eastern World, especially in the North American continent. Blue or purple in colour, the fruit is oval with bulging ventral side and compressed bilaterally. Flesh is firm and thick. The distinguishing feature of this group of plums is the high sugar content, which makes them suitable for drying without removal of the pit. Thus giving rise to the age-old adage that all prunes are plums but all plums are not prunes.

Reine Claude Gages

This group classified by botanists as either subspecies or distinctly different species comprises of a number of round, very high quality, golden or green plums. Botanist called them in different species *e.g., P. domestica cerola, P. italica* and *P. claudiana*. The exact origin of their group is still unknown.

Lombard Plums

The fruit is purplish red and of good-quality. Important cultivars of this group are Lombard, Bradshaw, Pond and Victoria.

Asiatic Species

Two *Prunus* species; *P. saliciana* and *P. simonii* belong to Asiatic species. *P. Saliciana* (Syn. *P. triflora, P. Japonica, P. hattan, P. ichangana, P. botan, P. masu*) known as the Japanese plum. It is next in importance to the large fruited European plums. It is actually believed to have originated in China. It was introduced in Japan some 200 to 400 years ago, from where it got disseminated around the world. Hence, the name 'Japanese plum'. The species is more vigorous. productive, precocious and resistant to diseases than the *Prunus domestica.*

It is characterized by medium sized sharp pointed leaves that are free from hair on the lower surface. It comes into bloom quite early and thus is a leader for early market. but this feature also makes it very susceptible to spring frosts. The fruits of Japanese cultivars are usually large and heart shaped often with a pronounced apex which distinguishes them from other types of plums. A few cultivars are however oblate or round. Tree growth habit varies considerably amongst cultivars. Some are low and spreading while others exhibit an extremely upright growth. Flowers are produced three in a bud on either one-year-old shoot or on spur. Fruit set is heavier and more consistent on spurs.

American Species

The important American species are *P. americana, P. nigra, P. angustifolia, P. hortulana* and *P. munsoniana*. *P.americana* known as common wild plum. Fruit quality is poor, small and spreading tree. The cultivated varieties of *P. americana* are

Pipestone, Tecumesh, Toka, Underwood and Waneta. *P. nigra*, known as Canadian plum. Fruit is oblong, orange red in colour; the stone is large and compressed. *P. angustifolia*, known as Chickasaw plum is a small bushy tree, fruit is small, cherry-like, red or yellow or yellow dotted. *P. hortulana*, known as hortulana plum. Tall tree, fruits are oblong, red to yellow, white dotted. *P. munsoniana* known as wild goose plum. Tall trees, fruit is oval, bright red or yellowish, marked with white dots.

Prunus insititia

This is a small fruited European plum. It is hexaploid like, the *Prunus domestica* to which it appears to be related. It grows wild through Europe and western Asia. Plums of this species are commonly known as the damsons, bullaces and mirabelles and extensively grown in western Britain. Damsons are believed to be native of the area around Damascus, where seeds of this species have been found in ancient ruins. The variability seen in this species is much less as compared with *Prunus domestica*. The fruits are small and purple (in damsons) or yellow (in mirabelles). The damsons are highly prized for preserves and jam, while mirabelles have an excellent flavour. The flavour is retained even after preservation for several years. The plants of this species are compact and form excellent hedgerows or shelter belts. Their silver-grey blossoms and fruits are very ornamental. In spite of the fact that, this is a very hardy species, it has remained largely neglected.

Cultivars

The cultivars of plum can be broadly classified into three categories. The first and the foremost are the European plums belonging to *Prunus domestica*. The next in importance are the Japanese plums belonging to *Prunus salicina* and its hybrids with various North American species. The Damson plums belonging to *Prunus insititia* are of limited importance. Some of the important plum cultivars are being briefly described here:

Italian

A prune, it is also known as French Fellmbery Agen petite. It was probably originated at Milan, Italy around 1800 century. The fruit is medium in size, oval and dark blue with a heavy bloom. The flesh is firm, high in sugar content and good in flavour. Several early maturing mutants, such as, De Moris, Greata and Richards have been selected and grown commercially. These strains mature 15 days earlier than the Italian but they are less productive.

Giant Prune

A prune evolved by Luther Burbank of California from Ponds seedling. It blooms late and is winter-hardy.

President

A prune raised by Thomas Rivers, it is an excellent dessert plum, being large, rich purple colour with its deep yellow flesh, juicy and sweet. Makes a large spreading tree. It blooms early and requires a pollenizer.

Other important prune cultivars are Imperial, Epineuse and Sugar Prune.

Green Gage

It is an old cultivar of the Reine Claude group. The fruit is medium in size, roundish with greenish-yellow skin. The flesh is firm, freestone and the flavour is sweet and very good in quality. It tends to overset and bears biennially.

Early Transparent Gage

A Reine Claude cultivar, it is a regular and heavy cropper, prolific bearer with delicious fruits which are rather small and roundish. The colour of the fruit is pale yellow with red dots. The fruits ripen in the third week of July.

Reine Claude

It is often described as a first rate plum with exquisite flavour (Genders, 1973). The tree is compact, self-fertile and blooms very late. The richly flavoured fruit is large and almost orange in colour, speckled with white.

Golden Transparent

Like all the 'Transparent' raised by Thomas Rivers it makes a dwarf tree, is self-fertile and blooms late. The fruit is very delicious, and is late in maturity. It is a Reine Claude plum.

Denniston's Superb

Though included in the group this cultivar is not truly a Reine Claude plum. It is of hybrid origin which was developed in 1835. It, however, possesses the true gage flavour and is extremely hardy and fertile. It is a good pollenizer for a number of other cultivars. It blooms in early mid-season and can set heavy crops without a pollenizer. The tree is vigorous, the fruits are green, flushed with crimson and ripens early.

Golden Drop

A yellow egg plum, it blooms early and requires frost protection. It bears large fruits of quite exceptional flavour, pale yellow speckled with crimson. The fruit has rich apricot flavour and characteristic sweetness. It makes a spreading tree and is a shy bearer. It requires an early season pollenizer.

Pond

A lombard plum. It makes a large tree and is valuable for cold situation since it blooms late. The rose crimson fruit is large and handsome and of good dessert quality. It does well on calcareous soils.

Victoria

It was discovered in a Sussex cottage garden more than a century ago. It is the most widely grown cultivar of all the *Prunus domestica* plums. The tree is very vigorous and extremely self-fertile. It is frost resistant and a prolific bearer. The fruit is used for all purposes; however, it is highly susceptible to silver leaf disease. It belongs to the Lombard group of plums.

European Plum

Japanese Plum

Mariposa

Plum Orchard

Plum Pox

President

Plum Trees in Flowering

Santa Rosa

Bradshaw

A lombard plum, the fruit is medium large, oblong, purplish-red, attractive and medium soft. The stone is semi-cling to clingstone. The quality of the fruit is fair to good.

Beauty

A Japanese plum, it makes an upright tree. It does not require pollenizers, but cross-pollination may increase set in some areas. The fruit is heart-shaped, clingstone, greenish-yellow to bright crimson and the flesh colour is amber streaked with scarlet. The cultivar is regular and prolific bearer.

Burmosa

It matures later than Beauty and requires cross-pollination for adequate fruit set. The tree is of spreading type, the fruit is roundish, yellow with red blush and the flesh light amber in colour.

Durado

The tree exhibits a spreading growth habit requiring cross-pollination. The fruit is flat to oval, purple with greenish amber flesh. This is a *Prunus salicina* plum.

June Blood

The fruit is medium to large, roundish, conical and dark red. The flesh is firm, juicy, clingstone and fair in quality. This is also a *Prunus salicina* plum.

Kelsey

A Japanese plum, it produces a tree of moderate vigour. The fruit is heart shaped with greenish yellow skin and flesh. The flesh is juicy, firm and of good quality.

Santa Rosa

A Japanese plum, it produces an upright tree. The fruit is large purplish crimson in colour. The flesh is amber with red near the skin. In India this cultivar has established itself primarily because of its self-fruitfulness, prolific bearing and characteristic flavour, not withstanding the offending sour taste of the relatively thick skin. A number of early and late maturing strains of this cultivar have been developed like the July Santa Rosa and Late Santa Rosa.

Mariposa

An upright growing tree that requires cross-pollination for adequate fruit set. The fruit is heart shaped with greenish-yellow fruit skin mottled with red. The flesh is red in colour of fair quality, belongs to *P. salicina*.

Starking Delicious

Developed by Ruth J. Johnson in 1931. The tree is upright, hardy and productive. The fruit is medium large with dark red skin and blood-red flesh. The quality is good. This is a very good commercial Japanese type plum. It requires a *Prunus salicina* pollenizer.

Damson (*Prunus insititia*)

It is recommended for cultivation in Canada. It is a hardy cultivar and bears small roundish dark blue, semi-clingstone fruits. The flavour is poor in fresh fruit. but excellent in jam. It is self-fruitful.

Bradley's King (*Prunus insititia*)

An extremely hardy variety. It blooms late and is a heavy bearer. It makes a vigorous tree and unlike other damsons. The wood is not brittle. The fruit is comparatively large, richly flavoured and dark crimson in colour. It is self-fruitful.

Farleigh Prolific (*Prunus insititia*)

This is the most prolific hearer of all if given a pollenizer like Bradley's King. It is also known as Crittendens or the Cluster Damson. The fruits hang in clusters. Tapering in shape and coal-black in colour and make a superb jam. The tree is small and compact.

Shropshire Prune

It makes slender tree those crops only lightly unless planted on a heavy loam soil and well supplied with nitrogen. It bears large oval fruit with an outstanding flavour. It is also known as Westmorland Damson.

New Varieties and Hybrids

Black Delight

An early ripening, dark red skin, yellow flesh, Japanese plum developed from California. Fruit medium, oblate, glabrous, flesh yellow, free stone, regular bearer and self-infertile.

Black Glow

A hybrid of Suplumsix X O.P, developed at the University of Bologna, Italy and introduced in 2002. It is a mid-season, deep blackish-purple skin, yellow flesh Japanese plum. Fruit medium to large, flesh yellow, early bearing and self-infertile.

Black Sunrise

A hybrid of Suplumsix X Suplumthirteen, also developed at University of Bologna, Italy and introduced in 2002. A mid-season, purplish brown skin, yellow flesh Japanese plum. Fruits medium, oblate, symmetrical, glabrous, very juicy, very productive and self-infertile.

Candy Stripe

This hybrid (King David x 4G 1180 plumcot) X Autumn Giant was developed at California. This is a mid-season, yellow-red skin, yellow flesh interspecific plum. Fruit medium, globose, glabrous, regular bearer, productive and self-interfile. Chilling requirement is about 450 to 500 h.

Constanza

A very late ripening, black skin, yellow-green flesh Japanese plum hybrid (Larry Anne x Suplumsix) developed from Santiago, Chile. Fruit medium large, round to slightly oblate, may lenticels, russet with tendency to crack, semi clingstone and self-infertile.

Prune 29

A late-season, purple-red skin, greenish-yellow flesh, European plum hybrid (French Improved x Stanley) developed from University of Bologna, Italy and released in 1996. Fruit medium, elliptical, good taste, suitable both for fresh fruit and processing. Tree medium-high vigour, average production and usually self-infertile.

Shezifon Kerassia

An early season, small-fruited dark blue, hexaploid, plum derived from *P. cereasia* (land race sweet Kerassia) from Israel and released in 2005. Fruit very small, globose, almost free stone, heavy producer, self-fertile, chilling requirement 590 h. and ripens with Santa Rosa.

Aitianli

A hybrid of Julian 6 x Formosa developed in China (Liu-HaiRong, 2000). It exhibited winter hardiness, fruits large, skin red, flesh yellow, fine and juicy, without fibres. Trees are precocious and productive.

Tezaohong

Another Chinese cultivar introduced in 1991. After 7 year trial, it proved immune to plum red spot (*Polystigma rubrum*) and highly resistant to aphids and plum fruit moth (*Cydia funebrana*). Tezaohong is very productive, large fruit (78-120 g), red skin, flesh yellow, crispy, juicy, strong aroma and good eating quality (Zhang-JunSheng, 2000).

Alge

Developed from a selfed seedling of cv. Ulen Renclod. Alge is productive, moderate vigour, moderate resistance to disease and cold, and is partially self fertile. Fruits medium in size, round-elliptical shape, skin greenish-yellow to red-blue, flesh yellow, tasty, sweet and juicy with soft texture (Lukosevicius, 1999)

Summer Fantasia

A new Japanese plum, originated from a cross between Soldam and Taiyo developed at Republic of Korea (Jun *et al.*, 2015). It is a mid-season Japanese plum cultivar that produce round, firm and very sweet fruit for fresh market. It produces higher yield than Oishiwase, Santa Rosa and Soldam.

Climate

Plums because of their wide variability between different species, exhibit considerable varietal differences in their climatic requirements. Thus plums are being grown in areas extending from the subtropical plains to the temperate high hills. A sizeable portion of the commercial cultivation of plums in India is located in the low and mid hills of the Himalyan range. This region is not particularly suited for apple cultivation. because of lesser chilling hours experienced in these areas. In general. European plums require 800 to 1500 hours below 7.2°C of winter chilling for satisfactory budbreak in the spring. The cultivars of the Japanese plum require comparatively less of winter chilling (about 100 and 800 hours below 7.2°C). (Griesbagh, 2007). Very low-chill Japanese plums have been developed in Florida, California, Australia, Taiwan, South Africa and Argentina requiring <450 chill hours (below 7°C). Some of the low-chill cultivars are Golden Talisma, Carmesim, Pluma-7, Harry Pickstone, Gigaglia, Gulfbeauty *etc.* Most of the Japanese plum cultivars are resistant to winter cold and thus can be grown in more northerly regions of the Northern Hemisphere.

Plums bloom early in the season and hence are prone to spring frost injury. Selection of relatively higher site with good air drainage would be desirable. Northern slopes should generally be avoided for plum planting except for the Japanese plum cultivars. The bloom period of the latter tends to get delayed on the northern slopes thus helping them to escape spring frost injury. A gentle slope is preferable rather than an enclosed plain area. This would ensure adequate air and water drainage. Presence of a large body of water in the vicinity would be advantageous since water helps to moderate the summer heat and winter cold.

Availability of ample good quality irrigation water is of primary importance while selecting a site for a plum plantation. Plum trees will fail to produce commercially acceptable crops if they suffer from water deficits. There are. however, cultivars like the Renklod Althanna (Althann's Gage) which are reported to be highly drought tolerant (Gnezdilova and Doroshenko, 1984).

Soil

The plum is not very fastidious in its soil requirements. Generally deep sandy loam soils with good internal drainage and freedom from alkalinity and salinity are considered best. (La Rue and Gerdts, 1976). European plums, however, have been reported to perform better on heavy loam rich clayey soils (Genders, 1973). For prunes, soils particularly rich in potassium should be preferred as these are heavy potash feeders. It prefers a soil with a pH of 5.5-6.5. Plum trees are the most tolerant of all stone fruits with regard to heavy soil and water logging and they tolerate drought better than do peach trees (Griesbach, 2007).

Rootstocks

Characteristics of Rootstocks

1. Myrobalan

Myrobalan (*Prunus cerasifera*) is the most widely used rootstock, particularly for the European plums (Hartmann and Kester, 1972) though it is compatible with the Japanese plum cultivars also (Kishore and Randhawa, 1983). It is propagated from seed and grows well in fairly heavy wet soils (Adriance and Brisan, 1967) and is also resistant to drought (Okie, 1987). It is susceptible to oak root fungus

(*Armillariella mellea*) but resistant to crown rot. A number of clonal selections with some distinctive feature have been made.

MYROBALAN 29C: It is a vigorous stock which may delay fruit ripening. The root anchorage is not good during the early years of orchard life and may require tree supports in windy location. Uptake of Ca and K is good. It does not perform well on extremely heavy soils, though it can withstand waterlogging due to high rainfall or heavy irrigations. It is immune to the root-knot nematodes (*Meloidogyne incognita* and *Meloidogyne javanica*) but like the seedlings and other Myrobalan selections, it is susceptible to the root-lesion nematode (*Pratylenchus vulnus*).

MYROBALAN B: This is one of the six clonal selections from Myrobalan seedlings made at the East Malling Research Station, UK (now rechristened Horticultural Research Institute, UK). Myrobalan-B (green-wooded) has proved outstanding (Tukey, 1964). It is now recognized as the standard clonal rootstock for vigorous plum trees. It tends to delay ripening and is a good co-absorber (Okie, 1987). The other Myrobalan clonal selections made were Myrobalan-A (narrow-leaved), -C (broad-leaved), -D (yellow-leaved) and -E. Clonal selections from Myrobalan seedlings have also been made at some other research centres of Europe.

MYROBALAN GF 31: It produces a vigorous tree and can easily be propagated through cuttings. It is precocious and performs well even on dry stony soils.

MYROBALAN 2-7: A vigorous, tolerant to drought rootstock exhibits very good K absorption.

MYROBALAN 5-Q: The rootstock tends to delay ripening but has not been fully tested.

2. Marianna

Like the Myrobalan this is a vigorous rootstock, but it is characterized by its resistance to nematodes, crown gall and the oak root fungus and is easy to propagate through cuttings. It originated in Marianna, Texas, USA in 1870s. Botanically, it is a cross between Myrobalan (*Prunus cerasifera*) and *Prunus munsoniana* and is a hexaploid (2n = 48). Some clonal selections made from this stock are enumerated below:

MARIANNA 2624: It produces semi-vigorous trees resistant to water logging and root-knot nematodes. It is, however, susceptible to bacterial canker and is unsuitable for heavy soils. Nitrogen is efficiently utilized by this stock.

MARIANNA 4001: The rootstock produces vigorous trees with very good anchorage and is reported to be tolerant to drought (Okie, 1987).

MARIANNA GF 8-1: Though the trees produced on this stock are vigorous, yet they are precocious. It is resistant to water logging, viruses and root-knot nematode. Okie (1987) has described it as suitable for both heavy and sandy soils and is also tolerant to high soil pH.

BUCK PLUM (Morrison's plum stock): This is a vigorous rootstock compatible with all Japanese and most European plums. It is resistant to crown gall.

3. *Prunus domestica* Selections

In Europe, selections from *Prunus domestica* have been widely used. Following are some of the distinguishing features of these selections.

ACKERMANN'S (Marunke): A selection made at the Ackermann-Torgace nursery in Thuringen, Germany, where it is found wild (Tukey, 1964). It produces a semi-vigorous tree and hastens fruit ripening. However, its productivity is low.

BROMPTON: It is of English origin and is easy to propagate through hardwood cuttings. It produces a semi-vigorous tree and may delay fruit ripening. It is reported to be incompatible with some prune cultivars.

BRUSSEL: Developed in Holland, it has been widely used in Europe. It is propagated easily by layering and makes a large tree. This rootstock, however, is now no longer recommended because of its susceptibility to diseases and other undesirable characters (Tukey, 1964).

COMMON PLUM: This stock has been used since long in England. Botanically it has not been assigned to any single species (Tukey, 1964), but is found fairly true to type. It can be propagated easily through both layers and hardwood cuttings. It is not compatible with all the plum cultivars, but when compatible it is of value as a semi-dwarfing plum rootstock.

PERSHORE: It is a yellow egg cultivar of the European plum. Propagation is done by layers or suckers. It is still recommended and produces a medium or intermediate sized tree.

NARROW LEAVED SHINY MUSSEL: It produces trees of intermediate size. It is, however, devoid of any special characteristics to sustain its continued recommendation as rootstock.

GROSSE GRUNE REINE CLAUDE (Large Grune Reine Claude): It is used by nurserymen in Europe and makes a medium-sized to vigorous tree.

HUTTNER IV: This is a selection from Germany which makes a medium-sized tree.

PRUNE GF 43: This is reported to be a dwarfing rootstock for plum and may also delay fruit ripening.

EM 478 (*Prunus domestica* x *Prunus* sp.): This is a selection made at the East Malling Research Station, England, from a cross between Brompton and Common Plum in 1933. It is reported to be semi-dwarfing but has not been fully evaluated.

DAMAS GF 1869: It originated from a cross between *Prunus domestica* and *Prunus spinosa.* It is reported to have a 5n chromosome number (Okie, 1987). It is semi-dwarfing, winter-hardy, tolerant to high pH, waterlogging and resistant to bacterial canker.

4. *Prunus insititia* Selections

ST. JULIEN STOCKS: St. Julien plums mentioned in old horticultural literature, closely related to the damsons, and consitituting a division within *Prunus insititia.* St. Juliens have long been recognized as the dwarfing stock for plums. They are propagated from seed as well as by layers and cuttings. Many clonal selections have been made and the more important ones are enumerated below:

ST. JULIEN A: It is of value as a semi-dwarfing rootstock for plum which is precocious and exhibits good Ca absorption.

ST. JULIEN K: This is a very dwarfing plum rootstock but has not been fully tested as yet.

ST. JULIEN GF 655-2: It produces a semi-dwarfing tree with a very good productivity rating (Okie, 1987). It is resistant to bacterial canker and viruses but is susceptible to high soil pH.

ST. JULIEN HYBRJD 1: Selected from a cross between *Prunus insititia* and *Prunus domestica.* It produces a vigorous tree.

ST. JULIEN HYBRID 2: Also selected from a cross between *Prunus insititia* and *Prunus domestica,* it is a semi-dwarfing rootstock. It is unsuitable for light stony soils.

ST. JULIEN W 61: A promising semi-vigorous rootstock which can be vegetatively propagated by cutting under mist. Trees on it yield high with large fruits (Riesen and Zbinden, 1986).

BLACK DAMES: It is also known as the Black Damascena and resembles St. Julien. It can be propagated by seed as well as by layers and cuttings. A number of clonal selections have been made, which are identified by the suffixes' A', 'B', 'C', 'D' or 'E'. Of these the Black Dames C is considered of some value (Tukey, 1964). It produces a vigorous to an intermediate tree and may delay fruit ripening (Okie, 1987; Tukey, 1964).

COMMON MUSSEL: It can easily be propagated by root-cuttings. It makes a medium-sized tree being unsuitable for dry soils and tends to delay fruit ripening.

PIXY: This is one of the recently developed plum rootstocks for which *in vitro* micropropagation techniques have also been evolved (Jones and Hopgood, 1988). The rootstock is dwarfing, induces precocity and hastens fruit ripening. However, the fruit size remains small. Pixy has proved compatible with most of the European plum cultivars. It has strong resistance to bacterial canker but not to replant disease, silver-leaf or plum pox.

5. American Plum Rootstocks

Prunus americana: The rootstocks have been reported (Tukey. 1964) useful as moderately dwarfing rootstocks. It is compatible with a wide range of cultivars of European prune, damsons, Japanese plum and the native American types. However, the scion tends to overgrow the rootstock. The stionic region is brittle, often succumbing to strong winds. Propagated from seed, it is considered useful where winter-hardiness is major consideration.

BEACH PLUM (*Prunus maritima*): It has been classified as a very dwarfing stock (Okie, 1987). Its productivity is low and anchorage poor. It is propagated from seed.

SAND CHERRIES: There are a numerous species of sand cherries found on the North American continent. Of these, two are important as rootstocks from plums.

Prunus pumila has been found to produce dwarf trees that are precocious and highly productive. The western sand cherry (*Prunus besseyi*) is a very dwarfing rootstock for plum (Tukey, 1964). The productivity of plum trees on *Prunus besseyi* has been rated as good (Okie, 1987). It is sand cherry but root anchorage is poor.

NANKING CHERRY (*Prunus tomentosa*): This is also a very dwarfing rootstock which is resistant to drought and frost. It can easily be propagated by hardwood cuttings. However, its compatibility with the cultivars is reported to be highly uncertain (Tukey, 1964).

6. Romanian Rootstocks

ROSIOR VARATIC (*Prunus domestica*): It is a self-sterile seed propagated stock. Seedlings are more uniform than those of Myrobalan (*P. cerasifera*). Trees are efficient producers (Stefan *et al.*, 1968) and well-adapted to heavy, calcareous soils (Tudosescu *et al.*, 1974).

COROCODUS 163 (*P. cerasifera*): It is quite vigorous and is resistant to low winter temperature. It can be propagated easily by hardwood cuttings (Parnia *et al.*, 1988).

OTESANI 8 (*P. domestica*): It is a dwarfing rootstock similar to pixy, propagated readily by hardwood cuttings. It is also resistant to low temperature and is well adapted to clayey soils.

PORUMBAR DE LASI (*P. spinosa*): Trees are 2.3m in the 10th year in the orchard. Seedlings are compatible with all Romanian commercial cultivars. When using this stock, orchard density could be as high as 1250 trees/ha, with high yield and good quality fruit (Parnia *et al.*, 1988).

7. Other Species and Hybrids

Almond (*Prunus amygdalus*) and peach (*Prunus persica*) have also been used as rootstock for plum. Clonal selections of almond X peach hybrids have been made. Some of the characteristic features of these rootstocks are detailed below:

BITTER ALMOND (*Prunus amygdalus*): It produces a semi-vigorous tree. Propagated by seed, it is tolerant to boron toxicity and resistant to drought. It is incompatible with cv. Santa Rosa.

PEACH (*Prunus persica*): It is a semi-vigorous rootstock for plum. Mainly seedlings of cultivars Halford, Lovell, Nemagaurd and S-37 are used.

GF 677: A selection of *Prunus amygdalus* x *Prunus persica*. It is vigorous and resistant to drought but susceptible to waterlogging and root-knot nematode. Recently it was reported as resistance to natural Plum Pox Virus (PPV) infection and recommended to use in region where PPV is a threat (Polak and Kominek, 2014).

GF 557: Also a selection from *Prunus amygdalus* x *Prunus persica* crosses which produces a vigorous tree, resistant to drought, root-knot nematode, and high pH. It is, however very sensitive to waterlogging.

ISHTARA, (GF 322 x 871/1): It is a hybrid of plum cv. Belsiana with Myrobalan plum. It produces semi-vigorous trees that are precocious giving high yield of large sized fruits (Hartman, 1987). It can be propagated by soft or hardwood cuttings.

Almond rootstock compatible with Japanese Plums.

FELINEM: Almond X peach hybrid rootstock for almond developed at Spain, adapted to calcereous soils, resistant to broad range nematodes and vigorous rootstock.

GARNEM: Almond X peach hybrid rootstock for almond. Similar characters like Felinem.

At East Malling Research Station, UK, Atkinson *et al.* (1984) observed that four-year-old Victoria and Marjorie's Seedling plum on Pershore rootstocks were 22 per cent and 32 per cent larger, respectively than similar trees on St. Julien A, and 160 per cent and 196 per cent larger than on Pixy (Table 6.4). They also observed symptoms of incompatibility on *Prunus tomentosa* and Victoria trees on *P. pumila.* Poor compatibility was also observed where used as interstocks between St. Julien A rootstock and either Opal or Czar scions.

Table 6.4: Size and Yield of 4-year-old Victoria and Marjorie's Seedling Plums on Five Rootstocks*

Rootstock	*Victoria*			*Marjorie's Seedling*		
	Tree Head Volume (m^3)	*Yield/Tree (kg)*	*Fruit Size (g)*	*Tree Head Volume (m^3)*	*Yield/tree (kg)*	*Fruit Size (g)*
St. Julien A	22.0	18.9	40.1	18.1	5.2	51.1
Pershore	26.9	31.1	39.1	24.0	9.1	51.3
Pixy	13.4	16.9	30.4	s.:	3.6	44.1
P. pumila	9.0	8.1	37.0	12.3	5.4	37.8
P. tomentosa	3.1	1.9	20.6	1.2	3.4	49.7

* Atkinson *et al.* (1984).

Kosina (2011) studied the performance of plum rootstocks Fereley, Ishtava, Pixy, CF655/2 and St. Julian A using scion cultivars Stanley, Cacanska Lepotica and Valjevka for 11 years. The highest yield efficiency with all scion cultivars were observed on Fereley and Ishtava, while Pixy showed the lowest productivity with all scion cultivars. The growth vigour of trees trained as central leader was larger than those trained as a spindle. Embree *et al.* (1999) working with European plum cultivars and selections developed at Ontario, Canada; evaluated on a range of rootstocks of differing vigour. They concluded that the cultivar Vision was very productive for which tree size can be controlled by the rootstocks Pixy and St. Julian A.

Propagation

Plums do not come true to type from seed and hence its cultivars are propagated by budding on the seedlings or vegetatively propagated rootstocks obtained by means of layering or cutting.

Mirabdulbaghi and Zarghami (2011) recorded maximum rooting (47 per cent) of a natural plum X apricot hybrid by using hardwood cutting treated with IBA at 3000 ppm. Bal and Sandhawalia (2000) noted higher rooting in Japanese plum cvs. Alu Bokhara and Amritsari by using basal cuttings collected on 15 December and treated with IBA at 3000-4000 ppm. Samoschechenkov and Tikhomirov (2000) working with different substrate for rooting green cuttings of plum recommended that moss should be added to any rooting substrate (perlite, peat, sand, sawadust) for higher rooting and peat should be excluded from surface substrates in direct contact with the cuttings.Howard and Nahlawi (1969) stated that rooting occurred most readily during autumn and again in late winter and spring. Cuttings made from thin shoots generally rooted more readily than those from thick ones. Some clones, including Myrobalan B and St. Julien A, respond to increasing concentrations of IBA to an optimum at 5000ppm and to increasing bottom heat to about 20°C. Others, including Brompton and Pershore, were least responsive to both of these factors (Table 6.5). Ozkarkas and Balkan (1999) suggested raising of plum seedling in polythene bags for grafting or budding. Seedlings raised in polythene bags are free from nematodes and other soil diseases; their roots are not damaged during transportation, they can be planted at any time of the year, and more seedlings can be produced per unit area. They also suggested using media containing ½ peat +1/2 sand mixture 4.5 liter bag size. Morini (1984) reported a novel and rapid method of propagating plum grafts. Planting of Myrobaian cuttings in November after IBA treatment (2500 ppm) and bottom heat (21°C for 15 days) and then Kerf-grafting them in the field in February gave a success rate of 73.5 per cent. The plants were ready for planting out in the orchard after only nine months as compared with two years required for the traditionally grafted trees. However, the trees so produced are much smaller and may not do well in locations where the climatic conditions are not favourable. Plums are mostly propagated by T-budding in autumn or in spring (Hartmann and Kester, 1972), on rootstocks of seedling origin or clonally propagated ones. Chip budding is now considered to be better than T-budding as it gives better take and can be carried over a longer period (Kurbanova, 1988).

Table 6.5: Rooting Response of Five Clones to IBA Concentrations*

Clones	*IBA (ppm)*				
	0	*625*	*1250*	*2500*	*5000*
	(percent rooting)				
Spring 1967 at 21°C					
Myrobalan B	5	71	78	91	76
Autumn 1967 at 19°C Myrobalan B	0	17	28	50	71
St. Julien A	0	20	19	54	42
8.21	0	7	30	31	22
Brompton	1	3	4	4	9
Pershore	1	1	1	4	1

Per cent rooting has been obtained by the back-transformation of angular transformed data.

* Howard and Nahlawi (1969).

In vitro micropropagation techniques for large-scale multiplication of plum rootstocks and cultivars have also been developed (Jones and Hopgood, 1979; Baleriola-Lucas and Mullins, 1982). Howard *et al.* (1989) reported long term improvement in rooting of plum rootstock cv. Pixy cuttings taken from stock plant hedges of *in vitro* origin. Such rejuvenation was an apparently sustained even after nine years of establishment of cultured plants *in vivo*. Vorpi *et al.* (2011) described a mircropropagation protocol of plum (*P. domestica*) cv. Shengjine by using single-node and shoot apical explants from plantlets grown in the field. Shoot proliferation was obtained on modified MS medium supplemented with 0.5 mg GA_3, 0.1mg IAA and different concentrations of BAP (0.5, 1, 1.5 and 2mg/l). Explants were rooted on a medium similar to that for shoot development, but without cytokinins and with IBA at 0.5, 1 and 1.5mg/l. Shoot proliferation and shoot length was found to improve by the addition of crude extracts from shoot apical of the same plants, collected in March, to the nutrient media at 30mg/l. Rooted explants were transferred to pots containing equal volumes of peat and perlite, about 75 per cent survived. Zhou-HouCheng *et al.* (2011) developed a complete protocol for adventitious shoot regeneration *in vitro* from the leaves of plum rootstock Mariana (*P. cerasifera* X *P. munsoniana*). They used the first leaves along with their petioles from *in vitro*-grown shoots as explants. They obtained significantly higher regeneration percentage by using ½ MS or WPM medium than MS or SH medium. Explants of leaves with petiole were found more effective than leaves without petiole or petiole explants in terms of regeneration. The highest regeneration percentage of 81.7 per cent and a mean of 7.46 +1.38 shoos were obtained on ½ MS medium supplemented with 2.0 mg/l thidiazuron, 0.1 mg/l IBA, 0.25 per cent agar and 3 per cent sugar. Adventitious shoots were rooted and the highest rooting percentage (96.7 per cent) was obtained on ½ MS medium supplemented with 0.5-1.0 mg/l IBA. The rooted plantlets survived after acclimatization in the greenhouse.

Planting

For the European plums two-year-old feathered trees give a better establishment than one-year-old whips. For the Japanese cultivar (*Prunus salicina* and its hybrids) one-year-old whips are quite suitable. Spring planting (before bud break) is usually recommended, especially for areas with extreme winter conditions in order to avoid winter injury. Autumn planting however, may be undertaken in areas which experience only mild temperatures during the winter.

The plants on arrival from the nursery should be planted immediately in their permanent position in the orchard. If some delay is anticipated then the plants on receipt should be 'healed in' (La Rue and Gerdts, 1976). This consists of planting the plants in a shallow trench with their roots covered with moist soil to prevent drying.

If hardpan, plough pan or any such shallow-soil compaction layers exist then these must be broken before undertaking plantation work. The ground should be well worked at least down to the planting depth. It would be better if planting is done in previously dug and filled pits large enough to accommodate the root system in its naturally spread out position and deep enough to hold the plant at the same depth as they grew in the nursery (La Rue and Gerdts, 1976). Though, Teskey and

Shoemaker (1972) recommend setting the plants slightly deeper than their position in the nursery, if the holes are much deeper than the root system, undesirable tree setting will occur. Before setting the plants in their permanent orchard position all broken roots and also the long ones are trimmed to fit in the hole rather than to curl around. When the roots have been well spread out, they are covered with soil and farmyard manure mix (1: 1). Soil around the plant firmly consolidated to remove all air pockets. Finally a layer of loose soil is spread on top to act as mulch.

Trees should normally be headed back at planting time to 60 cm. Large side branches below this point should be removed or clippedback to one bud. The trunk should be painted white up to 15 cm from the headed tree top. Ordinary latex paint or lime mixed with linseed oil may be used. This will prevent sunburn and the flat headed borer damage. Alternatively, cardboard tree wraps can also be used.

Planting Distance

A square planting of 6 m rows with trees 6 m apart in the rows is considered ideal for almost all plum cultivars, though shading of the lower wood does eventually occur as plantations age. A closer spacing of 5.4 x 5.4 m may be adopted on very sandy soil. A few plum cultivars with a spreading type of growth habit have been reported to produce well when planted 6.6 x 6.6 m or 7.2 x 7.2 m on deep sandy loam soils (La Rue and Gredts, 1976). With the increasing availability of dwarfing rootstocks and efficient growth regulators high density plantings are being extensively experimented with. Koroid (1987) has reported yield increase up to 168 per cent by reducing the planting distance to 6.2 m and training trees on Tatura trellis. Other high density planting systems consist of free standing trees planted at closer giving a tree stand of 6666 trees/ha (Cepoiu, 1987). The trees for such high density plantings may be obtained by grafting the scions onto dwarfing rootstocks. Alternatively, scion cultivars have also been multiplied by hardwood cuttings and planted as own-root trees (Cepoiu, 1987). Mika and Buler (2011) developed an intensive plum orchard model. Six plum cultivars were grafted on semi-dwarfing rootstock 'Prune Wangenheim' (*P. domestica*) and planted at three desities (1000, 1250 and 1666 trees/ha). A new training system; central leader spindle was applied. In this system leader was not headed back after planting and summer training treatments were performed. From the third year onwards renewal pruning was done after harvesting. The new training and pruning system resulted in very rapid tree growth, much young wood, fruit bud formation on young wood and early bearing. Tree was found suitable for hand and mechanical harvesting within four years of planting. Miletic *et al.* (2011) working with two plum cultivars Cacanska Rodna and Staley planted at different plant densities of 2500, 1666, 1250 and 500 trees/ha observed the highest yield of 11.5 t in cv. Cacanska Rodna at a plant density 2500/ha compared with 1.48 t in control (500 trees/ha). In cv. Stanley, the highest yield of 7.56 t was however, noted at a density of 1250 trees/ha.

Training and Pruning

The plum trees vary greatly in growth and fruiting habit which is true even for cultivars belonging to the same species. The pruning procedures for individual cultivars will, therefore, have to be accordingly different.

Basically there are two distinct phases in the pruning of fruit trees, *viz.* training and maintenance. A young tree is pruned (or trained) to provide structural strength and maximum fruiting area and fruitfulness at an early age. For plum tree proper care taken at this stage is crucial, since unlike the peaches, plums do not respond favourably to heavy renewal pruning (Christopher, 1976). Plum trees are usually trained to modified central leader system or the open centre system. The Japanese plum trees are particularly more adaptable to the open centre system. The trees are usually obtained from the nursery as l-year-old whips or 2-year-old branched trees. The former are simply headed back to the desired height of about 60 cm. Heading back at a height lower than this usually results in trees branching too close to the ground. In case of older branched trees obtained from the nursery, not more than 3-5 appropriately placed existing branches may be selected to develop as scaffold branches and if the trees sought to be trained to the modified leader system then keep the central growth dominant over the scaffold branches. Primary branches should be spaced 15 to 20 cm apart along the trunk, but this may not always be possible because most branches arise near the point of heading (La Rue and Gerdts, 1976). Secondary branches are to be selected during the second, third and fourth dormant seasons. At the end of four years of growth and pruning, 7 to 9 well-spaced secondary branches are obtained, which do not hinder the light penetration into the centre of the tree and thereby stimulate fruiting there. Mika *et al.* (1999) in their experiment with plum cvs. Cacanska Rodna, Cacanska Rana, Cacanska, Najbolja, Sanctus Hubertus, Herman, Valor, Empress, Oneida and Bluefire grafted on *P. divaricata* (Myroblan seedlings) planted at a distance of 3.5 x 2.0m and trained as a spindle, hedgerow or central axis system recorded higher yield in trees trained as a spindle than those trained as hedgerow. In the fifth year after planting, yield per hectare ranged from 30 to 44 t for hedgerow trees and from 41-66 t for spindles. Sosna (2008) conducted field experiment in Poland to evaluate the yield performance of young plum (cvs. Hanita and Elena) trees based on the training system used. The trees were trained as near-natural crown shape (spacing at 4 x 3 m) and spindle (spacing at 4 x 2.5 m) in the first year after planting. He observed the highest yield of 38.6 kg/tree in cv. Elena in the fifth year by training as near-natural system. The trees of Elena grew more vigorously and their fruits were significantly smaller compared to Hanita.

Pruning in the formative years or during the pre bearing period is usually corrective pruning to increase light penetration. Heavy heading back of the shoots should be avoided as this would only encourage development of long upright shoots and too dense tops. Only light thinning should be adopted in order to admit sunlight, facilitate spraying operations and to develop a healthy spur system (Teskey and Shoemaker, 1972). However, in areas with warm dry summers and where irrigation is limited heavier pruning may have to be resorted to, so as to enable the tree to produce large fruits. Since, heavier pruning thins out fruit buds resulting in reduced crop load but increased fruit size (Magness, 1941). In bearing plum trees, pruning is directed towards increasing the size of fruits, preventing the breakage of branches as a result of too heavy crops and promoting continuous new growth.

Most of the plum crops are borne on vigorous spurs on 2-year-old wood. The fruit spurs of European plums differ from the Japanese types in that these are more in number and branched (Teskey and Shoemaker, 1972). Another fundamental difference between the two plum types is that the Japanese plums tend to overbear and may set such heavy crops that fruit size is considerably reduced. Keeping this in view, the Japanese plums require heavier annual pruning as compared to the European plums. This is however, not universally true for all Japanese plum cultivars, Santa Rosa being one such exception (Christopher, 1976).

The life-span of the plum spurs is believed to be 5-8 years, while it may be necessary to prune for some spur renewal each season, removal of most of the new growth (75 to 85 per cent) is recommended (La Rue and Gerdts, 1976). The extent of pruning ought to be so regulated as to induce an annual extension growth of 25 to 50 cm in the young bearing trees of both the European and Japanese plums. Fifteen cm of annual extention growth is considered adequate for the older trees of European plums. The older trees of Japanese plums, however, would require 25-30 cm of annual extension growth for proper fruiting (Teskey and Shoekmaker, 1972). Insufficient thinning of the interior of the canopy can result in a condition in which the bearing area is concentrated on the periphery of the tree canopy, while the interior of the tree remains unproductive. Kumar and Thakur (2012) suggested 25 percent head back pruning which showed good production and fruit quality of plum cv. Santa Rosa, Increasing the pruning intensity to 75 percent head back improved fruit quality but the production decreased.

Nutrition

Role of Nutrients

Nitrogen

Plum trees require a fairly large amount of N for growth and fruiting. Deficiency of N caused decrease in plant vigour, early maturity of fruit and reduction in yield (Sandvad, 1962). The cvs. Anna Spath and Vinete Romenesti produced maximum yield with 50 g N/tree (Negvila *et al.*, 1969). Pejkic and Boskovic (1985) reported that trees with optimal leaf N (2.68 per cent) content showed 39.43 per cent flower fertility and a yield of 47.0 kg/tree compared with 11.16 per cent flower fertility and a yield of 8.5 kg/tree with leaf N content of 1.67 per cent. Application N in the form of calcium nitrate at 500g N/tree or urea + lime 50 per cent at 500g N/tree increased fruit size and weight and improved fruit quality of plum cv. Santa Rosa (Saini *et al.*, 2013). Detzel *et al.* (2000) working with plum cv Ersinger observed that an increase dose of N Fertilization (80 kg/ha in March and 20 kg/ha in June) caused increased fertilization. The development of embryo sac and haustorium was accelerated and caused a significant increase in the relative fertility. In Cacak, Serbia, Pesakovic *et al.* (2008) recorded maximum yield of plum cv Stanley by application of 600 kg/ha of NPK 8:16:24+3 per cent MgO.

Phosphorus

Nyhlen (1965) suggested application of phosphate fertilizer before planting. Deficiency of P caused reduction in growth and tip burn of leaf (Ahmed and Twefik, 1962).

Potassium

Deficiency of K may cause crop loss up to 90 per cent. The deficiency may be corrected by soil application of muriate of potash (8 kg/tree) or potassium sulphate (9.5 kg/tree) and a single application is considered sufficient for 3-4 years (Olson *et al.*, 1987).

Calcium

Although plum trees require very little amount of calcium (Davis, 1934); a leaf Ca content of 2.0 per cent was found optimum for good yield of Stanley plum. Polish workers reported Ca concentrations of 5.7-6.6 and 8.5-9.1 mg/100g fresh weight in Stanley and Dabrowicka prune, respectively (Plich and Wojcik, 2002), Posthawest treatment of Japanese plums with calcium chloride solution reduced the physiological response to mechanical damage (Serrano *et al.*, 2004)

Magnesium

Magnesium deficiency resulted pale colouration of leaves and interveinal necrosis. Premature fruit drop also occurs due to Mg deficiency. Spraying of magnesium sulphate at 2 Ib/100 gallon after flowering and then at monthly interval was found effective to control the deficiency (Bowen, 1956).

Boron

European plums showed brown shunken areas in the flesh of the fruit as a single small spot or on the whole fruit (Hansen and Proebsting, 1949). While the heat spot, Kelsey spot or drouth spot was found in Japanese plum (Proebsting, 1937) due to B deficiency, Hansen and Proebsting (1949) suggested soil application of 0.5 lb borax in the late summer or fall to correct the deficiency.

Iron

Iron deficiency showed yellowing of leaves which finally turned ivory or even white (Linder and Harley, 1944). Yoshikawa *et al.* (1982) suggested injection of ferrus sulphate onto tree to correct the deficiency. The treatment also caused marked increase in fruit set in the following year.

Zinc

Deficiency of Zn showed chlorosis of leaves and crinkling of leaf margins (Beyers, 1969). Foliar spraying of zinc oxide or zinc oxysulphate was suggested to control the Zn deficiency.

Copper

Exanthema or die back of twigs was found due to Cu deficiency (Hoagland, 1941). Spraying of copper oxysulphate or copper oxychloride with hydrated lime was recommended to control the Cu deficiency (Beyers and Terblanche, 1971).

Fertilizer Recommendations

According to a long standing recommendation in Himachal Pradesh, India (Harbans Singh's undated farmers bulletin-2) newly planted plum trees should be given a dose of 450g calcium ammonium nitrate (CAN), 190g of superphosphate (SP) and 150 g of muriate of potash (MOP) besides 10 kg of compost. Thereafter, the dosage of these fertilizers is recommended to be increased at the rate of 450 g of CAN, 190 g of SP and 150 g of MOP for each complete year of plum trees in the orchard. The fertilizer dose being stabilized at the age of six years. Bajwa and Misra (1970) moderated these rates to 120 g of CAN, 95 g of SP and 60 g of MOP. Teskey and Shoemaker (1972) recommended application of 224 to 336 kg/ha of complete fertilizer in plum orchard. The fertilizer dose being further supplemented with additional potash for prune trees during years of heavy cropping to prevent decline.

Foliar application of urea (0.5 per cent to 2.5 per cent) for correcting nitrogen deficiency have been reported (Chandra *et al.*, 1988; Mishra, 1984), however, La Rue and Gerdts (1976) did not recommend foliar sprays as plum trees absorbed little nitrogen by this method of application. Foliar application of chelated mixture of Fe, Zn and Mn at 0.05 per cent in combination with GA_3 at 20 ppm increased fruit set, yield and improved fruit quality of plum cv. Hollywood growing under loamy soil condition of Giza, Egypt (Hassan *et al.*, 2010)

Table 6.6: Optimum Levels of Nutrient Element in Leaves of European and Japanese Plums

Element	*Optimum Range*	
	For European Plum (1)	*For Japanese Plum (2)*
N per cent	2.1 - 3.0	2.3 - 2.8
P per cent	0.15 - 0.30	0.1 -0.3
K per cent	2.0 - 3.5	< 1.0
Ca per cent	2.0 - 4.0	< 1.0
Mg per cent	0.2 - 4.0	< 0.25
Mn ppm	50 -150	20
Fe ppm	N.E	N.E.
B ppm	30 - 56	30 - 60
Cu ppm	5 - 10	4
Zn ppm	NE	> 15
Mo ppm	NE	NE
S per cent	0.2 - 0.7	NE
Na ppm	< 0.18	< 0.2
C1 per cent	< 0.69	< 0.3

NE = Not estimated

1. Shear and Faust (1980); 2. La Rue and Gerdts (1976)

Tissue Analysis

It is very difficult to generalize fertilizer and manurial schedule for fruit tree plantations. Particularly for those located on highly undulating mountaineous terrain. Such a topography *per se* indicates a highly heterogenous soil structure and fertility. As such a truelly effective schedule for a plum plantation ought to be arrived at after a thorough soil and leaf analysis. The optimum ranges of the various nutrient mineral elements have been arrived at and are presented in Table 6.6.

Leaf analysis has proved to be the most important guide to determine deficiencies and excesses in fruit tree nutrition. Like the other deciduous fruit trees, plum requires adequate amount of all major elements including nitrogen, phosphorus, potassium, magnesium, calcium and sulphur and the microelements like zinc, manganese, iron, boron, and copper. The most suitable time for sampling of leaves for foliar diagnosis has been determined to be during the second-half of June (Badyal and Kar, 1985) or the month of July (Sanchez-Alonso and Lachica, 1987). According to Vitanova (1986b), an eight- to eleven-year-old plum plantation annually removed various nutrients at the following rates:

Removal of Nutrients in kg/ha

N	24.63 - 62.58
P_2O_5	4.68 -13.12
K_2O	23.27 -61.84
Ca	3.91 -11.59
Mg	2.07 -6.73

This nutrient removal was in the form of fruits, fallen leaves and pruning wood. Vitanova (1986a) had also reported a negative correlation between N:K ratio in leaf and yield. An N:K ratio of more than 2 decreased yield in the following year whereas a ratio of 1.2 to 1.6 increased it. Cmelik (2000) studied the seasonal changes in calcium concentration of bearing plum shoots of cv. Bistrica. The study showed that seasonal changes in Ca concentration in particular plum shoots was quite different. Decreasing Ca concentration was observed in the vegetative and generative buds in the period from the beginning of vegetative growth to full bloom and in the fruits from when the shoot growth ceased up to fruit ripening. However, Ca concentration increased during growth in the leaf blade, leaf petiole and in shoot bark. Ca concentration in the woody tissues of shoots did not change over the growing season. In the bark of the previous year shoots, Ca concentration was gradually increased from the beginning of growth until full bloom. Vitanova and Bozhkova (1999) after studying the dynamics of the content of nitrogen and of some ash elements in the leaves of two Myrobalam plum cultivars stated that the concentrations of leaf nutrients were more stable from the middle of July to the middle of August. They concluded that this one-month period was the most suitable for sampling leaves for plant analysis.

Irrigation

Plum tress must have adequate amount of water available throughout the growing season in order to attain proper fruit size and induce sufficient new growth and flower bud differentiation. It is however, not necessary to maintain soil moisture far above the permanent wilting point, as no advantage in terms of greater proportion of large sized fruit is obtained (Teskey and Shoemaker, 1972). If however, readily available moisture becomes a limiting factor before the fruits are fully grown then the proportion of large sized fruits may be decreased (Hendrickson, 1948). La Rue and Gredts (1976) stated that plums require readily available moisture through harvest with decreasing amount for the rest of the season. Farmers tend to over-irrigate plum trees which is though not of any particular benefit, some tree losses can occur, especially if peach rootstock has been used. On the other hand, young trees are particularly sensitive to improper scheduling of irrigation. The root system of such plants is very limited and may experience dry condition in the small volume of soil occupied. Even though surrounding soil is wet. Such desiccations for short periods will result in erratic growth flushes and less total growth for the season. The irrigation requirement of plum trees has been estimated to be about 36 acre-inches. Treder *et al.* (1999) studied the effect of drip irrigation on plum cv. Valor grafted on Myrobalan and Wangenheim Prune. Drip irrigation significantly increased the tree growth, yield and fruit quality however the effect depended on weather conditions in the individual years. Lakatos *et al.* (2012) stated that micro sprayed irrigation significantly decreased the plants surface temperature and air temperature which helps in frost protection. It also delays flowering and improves fruit quality.

Regulated deficit irrigation (RDI) has been proven as a useful irrigation strategy in a wide range of fruit crops including plum to deal with water scarcity. However, long-term RDI techniques in developing orchards can deduce tree growth resulting in a loss of tree yield capacity. Strategies that could be used to quickly recover trees from the possible carry-over effects of long-term deficit irrigation are regulate crop load. Tree crop load is a determinant factor of the carbohydrates partitioning between fruit and vegetative sinks, being fruit sink strength higher than that of vegetative organs. The recovery of a young Japanese plum cv. Black Gold orchards after seven seasons under RDI was studied by testing combinations of two crop load levels (medium and low) and two drip irrigation regimes (100 and 133 per cent ETc) and two different number of emitters per tree (6 and 8) by Intrigliolo *et al.* (2014) at Spain. Results showed that trees watered at 133 per cent ETc and thinned at low cropload, which underwent a very severe RDI strategy previous to this experiment, were the trees that shared the greatest recovery in tree size. This work shows that crop load regulation is a useful tool to quickly recover plum trees from the detrimental effects of long term deficit irrigation.

Soil Management

A combination of cultivation and cover crops is generally recommended for the plum orchard. Teskey and Shoemaker (1972) consider plum trees performance better under sod and mulch management compared with that of peach and cherry trees. In Himachal Pradesh, India, plum trees, as is the case with other temperate fruit crops are grown under a semi-non-cultivation system. The basins of the trees, extending up to the canopy girth are kept relatively clean usually by hand weeding and the rest of the orchard floor is kept under permanent sod with occasional mowing/cuttings to cater to the requirement of milch and other cattle maintained in the farm. LaRue and Gerdts (1976) also recommend semi-non-cultivation as an ideal system for plum orchards. According to the system-described by them a 1 to 1.5 m strip with 15 cm high berms along the tree rows is kept clean by occasional cultivation or once-a-year pre-emergence herbicide application. Weeds or planted permanent cover crops, such as, the self-seeding clovers are regularly mowed.

Both clean cultivation, or bare no-cultivation and cover cropping have their disadvantages. Bare orchard floor tends to deprive the soils of its humus (Teskey and Shoemaker, 1972) and can experience water penetration problems (La Rue and Gerdts, 1976). Cover crops, though can improve soil structure, but orchards operations are hindered and they tend to make the orchard cooler during the bloom period as they accumulate less heat during the day and lose heat faster at night (La Rue and Gerdts, 1976).

Weed Control

Control of weeds in plum orchards is an important and essential orchard management practice. Ognjanovic and Saric (1988) working at Yugoslavia found that ploughing + discing + harrowing favoured the appearance of therophytes, particularly *Chenopodium* spp., *Amaranthus retroflexus, Echinochloa crus-galli, Setaria glauca, Atriplex patula* and *Daucus carota.* Discing and rotary cultivation favoured a different weed flora, mainly species that propagate vegetatively, including *Sorghum halepense, Cirsium arvense* and *Convolvulus arvensis.* They reported that use of simazine, dichlobenil, amitrole and amitrole + diuron + MCPA controlled weeds, including perennials, well without injury to fruit trees. Annual grasses and broad-leaved weeds in 1- to 3-year-old plums cvs. Kustendilska, Stanley, Strinova and Gabrovska were well controlled by 0.15 and 0.30 kg of imazaquin/ha. The addition of 1.32 kg pendimethalin/ha enhanced the activity of 0.15 kg imazaquin and showed season-long weed control (Nikolova and Baeva, 1987). In plum orchards of Himachal Pradesh, India, several monocot weeds like *Cyperus rotundus, Commelina nudiflora, Digitaria sanguinalis, Setaria glauca, etc.,* and dicots like *Amaranthus viridis, Chenopodium album, Medicago* sp., *etc.,* were recorded (Bhan *et al.,* 1983). Bromacil and atrazine at 2.5 to 7.5 kg/ha were found effective to reduce the weed population. Bhutani and Bhatia (1986) suggested use of terbacil (3.0 kg/ha) or simazine (4.0 kg/ha) in plum orchards, as this herbicides also increased leaf N, K and Ca contents. Both diuron and atrazinc at 4 kg/ha proved effective in controlling weeds in 12-year-old Santa Rosa plum orchards (Khokhar *et al.,* 1986). The selectivity of new formulations of glyphosate (0.587 to 1.440 kg/ha) and sulfosate [glyphosate] (0.480 to 1.440 kg/ha) to plum (*P. salicina*) cv. Carmesin was tested by Oliveira-Junior *et al.* (2000) at Brazil. Post-emergence spraying of these herbicides between crop rows, even at highest rates, did not cause any injury sysmptoms on plums. Perennial weeds such as *Paspalum conspersum, Cyperus ferax* and *Boerhavia decumbens* were controlled effectively only with the highest rate of each herbicide, whereas *Spermacoce latifolia* and *Bromus catharticup* control was achieved with lower rates. Soil herbicide flumioxazin (Pledge 50 WP) at 4 kg/ha was found effective in young plum plantation when applied in the in-row strip. It control in-row weeds and the posteffect lastsed about 5 months (Rankova and Popev, 2011)

Flowering

Diploid plum species such as *P. salicina* and *P. cerasifera* are mostly self-incompatible, while fertility relations of the hexaploid European plum (*P. domestica*) cultivars vary between self-fertile, partially self-fertile and self incompatible (Szabo and Nyeki, 1996). European plum (*P. domestica*) and Japanese plum (*P. salicina*) cultivars are also classified as early-medium or late-flowering (Nyeki *et al.*, 1985). Vitkovskii (1986) in studying of over 1000 cultivars and hybrids representing several plum species, classified into 5 basic types with petal shape being the most polymorphic trait. Gavrilina (1985) working with 15 cultivars in Leningrad province classified the cultivars into 3 groups according to the date of differentiation of the floral primordia (10-13 July, 18-23 July and 27 July to 6 August) and in respect of mean duration of the period of floral development (to pistil formation), with Evraziya 21, Kornpotnaya, Renklod Pushkinogorskii, Pamyat, Tirniryazeva, Ternosliv Dubovskii, Ternosliv Osennii 7, Smuglyanka and Iskra having the shortest period (66-71 days) and Renklod Zheityi Mestnyi the longest (82 days). Iliev (1985) stated that flowering of plum in the Dryanova area of Bulgaria lasted on an average for 4 weeks. In an investigation with more than 60 cultivars, he found that 54 trees were in full bloom for 6-8 days. In India, Sud (1979) found that the flowering varied from 11-14 days, which started on 5 March and ended on 21 March. In UK, spraying with ethephon at 250, 500 mg/I in late September was found to delay the flowering by 4 and 7 days, respectively, of plum cvs. Victoria and Czar (Webster, 1986).

Pollination

Most European and Japanese plum cultivars grown commercially require pollination to set satisfactory crops. For a cultivar to be classified as self-fruitful, it should exhibit an average of 30 per cent fruit set which is considered adequate for a commercial set (La Rue and Gerdts, 1976). The cultivars that are self-unfruitful set only 1 to 2 per cent fruit. These cultivars must be interplanted with pollenizing cultivars. Most Japanese plum cultivars are self-unfruitful and require cross-pollination (Arora and Singh, 1987a). Some cultivars like the Santa Rosa and Beauty are partially self-fruitful, and may bear heavier crops when planted among other cultivars.

Both in the *domestica* and *salicina* plum; self-fertile, self-sterile and partially self-fertile cultivars are known. Cross-incompatibility also occurs both in the *domestica* (Childers, 1973) and *salicina* cultivars (Theile and Strydom, 1964), but is not very common. Iliev (1985) distinguished three groups in respect of self-fertility: (i) the self-sterile Zh'lta Edra (Yellow Large), Zelena Renkloda (Green Gage), Moldavskaya Rannyaya (Moldavian Early), Pacific and Rotunda (ii) the partially self-fertile Imperial and Stanley and (iii) the self-fertile Rimska, Blufre, Opal, Paczelt, Pozegaca, Sopernitsa and Renklod Khramovykh. The last four cultivars were found capable of self-pollination without the aid of insect pollinators. Gavrilina (1986) observed high self-fertility in Iskra, Pamyat' Timiryazeva, Kreek and Mirapel' Krasnopakharskaya. These had a fruit set of 18.3-22.3 per cent under natural and 10.1-30.9 per cent under artificial self-pollination compared with 29.3-47.5 per cent under open-pollination. The cvs. Early Round and Alucha Black were found to be self-compatible (10.8-27.1 per cent fruitset). The highly compatible pollen helped weakly compatible or incompatible pollen to fertilize some of the ovules, presumably by the exchange of pollen growth substances (Arora and Singh, 1987b).

Self-fruitfullness is not a constantly stable character and may vary from year to year and area to area. The cultivar Santa Rosa, for example, is considered self-fruitful in India and in.some areas of the USA, where it gives heavy yield when planted in solid blocks. However, in some northerly located areas of California and Canada, Santa Rosa requires cross-pollination by bees from other cultivars.

The germination and development of the pollen tube are greatly affected by the prevailing temperature during the bloom period. Pollen tubes take only 4 days to reach the embryo sac at 20° to 21°C, while at 10° to 11°C they take 8 days. The receptability of the styles is maintained up to a maximum of 6 days (Lee and Benemann, 1981). Low temperatures at blooming time are thus highly detrimental to fruit set. The growth of the pollen tube is promoted by the presence of a large number of pollen grains on the stigma (Gur, 1986). Inefficient pollenizers probably due to low germination of their pollen are quite common among *salicina* plum cultivars (Cobianchi, *et al.*, 1978).

Many factors are to be taken into consideration while selecting the right combination of cultivars to ensure consistent fruit set. Cultivars must bloom at about the same time and provide adequate pollen capable of setting fruit in order to be good pollenizer. Warm weather in February following adequate chilling in winter usually results in a relatively short bloom period during which the blooming of most cultivars overlaps. However, following mild winters blossoming is often delayed and the bloom period greatly lengthened. In such a situation certain cultivars may not cross pollinate because their bloom is delayed too long to coincide with cultivars having a shorter chilling requirement. This is further complicated by the fact that certain varietal combinations are ineffective as pollenizers for each other, though they are good pollenizers for other cultivars. Certain other varietal combinations are not fully inter-compatible. Cultivar Laroda, for instance, can effectively pollenize cv. EI Dorado but *vice-versa* is not true.

Other factors that are to be taken into consideration while choosing an appropriate combination are the commercial value of each cultivar, time of fruit maturity, area where grown, bee activity and proximity between trees of the pollenizing variety and that to be pollenized. Excessive pollination leads to excessive fruit set resulting in high thinning costs. Slight under-pollination often minimizes thinning cost yet provides adequate crops of good-sized fruits (La Rue and Gerdts, 1976).

Honey-bees are the single most important pollinating agent in a plum orchard. Though in some orchards sufficient wild bees may be present yet for consistently good yield honey-bee hives should be placed at strategic locations throughout the orchard. The most effective pollination occurs within the radius of 100 m from the colonies. Since it is not always possible to space the hives in overlapping 100 m circles, in practice the hives may be scattered in groups of 5 to 10 on the orchard periphery. In large orchards, however, some colonies may have to be placed within the orchard. Elimination of blooming weeds and cover crops is essential for effective pollination.

Fruit Set

Some Japanese plum cultivars are particularly prone to erratic fruit set showing very low or even no fruit set. High incidence of premature ovule degeneration was reported to be the cause of poor fruit set in some cultivars (Guerra *et al.*, 2011) Nakajo *et al.* (2014) made histological study of ovule development in triploid Japanese plum cv Kiyo. They observed that even during the flowering time the embryo sac of 60 per cent or more remained undifferentiated. A higher rate of embryo sacs with developmental delay/or degeneration, and embryo sacs without a zygote and/or endosperm nucleus were also recorded and were identified as the cause of poor fruit set.

The fruit set varies with the cultivars and depends largely on the environmental factors. A fruit setting of 4 per cent to 8 per cent with heavy bloom and 12 per cent to 18 per cent with light bloom is considered satisfactory (Thompson and Liu, 1972). Under open-pollination the fruit set was found 9.9 per cent in cv. Blufre and as high as 33.4 per cent in Imperial (Iliev, 1985). Ughini and Roversi (1987) found a considerable higher percentage of fruit set on 1-than on 2-year-old shoots. The fruit set was found unaffected by temperature during flowering but the occurrence of rain during the hours of bee flight (0900 to 1700 h) reduced it. In drought susceptible cultivars, increased leaf and fruit drop under unfavourable conditions was found associated with decrease in the nucleic acid and protein N contents of the fruit and leaves and the rapid formation of abscission layers (Goncharova *et al.*, 1985). Spraying of gibberellic acid (GA_3) to open pollinated plum was found to increase final set and yield (Ugolik, 1966). Sprays containing GA_3, auxin and MN'-diphenylurea (DPU), a substance showing cytokin-type activity, have also induced large yield increase of Victoria plum (Webster and Goldwin, 1978, 1981). Excessive fruitlet abscission in late May and June is frequently the cause of poor Victoria plum yield in the UK. Webster and Treharne (1983) reported that spraying of GA_3 + 2, 4, 5-TP at 4 and 6 weeks after 50 per cent petaffall considerably reduced fruitlet abscission. Rangelov *et al.* (1985) found that application of GA_3 at 50 and 100 ppm with 2,4, 5-TP at 10 ppm on the 10th day after full bloom increased fruit set of cv. Burja from 3.1 to 17.7 and 6.5 to 23.0 fruits/100 flowers for the two concentrations, respectively. In Sangold plums, paclobutrazol application once at 500 µg/ml in autumn or at 250 µg/ml in autumn followed by spring application enhanced fruit set and increased yield (Strydom and Honeyborne, 1986).

Fruit Thinning

Some plum cultivars tend to overbear (Teskey and Shoemaker, 1972) particularly the Japanese plum. Removal of part of the crop enables the remaining crop to receive a greater share of the food manufactured by the leaves. Thinning also results in removal of the undesirable fruits and thus reduces limb breakage. Early thinning tends to increase average fruit weight and hasten maturity without affecting total yield (Belmans and Keulemans, 1987). Heavy crop of plum trees in one year showed marked effect on flowering, tree growth and fruit soluble solids content in the following year (Couranjou, 1970, Kwong, 1973). The fruit size of Japanese (Bajwa and Singh, 1965, 1970) and European type cultivars (Painter, 1930) were found to increase by thining of fruits. Wells and Bukovac (1978) reported that fruit thinning significantly increased fruit size, soluble solids and colour, and markedly decreased yield in Stanley plum (Table 6.7). European type cultivars appear to be more efficient than Japanese type cultivars in sizing fruits borne in clusters, and thus require less thinning (Allen, 1939). Severe fruit thinning has become a common practice for *salicina* plums (Gur, 1986). La Rue and Gerdts (1976) advocated hand thinning, though the possiblity of using mechanical shakers followed by selective, touch up hand thinning is also recognized. Seehuber *et al.* (2011) evaluated mechanical thinning of European plum cv. Ortenauer in Germany. They reported that thinning of flowers by using a rotor (Speed 200-500 rpm) after 30 days of full bloom thinned 1/3 flowers even with the slowest rotor speed of 300 rpm. The number of fruit per bunch reduced significantly from 152 to 67-76, equivalent to a (source: sink) leaf: fruit ratio of 5:1. Mechanical thinning also enlarged fruit mass from 28g in the un-thinned control to 30-32g with rotor speed of 400 or 500 rpm. Thinning of plums by the use of chemicals has also been recommended. Batjer (1965) suggested a spray of 4, 6-dinitro-ortho-cresol (DNOC) 42 to 85 ml per 100 litre, quite in accordance with the recommendation of Teskey and Shoemaker (1972) who suggested the use of dinitro 318 at ½ pint per 100 gallon (or 63.33 ml per 100 litre). Other chemicals that have been suggested for use as plum fruit thinners are 2, 4, 5-trichlorophenoxy acetic acid with or without gibberellic acid (Paunovic and Ogasonovic, 1979), naphthaleneacetic acid (Lal and Thakur, 1979), 3-chlorophenoxypropionic acid or 3-chlorophenoxy propionamide or a mixture of both (Webster, 1980; Beutel, 1969) and (2-chloroethyl) phosphonic acid (Martin, *et al.*, 1975). However, response to these chemicals is quite uncertain (Dodd, 1967; Webster, 1980) and none can be recommended for use on all the cultivars of plum. The spray should be applied slightly before full bloom to obtain adequate thinning. In case inclement weather occurs during the pre-bloom

Table 6.7: Effect of Fruit Thinning on Yield, Fruit Size and Quality of Stanley Plum*

Observation	*Degree of Thinning*		
	Light	*Medium*	*Heavy*
Fruit load (fruits /spur)^z	1.9	1.1	05
Yield /tree (kg)	36.5 b^Y	42.4 b	14.4 a
Yield (g /cm^{2}trunk cross-sectional area)	647 c	473 b	233 a
Fruit cheek dia (mm)	32.8 a	34.6 ab	35.8 b
Fruit wt (g/fruit)	25.2 a	28.3 ab	31.1 b
Soluble solids (per cent)	10.8 a	11.2 a	\2.1 b
Fruit color (absorbance 515 nm)	0.37 a	0.36 a	0.44 b

Z: The thinning treatments were established during late June drop on June 24, 1976 (77 days before harvest); Y: Mean separation in rows by Tukey's w test, 5 per cent level

* Wells and Bukovac (1978).

period, which could quantitatively affect the full bloom and subsequent fruit set, the spray may be delayed until a day or two after full bloom.

Lupescu and Tertecel (1985) in trials with 11-year-old Stanley plum, sprayed Flordimex (ethephon) at 200 or 400 ppm or NAA at 20 or 10 ppm, 2 weeks after full bloom. They found that ethephon at 200 ppm produced yields of 28-29 tonnes/ha compared with 23-27 tonnes/ha in the control. Ethephon also increased fruit weight by 3-4 g compared with control while NAA overthinned and reduced yields by 3-7 tonnes/ha below those of control. Webster (1984) noted severe fruitlet abscission by treatment with paclobutrazol (PP 333) treatment on 2nd June and yields were found to reduce to 17 per cent and 9 per cent of unsprayed controls by the 1000 and 2000 mg/l treatments, respectively, whilst fruit size was increased by 37 per cent and 28 per cent (Table 6.8). Treatments applied earlier in the season has smaller effects on fruitlet retention but all improved fruit size compared with untreated trees.

Table 6.8: Effect of Paclobutrazol Sprays on Yields and Fruit Size of Victoria Plum*

	Control	*Date of Sprays and Concentrations (mg/l)*							
		21st April		*12th May*		*2nd June*		*23rd June*	
		1000	*2000*	*1000*	*2000*	*1000*	*2000*	*1000*	*2000*
Yield (kg / tree)	164	139	131	106	34	29	15	139	148
Mean fruit weight (g)	35.8	40.8	44.3	41.5	49.1	49.1	46.0	34.4	32.4

* Webster (1984).

Proper timing of the thinning is crucial for optimum crop development at lowest cost. Thinning too early, besides being labour intensive, involves risk because of the possible fruit drop due to hails, frosts and the pre-harvest drop. Too late thinning would result in only partial increase in fruit size. The operation should be completed before the pit-hardening stage. Thinning after this stage does not produce maximum fruit size (La Rue and Gerdts, 1976).

Fruit Growth and Development

In all the *Prunus* species the pattern of growth of fruit follows the double sigmoid curve (Thompson and Liu, 1972; Ugolik, 1970). The first zygotic development may be evident only after 5 to 19 days after fertilization (Cerkova and Skipina, 1966). According to Sterling (1953), during the subsequent development of the fruit the outer wall of the ovary gives rise to three distinct layers of the pit. Two hypodermal layers in the ovary produce the skin of the fruit which consists of the cuticle, the epidermis and a few layers of collenchymatous cells. The flesh contains most of the vascular bundles of the fruit which are embedded in the lignified tissue of the pit.

Prior to fertilization, one of the two ovules of the plum fruit aborts. The side with the persistent ovule develops more rapidly and may ripen somewhat earlier than the opposite side. The seed is attached to the more developed side of the carpel. This probably explains the asymmetry in fruit development (Tukey, 1936; Stosser and Hartman, 1982). The fresh weight of fruit and size increased throughout the growing season (Josan and Chohan, 1982); whilst the weight of ethanol insoluble material in Victoria plum was found to increase to a peak and then declined (Boothby, 1983). The sugar content of fruit increased gradually and rapidly during ripening. Sorbitol also rose gradually and then sharply while the N, P, K, Ca and Mg contents in fruits were found maximum in newly set fruit decreased at ripening and stabilized in ripe fruits (Pangelova-Shchrkova and Vitanova, 1987a). In cvs. Stanley and Strinava, intensive pericarp growth was found before the start of ripening, whereas the seeds had a high growth rate from the start and reached maximal size about a month before ripening. Dry matter content increased gradually then more sharply at ripening. The organic acid content increased initially but decreased during fruit development (Pangelova-Shchrkova and Vitanova, 1987b). During maturation glucose and galactose were found to be the major sugar, sucrose content being very low; however, sucrose accumulated rapidly during the last few weeks before harvest, while fructose and glucose decreased (Vangdal, 1982). Determination of bioactive GA content during plum fruit ontogeny revealed that GA_1 and GA_4 are critical for fruit growth and development. Characterization of several genes involved in GA signaling showed that their transcriptional regulation are generally GA dependent, confirming their involvement in GA signalling. Based on these results, El-Sharkaway *et al.* (2014) developed a model which presented how the potential association between GA and other hormones may contribute to fruit development, Ps1G ID1 protein structure, Y_2H and BiFC assays indicated that plum GA receptors can form a complex with AtDELLA-receptors in a GA-dependent manner. Moreover, phenotypical-, molecular- and GA-analyses of various Arabidopois background ectopically expressing Ps1G ID1 sequences provide evidence on their role as active GA signaling components that mediate GA-responsivenes. The findings of El-Shavkaesy *et al.* (2014) support the critical contribution of GA alone or in association with other hormones in mediating plum fruit growth and development.

Maturity

Plum should be picked when mature enough to continue its ripening process off the tree. The mature fruit is not eaten ripe at time of harvest, but it can attain the desirable quality following a postharvest ripening period. The appropriate maturity standards vary with the cultivar and the intended use of the fruit. Plums grown for distant markets should be picked when firm ripe. At this stage the fruit is less likely to be injured during handling and storage. Optimum picking maturity varies from cultivar to cultivar. It has been found to be strongly correlated with cummulative day temperature (Gigleux-Spitz, 1988, Sud and Kar, 1985). Chopra, *et al.* (1986) had sought to standardize picking dates on the basis of days from full bloom (DFFB). They determined 94± 3 DFFB as optimum for picking cv. Santa Rosa, however, such predictions would be valid for areas with relatively constant summer temperature or for species in which very high summer temperatures do not hasten fruit maturity. With prunes, however, the ripening process is considerably delayed by temperatures above 40°C (Baker and Brooks, 1944). In spite of this, counting the days between full bloom and maturity have been recommended for several *domestica* and *salicina* cultivars (Samish, 1948; Westwood, 1978; Chopra *et al.*, 1986). Josan and

Chohan (1982) suggested harvesting of Kala Amritsari plum around 4th June. *i.e.*, after 84 days of fruit set under subtropical conditions at Abohar, India. Some of the indices of maturity that are considered more reliable are:

(i) **Per cent fruit surface colouration** : In cv. Santa Rosa, fruit colour and firmness were found to be the best indices for picking maturity. With Hary Pickstone fruit colour was found only reliable index and for Sangold colour and TSS. In cv. Gaviota, colour, firmness and TSS were found equally reliable indices (Torman and Van Zyl, 1982). Borsheim (1980) stated that the differences in colour between the cultivars were due to differences in pigment concentration rather than pigment composition. However, fruit colour was found affected by the fruit load on the tree and was related to fruit size. and it varied with harvesting date (Gigleux-Spitz, 1988).

(ii) **Spectrophotometric method**: This has been successfully applied to plums (Krivoshiev and Chalukova, 1981). Borsheim (1980) found a strong correlation between visual assessment of fruit maturity and assessment by colorimetric determination.

(iii) **Fruit flesh firmness as determined by a standard pressure tester** (Westwood, 1978): Recommended pressures for plums range between 4.1 to 0.1 kg varying with cultivar.

(iv) Total soluble solids (TSS) content of the juice has also been suggested as a reliable index for determining fruit maturity. However, the TSS values varied widely among cultivars (Davey, 1978). Josan and Chohan (1982) suggested harvesting of Kala Amritsari plum when the TSS value is around 13.0 per cent,

(v) **TSS/acid ratio**: The ratio between the two has been considered a more reliable criteria for determining fruit maturity than either of them alone. Increasing evidence of consumer research warrants more reliance of soluble solids content, titratable acidity and their ratio for good consumer experience (Crisosto *et al.*, 2007). In South Africa and Australia, the harvest maturity is determined on the basis of a combination of parameters such as skin colour, firmness and soluble solids content. The harvest maturity standards for some commercial cultivars of plums exported from Western Australia (Ward and Melvin-Carter, 2001) are listed in Table 6.9.

Maturity of plum could be enhanced by applying ethephon to developing fruits. Cassibba *et al.* (1979) studying with Stanley plum in 3 localities at Italy reported that application of ethephon at 200 or 250 ppm when the fruits were 30-50 per cent coloured, advanced commercial maturity by 7-12 days without, affecting quality. In Santa Rosa plums, spraying of ethephon at 80 ppm 3 weeks before maturity, advanced the anthocyanin development by about 10 days and was found accompanied by a rapid increase in total sugar content (Lee and Lee, 1980). However, Testoni and Spada (1981) found increase fruit drop, reduction in eventual colouring, size, firmness and soluble solids by application of ethephon on trees previously treated with daminozide.

Table 6.9: Harvest Maturity Standards for Export Plums in Western Australia*

Cultivar	*Colour Description*	*Firmness Range (kg)*	*Minimum SSC (per cent)*
Blackamber	Deep burgundy to purple	6.0 - 7.0	11.0
Amber Jewel	Dark red with remainder yellow/green	5.0 - 6.0	19.0
Santa Rosa	50 per cent red to full red over yellow/ green	4.5 - 7.0	10.5
Laroda	Dark red with remainder yellow/green to yellow	6.0 -7.5	13.0
Friar	Dark red to burgundy	8.0 - 9.0	13.0
Tegan Blue	Deep maroon to purple	7.0 - 8.0	15.0
Radiance	Yellow with red blush	4.0 - 6.0	16.0
Autumn Giant	Bright red to dark red	7.0 - 8.0	13.5

* Ward and Melvin-Carter (2001).

Harvesting

Plums are very perishable and delicate and hence must be picked with care. They are picked by hand into buckets or baskets with padded liners. So far as possible stems (peduncles) should be allowed to remain attached to the fruit (La Rue and Gerdts, 1976). Mechanical harvesting as yet has not been found feasible with plum (Kader, 1983). Precaution must be taken not to puncture the fruit with finger nails, *etc.* The pouring of fruit from one container into another should also be done very carefully and slowly with minimum damage. Care should also be taken to prevent direct exposure of the fruit to sun after harvesting.

Plums are packed in shallow crates about 4 to 5 inches deep, packing not more than three layers of fruit. After packing, the fruit should be cooled immediately to 0°C. (La Rue and Gerdts, 1976; Sekse, 1989), which would stop ripening process in plums for a period of approximately 12 days (Teskey and Shoemaker, 1972).

Storage

Cold Storage

Most of the plum cultivars can be stored at 0°C for 3-4 weeks. However, plums or prunes are susceptible to cold storage disorder when stored at 5°C than at 0°C. Storage disorders symptoms include flesh browning, gel breakdown, mealiness, flesh translucency, red pigment accumulation, over ripening, and loss of flavour (Dodd, 1984, Crisosto *et al.*, 1999). Cultivars Elena, Haganta, Hanita, Jojo, Presenta, President, Topgigant and Tophit can be stored at 1°C for 30-40 days. The cvs. Presnta and Elena can be stored even until 60 days at 1°C (Blazek and Pistekova, 2011). The cv. Reine Claude D'Althan can be stored at 5°C and 90-98 per cent RH, whereas cv. Belle de Louvin loss firmness rapidly and showed poor flavour at this temperature (Peirs *et al.*, 2000). Cacanska Najbolja plum can be stored at -0.5°C for upto eight weeks with minimal internal breakdown and only moderate loss of firmness (Plich, 2008).

CA Storage

Carbon dioxide concentrations of 10 per cent or above have been found to delay the onset of ripening of plums and to lengthen the subsequent ripening period (Allen, 1939). However, box liners made of polythene film to provide modified atmosphere have not proved useful for several plum cultivars (Huelin and Tindale,1940). Melvin Couey (1960) stated that temperature near 1°C were required for optimum storage of Eldorado plums. Fruits held at 3° to 7°C softened and coloured during storage and developed gelatinous texture which was very undesirable. Plums stored at a lower than normal temperature (-2°C) for 1-1.5 months showed no reduction in their organoleptic and marketable qualities. Storage life was greatly extended with a combination of CA storage and at this temperature of -2°C (Gajdaj, 1986). The reported beneficial effects of CA/MA in European and Japanese plums include delayed fruit ripening, reduced rates of respiration and ethylene production, inhibition of fruit softening and skin colour changes and reduction of incidence and severity of chilling injury. CA is commercially used for long distance transport of plums. Evolution of CO_2 from fruits during storage was found to be strongly temperature dependent and distinct differences in postharvest respiration rate were noted among the cultivars. At low storage temperatures (0° and 4°C) fruit rot development was distinctly slower than at higher temperatures (8-12°C) and slightly unripe fruits had better resistance to fungus rot infection than had tree ripe fruits (Sekse, 1989). Streif (1987) stored firm, ripe and unblemished plums at 2°C under controlled atmosphere (CA) with 3, 6, 12, 15, 20 or 25 per cent CO_2 and 16-21 per cent O_2 for 2, 4 or 10 days before being transfered to a cold store (2°C) without CA. Fruit firmness in CA with 3-15 per cent CO_2 and 16-21 per cent O_2 was found similar to that of control fruit after 4 weeks storage, with 20 per cent and 25 per cent CO_2 there was only a slight positive effect over controls. In second experiment, the fruits were stored at 2°C with 3, 6 or 12 per cent CO_2 and 2 per cent O_2 or 25 per cent CO_2 and 18 per cent O_2 throughout the storage period and he found that highest CO_2 and lowest O_2 concentrations gave the best fruit firmness, least change in fruit colour, best flavour and the smallest increase in sugar content after 4 weeks' storage.

Tsuji *et al.* (1983) reported that anthocyanin content did not vary during storage at 3°C, but in fruits stored at 20°C. Naichenko and Romanshchak (1984) suggested the optimum storage conditions for plum cv. Vengerka Obyknovennaya in CA with 5 per cent CO_2 + 3 per cent O_2 + 92 per cent N_2 at -2°C in which the fruit stored well for 125 days, compared with 19 days at 0 or -2°C only.

Extensive low temperature breakdown during cold storage of plums has been observed particularly when plums are stored at -0.5°C. This low temperature damage consists of jellying of the flesh and is due to damage of the mechanism by which insoluble pectin is transformed to soluble pectin (Gur, 1986). Low temperature damage can be prevented by (i) controlled atmosphere (CA) storage at 2 per cent to 5 per cent oxygen and 1 per cent to 8 per cent carbon dioxide and 4.4°C, alternatively the fuit may be stored in polyethylene wrappers wherein a micro environment of 10 per cent to 16 per cent oxygen and 6 per cent to 7.8 per cent carbon dioxide at 4.4°C created; (ii) delaying storage at -0.5°C by 3 days during which period the fruit is held at ordinary room temperature; or (iii) by interrupting cold storage for 2 days and holding the fruit at 18.3°C (Gur, 1986). Low fruit N content in October and high fruit Ca and Mg contents at harvest showed a low incidence of disorder in storage of plum cvs. Santa Rosa, Hary Pickstone and Sangold (Kotze *et al.*,1989).

Other Treatments

Effect of spraying of amino ethoxy vinylglycine (AVG) and methyl jasmonate (MeJA) at 2 weeks ahead of estimated harvest date on plum cvs. Black Amber, Black Beauty and Fortune was tested by Ozturk *et al.* (2013). Fruits were stored at 0° and 5°C and 90±5 per cent RH for 4 weeks. AVG treatment resulted in significant increase in flesh firmness at the end of storage. MeJA increased significantly the firmess of Black Beauty and Fortune fruit at the end of storage. AVG treatment increased the contents of chlorogenic and *p*-coumaric acid in Black Amber fruit during storage. MeJA treatment increases the content of rutin, ferulic acid and naringenin in the fruits of Black Beauty and Fortune.

Post harvest application of 1-MCP at 1µl/l delayed skin colour development, decay and firmness of Japanese plum cv. Formosa at storage (0°C for 25 days). The treatment also inhibited chilling injury (Bae RoNa, 2011). Treatment with 1-MCP (1µl/l) preserve fruit quality during storage at – 0.5°C for 30 days of plum cv. Laetitia and the effect of 1-MCP persists even after removal fruit from cold storage (Alves *et al.*, 2010) Erkan and Eski (2012) reported that combined treatment of nonperforated modified atmosphere packaging and 1-MCP (500 ppb) of Japanese plum cvs. Autumn Giant and Black Beauty effectively decreased weight loss, softening of fruit, increasing shelf life and maintain higher overall fruit quality at storage (0°C for 60 days). JiHua *et al.* (2012) suggested use of film packing (30µm), ethylene absorbent and 1-MCP (0.5µm/l) treatment after harvest for extended storage life of Angeleno plum.

Ripening

In California, plums are usually harvested for eastern shipment before they are fully ripe. It has been observed that some cultivars will not ripen normally at temperatures of 80°F or higher (Pentzer and Allen, 1944). Davis and Boyes (1940) and Smith (1939) were able to stimulate the ripening of plums with ethylene and acetylene. Uota (1955) made quantitative determinations of carbon dioxide, ethylene and other oxidizable volatiles on three cultivars of plums at different temperatures. 'Beauty' plums ripened normally at 90°F, had a higher respiration rate and produced more ethylene than at lower temperatures. 'Santa Rosa' plums had a higher rate of respiration at 90°F than at lower temperature but the fruit did not colour intensely at that temperature although the ripening appeared to be normal. Ethylene production was found almost completely inhibited at 90°F in 'Durate' plums unless first treated with ethylene and the fruit did not ripen normally at that temperature.' although the rate of respiration was highest at 90°F.

Most of the Japanese plums are climacteric. However, few cultivars are classified as suppressed-climacteric, that are unable to biosynthesize ethylene in a sufficient quantity to induce a climacteric rise in respiration and ethylene production (Diaz-

Mula *et al.*, 2009). Some of the reported climacteric cultivars are Blackamber, Amba Jewel, Gulfruby, Beauty and Tegan Blue while Angeleno, Shiro, Songold, Golden Japan and Rubyred have been reported to be suppressed climacteric type (Abdi *et al.*, 1998; Khan and Singh, 2007; Martinez Romero *et al.*, 2003). Kitamura *et al.* (1983) in Japan investigated the ripening changes of 3 plum cultivars, the fruits of which were picked 2-6 times at different stages of maturity and stored at 20°C. In the early maturing cultivars (Ooishi Wase and Beauty), the rise in respiration was found to occur earlier and progressed more rapidly than it did in the late cv. Santa Rosa. In the early cultivars climacteric rise advanced rapidly with red colour development and softening of the flesh, while in Santa Rosa the climacteric rise and flesh softening were much slower than colour development. Free ABA content of attached fruit was found to increase with the rise in respiration rate during ripening on the tree. Measurements of CO_2 and ethylene evolution from cv. Victoria plums during the last part of growing season showed typical climacteric respiration patterns. The amount of CO_2 evolved decreased from approximately 50 mg/kg/h in late August to a minimum of less than 5 mg/kg/h at the optimum harvest date, then increased sharply to approximately the level first measured (Sekse, 1988).

Vangdal (1981) in trials with fruits of cvs. Mallard, Opal, Oullins (Oullins Gage), Rivers Early Prolific (Early Rivers) and Victoria which were harvested at 4 stages of maturity and stored at 2° or 20°C found that, during ripening at either temperature, fruit colour increased and firmness and acidity decreased. The TSS content rose in Mallard but not in other cultivars. Fruits of all cultivars harvested when immature improved in flavour during ripening but only Mallard fruits attained an acceptable flavour rating. Romojaro *et al.* (1979) reported that fruit chlorophyll levels decreased with ripening and they found increase in fruit carotenoid content during ripening in cv. Golden Japan but decreased in Reine Claude Verte. Lee and Lee (1980) stated that post harvest treatment of fruit with ethephon stimulated anthocyanin development but did not affect fruit composition. Kitamura *et al.* (1983) found no respiratory climacteric in fruits treated with ethylene at 500 ppm for 48 hours after harvest. A rise in ethylene production or earlier ripening was observed when the fruits were harvested at or after the minimum climacteric.

Pests

San Jane Scale (*Quadraspidiotus perniciosus*) [*Diaspidiotus perniciosus*]

This is a serious pest of plums and apples causing considerable economic losses (La Rue and Gerdts, 1976). The adults which feed on sap from limbs, twigs and fruits are small, circular and grey, on the fruit they cause spotting. The insects cause permanent injury to mature trees by weakening the fruiting branches and even the main scaffold branches. Injured limbs exhibit gum exudation, rough bark and die back. Dead leaves adhering to fruit spurs during the dormant season indicate presence of the scale. Effective control is usually obtained by dormant season universal oil sprays or by suitable insecticide treatment during the spring crawler hatch in early May. Application of white oil (oliocin) at 3.5 per cent when the new growth resumed in March showed effective control in Italy than traditional insecticides. Spraying of imidacloprid + oliocin at 2.5 per cent also found effective (Pasqualini and Civolani, 2000)

Peach Twig Borer (*Anarsia lineatella*)

A common pest, the borer attacks plum fruits and twig terminals. The cultivars ripening during the hatching period require utmost protection, since ripening fruits are very vulnerable. Egg hatch dates may vary with the prevailing season, but usually one brood occurs in mid-May and another in early July. Properly timed sprays can control the insects at these times (La Rue and Gerdts, 1976).

Plum Fruit Moth (*Laspeyresia pomonella, L. funebrana*)

This insect is prevalent in the cool and temperate regions. It is also a major pest of walnuts and apples. The larvae enter both immature and mature fruit and tunnel deeply into the fruit tissue. The entrance is usually surrounded by frass and droplets of gum (Heddergott, 1971). For effective control, services of a forecasting and warning system are necessary and knowledge of the location and specific life cycle of the pest are essential. Insecticides based on azinophos, bromophos, carbaryl, diaxinon, dimethoate, malathion, methidathion and propokur are effective against the plum fruit moth (codling moth). Charmillot *et al.* (1986) found effective control of the first generation larvae of the tortricid by applying fenoxycarb at the end of June and again 2 weeks later at a concentration of 0.03 per cent.

Mites

Three species of mites, *viz.*, *Tetranychus urticae* (the two spotted mite), *T. pacificus* (the Pacific mite) and *Panonychus ulmi* (the European red mite) have been reported to cause damage to the plum trees. These feed on leaves, and when present in large numbers can cause leaf browing and defoliation. The European red mite attack gives the trees a general light green or yellow-green appearance. In the plum orchards of Hungary, eight zoophagous mite species have been identified *viz. Amblyseius andersoni, Euseius finlandicus, Anthoseius occiduus, Kampimodromus aberrans, Paraseiulus triporus, Phytoseius macropilis* and *Typhloseiulus simplex* belonging to the family Phytoseiidae and *Zetzellia mali* belonging to the family Stigmaeidae (Hajdu, 2014). Control measures using acaricides should be timed to reduce population growth before serious leaf injury occurs (La Rue and Gerdts, 1976). Though dormant spray treatments are effective against the European red mite, but sometimes foliar acaricide applications are necessary. In Egypt, Nassar *et al.* (1982) found good control by using propargite (as 130 ml comite 75 per cent e.c. in 100 I water), dicofol (as 250 ml kelthane 18.5 per cent e.c.) binapacryl (as 100 ml morocide 40 per cent e.c.), fenbutatin oxide (as 150 ml torque 55 per cent e.c.), chlorobenzilate (as 150 ml akar 50 per cent e.c.) and a mixture (tedifol e.c. at 250 ml) as dicofol and tetradifon (as tedion 6 per cent) which reduced mite number in the field by 74.81-97.46 per cent.

Plum Sawflies

Two species have been recorded on the plum. These are the small plum sawfly (*Hoplocampa minuta*), the yellow plum sawfly (*H. flava*). Blossoms or young fruits are damaged by fissures in which eggs may be found. Plum sawflies lay their eggs on flower sepals. The white grubs that emerge, damage the young fruits by boring holes into them and feeding on the fruit tissue and kernel (Heddergott, 1971). Each grub damages 2-5 fruits. Both species mainly are found together, but *H. minuta* is considered to be dominant and contributes to the most of the damage caused. If conditions are favourable and no control measures are taken, in could result in almost complete destruction of crop. Wildbolz and Staub (1984) have recommended the use of white traps for monitoring the flight intensities. A total catch of 100-120 insects has been considered the provisional threshold counts beyond which chemical control measures should be adopted. Spray just after petal fall with insecticides based on carbaryl, demeton, dichlorvos, dimethoate, formathion, methidathion or promecarb has proved effective (Heddergott, 1971). Broad-Spectrum pyrethroids or neo-nicotinoids with contact mode of action are used to reduce sawfly damage. One or two insecticide treatments-before the flowering and during the petal fall are usually applied against plum sawflies (Tamosiunas *et al.*, 2014)

Nematodes

Nematodes are small transparent worm-like organisms which attack roots, and thereby cause reduced tree growth. The nematode, attacking plum roots are *Meloidogyne* sp. (root-knot nematode) and *Pratylenchus* sp. (lesion nematodes) and *Criconemoides* sp. (ring nematodes). Nematode infestation also predisposes the trees to various soil-borne diseases. Proper rootstock selection can reduce the damage caused by the root-knot nematodes to which Mariana 2624, Myrobalan 29C and Nemagaurd are resistant. However, for the damage caused by the other two groups of nematodes, pre-plant and post-plant soil fumigation is essential.

Diseases

Bacterial Canker (*Xanthomonas pruni* and/or *Pseudomonas syringae*)

The bacteria form over-wintering cankers on the twigs. The leaf is invaded through stomata to cause irregular circular shaped lesions. The infected area dies and falls out, thereby producing a shot-hole appearance. Sometimes infected leaves become yellow and fall from the tree, this is accentuated by nitrogen deficiency and infertile light sandy soils. This is particularly so when the causal organism is *Xanthomonas pruni* (McNew 1953). Infected trees become chlorotic, wilted or fail to leaf out properly in the spring. Dissection of the diseased limbs reveals dark sour smelling cankers (La Rue and Gerdts, 1976). These may sometime girdle limbs or entire trunks and cause death of the tree parts above the girdle. The leaves and flowers are particularly susceptible when rains and low-temperature occur during the blossom and leafing out periods. This results in leaf and blossom blast when the disease is also known as bacterial blast. There is no satisfactory control against this disease though pre-plant soil fumigation followed by annual post-plant application of Bordeaux sprays at leaf-fall provide some degree of protection. Delaying the pruning until late winter is also reported to be helpful. (La Rue and Gerdts, 1976).

Oak Root Fungus

The disease is caused by *Armillarella mellea,* and is responsible for large annual losses in most cultivars of tree fruits The fungus infects the roots and lower part of the trunk. The disease spreads in a circular pattern in the orchard. Infected trees show a progressive yellowing and wilting of the foliage and a stunted growth because of trunk girdling and death of the root system. The crown and foot of infected trees exhibit white fan-shaped mycelial growth between the wood and bark. Myrobalan 29 and Mariana 2624 show more resistance to the fungus than other plum rootstocks but even these can be severely injured or killed by the disease. Soil fumigation in autumn after fruit harvest with carbon bisulphate is recommended for control. The fumigant should not be applied closer than 2 to 2.7 m from any living tree as the material is toxic to living roots.

Brown Rot

It is caused by the fungus *Monilinia fructicola* which attacks peaches, apricots and cherries besides plums in orchard, and during transit, storage and market. In the orchard the disease may occur on leaves, blossoms (blossom blight), fruits, twigs (twig blight) and limbs (brown rot canker) (Ramsey and Smith, 1953). The relative prevalence of plum curculio and the oriental fruit moth are also important factors in the seasonal severity of the disease. The disease becomes active as the fruit begins to ripen and first shows as brown spots on the fruit surface. As the rot progresses, greyish spore masses are evident on the older rotted tissue (La Rue and Gerdts, 1976). Preharvest sprays followed by postharvest treatments are recommended. With highly susceptible cultivars like the Santa Rosa and Wickson, spring blossom sprays may also be necessary. Diseased and mummified fruits on trees as well as from orchard floor should be collected and burnt.

Crown Rot (*Phytophthora* spp.)

Primarily a disease of the young trees, its onset begins during extended periods of high soil moisture. Cankers develop on the bark below the ground and infected trees appear pale after leafing. If the cankers girdle the trunk, the tree will eventually die. There is no satisfactory control measure once the disease appears. However, it can be prevented by planting the trees at the same depth at which they grew in the nursery, or even higher (graft union should be kept out 30-45 cm above the soil level). Ensuring proper drainage in the tree basins is crucial for preventing outbreak of the disease. In poorly drained orchard sites, diversion channels may be provided to divert the run-off from adjoining areas. Plum rootstocks which exhibit greater resistance to the fungus should be preferred instead of peach in heavier soil conditions (La Rue and Gerdts, 1976). Pit disinfection at the time of planting should be done. This can be done by burning stubble in the pit and or mix soil with zinc sulphate, copper sulphate and lime (5: 1 :4) which should be mixed in the back fill soil. Bordeaux paste (2 kg copper sulphate and 3 kg quick lime dissolved separately each in 15 litres of water and then simultaneously poured into a third vessel) should be applied on the tree trunks before the onset of the rainy season.

Cytospora Canker (*Cytospora leucostoma*)

It is prevalent on the European plums, while Japanese plums are resistant to the disease. The fungus usually infects limb areas weakened by sunburn, over-cropping and scale insects. Cankers develop in the bark along main scaffolds resulting in girdling and sudden collapse (La Rue and Gerdts, 1976). Sunburn of the main scaffolds can be prevented by taking care not to remove the inside wood of the tree while undertaking annual pruning. Shot-hole borers are believed to spread the disease. Elimination of their sources in the orchard vicinity helps in checking the spread of the disease. Good cultural practices such as adequate fertilization, irrigation and scale insect control are important in preventing Cytospora canker. Application of Bordeaux paste on the weakened limbs also helps in preventing the occurrence of the disease.

Blossom Blight (*Sclerotinia sclerotiorum*)

Blossom blight of Japanese plum was reported from Chile (Ferrada *et al.*, 2014), Disease prevalence of 8 per cent and 1 per cent were estimated in 2012 and 2013, respectively. Early symptoms appeared as small pale-brown necrotic lesions on the petals that eventually affected the entire flower.

Plum Pocket (*Taphrina pruni*)

The disease affects the fruits which become greatly enlarged, distorted and hollow like a bladder. It is readily controlled by any good fungicidal spray as for example those containing copper oxychloride (Teskey and Shoemaker, 1972).

Viral Diseases

Sharka or Plum Pox

Sharka or plum pox disease is one of the most devastating viral diseases of stone fruits. The disease is caused by *plum pox virus* (PPV), a member of *Pllyvirus* genus in *Potyviridae* family (Sochor *et al.*, 2012). PPV usually affects both leaves and fruit. The symptoms include chlorotic spots or rings, oak-leave patterns and vein clearing on leaves. In fruits, it shows shallow rings and arabesque depressions, sometimes with brownish or reddish necrotic flesh. In addition fruits may drop prematurely (Paprstein *et al.*, 2007). The spread of the disease can be checked by removing and burning the infected trees as soon as they are detected (Heddergott, 1971). Replanting should be done with only virus-indexed trees. Control of aphids as virus-vectors by the application of organophosphorus insecticides is also helpful in checking the spread of the disease.

Spot real-time reverse transcriptase polymerase chain reaction (RT-PCR) was found more sensitive while double antibody sandwich indirect enzyme-linked immunosorbent assay (DASI-ELISA) using the PPV-specific monoclonal antibody 5B-IVIA/AMR was found more specific for PPV detection (Vidal *et al.*, 2012)

Plum Line Pattern

This disease is transmitted through grafting, and affects the foliage only (Cochran and Reeves, 1953; Heddergott, 1971). The symptoms on the leaf are quite similar to the Sharka except that they are more sharply defined and frequently follow the vein structure. The infected trees remain stunted, their leaf size is reduced and trunk growth is restricted. Fruit TSS, vitamin C content, acidity and total sugars reduced (Kanaujia *et al.*, 1985). The outbreak of the disease can be prevented by planting only certified virus-indexed trees (indicator plant-Myrobalan plum seedlings). Infected saplings can be made virus free by heat therapy at 37°C for 3-4 weeks followed by tip propagation (Heddergott, 1971; Styer and Chin, 1983).

Prune Dwark

The virus is transmitted through budding and grafting material. The leaves become small, narrow, rugose, thickened, distorted, sometimes exhibiting mosaic mottle. Terminal growth is reduced, internodes become shortened and yield is poor. The fruits become abnormally enlarged. No control measure is presently available for the diseased plants. The spread of the virus can, however, be checked by planting only indexed plants.

'Stanley' Prune Decline

Constriction at the graft union and decline of.cv. Stanley prune has been reported to be associated with the tomato ring spot virus (TRSV). The virus has been consistently detected in *Prunus cerasifera* and other rootstocks and cultivars of *Prunus domestica* that exhibited the symptoms (Cummins and Gonsalves, 1986). Growing 'Stanley' prune trees on their own roots appears to effectively check the disease. Cummins and Gonsalves (1986) observed that own-rooted 'Stanley' prune trees had remained free of natural infection for 14 years while 36 per cent of those on rootstocks had become infected.

Breeding and Improvement

The basic objective of any fruit tree breeding programme is the development of a desirable fruit which can be grown successfully in a particular locality and which can be marketed profitably. For plums the desirable fruit must have an attractive appearance, adequate size and firmness and acceptable flavour and texture. To be grown successfully the tree must be productive, resistant or tolerant to problems which may be location specific. Thus cold-hardiness would be important for the northern regions while low chilling requirement for buds would be a primary concern for plum breeders of the southern regions in India.

As to what constitutes desirable fruit varies greatly from one region to another. In the United States the breeders are concerned with developing cultivars that are firm enough to withstand long distance transportation, coupled with maturity dates suitably spaced that make shipping plums available at weekly interval throughout the fruiting season starting from April and extending up to October (Weinberger, 1975). Other breeding objectives vary with the location specific problems. Thus for example in Washington state breeders are looking for cultivars that are free from the physiological problems to which the cv. Italian Prune is susceptible. Resistance to leaf curl and premature fruit drop and problems of softening, internal browning and poor quality are other areas of concern. In Florida state, the main plum breeding objectives are to develop desirable plum with low chilling requirement and resistance to canker caused by *Pseudomonas syringae* (Sherman and Sharpe, 1970).

The inheritance of some of the characters of *Prunus salicina* has been worked out. The time of ripening and size of the fruit are quantitatively inherited. Shape of the fruit is controlled by multiple factors with neither round nor ovate shape being dominant. Yellow skin colour is recessive to black, red or purple which are, however, quantitatively inherited. Similarly, red flesh colour is dominant over yellow. The freestone character is apparently recessive to clingstone. The degree of clinginess being affected by fruit maturity and firmness of the flesh (Weinberger and Thompson, 1962).

In *Prunus domestica,* oval fruit shape has been reported to be dominant over round fruit shape; yellow skin colour recessive to red colour; thick fruit bloom dominant over thin fruit bloom and freestone recessive to clingstone.

Prunus nigra and *Prunus americana* have been used for increasing the winter-hardiness of the cultivated plums. Of these two *Prunus nigra* has been reported to transmit greater hardiness than *Prunus americana.* European plum (*P. domestica*) have high degree of heterozygosity and because of hexaploid nature, it is very difficult to investigate the inheritance of an individual trait. Most European plums have a moderate to high chilling requirement which is exploited to develop frost resistance. Interspecific hybridization with *P. salicina* cultivars may be tried to develop low chilling cultivars. Similarly interspecific hybridization with *P. microcarpa* may be useful to develop drought and heat resistance cultivars. Cultivars with high and regular yield are the demand of the growers. Good donors for precocity and high yield are Stanley, Verity, Cacanska Lepotica and Cacanska rodna. In European plum, a wide range of different fruit shapes exists. Oval fruit shape was reported to be dominant over round shape. Like shape, wide variations exists in fruit colour. In middle Europe, blue coloured fruits are preferred. As dark skin are often heterozygote concerning the trait, descendants originating from a crossing of two dark skin cultivars can show red, green or yellow coloured progeny. The flesh colour trait is variable. A good donors for orange flesh are the cvs. Hanita and Veeblu while cvs. Queen Garnet and Rubysweet are the good donors for deep red colour of flesh.

Selection of parents in any breeding programme is of prime importance. They should be strong in the particular traits desired in they progeny. The idea is to combine the desirable traits occurring separately in a single individual. However, since plums are highly heterozygous large progenies have to be raised and evaluated. According to one estimate at least 5000 seedlings would have to be raised and screened in order to have some chance of attaining success (Weinberger, 1975).

In Europe, the main breeding objectives are (i) fruit quality (ii) ripening time, (iii) resistance to diseases and pest specially sharka and (iv) climatic robustness. The main activities (80 per cent) is on *P. domestica*. Fruit Research Institute Cacak, Serbia is working on improvement of old cv. Pozegaca (increasing fruit size, repening time and resistance to PPV). The main objectives of fruit beeders at University Hohenheim, Stuttgart, Germany is to develop cultivars with superior fruit quality and PPV resistance by hypersensitivity. The INRA, Bordeaux, France is working on PPV resistance by transgenic plants. The Universities of Italy at Bologna, Firenze, and Forli are working on both Japanese and European plums for improvement of quality and resistance to PPV. Research station at East Malling, UK focused in developing late blooming, frost and PPV resistance cultivars. Until 2011, 184 plum cultivars has been released from Europe of which 38 from Romania, 28 from Germany, 18 from Italy and 16 from Estonia. The important European plum cultivars released from Italy include Liable, Sugar top, Prugna 29 and Maria Novella. The cultivars Hanka, Jojo, Present and Haganta was released from Germany. In Romania, since 1960 plum breeding is being carried out in different universities and institutes. The general objectives between 1960-1980 was for improvement of old cultivars Tuleu Gras, Vinete Romanesti and Grase Romanesti; between 1980-2000, it was the improvement of fruit quality for fresh market, extension of ripening season and tolerance to PPV while from 2000 the focus is on tolerance to PPV, improvement of fruit size and self fertility. The cultivars used as genitors for tolerance to PPV were Oneida, Kirke, Grase de Becz and Grase Pesteana; for fruit quality Grand Prize, Vision Oneida, Cacanska Rodna, Valcean, Grase Romanesti; for yield capacity Stanley, Anna Spath, Bluefree and Standard; for self fertility Lalomita, Stanley, Anna Spath and Bluefree and for earliness Early Rivers, Ruth, Gerstetler, Diana and Lalomita. Thirty eight cultivars have been developed of which important cultivars are (1) Roman (Tuleu gras x Early Rivers)-a medium ripening (August), large fruited (>408) and PPV tolerant cultivar; (2) Agent (open pollinated seedling selection)-medium to late (end August), higher sugar content and tolerant to PPV; (3) Robelle (Stanley x Valcean)-medium ripening (August), very large fruit (>50g) and tolerant to PPV; (4) Ivan (Tuleu gras x Vanat de Italia) – medium ripening (early August) and medium fruit (40g); (5) Romaner (Tuleu gras x Renclod Althan) – medium ripening (August), very large fruit (>50 g) and tolerant to PPV; (6) Zamfira (Anna Spath x Renclod Althan) – late ripening (beginning of September), medium fruit (45g) and tolerant to PPV; (7) Andreea- medium – late ripening (end of August), high sugar content and tolerant to PPV.

California is the world leader in dried plum production providing 99 per cent of the US supply and 40-60 per cent of the world supply. In 2011, California exported over 68,000 tons of dried plum to 44 countries. The cv. Improved French which was brought to California in 1856 represents 98 per cent of the total dried plum acerage in California (Scorza *et al.*, 2013). Because of the preponderance of a single cultivar, the entire industry must harvest and dehydrate the crop within the span of few weeks. The plum industry of California identified breeding priorities for new cultivars that focus on tolerance to pests and diseases, improved consumer traits such as flavour and nutritive value, and cost saving characters such as lower drying cost and reduced pruning. High sugar levels are a major need identified by the industry in terms of increased consumer preference and due to the fact that higher sugar levels decreases the drying time. The University of California, Davis Dried Plum Development and Evaluation programme, initiated in 1975 developed two cultivars 'Sutter' and 'Muir Beauty', but the dried plum industry remains overwhelmingly dependent upon a single cultivar 'Improved French'. One of the particular valnerabilities of this crops is its susceptibility to PPV. In an effort to develop new approaches to plum improvement the USDA-ARS Appalachian Fruit Research Station fruit breeding programme in collaboration with partners in the U.S. and Europe have developed a genetic engineering (GE) approach to target resistance to PPV. This programme has resulted in the development of a GE plum

'HoneySweet' which has been tested for 15 years in the E.U. and U.S. and is highly resistant to PPV. 'HoneySweet' has received full regulatory approval in the U.S. It represents a new source of PPV resistance for growers and it can be used by breeders to develop additional resistant cultivar/rootstock (Scorza *et al.*, 2013)

Plum breeding programme at Brazil resulted release of several low-chilling Japanese plum *viz.* Carmeslin (1973), Rosa Paulista (1978), Grancuore (1978), Gema-de-Ouro (1975), Golden Talisma (1979), Rosa Mineira (1983), Januaria (1985), Kelsey 31 (1987), Centenari (1989) *etc.* Most of these are seedlings of 'Kelsey Paulista' which is a local selection of 'Kelsey' (Ojima *et al.,* 1992). Plum breeding programme at Chapingo, Mexico was initiated in the early 1980s. The major goal is to develop of low chill, large, firm-fleshed cultivars for subtropical climates. Two low chill cultivars 'Cirena 1' and 'Corazon Roja' were released. In South Africa, plum breeding programme was initiated in 1960s. The goal is development of large-fruited plums with resistance to bacterial spot and bacterial canker and the ability to store without internal breakdown. Storage ability of four weeks is crucial to export the fruit by ship. Plums also need a moderate chilling requirement to enhance crop set in lower chilling areas of South Africa. The breeding programme resulted in release of several good cultivars like Songold, Harry Pickstone, Reubennel, Redgold, Laetitia, Celebration, Sapphire, Souvenir, Pioneer, Atlantic, Lady Red *etc.* Celebration is their first black-skinned cultivar.

The plum breeding programme in Japan resulted release of cultivars Ohishi-Nakate, Ohishi Wase, Whte Queen, Beniryouzen, Royozenwase, Shichiro, Pararu, Tsukuba No. 4, Tsukuba No. 5 (Sordum x Nishida).

Genetic Transformation

Genetic transformation in European plum has been established through regeneration of the cotyledons and hypocotyls sections of mature embryos, immature embryos and embryo axes. Scorza *et al.* (1994) used medium supplemental with thidizuron (TDZ) at 7.5 µM in combination with IBA at 0.25 µM for shoot regeneration with a transformation efficiency of 1.2 per cent. Petri *et al.* (2012) reported 42 per cent increase transformation efficiency by using 2, 4-dichlorophenoxyacetic acid (2, 4-D) at 9 µM for three days during co-cultivation, followed by a regeneration medium containing TDZ (7.5 µM) and kanamycin at 80 mg/l as selective agent. Using this protocol Srinivasan *et al.* (2012) successfully transformed European plum with *Pt* FT1 gene. Mikhailov and Dolgov (2012) transformed leaves of plum cv. Startovaya by using *Agrobacterium tumefaciens* Strain AGLO: pcamPPVRNAi carrying the hygromycin phosphotransferase gene (hpt), the gus gene and self-complementary sequences of fragment of PPV-CP gene separated by an entron to produce a 'hairpin' RNA (hp-RNA) structure. European plum plants with high expression of the flowering locus T1 (FT1) from *Populus trichocarpa* showed a continuous reproductive phenotype (flower and fruit) under greenhouse conditions, and during spring and autumn in the field (Srinivasan *et al.*, 2012)

European plum cv Honey Sweet (developed from C5 transgenic line) was the first transgenic line developed in which plum pox virus (PPV) resistance trait is heritable and stable. Virus resistance is due to the post transcriptional gene silencing

(PTG) given by the insertion of a transgene-multicopy complex and aberrant copies of the virus coat protein (PPV-CP). The stability of 'C5' has been evaluated under field conditions in different countries, showing that it can spend several years without becoming infected (Scorza *et al.*, 2001). Moreover, 'C5' trees exposed to natural aphid vectors in the field remained unaffected after 4 years (Jarosova *et al.*, 2010). Field testing for over 15 years in Europe has demonstrated that the stable and durable PPV resistance of 'HoneySweet', resistance is based on gene silencing whereby the inserted gene induce a natural plant defense mechanism against viruses. This resistance has been transferred to seedlings through cross hybridization as a single locus dominant trait making it useful as a parent for developing new plum varieties for specific growing areas and markets. HoneySweet plums are of high quality and compare well to the quality and nutritional value of conventional plums (Ravelonandro *et al.*, 2013).

References

Abdi, N., McGlasson, W.B., Hlford, P., Williams, M. and Mizrahi, Y. (1988). Postharvest Biol. Technol., 14: 29-39.

Adriance, G.W. and Brisan, F.R. (1967). Propagation of Horticultural Crops, Tata McGraw Hill Ltd.

Ahmed, M.B. and Twefik, M. (1962). Proc. 16th Int. Hort. Congr. Brussels. 1: 197-98.

Allen, F.W. (1939). Proc. Amer. Soc. Hort. Sci., 37: 467-72.

Alves, E, de, O., Steffins, C. A., Amarante, C.V. T. do., Weber A., Miqueloto, A., and Brackmann, A. (2010). Ciencia Rural, 40: 30-36.

Anonymous, (1969). The Wealth of India. Publications and Information Directorate. CSIR. New Deihl, pp. 250-82.

Arora, R.L. and Singh, R. (1987a). Indian J. Hort., 44: 143-47.

Arora, R.L. and Singh, R. (1987b). Incompatibility News Letter. 19: 4-9.

Atkinson, D., Thomas. C.M., White. G.C., Crisp, C.M., Taylor, N. and Clay, D.V. (1984). Rept. East Malling Res. Stn. for 1983. p. 40.

Badyal, J.M. and Kar, P.L. (1985). Haryana J. Hort. Sci., 14: 41-47.

Bae, RoNa, Lee, JiHyun and Lee, Seungkoo (2011). Korean J. Hort. Sci. Technol., 29: 592-99.

Bajwa, M.S. and Misra, K.A. (1970). Punjab Hort. J., 10: 19-20.

Bajwa, M.S. and Singh, K.K. (1965) J. Res. PAU. Ludhiana, 2: 187-90.

Bajwa, M.S. and Singh, K.K. (1970). Punjab Hort. J., 10: 29-41.

Baker, G.A.A. and Brooks, R.M. (1944). Proc. Amer. Soc. Hort. Sci., 45: 95-104.

Bal, J.S. and Sandhawalia, S.S. (2000). Acta Hort., 517: 151-58.

Baleriola-Lucas, C. and Mullins, M.G. (1982). Res. Rpt. Dept. Agronomy and Hort. Sci. University of Sydney, No. 10, 34.

Batjer, L.P. (1965). Agric. Inform. Bull No. 289. USDA.

Belmans, K. and Keulemans, J. (1987). Boer en de Tuinder, 93 : 23-24.

Beutel, J. (1969). Calif. Agric., 23 : 35-40.

Beyers, E. (1969). Dec. Fr. Grower, 19: 35-40.

Beyers, E. and Terblanche, J.H. (1971). Dec. Fr. Grower, 21: 199-202.

Bhan, S., Bhutani, V.P. and Chopra, S.K. (1983). Punjab Hort. J., 22: 181-85.

Bhutani, V.P. and Bhatia, H.S. (1986). In. Adv. Res. on Temp. Fruits. Proc. Nat. Symp. Temp. Fruit. pp. 199-202.

Blazek, J. and Pistekova, I. (2011). Vedecke Prace Ovocnarske, 22: 115-27.

Boothby, D. (1983). J. Sci. Fd. Agric., 34: 1117-22.

Borsheim, A. (1980). Meidinger fra Norges Landbrukshogskole, 59: 24.

Bowen, T.J. (1956). Qld. Agric. J., 82: 373-76.

Cassibba, L., Schiraparelli, A. and Peyron, F. (1979). Frutticoltura, 41: 51-56.

Cepoiu, N. (1987). Lucari Stiintifice Insrirutul Agronomic 'Nicolae Balcescu', B. (Horticulturae), 30: 61-65.

Cerkova, L.L. and Skipina, K.P. (1966). Biol. Nauk, 9: 3

Chandra, A., Gupta, S.P., Yamdagini, R. and Singh. D.(1988). Haryana Agric. Univ. J. Res., 18 : 53-55.

Charmillol, P.J., Bloesch, B., Frischknecht, M.L., Gingins, E. and Hohn, H. (1986). Mitteilungen des Schweizerische Entomologischen Gesellschaft, 59: 39-46.

Childers, N.F. (1973). Modern Fruit Science, Hortic. Pub. Rutgers Univ., New Burnswick. NJ. pp. 393-95.

Chopra. S.K., Misra, S.S., Bhutani, V.P. and Kashyap, A.S. (1986). In: Advances in Research on Temperate Fruits.Proc. Nat. Sym. on Temp. Fr., Dr. Y.S. Parmar Univ. Hort. and Forestry, Solan, India, pp. 319-25.

Christopher, E.P. (1976). In: The Pruning Manual. Macmillan Co. Inc. NY, p. 320.

Cmelik, Z. (2000). Agric. Conspectus. Scientificus Poljoprivredna D., Znastvena Somtra, 65: 71-78

Cobianchi, D., Faedi, W., Rivalta, L. and Batelli, T. (1978). Riv. Ortofruttic. Ital., 62: 552-62.

Cochran, L.C. and Reeves, E.L. (1953). In: Plant Diseases. The Yearbook of Agriculture. USDA, p 717.

Couranjou, J. (1970). Ann. Amel. Plantes, 20: 297-99.

Crane, M.B. and Brown, A.G. (1939). J. Pomol. Hort. Sci., 17: 51-56.

Crisosto, C.H., Crisosto, G.M., Echeverria, G. and Puy, J. (2007). Postharvest Biol. Technol., 44: 271-76.

Crisosto,C.H., Garner, D., Cid, L. and Day, K. (1999). Calif Agric., 53: 33-36.

Cummins, J.M. and Gonsalves, D. (1986). J. Amer. Soc. Hort. Sci., 111: 314-18.

Davey, P.C. (1978). Farmers News Letter, 1390: 24-26.

Davis, L.D. (1934). Proc. Amer. Soc. Hort. Sci., 32: 125-30.

Davis, R. and Boyes, W.W. (1940). South Africa Dept. Agric. Forestry Rept. Low Temp. Res. Lab. for 1938, pp. 25-30.

Detzel, B.M.P., Hartman, W. and Stosoer, R. (2000). Rivista de Fruiticultura e. di. Ortofloricoltura, 62: 83-86

Diaz-Mula, H.M., Zapata, P.J., Guillen, F., Martinez-Romero, D., Castillo, S., Serrano, M. and Valero, D. (2009). Postharvest Biol. Technol., 51: 354-63.

Dodd, M.C. (1984). Dec. Fr. Grower, 34: 255-56.

El-Sharkawy, I., Sherif, S., Kayal, W.E., Mahboob, A., Abubaker, K., Pratibha Ravindran, Joythi Prakash, P.A., Kumar, P.P. and Subramanian, J. (2014). Pl. Mol. Biol., 84: 399-413.

Embree, C.G., Tehrani, G. and McRae, K.B. (1999). Canadian J. Pl. Sci., 79: 611-15.

Endlich, Von J. and Murawski, H. (1962). Zuchter, 32: 121-33.

Enikeev, H.K. (1968). Vestn. Selskohoz. Nauka., 8: 54-58.

Erkan, M. and Eski, H. (2012). Turkish J. Agric. Forestry, 36: 563-75.

Ferrada, E.E., Diaz, G.A., Zoffoli, J.P. and Latorre, B.A. (2014). Pl. Disease, 98: 695

Gajdaj, G.S. (1986). Kholodil'naya Tekhnika, 6: 20-22.

Gavrilina, Z.M. (1985). Trudy po Prikladnoi Botanike, Genetike i. Selektsii, 76 : 49-55.

Gavrilina, Z.V. (1986). Sbornik Nauchnykh Trudov po Prinklandnii Botanike, Genctike i Selcktsii, 99: 54-57.

Gavrilovic, M. and Paunovic, A.S. (1962). Proc.16th Int. Hort. Cong, Brussels, Belgium, pp. 64-71.

Genders, R. (1973). In: Planting Fruit Trees. Roberthale and Co., UK, p 138.

Gigleux-Spitz, C. (1988). Infor. Centre Technique Interprifesionnel des Fruits et Legumes, France, 42: 11-14.

Gnezdilova, E.M. and Doroshenko, A.F. (1984). In : Intensiv Sadovod. Stavropol. Kraya. Stavropol., USSR, pp. 16-20.

Goncharova, E.A., Eremin, G.V. and Magomedova, R.A. (1985). Sbornik Nauchnykh Trudov po Prinkladnci Botanike, Genctike i Selektsii, 97: 49-54.

Griesbach, J. (2007). Growing Temperate Fruit Trees in Kenya. World Agroforestry Centre, Nairobi, pp. 68-72.

Guerra, M.E., Wunsch, A., Lopez-Corrales, M. and Rodrigo, J. (2011). J. Amer, Soc. Hort. Sci., 136: 375-81.

Gur, A. (1986). CRC Handbook of Fruit Set and Development (ed. S.P. Monselise) pp. 401-18. CRC Press Inc., Boca Raton, Florida.

Hajdu, Z. (2014). Ph.D. Thesis submitted to Corvinus University of Budapest, Hungary.

Hansen, C. J., and Proebsting, E.L. (1949). Proc. Amer. Soc. Hort. Sci., 53: 13-20.

Hartman, W. (1987). Obstbau., 12: 304-305.

Hartmann, H.T. and Kester, D.E. (1972). In: Plant Propagation. Prentice Hall of India, Pvt. Ltd., p. 702.

Hassan, H.S.A., Sarrwy, S.M.A. and Mostafa, E.A.M. (2010). Agric. Biol. J.N.Am., 1: 638-43.

Heddergott, H. (1971). In: Crop Protection Guide. Landwirtschaftsverlag Gmbtt, Hiltrup. FRG. pp. 1-255.

Hedrick, U.P. (1911). The Plums of New York, Rept. N.Y. State Agric. Exp. Stn., pp 1-6.

Hendrickson, A.H. (1948). Proc. Amer. Soc. Hort. Sci., 51: 234-38.

Hoagland, D.R. (1941). Proc. Amer. Soc. Hort. Sci., 38: 8-12.

Howard, B.H. and Nahlawi, W. (1969). J. Hort. Sci., 44: 303-10.

Howard, B.H., Jones, O.P. and Vasek, J. (1989). J. Hort. Sci., 64: 147-56.

Huelin, F.E. and Tindale, G.B. (1940). J. Agr. Victoria, May 1940. pp. 247-253.

Iliev, P. (1985). Rasteniev dni Nauki, 22: 85-92.

Intrigliolo, D.S., Ballester, C. and Castel, J.R. (2014). Ag. Water Management, 132: 23-29

Jaiswal, R., Karakose, H., Ruhmann, S., Goldner, K., Neumuller, M., Treuttev, D. and Kuhnert, N. (2013). J. Agric Fd. Chem., 61: 12020-31.

Jarosova, J., Gudious, S., Polak, J., Ravelonandro, M., Scorza, R. and Kumar, J. K. (2010). Julius Kuhn Arch, 427: 330-33.

Ji, Hua, Liu, Yuan, Guan, Junteng, Feng, YunXiao and Sun, Yulong, (2012). J. Fruit. Sci., 29: 404-408

Jones, O.P. and Hopgood, M.E. (1979). J. Hort. Sci., 52: 235-38.

Jones, O.P. and Hopgood, M.E. (1988). J. Hort. Sci., 63: 63-66.

Josan, J.S. and Chohan, G.S. (1982). Progressive Hort., 14: 11-13.

Jun, J.H., Nam, E.Y., Chung, K.H., Yun, S.K., Kwon, J.H., Yun, I.K., Kwack, Y.B., Kim, S.J. and Cho, K.H. (2015). HortSci., 50: 932-33.

Kader, A.A. (1983). HortSci., 18: 409-11.

Kanaujia, J.P., Singh, R.N., Singh, G. and Bhandari, T.P.S. (1985). Progressive Hort., 17 : 69-71.

Khan, A.S. and Singh, Z. (2007). Postharvest Biol. Tehnol., 43: 298-306.

Khokhar, U.V., Bhutani, V.P. and Kar, P.L. (1986). In: Adv. Res. on Temp. Fruits Proc. Nat. Symp. Temp. Fruit, pp. 193-197.

Kishore, D.K. and Randhawa, S.S. (1983). Scientia Hort., 19: 251-55.

Kitamura, T., Itamura, H. and Fukushima, T. (1983). J. Jap. Soc. Hort. Sci., 52: 325-31.

Knight, R.L. (1969). Prunus, Tech. Commun. Common. Bur. Hort. Plantation Crops, 31: 649.

Kolesnikov, V.A. (1966). Fruit Biology. Mir Publishers Moscow. pp. 1-338.

Koroid, A.S. (1987). Sadovodstvo, 3: 16-17.

Kosina, J. (2011). Vedecke- Prace-Ovocnarske, 22: 143-49.

Kotze, W.A.G., Notte, S.H., Dodd, M.C., Gurgen. K.H. and Crouse. K. (1989). Dec. Fruit Grower, 39: 64-68.

Krivoshiev, G. and Chalukova, R. (1981). Acta Aliment., 10: 45-60.

Kumar, J. and Thakur, D. (2012). Asian J. Hort., 7: 484-87

Kurbanova, R.I. (1988). Sadovodstvo i Vinogradarstvo, No.6. p. 21.

Kwong, S.S. (1973) Proc. Amer. Soc. Hort. Sci., 98 : 72-74.

La Rue, J.H. and Gerdts, M.H. (1976). Commercial Plum Growing in California. Leaflet No. 2458. Div. of Agricultural Sciences, University of California.

Lakatos, L., Zyromski, A. and Bimak-Pierog, M. (2012). J. Water and Land Dev., 16: 29-34

Lal, B.B. and Thakur, D.R. (1979). Haryana J. Hort. Sci., 7: 21-24.

Lee, C.L. and Denemann, G. (1981). Gartenbau.. 46: 217-23.

Lee, J.C. and Lee, Y. B. (1980). J. Korean Soc. Hort. Sci., 21: 36-41.

Linder, R.C. and Harley, C.P. (1944). Plant Physiol., 19: 420-39.

Liu-HaiRong (2000). China Fruits, 1: 3-4.

Lukosevicius, A. (1999). Sodininkyste-ir-DarZininkyste, 18: 3-12.

Lupescu, F. and Tertecel, M.(1985). Lucrari Stiintifice. Institutul Agronomic Nicolae Balcescu B (Horticulturae), 28: 41-45.

Magness, J.R. (1941). Pruning Hardy Fruit Plants. USDA Agric. Bull. No. 1870.

Martin, G.C., Fitch L.B., Sibbetr, G.S., Cornill, G.L. and Ramos. D.E. (1975). J. Amer. Soc. Hort. Sci., 110: 90-93.

Martinez-Romero D., Dupille, Z., Guillen, F., Valverde, J.M., Serrano, M. and Valero, D. (2003). J. Agric. Fd. Chem., 51: 4680-86.

McNew, G.L. (1953). Plant Diseases. The Yearbook of Agriculture. USDA. pp. 108-109.

Melvin Couey, H. (1960). Proc. Amer. Soc. Hort. Sci., 75: 207-15.

Mika, A. and Buler, Z. (2001). Adv. Hort. Sci., 25: 193-98.

Mika, A., Buler, Z. and Chlebowska, D. (1999). J. Fruit Ornamental Pl. Res., 7: 61-69.

Mikhailov, R.V. and Dolgov, S.V. (2012). Acta Hort., 941: 85-90.

Miletic, R., Lukovic, J., Paunovic, S.M. and Karaklajic-Stajic, Z. (2011). Vocarstvo, 45: 123-28.

Mirabdulbaghi, M. and Zarghami, R. (2011). Acta Hort., 923: 117-20.

Mishra, K.A. (1984). Agric. Sci. Digest. 4: 226-27.

Morini. S. ((1984). J. Hort. Sci., 59 : 287-94.

Naichenko,V.M.and Romanshchak, S.P.(1984).Fiziologiya i Biokhirruya Kul'turnykh Rastenii. 16: 143-148.

Nakajo, Y., Obayashi, S., Yahata, M., Nagashima, Y., Naruse, H., Masuda, Y., Mukai, H., Harada, H. and Takagi, T. (2014). Hort. Res. Japan, 13: 107-11.

Nassar, M.E., E-Halawany, M.E. and Kandeal, M. M.H. (1982). Agric. Res. Rev., 60: 275-80.

Negvila, A., Molitiu, I. and Lupescu, F. (1969). Lucrari Stiintifice. Institutul Agronomic 'Nicolae Balcescu B., 12 : 215-21.

Nyeki, J., Szabo, Z., Toth, F. and Pete, A. (1985). Kertgazdaasag, 17: 35-63.

Nyhlen, A. (1965). Landbr Hogsk. Medd., Upsala, Ser. A., 35: 19.

Ognjanovic, R. and Saric, T. (1988) Fragmenta Herbologica Jugoslavica, 17: 101-14.

Ojima, M., Campo-Dall Orto, F.A., Barbosa, W. and Rigitano, O. (1992). Inst. Agron, Campinas Bul.Tech., 144.

Okie, W.R. (1987). In: Rootstocks for Fruit Crops. (eds. R.C. Rom and R.F. Carlson) pp. 321-60.

Okie, W.R. and Weinberger, J.H. (1996) Plum. In: Fruit Breeding. Vol. I. Tree and Tropical Fruits (eds. J. Janick and J.N. Moore), Wiley, New York, pp. 559-607.

Oliveira-Junior, R.S. de., Fornarolli, D.A. and Chehata, A.N. (2000). Rivista Brasileira de Fruticultura, 12: 134-36.

Olson, W.H., Uriu, K., Carlson, R.M., Krueger, W.H. and Perrson, J. (1987). Calif. Agric., 41: 20-21.

Os, J., Wilska-Jeszka, J. and Pawlak, M. (2000). Polish J. Fd. Nutr. Sci., 9: 35-38.

Ozkarakas, I. and Balkan, C. (1999). Anadolu, 9: 1-11.

Oztruk, B., Kucuker, E., Karaman, S., Yldz, K. and Klc, K. (2013). Int. J. Fd. Eng., 9: 421-32.

Painter, A.C. (1930). Rept. East Malling Res. Stn. for 1929, pp. 68-73.

Pangelova-Shchrkova, I. and Vitanova, I. (1987a). Fiziologiya na Rasteniyata, 13: 34-42.

Pangelova-Shchrkova, I. and Vitanova, I. (1987b). Fiziologiya na Rasteniyata. 13: 61-67.

Paprstein, F., Karesova, R. and Navratil, M. (2007). Proc. 1 8th Int. Symp. Plum and Prune Genet. Breed. Pomol., pp. 255-57.

Parnia, P., Mladin, Gh., Dutu, I. and Stanciu, N. (1988). HortSci., 23: 107-109.

Pasqualini, E. and Civolani, S. (2000). Informature Agrario, 56: 93-95.

Paunovic, S.A. and Ogasonovic, D. (1979). Jugoslovensko Vocarstvo, 10: 491-99.

Pcjkic, B. and Boskovic, R. (1985) Jugoslovensko Vocarstvo, 19: 439-45.

Peirs, A., Parmentier, V., Wustenberghs, H. and Keulemans, J. (2000). Acta Hort., 518: 145-50.

Pentzer, W.T. and Allen, F.W. (1944). Proc. Amer. Soc. Hort. Sci., 44: 148-56.

Peskabovic, M., Djukic, D., Mandic, L. and Miletic, R. (2008). Cereal Res. Comm., 36: 675-81.

Petri, C., Webb, K., Hily, J.M., Dardick, C. and Scorza, R. (2008). Mol. Breed., 22: 581-91.

Plich, H. (2008). Acta Hort., 768: 135-42.

Plich, H. and Wojcik, P. (2002). Acta Hort., 594: 445-51.

Polak, J. and Kominek, P. (2014). Canadian J. Pathol., 36: 116-20.

Proebsting, E.L. (1937). Proc. Amer. Soc. Hort. Sci., 34: 272-74.

Ramming, D.W. and Cociu, V. (1990). Acta Hort., 290: 1-63

Ramsey, G.B. and Smith, M.A. (1953). In: Plant Diseases, The Agriculture Yearbook USDA. pp. 809-16.

Rangelov, B., Mondishka, P. and Nesheva, M. (1985). Proc. 4th Int. Symp. PI. Growth Regulators, Bulgaria, Part 2.

Rankova, Z. and Popov, S. (2011). Acta Agriculturae Servica, 16: 51-57.

Ravelonandro, M., Scorza, R., Polak, J., Callahan, A., Krska, B., Kundu, J. and Brirard, P. (2013). Fd. Nutr. Sci., 4: 45-49

Rehder, A. (1967). Manual of Cultivated Trees and Shrubs. 2nd. ed. Macmillan. New York.

Riesen, W. and Zbinden, W. (1986). Schweizerishe Zeitschrift fur Obst-und Weinbau,122: 197-202.

Robertson, J.A., Meredith, F.I. and Senter, S.D. (1992). J. Sci. Fd. Agric., 60: 339-47.

Romojaro, F., Banet, E. and Llorente, S. (1979). Anales de Edalol. Y. Agrobiol, 38: 1815-25.

Saini, P., Pandey, A. and Rehalia, A.S. (2013). Curr. Adv. Agric. Sci., 5: 132-34.

Salesses, G. (1977). Ann. Arnel. Plantes, 27: 235-43.

Samish, R.M. (1948). Agric. Res. Stn. Rehovot. Bull, No. 50 p. 92.

Samoshchenkov, E.G and Tikhomirov, V.A. (2000) Sadovodstvo i vinogradarstvo, 1: 9-11.

Sanchez-A1onso, F. and Lachica, M. (1987). Comm. Soil Sci. PI. Analysis, 18: 31-43.

Sandvad, K. (1962). Tidsskr Plantevl., 60: 609-42.

Scorza, R., Sallshan, A.M., Dardick, C.D., Srinivasan, C., Dejong, T.M., Abott, A.G. and Ravelonandro, M. (2013). Acta Hort., 985: 111-18.

Scorza, R., Callahan, A., Levy, L., Damsteegt, V., Webb, K. and Ravelonandro, M. (2001). Transgenetic Res., 10: 201-209

Scorza, R., Levy, L., Damsteegt, V., Callahan, A., Webb, K. and Ravelonandro, M. (1994). Focus 20 (Suppl.): 12-13

Seehuber, C., Damerow, L. and Blanue, M. (2011). Plant Growth Reg., 65: 335-41.

Sekse, L. (1988). Acta Agric. Scandinavica, 38: 317-20.

Sekse, L. (1989). Acta Agric. Scandinavica, 39: 15-21.

Serrano, M., Martinez-Romero, D., Castilo, S., Guillen, F. and Valero, D. (2004). Postharvest Biol. Technol., 34: 155-67.

Shear, C.B. and Faust, M. (1980). Hort. Rev., 2: 3-4.

Sherman, W.B. and Sharpe, R.H. (1970). Fruit Var. Hort. Dig., 24: 3-4.

Singh, S.P. and Singh, Z. (2008). J. Amer. Pom. Soc., 62: 185-90.

Singh, S.P., Singh, Z. and Swinny, E.E. (2009). Int. J. Fd. Sci. Tech., 44: 1973-82.

Smith, W.H. (1939). Great Britain Dept. Sci. Ind. Res. Rept. for 1938. pp. 165-69.

Sochor, J., Babula, P., Adam, V., Krska, B. and Kizek, R. (2012). Viruses, 4: 2853-901.

Sosna, I. (2008). Zeszytu Naukowe Instytutu Sadownictwa I Kwiaciarstwa w Skierniewicach, 16: 103-11.

Srinivasn, C., Dadrick, C., Callahan, A. and Scorza, R. (2012). PLOS 1 Biol., 7: 40715

Stefan, N., Liacu, A. and Tudosescu, O. (1968). Acta Hort., 10: 265-75.

Sterling, C. (1953). Bull. Torrey Bot. Club., 80: 475-77.

Stosser, and Hartman, W. (1982). In: Beziehung zur Fruchtform. Min. losterneuburg Reb. Wein. Obstbau. Fruchteverwertung, 32: 124-30.

Streit, J. (1987). Erwerbsobstbau, 29: 172-74.

Strydom, D.K. and Honeyborne, G.E. (1986). Dec. Fruit Grower, 36: 259-60.

Styer, D. J. and Chin, C.K. (1983). Hort. Rev., 5: 221-27.

Sud, G. (1979). Indian J. Hort., 36: 423-42.

Sud, G. and Kar, P.L. (1985). Himachal J. Agric. Res., 11: 34-39.

Szabo, Z. and Nyeki, J. (1996). Acta Hort., 423: 23-29.

Tamosiunas, R., Valiuskaite, A., Surviliene, E. and Rasiukevictute, N. (2014). Zemdirbyste Agric., 101: 91-100.

Teskey, B.J.E. and Shoemaker, J.S. (1972). In: Tree Fruit Production. The AVI Puhlishing Co. Inc., pp. 294-314.

Testoni, A. and Spada, G. (1981). Annali dell Istituto Sperimcntalc per la Valorizzasione Technologin dei Prodotti Agricoli, 12: 133-38.

Thiele, I. and Strydom, D.K. (1964). S. Affr. J. Agric. Sci., 7: 165-68.

Thomas, H.E., Wilhum, S. and MacLean, N.A. (1953). In: Plant Disease. The Yearbook of Agriculture, USDA. pp. 702-03.

Thompson, M.M. and Liu, L.J. (1972). J. Amer. Soc. Hort. Sci., 97: 193-97.

Tormann, H. and Van Zyl, H.J. (1982). Dec. Fruit Grower, 32: 381-41.

Treder, W., Grzyb, Z. and Rozpara, E. (1999). Acta Agrobotanica, 52 : 95-101.

Tsuji, M., Harakawa, M. and Komiyama, Y. (1983). J. Jap. Soc. Fd. Sci. Tech., 30: 688-92.

Tudosescu, O., Stefan, N., Liacu, A., Parnia, P., Prica, I., Manughevici, S., Casavcla, I., Bodi, G., Dumitrcscu, D., Cvasnii, M., Mites. N., Cojeneanu, I., Modoran, K. and Movileanu, M. (1974). In: Portaltoi Principalelor Specii de Ponii Fruitiferi. pp. 171-94.

Tukey, H.B. (1936). Science, 84: 513-15.

Tukey, H.B. (1964). Dwarfed Fruit Trees. Macmillan Co. New York.

Ughini, V. and Roversi, A. (1987). Annali delia Facolta di Agraria. Univ. Cattolica del Sacro Cuore, Milano., 27: 31-43.

Ugolik, M. (1966). Byulleten Glavnogo Botanicheskogo Sada, 61: 60-66.

Ugolik, M. (1970). In: Physiologische Probleme in Obstbau., pp. 265-77.

Uota, M. (1955). Proc. Amer. Soc. Hort. Sci., 65: 231-43.

Vangdal, E. (1981). Forskning Og Forsok i Landbruket, 32: 13-20.

Vangdal, E. (1982). Meidinger Fra. Norges Landbruks-shogskole, 61: 39

Venter, A., Joubert, E. and Beer, D.de (2014). Molecules, 19: 3084-109.

Vidal, E., Moreno, A., Bertolini, E. and Cambra, M. (2012). Pl. Pathol., 61: 413-22.

Vitanova, I. (1986a). Rastenivdni Nauki, 23: 102-09.

Vitanova, I. (1986b). Rastenivdni Nauki, 23: 98-105.

Vitanova, I. and Bozhkova, V. (1999). Rastenievdni Nauki, 36: 384-87.

Vitkovskii, V.L. (1986). Sbornik Nauchnykh Trudoy po Prikladnoi Botanike, Genetikeii Selektsii, 99: 31-35.

Vorpsi, V., Rama, P. and Vuksani, G. (2011). Albanian J. Agric. Sci., 10: 23-26.

Ward, G. and Melvin-Carter, E. (2001). Strategies to reduce losses from internal breakdown in plums. Hort. Aust. Project No. SF 00024: 1-66.

Watkins, R. (1976). In: Evolution of Crop Plants (ed. N.W. Simmonds) Longmans, London, pp. 243-46.

Webster, A. D. (1986). Acta Hort., 179: 171-72.

Webster, A.D. (1980). J. Hort. Sci., 55: 19-26.

Webster, A.D. (1984). Rept. East Malling Res. Stn., for 1983, p. 39.

Webster, A.D. and Goldwin, G.K. (1978). J. Hort. Sci., 53: 123-129.

Webster, A.D. and Goldwin, G.K. (1981). J. Hort. Sci., 56: 27-40.

Webster, A.D. and Treharne, K.J. (1983). Acta Hort., 137: 297-305.

Weinberger, J.H. (1975). In. Advances in Fruit Breeding. (eds. J. Janick and J.N. Moore) Purduc. Univ. Press. USA. pp. 336-47.

Weinberger, J.H. and Thompson, L.A. (1962). Proc. Amer. Soc. Hort. Sci., 81: 172-79.

Wells, J.M. and Bukovac, M.J. (1978). J. Amer. Soc. Hort. Sci., 103: 612-16.

Werneck, H.L. (1959). Ann. Bot., 33: 19-33.

Westwood, M.N. (1978). Temperate Zone Pomology, W.H. Freeman, San Francisco, pp, 233-35.

Wildbolz, T. and Staub, A. (1984). Schweizerische Zeitschrift fur Obst und Weinbau, 120: 228-32.

Wills, B.H., Scriven, F.M. and Greenfield, H. (1983). J. Sci. Fd. Agric., 34: 1383-89.

Yoshikawa, F.T., Reil, W.O. and Stromberg, L.K. (1982). Calif. Agric., 23: 13.

Zhang-JunSheng, (2000). China Fruits, 2: 51-52.

Zhout HouCheng, Zhao Xia, Li Ming and Guo AiGuang, (2011). Acta Bot. Boreali Occidentalia Sinica, 31: 2113-18.

Index

B

C

D

E

F

G

Q

R

S

T

V

W

X

Y

Z

Apple Orchard p.13

Bitter Pit p.13

Columnar Apple Trees p.13

Gala p.13

Fire Blight p.13

Discovery p.13

Gloster p. 14

Greensleeves p. 14

Granny Smith p. 14

Golden Delicious p. 14

Jonagold p. 14

Kalei p. 14

Leaf Miner p.15

Red Delicious p. 15

Red Fuji p.15

Russeting in Apple p. 15

Watercore p.15

Central Leader p. 16

Modified Cenreal Leader p. 16

Cordon p. 16

Espailer p. 16

Open Centre p. 16

Apricot Orchard p.181

Brown Rot p.181

Blight p.181

Harcot p.181

Goldbar-150x150 p.181

Moorpark p.181

New Castle p. 182

Tilton p. 182

Wenatchee p. 182

Bacterial Canker p. 235

Bing p. 235

Black Cherry Aphid p. 235

Cherry Orchard p. 235

Compact Lambert p. 235

Compact Stella p. 235

Early Rivers p. 236

Montmorency p. 236

Grace Star p. 236

Sam p. 236

August Prince p. 303

Autumn Crunch p. 303

Babcock p. 303

Bacterial spot p. 303

Brown Rot p. 303

Elberta p. 303

Peach Flowering p.304

Peach Leaf Curl p.304

Peach Leaf Curl p.304

Peach Orchard p.304

Sweet Scarlet p.304

Spring Crest p.304

Concorde p.407

D'Anjou-Pear-450w p.407

Corck Spot p.407

Conference p.407

Le Conte p.407

Packham p.407

Eldorado p. 408

Pear Scab p. 408